Mathematical and Scientific Analysis Handbook

Mathematical and Scientific Analysis Handbook

Edited by **Lawrence Bech**

Clanrye International

New Jersey

Published by Clanrye International,
55 Van Reypen Street,
Jersey City, NJ 07306, USA
www.clanryeinternational.com

Mathematical and Scientific Analysis Handbook
Edited by Lawrence Bech

International Standard Book Number: 978-1-63240-336-0 (Hardback)

Printed in the United States of America.

Contents

Preface

The dictionary defines 'analysis' as a 'detailed examination of the elements or structure of something'. In other words, analysis can refer to and be relevant to anything from chemistry to rocket science and from business to military.

It essentially refers to the process of breaking down any and all forms of complex data into simpler form for better absorption and understanding. Analysis can refer to any field of science or arts. With technological progress, even the field of analysis is broadening. Some of the prominent names, who have contributed significantly to this field, include René Descartes, Galileo Galilee, and Isaac Newton among numerous others.

From behavioural analysis to chemical analysis, there are specialists in all fields working tirelessly to improve and advance their respective spheres and bring about changes that can improve life.

This book contains pieces and contributions from some highly renowned analysts in their chosen fields and looks into the progress that they have made over the years. This book will be relevant and important to everyone, given its all pervading relevance to a vast spectrum of topics.

I wish to thank all the contributing authors who not only shared their expertise in this book but also followed the time parameters set for this publication. I also want to thank my publisher for believing in me and giving me this unmatched opportunity. Lastly, I would like to dedicate this book to my family, without their constant support I would not have been able to complete this project.

Editor

Integral Estimates for the Potential Operator on Differential Forms

Casey Johnson and Shusen Ding

Department of Mathematics, Seattle University, Seattle, WA 98122, USA

Correspondence should be addressed to Shusen Ding; sding@seattleu.edu

Academic Editor: Tohru Ozawa

We develop the local inequalities with new weights for the potential operator applied to differential forms. We also prove the global weighted norm inequalities for the potential operator in averaging domains and explore applications of our new results.

1. Introduction

This paper deals with the weighted estimates for the potential operator applied to differential forms. Throughout this paper, Ω will denote an open subset of $\mathbb{R}^n$, $n \geq 2$, and $\mathbb{R} = \mathbb{R}^1$. Let $e_1 = (1, 0, \ldots, 0)$, $e_2 = (0, 1, \ldots, 0)$,..., $e_n = (0, 0, \ldots, 1)$ be the standard unit basis of $\mathbb{R}^n$. For $l = 0, 1, \ldots, n$, the linear space of l-vectors, spanned by the exterior products $e_I = e_{i_1} \wedge e_{i_2} \wedge \cdots \wedge e_{i_l}$, corresponding to all ordered l-tuples $I = (i_1, i_2, \ldots, i_l)$, $1 \leq i_1 < i_2 < \cdots < i_l \leq n$, is denoted by $\wedge^l = \wedge^l(\mathbb{R}^n)$. The Grassman algebra

$$\wedge = \wedge(\mathbb{R}^n) = \bigoplus_{l=0}^{n} \wedge^l (\mathbb{R}^n) \tag{1}$$

is a graded algebra with respect to the exterior products. For $\alpha = \sum \alpha^I e_I \in \wedge$ and $\beta = \sum \beta^I e_I \in \wedge$, the inner product in $\wedge$ is given by

$$\langle \alpha, \beta \rangle = \sum \alpha^I \beta^I \tag{2}$$

with summation over all l-tuples $I = (i_1, i_2, \ldots, i_l)$ and all integers $l = 0, 1, \ldots, n$. We should also notice that $dx_i \wedge dx_j = -dx_j \wedge dx_i$, $i \neq j$, and $dx_i \wedge dx_i = 0$.

Assume that $B \subset \mathbb{R}^n$ is a ball and σB is the ball with the same center as B and with $\operatorname{diam}(\sigma B) = \sigma \operatorname{diam}(B)$. Differential forms are extensions of functions defined in $\mathbb{R}^n$. A function $f(x_1, \ldots, x_n)$ in $\mathbb{R}^n$ is called a 0-form. A differential l-form is of the form

$$\begin{aligned} w(x) &= \sum_I w_I(x)\, dx_I \\ &= \sum w_{i_1 i_2 \cdots i_l}(x)\, dx_{i_1} \wedge dx_{i_2} \wedge \cdots \wedge dx_{i_l} \end{aligned} \tag{3}$$

in $\mathbb{R}^n$. Differential forms have become invaluable tools for many fields of sciences and engineering, including theoretical physics, general relativity, potential theory, and electromagnetism. They can be used to describe various systems of PDEs and to express different geometrical structures on manifolds. Many interesting and useful results about the differential forms have been obtained during recent years; particularly, for the differential forms satisfying some version of A-hrmonic equation, see [1–8]. The n-dimensional Lebesgue measure of a set $E \subseteq \mathbb{R}^n$ is denoted by $|E|$. We call w a weight if $w \in L^1_{\text{loc}}(\mathbb{R}^n)$ and $w > 0$ a.e. For $0 < p < \infty$, we denote the weighted L^p norm of a measurable function f over E by

$$\|f\|_{p,E,w^\alpha} = \left(\int_E |f(x)|^p w^\alpha dx \right)^{1/p} \tag{4}$$

if the above integral exists. Here α is a real number. It should be noticed that the Hodge star operator can be defined equivalently as follows.

Definition 1. If $\omega = \alpha_{i_1,i_2,\ldots,i_k}(x_1, x_2, \ldots, x_n)dx_{i_1} \wedge dx_{i_2} \wedge \cdots \wedge dx_{i_k} = \alpha_I dx_I$, $i_1 < i_2 < \cdots < i_k$, is a differential k-form, then

$$\star\omega = \star\alpha_{i_1 i_2 \cdots i_k} dx_{i_1} \wedge dx_{i_2} \wedge \cdots \wedge dx_{i_k} = (-1)^{\sum(I)} \alpha_I dx_J, \quad (5)$$

where $I = (i_1, i_2, \ldots, i_k)$, $J = \{1, 2, \ldots, n\} - I$, and

$$\sum(I) = \frac{k(k+1)}{2} + \sum_{j=1}^{k} i_j. \quad (6)$$

The following $A(\alpha, \beta, \gamma; E)$-weights were introduced in [8].

Definition 2. One says that a measurable function $g(x)$ defined on a subset $E \subset \mathbb{R}^n$ satisfies the $A(\alpha, \beta, \gamma; E)$-condition for some positive constants α, β, γ, writes $g(x) \in A(\alpha, \beta, \gamma; E)$ if $g(x) > 0$ a.e., and writes

$$\sup_B \left(\frac{1}{|B|}\int_B g^\alpha dx\right)\left(\frac{1}{|B|}\int_B g^{-\beta} dx\right)^{\gamma/\beta} < \infty, \quad (7)$$

where the supremum is over all balls $B \subset E$. One says that $g(x)$ satisfies the $A(\alpha, \beta; E)$-condition if (7) holds for $\gamma = 1$ and write $g(x) \in A(\alpha, \beta; E) = A(\alpha, \beta, 1; E)$.

Notice that there are three parameters in the definition of the $A(\alpha, \beta, \gamma; E)$-class. We obtain some existing weighted classes if we choose some particular values for these parameters. For example, it is easy to see that the $A(\alpha, \beta, \gamma; E)$-class reduces to the usual $A_r(E)$-class if $\alpha = \gamma = 1$ and $\beta = 1/(r-1)$.

Recently, Bi extended the definition of the potential operator to the case of differential forms; see [2]. For any differential k-form $w(x)$, the potential operator P is defined by

$$Pw(x) = \sum_I \int_E K(x, y) w_I(y)\, dy\, dx_I, \quad (8)$$

where the kernel $K(x, y)$ is a nonnegative measurable function defined for $x \neq y$ and the summation is over all ordered k-tuples I. The $k = 0$ case reduces to the usual potential operator,

$$Pf(x) = \int_E K(x, y) f(y)\, dy, \quad (9)$$

where $f(x)$ is a function defined on $E \subset \mathbb{R}^n$. A kernel K on $\mathbb{R}^n \times \mathbb{R}^n$ is said to satisfy the standard estimates if there are δ, $0 < \delta \leq 1$, and constant C such that, for all distinct points x and y in $\mathbb{R}^n$ and all z with $|x - z| < (1/2)|x - y|$, the kernel K satisfies (i) $K(x, y) \leq C|x - y|^{-n}$, (ii) $|K(x, y) - K(z, y)| \leq C|x - z|^\delta |x - y|^{-n-\delta}$, and (iii) $|K(y, x) - K(y, z)| \leq C|x - z|^\delta |x - y|^{-n-\delta}$. In this paper, we always assume that P is the potential operator with the kernel $K(x, y)$ satisfying the above condition (i) in the standard estimates. In [2], Bi proved the following inequality for the potential operator:

$$\|P(u) - (P(u))_E\|_{p,E} \leq C|E| \operatorname{diam}(E) \|u\|_{p,E}, \quad (10)$$

where $u \in D'(E, \wedge^k)$, $k = 0, 1, \ldots, n-1$, is a differential form defined in a bounded and convex domain E.

Definition 3. One says that a differential form $u \in \wedge^l(E)$ belongs to the $\mathrm{WRH}(\wedge^l, E)$-class and writes $u \in \mathrm{WRH}(\wedge^l, E)$, $l = 0, 1, 2, \ldots, n$, if for any constants $0 < s, t < \infty$, the inequality

$$\|u\|_{s,B} \leq C|B|^{(t-s)/st} \|u\|_{t,\sigma B} \quad (11)$$

holds for any ball B with $\sigma B \subset \Omega$, where $\sigma > 1$ and $C > 0$ are constants.

From [1], we know that any solution of A-harmonic equations satisfies (11). Hence, the $\mathrm{WRH}(\wedge^l, \Omega)$-class is a large class of differential forms. We will use the following Hölder inequality repeatedly in this paper.

Lemma 4. *Let both f and g be measurable functions in $\mathbb{R}^n$ and $p, q > 0$ and $1/s = (1/p) + (1/q)$. Then*

$$\left(\int_E |fg|^s dx\right)^{1/s} \leq \left(\int_E |f|^p dx\right)^{1/p} \left(\int_E |g|^q dx\right)^{1/q} \quad (12)$$

for any $E \subset \mathbb{R}^n$.

2. Local Inequalities

In this section, we will prove some local weighted norm inequalities for the potential operator.

Theorem 5. *Let P be the potential operator applied to a differential form $u \in \mathrm{WRH}(\wedge^l, \Omega)$, where $\Omega \subset \mathbb{R}^n$ is a domain. Assume that $w(x) \in A(\alpha, \beta; \Omega)$ with $\alpha, \beta > 0$. Then, there exists a constant C, independent of u, such that*

$$\|P(u) - (P(u))_B\|_{s,B,w} \leq C|B| \operatorname{diam}(B) \|u\|_{s,B,w} \quad (13)$$

for all balls B with $B \subset \Omega$, where $s > 1$ is a constant.

Proof. Let $t = s\alpha/(\alpha - 1)$, then $t > s$. Using Lemma 4 with $1/s = (1/t) + ((t-s)/st)$ yields

$$\begin{aligned} &\|P(u) - (P(u))_B\|_{s,b,w} \\ &= \left(\int_B |P(u) - (P(u))_B|^s w\, dx\right)^{1/s} \\ &= \left(\int_B \left(|P(u) - (P(u))|_B w^{1/s}\right)^s dx\right)^{1/s} \\ &\leq \left(\int_B |P(u) - (P(u))_B|^t dx\right)^{1/t} \\ &\quad \times \left(\int_B \left(w^{1/s}\right)^{st/(t-s)} dx\right)^{(t-s)/st} \\ &\leq C_1 \|B\| \operatorname{diam}(B) \|u\|_{t,B} \left(\int_B w^\alpha(x)\, dx\right)^{1/\alpha s}. \end{aligned} \quad (14)$$

Set $k = \beta s/(1 + \beta)$, then $0 < k < s$. Since u is in the $\mathrm{WRH}(\Omega)$ class,

$$\|u\|_{t,B} \leq C_2 \|B\|^{(k-t)/kt} \|u\|_{k,\sigma_1 B}, \quad (15)$$

where $\sigma_1 > 1$ in a constraint. Since u is in the WRH(Ω)-class again (note that $1/k = (1/s) + (s-k)/sk$),

$$\begin{aligned}
\|u\|_{k,\sigma_1 B} &= \left(\int_{\sigma_1 B}\left(|u|\, w^{1/s} s^{-1/s}\right)^k dx\right)^{1/k} \\
&\le \left(\int_{\sigma_1 B}\left(|u|\, w^{1/s} w^{-1/s}\right)^k dx\right)^{1/k} \\
&\quad\times \left(\int_{\sigma_1 B}\left(w^{-1/s}\right)^{ks/(s-k)} dx\right)^{(s-k)/ks} \\
&= \left(\int_{\sigma_1 B}|u|^s w\, dx\right)^{1/s} \\
&\quad\times \left(\int_{\sigma_1 B} w^{-k/(s-k)} dx\right)^{(s-k)/ks} \\
&= \left(\int_{\sigma_1 B}|u|^s w\, dx\right)^{1/s}\left(\int_{\sigma_1 B} w^{-\beta} dx\right)^{1/\beta s},
\end{aligned} \tag{16}$$

where we have used the following calculation:

$$\begin{aligned}
\frac{k}{s-k} &= \frac{\beta s/(1+\beta)}{s-(\beta s/(1+\beta))} = \frac{\beta/(1+\beta)}{a-(\beta/(1+\beta))} \\
&= \frac{\beta/(1+\beta)}{(1+\beta-\beta)/(1+\beta)} = \beta.
\end{aligned} \tag{17}$$

Combining (14), (15) and, (16) gives

$$\begin{aligned}
&\|P(u) - (P(u))_B\|_{s,B,w} \\
&\quad\le C_3 |B| \operatorname{diam}(B)\, |B|^{(k-t)/kt}\left(\int_{\sigma_1 B}|u|^s w\, dx\right)^{1/s} \\
&\quad\times \left(\int_B w^{\alpha} dx\right)^{1/\alpha s}\left(\int_{\sigma_1 B} w^{-\beta} dx\right)^{1/\beta s}.
\end{aligned} \tag{18}$$

Note that since $w \in A(\alpha, \beta, \alpha; \Omega)$, it follows that

$$\begin{aligned}
&\left(\int_B w^{\alpha} dx\right)^{1/\alpha s}\left(\int_{\sigma_1 B} w^{-\beta} dx\right)^{1/\beta s} \\
&= \left(\left(\int_{\sigma_1 B} w^{\alpha} dx\right)\left(\int_{\sigma_1 B} w^{-\beta} dx\right)^{\alpha/\beta}\right)^{1/\alpha s} \\
&= \left(|\sigma_1 B|^{1+(\alpha/\beta)}\left(\frac{1}{|B|}\int_{\sigma_1 B} w^{\alpha} dx\right)\left(\frac{1}{|B|}\int_{\sigma_1 B} w^{-\beta} dx\right)^{\alpha/\beta}\right)^{1/\alpha s} \\
&= C_4 |B|^{(1/\alpha s)+(1/\beta s)}.
\end{aligned} \tag{19}$$

Plugging (19) into (18), we have

$$\begin{aligned}
&\|P(U) - (P(u))_B\|_{s,B,w} \\
&\quad\le C_5 \operatorname{diam}(B)\, |B|^{1+(1/t)-(1/k)} \\
&\quad\times \left(\int_{\sigma_1 B}|u|^s w\, dx\right)^{1/s} C_4 |B|^{(1/\alpha s)+(1/\beta s)} \\
&\quad\le C_6 \operatorname{diam}(B)\, |B|^{1+(1/t)-(1/k)+(1/\alpha s)+(1/\beta s)} \\
&\quad\times \left(\int_{\sigma_1 B}|u|^s w\, dx\right)^{1/s}.
\end{aligned} \tag{20}$$

We should notice that

$$1 + \frac{1}{t} - \frac{1}{k} + \frac{1}{\alpha s} + \frac{1}{\beta s} = 1 + \frac{\alpha-1}{s\alpha} - \frac{1+\beta}{s\beta} + \frac{1}{\alpha s} + \frac{1}{\beta s} = 1. \tag{21}$$

Combining (20) and (21) gives us

$$\|P(u) - (P(u))_B\|_{s,B,w} \le C |B| \operatorname{diam}(B)\left(\int_{\sigma_1 B}|u|^s w\, dx\right)^{1/s}. \tag{22}$$

The proof of Theorem 5 has been completed. □

A continuously increasing function $\varphi : [0,\infty) \to [0,\infty)$ with $\varphi(0) = 0$ is called an Orlicz function. A convex Orlicz function φ is often called a Young function. The Orlicz space $L^{\varphi}(\Omega,\mu)$ consists of all measurable functions f on Ω such that $\int_{\Omega}\varphi(|f|/\lambda)d\mu < \infty$ for some $\lambda = \lambda(f) > 0$. If φ is a Young function, then

$$\|f\|_{\varphi(\Omega,\mu)} = \inf\left\{\lambda > 0 : \int_{\Omega}\varphi\left(\frac{|f|}{\lambda}\right) d\mu \le 1\right\} \tag{23}$$

defines a norm in $L^{\varphi}(\Omega,\mu)$, which is called the Orlicz norm or the Luxemburg norm.

Definition 6 (see [9]). One says that a Young function φ lies in the class $G(p,q,C)$, $1 \le p < q < \infty$, $C \ge 1$, if (i) $1/C \le \varphi(t^{1/p})/j(t) \le C$ and (ii) $1/C \le \varphi(t^{1/q})/h(t) \le C$ for all $t > 0$, where j is a convex increasing function and h is a concave increasing function on $[0,\infty)$.

From [9], each of φ, j, and h in the above definition is doubling in the sense that its values at t and $2t$ are uniformly comparable for all $t > 0$ and the consequent fact that

$$C_1 t^q \le h^{-1}(\varphi(t)) \le C_2 t^q, \qquad C_1 t^p \le j^{-1}(\varphi(t)) \le C_2 t^p, \tag{24}$$

where C_1 and C_2 are constants. Also, for all $1 \le p_1 < p < p_2$ and $\alpha \in \mathbb{R}$, the function $\varphi(t) = t^p \log_+^{\alpha} t$ belongs to $G(p_1, p_2, C)$ for some constant $C = C(p, \alpha, p_1, p_2)$. Here $\log_+(t)$ is defined by $\log_+(t) = 1$ for $t \le e$ and $\log_+(t) = \log(t)$ for $t > e$. Particularly, if $\alpha = 0$, we see that $\varphi(t) = t^p$ lies in $G(p_1, p_2, C)$, $1 \le p_1 < p < p_2$.

Theorem 7. *Let P be the potential operator applied to a differential form $u \in \mathrm{WRH}(\wedge^l, \Omega)$ and φ a Young function in the class $G(p, q, C)$, $1 \le p < q < \infty, C \le 1$, and where Ω is a bounded domain. Assume that $\varphi(|u|) \in L^1_{loc}(\Omega, \mu)$. Then, there exists a constant $M > 0$, independent of u, such that*

$$\int_B \varphi\left(\frac{|P(u) - (P(u))_B|}{k}\right) dx \le M \int_{\sigma B} \varphi\left(\frac{|u|}{k}\right) dx \qquad (25)$$

for all balls B with $B \subset \Omega$, where $\sigma > 1$ is a constant.

Proof. Using Jensen's inequality for h^{-1} and (24), we have

$$\begin{aligned}
&\int_B \varphi\left(\frac{|P(u) - (P(u))_B|}{k}\right) dx \\
&= h\left(h^{-1}\left(\int_B \varphi\left(\frac{|P(u) - (P(u))_B|}{k}\right) dx\right)\right) \\
&\le h\left(\int_B h^{-1}\left(\varphi\left(\frac{|P(u) - (P(u))_B|}{k}\right)\right) dx\right) \\
&\le h\left(C_1 \int_B \left(\frac{|P(u) - (P(u))_B|}{k}\right)^q dx\right) \\
&\le C_2 \varphi\left(\left(C_1 \int_B \left(\frac{|P(u) - (P(u))_B|}{k}\right)^q dx\right)^{1/q}\right) \\
&\le C_2 \varphi\left(\frac{1}{k}\left(C_1 \int_B (|P(u) - (P(u))_B|)^q dx\right)^{1/q}\right) \\
&\le C_3 \varphi\left(\frac{1}{k}\left(\int_B (|P(u) - (P(u))_B|)^q dx\right)^{1/q}\right).
\end{aligned} \qquad (26)$$

Since $u \in \mathrm{WRH}(\Omega)$, by (11) we obtain

$$\left(\int_B |u|^q dx\right)^{1/q} \le C_4 |B|^{(p-q)/pq} \left(\int_{\sigma B} |u|^p dx\right)^{1/p}, \qquad (27)$$

where $\sigma > 1$ is a constant. Using (13), (24), and Jensen's inequality,

$$\begin{aligned}
&\varphi\left(\frac{1}{k}\left(\int_B |P(u) - (P(u))_B|^q dx\right)^{1/q}\right) \\
&\le \varphi\left(\frac{1}{k}|B|^{1+(1/n)}\left(\int_B |u|^q dx\right)^{1/q}\right) \\
&\le \varphi\left(\frac{1}{k}|B|^{1+(1/n)} C_4 |B|^{(1/q)-(1/p)}\left(\int_{\sigma B} |u|^p dx\right)^{1/p}\right) \\
&= \varphi\left(\left(\frac{1}{k^p} C_4^p |B|^{p(1+(1/n)+(1/q)-(1/p))} \int_{\sigma B} |u|^p dx\right)^{1/p}\right) \\
&\le C_5 j\left(\frac{1}{k^p} C_4^p |B|^{p(1+(1/n)+(1/q)-(1/p))} \int_{\sigma B} |u|^p dx\right) \\
&\le C_5 j\left(\int_{\sigma B} \frac{1}{k^p} C_4^p |B|^{p(1+(1/n)+(1/q)-(1/p))} |u|^p dx\right) \\
&\le C_5 \int_{\sigma B} j\left(\frac{1}{k^p} C_4^p |B|^{p(1+(1/n)+(1/q)-(1/p))} |u|^p\right) dx.
\end{aligned} \qquad (28)$$

Note that $p \ge 1$, then $1 + (1/n) + (1/q) - (1/p) > 0$. Thus,

$$|B|^{1+(1/n)+(1/q)-(1/p)} \le |\Omega|^{1+(1/n)+(1/q)-(1/p)} \le C_6. \qquad (29)$$

Using the above inequality and (i) in Definition 6, we find that $j(t) \le C_7 \varphi(t^{1/p})$. Therefore,

$$\begin{aligned}
&\int_{\sigma B} j\left(\frac{1}{k^p} C_4^p |B|^{p(1+(1/n)+(1/q)-(1/p))} |u|^p\right) dx \\
&\le C_7 \int_{\sigma B} \varphi\left(\frac{1}{k} C_4 |B|^{1+(1/n)+(1/q)-(1/p)} |u|\right) dx \\
&\le C_7 \int_{\sigma B} \varphi\left(C_8 \frac{|u|}{k}\right) dx \\
&\le C_9 \int_{\sigma B} \varphi\left(\frac{|u|}{k}\right) dx.
\end{aligned} \qquad (30)$$

Combining (28) and (30) yields

$$\varphi\left(\frac{1}{k}\left(\int_B |P(u) - (P(u))_B|^q dx\right)^{1/q}\right) \le C_{10} \int_{\sigma B} \varphi\left(\frac{|u|}{k}\right) dx. \qquad (31)$$

Finally, substituting (31) into (26), we obtain

$$\int_B \varphi\left(\frac{|P(u) - (P(u))_B|}{k}\right) dx \le M \int_{\sigma B} \varphi\left(\frac{|u|}{k}\right) dx. \qquad (32)$$

The proof of Theorem 7 has been completed. □

3. Global Inequalities

In 1989, Staples introduced the following L^s-averaging domains in [10]. A proper subdomain $\Omega \subset \mathbb{R}^n$ is called an L^s-averaging domain, $s \ge 1$, if there exists a constant C such that

$$\left(\frac{1}{|\Omega|} \int_\Omega |u - u_\Omega|^s dx\right)^{1/s} \le C \sup_{B \subset \Omega} \left(\frac{1}{|\Omega|} \int_B |u - u_B|^s dx\right)^{1/s} \qquad (33)$$

for all $u \in L^s_{loc}(\Omega)$. Here the supremum is over all balls $B \subset \Omega$. The L^s-averaging domains were extended into the $L^s(\mu)$-averaging domains recently in [11]. We call a proper subdomain $\Omega \subset \mathbb{R}^n$ an $L^s(\mu)$-averaging domain, $s \geq 1$, if $\mu(\Omega) < \infty$ and there exists a constant C such that

$$\left(\frac{1}{\mu(\Omega)}\int_\Omega |u - u_{B_0}|^s d\mu\right)^{1/s} \leq C \sup_{4B\subset\Omega}\left(\frac{1}{\mu(B)}\int_B |u - u_B|^s d\mu\right)^{1/s} \tag{34}$$

for some ball $B_0 \subset \Omega$ and all $u \in L^s_{\text{loc}}(\Omega;\mu)$. The $L^s(\mu)$-averaging domain was genralized into the following $L^\varphi(\mu)$-averaging domain in [12].

Definition 8. Let φ be a continuous increasing convex function on $[0,\infty)$ with $\varphi(0) = 0$. One calls a proper subdomain $\Omega \subset \mathbb{R}^n$ an $L^\varphi(\mu)$-averaging domain, if $\mu(\Omega) < \infty$ and there exists a constant C such that

$$\frac{1}{\mu(\Omega)}\int_\Omega \varphi\left(\tau |u - u_{B_0}|\right) d\mu \leq C \sup_{4B\subset\Omega}\frac{1}{\mu(B)}\int_B \varphi\left(\sigma |u - u_B|\right) d\mu \tag{35}$$

for some ball $B_0 \subset \Omega$ and all u such that $\varphi(|u|) \in L^1_{loc}(\Omega;\mu)$, where the measure μ is defined by $d\mu = w(x)dx$, $w(x)$ is a weight, and τ, σ are constants with $0 < \tau < 1$, $0 < \sigma < 1$, and the supremum is over all balls $B \subset \Omega$ with $4B \subset \Omega$.

From the above definition, we see that $L^s(\mu)$-averaging domains are special $L^\varphi(\mu)$-averaging domains when $\varphi(t) = t^s$ in Definition 8.

Theorem 9. *Let φ be a Young function in the class $G(p,q,C)$, $1 \leq p < q < \infty$, $C \geq 1$; and $\Omega \subset \mathbb{R}^n$ any bounded $L^\varphi(\mu)$-averaging domain, and P the potential operator applied to a differential form $u \in \text{WRH}(\wedge^l,\Omega)$, $l = 1,2\ldots,n$. Assume that $\varphi(|u|) \in L^1(\Omega,\mu)$. Then, there exists a constant C, independent of u, such that*

$$\int_\Omega \varphi\left(\frac{1}{k}|P(u) - (P(u))_{B_0}|\right) d\mu \leq C\int_\Omega \varphi\left(\frac{|u|}{k}\right) d\mu, \tag{36}$$

where $B_0 \subset \Omega$ is some fixed ball.

Proof. From Definition 8, (25), and noticing that φ is doubling, we have

$$\begin{aligned}\int_\Omega \varphi\left(|P(u) - (P(u))_{B_0}|\right) d\mu &\leq C_1 \sup_{B\subset\Omega}\int_B \varphi\left(|P(u) - (P(u))_B|\right) d\mu\\ &\leq C_1 \sup_{B\subset\Omega}\left(C_2\int_{\sigma B}\varphi(|u|)\,d\mu\right)\\ &\leq C_1 \sup_{B\subset\Omega}\left(C_2\int_{\Omega}\varphi(|u|)\,d\mu\right)\\ &\leq C_3\int_\Omega \varphi(|u|)\,d\mu.\end{aligned} \tag{37}$$

We have completed the proof of Theorem 9. □

Choosing $\varphi(t) = t^s\log_+^\alpha t$ in Theorem 9, we obtain the following Poincaré inequalities with the $L^s(\log_+^\alpha L)$-norms.

Corollary 10. *Let $\varphi(t) = t^s\log_+^\alpha t$, $s \geq 1$, $\alpha \in \mathbb{R}$, and P the potential operator applied to a differential form $u \in \text{WRH}(\wedge^l,\Omega)$, $l = 1,2\ldots,n$. Assume that $\varphi(|u|) \in L^1(\Omega,\mu)$. Then, there exists a constant C, independent of u, such that*

$$\int_\Omega |P(u) - (P(u))_{B_0}|^s \log_+^\alpha\left(|P(u) - (P(u))_{B_0}|\right) d\mu \leq C\int_\Omega |u|^s\log_+^\alpha(|u|)\,d\mu \tag{38}$$

for any bounded $L^\varphi(\mu)$-averaging domain Ω and $B_0 \subset \Omega$ is some fixed ball.

Note that (38) can be written as the following version with the Luxemburg norm:

$$\|P(u) - (P(u))_{B_0}\|_{L^s(\log_+^\alpha L)(\Omega)} \leq C\|u\|_{L^s(\log_+^\alpha L)(\Omega)} \tag{39}$$

provided that the conditions in Corollary 10 are satisfied.

4. Applications

We have established the local and global weighted estimates for the potential operator applied to the differential forms in the $\text{WRH}(\wedge^l,\Omega)$-class. It is well known that any solution to A-harmonic equations belongs to the $\text{WRH}(\wedge^l,\Omega)$-class. Hence, our inequalities can be used to estimate solutions of A-harmonic equations. Next, as applications of the main theorems, we develop some estimates for the Jacobian $J(x,f)$ of a mapping $f : \Omega \to \mathbb{R}^n$, $f = (f^1,\ldots,f^n)$. We know that the Jacobian $J(x,f)$ of a mapping f is an n-form, specifically, $J(x,f)dx = df^1\wedge\cdots\wedge df^n$, where $dx = dx_1\wedge dx_2\wedge\cdots\wedge dx_n$. For example, let $f = (f^1,f^2)$ be a differential mapping in $\mathbb{R}^2$. Then,

$$\begin{aligned}J(x,f)\,dx\wedge dy &= \begin{vmatrix} f_x^1 & f_y^1\\ f_x^2 & f_y^2\end{vmatrix} dx\wedge dy\\ &= \left(f_x^1 f_y^2 - f_y^1 f_x^2\right) dx\wedge dy,\\ df^1\wedge df^2 &= \left(f_x^1 dx + f_y^1 dy\right)\wedge\left(f_x^2 dx + f_y^2 dy\right)\\ &= f_y^1 f_x^2 dy\wedge dx + f_x^1 f_y^2 dx\wedge dy\\ &= \left(f_x^1 f_y^2 - f_y^1 f_x^2\right) dx\wedge dy,\end{aligned} \tag{40}$$

where we have used the property $dx_i\wedge dx_j = -dx_j\wedge dx_i$ if $i \neq j$, and $dx_i\wedge dx_j = 0$ if $i = j$. Clearly, $J(x,f)dx\wedge dy = df^1\wedge df^2$.

Let $f : \Omega \to \mathbb{R}^n$, $f = (f^1,\ldots,f^n)$ be a mapping, whose distributional differential $\text{Df} = [\partial f^i/\partial x_j] : \Omega \to \text{GL}(n)$ is a locally integrable function on M with values in the space $\mathrm{GL}(n)$ of all $n \times n$-matrices. We use

$$J(x,f) = \det Df(x) = \begin{vmatrix} f^1_{x_1} & f^1_{x_2} & f^1_{x_3} & \cdots & f^1_{x_n} \\ f^2_{x_1} & f^2_{x_2} & f^2_{x_3} & \cdots & f^2_{x_n} \\ \vdots & \vdots & \vdots & \ddots & \vdots \\ f^n_{x_1} & f^n_{x_2} & f^n_{x_3} & \cdots & f^n_{x_n} \end{vmatrix} \quad (41)$$

to denote the Jacobian determinant of f. Assume that u is the subdeterminant of Jacobian $J(x,f)$, which is obtained by deleting the k rows and k columns, $k = 0, 1, \ldots, n-1$; that is,

$$u = J\left(x_{j_1}, x_{j_2}, \ldots, x_{j_{n-k}}; f^{i_1}, f^{i_2}, \ldots, f^{i_{n-k}}\right) = \begin{vmatrix} f^{i_1}_{x_{j_1}} & f^{i_1}_{x_{j_2}} & f^{i_1}_{x_{j_3}} & \cdots & f^{i_1}_{x_{j_{n-k}}} \\ f^{i_2}_{x_{j_1}} & f^{i_2}_{x_{j_2}} & f^{i_2}_{x_{j_3}} & \cdots & f^{i_2}_{x_{j_{n-k}}} \\ \vdots & \vdots & \vdots & \ddots & \vdots \\ f^{i_{n-k}}_{x_{j_1}} & f^{i_{n-k}}_{x_{j_2}} & f^{i_{n-k}}_{x_{j_3}} & \cdots & f^{i_{n-k}}_{x_{j_{n-k}}} \end{vmatrix}, \quad (42)$$

which is an $(n-k) \times (n-k)$ subdeterminant of $J(x,f)$, $\{i_1, i_2, \ldots, i_{n-k}\} \subset \{1, 2, \ldots, n\}$ and $\{j_1, j_2, \ldots, j_{n-k}\} \subset \{1, 2, \ldots, n\}$. Note that $J(x_{j_1}, x_{j_2}, \ldots, x_{j_{n-k}}; f^{i_1}, f^{i_2}, \ldots, f^{i_{n-k}})dx_{j_1} \wedge dx_{j_2} \wedge \cdots \wedge dx_{j_{n-k}}$ is an $(n-k)$-form. Thus, all estimates for differential forms are applicable to the $(n-k)$-form $J(x_{j_1}, x_{j_2}, \ldots, x_{j_{n-k}}; f^{i_1}, f^{i_2}, \ldots, f^{i_{n-k}})dx_{j_1} \wedge dx_{j_2} \wedge \cdots \wedge dx_{j_{n-k}}$. For example, choosing $u = J(x,f)dx$ and applying Theorems 7 and 9 to u, respectively, we have the following theorems.

Theorem 11. *Let φ be a Young function in the class $G(p,q,C)$, $1 \le p < q < \infty, C \le 1$. Let $f = (f^1, \ldots, f^n) : \Omega \to \mathbb{R}^n$ be a mapping such that $J(x,f)dx \in \mathrm{WRH}(\wedge^n, \Omega)$, where $J(x,f)$ is the Jacobian of the mapping f and $\Omega \subset \mathbb{R}^n$ is a bounded domain in $\mathbb{R}^n$. Assume that $\varphi(|J(x,f)|) \in L^1_{\mathrm{loc}}(\Omega, \mu)$. Then, there exists a constant C, independent of $J(x,f)$, such that*

$$\int_B \varphi\left(\frac{|P(J(x,f)) - (P(J(x,f)))_B|}{k} du\right) \le C \int_{\sigma B} \varphi\left(\frac{|J(x,f)|}{k}\right) du \quad (43)$$

for all balls $B \subset \Omega$ and some constant $\sigma > 1$.

Theorem 12. *Let φ be a Young function in the class $G(p,q,C)$, $1 \le p < q < \infty, C \le 1$. Let $f = (f^1, \ldots, f^n) : \Omega \to \mathbb{R}^n$ be a mapping such that $J(x,f)dx \in \mathrm{WRH}(\wedge^n, \Omega)$, where $J(x,f)$ is the Jacobian of the mapping f and $\Omega \subset \mathbb{R}^n$ is a bounded $L^\varphi(\mu)$-averaging domain in $\mathbb{R}^n$. Assume that $\varphi(|J(x,f)|) \in L^1(\Omega, \mu)$. Then, there exists a constant C, independent of $J(x,f)$, such that*

$$\int_\Omega \varphi\left(\frac{|P(J(x,f)) - (P(J(x,f)))_{B_0}|}{k} du\right) \le C \int_\Omega \varphi\left(\frac{|J(x,f)|}{k}\right) du \quad (44)$$

for some ball $B_0 \subset \Omega$.

References

[1] R. P. Agarwal, S. Ding, and C. A. Nolder, *Inequalities for Differential Forms*, Springer, New York, NY, USA, 2009.

[2] H. Bi, "Weighted inequalities for potential operators on differential forms," *Journal of Inequalities and Applications*, vol. 2010, Article ID 713625, 2010.

[3] H. Bi and Y. Xing, "Poincaré-type inequalities with $L^p(\log L)^\alpha$-norms for Green's operator," *Computers and Mathematics with Applications*, vol. 60, no. 10, pp. 2764–2770, 2010.

[4] B. Liu, "$A_r^\lambda(\Omega)$-weighted imbedding inequalities for A-harmonic tensors," *Journal of Mathematical Analysis and Applications*, vol. 273, no. 2, pp. 667–676, 2002.

[5] Y. Wang, "Two-weight Poincaré-type inequalities for differential forms in $L^s(\mu)$-averaging domains," *Applied Mathematics Letters*, vol. 20, no. 11, pp. 1161–1166, 2007.

[6] X. Yuming, "Weighted integral inequalities for solutions of the A-harmonic equation," *Journal of Mathematical Analysis and Applications*, vol. 279, no. 1, pp. 350–363, 2003.

[7] Y. Xing and Y. Wang, "Bmo and Lipschitz norm estimates for composite operators," *Potential Analysis*, vol. 31, no. 4, pp. 335–344, 2009.

[8] Y. Xing, "A new weight class and Poincaré inequalities with the Radon measures," *Journal of Inequalities and Applications*, vol. 2012, article 32, 2012.

[9] S. M. Buckley and P. Koskela, "Orlicz-hardy inequalities," *Illinois Journal of Mathematics*, vol. 48, no. 3, pp. 787–802, 2004.

[10] S. G. Staples, "L^p-averaging domains and the Poincaré inequality," *Annales Academiæ Scientiarum Fennicæ Mathematica*, vol. 14, pp. 103–127, 1989.

[11] S. Ding and C. A. Nolder, "$L^s(\mu)$-averaging domains," *Journal of Mathematical Analysis and Applications*, vol. 283, no. 1, pp. 85–99, 2003.

[12] S. Ding, "$L^\varphi(\mu)$-averaging domains and the quasi-hyperbolic metric," *Computers and Mathematics with Applications*, vol. 47, no. 10-11, pp. 1611–1618, 2004.

2

On a New I-Convergent Double-Sequence Space

Vakeel A. Khan and Nazneen Khan

Department of Mathematics, A.M.U., Aligarh 202002, India

Correspondence should be addressed to Vakeel A. Khan; vakhan@math.com

Academic Editor: Wen Xiu Ma

The sequence space BV_σ was introduced and studied by Mursaleen (1983) . In this article we introduce the sequence space ${}_2BV_\sigma^I$ and study some of its properties and inclusion relations.

1. Introduction and Preliminaries

Let $\mathbb{N}$, $\mathbb{R}$, and $\mathbb{C}$ be the sets of all natural, real, and complex numbers, respectively. We write

$$\omega = \{x = (x_k) : x_k \in \mathbb{C}\}, \tag{1}$$

showing the space of all real or complex sequences.

Definition 1. A double sequence of complex numbers is defined as a function $x : \mathbb{N} \times \mathbb{N} \to \mathbb{C}$. We denote a double sequence as (x_{ij}) where the two subscripts run through the sequence of natural numbers independent of each other [1]. A number $a \in \mathbb{C}$ is called a double limit of a double sequence (x_{ij}) if for every $\epsilon > 0$ there exists some $N = N(\epsilon) \in \mathbb{N}$ such that

$$\left|(x_{ij}) - a\right| < \epsilon, \quad \forall i, j \geq N, \tag{2}$$

(see [2]).

Let l_∞ and c denote the Banach spaces of bounded and convergent sequences, respectively, with norm $\|x\|_\infty = \sup_k |x_k|$. Let v denote the space of sequences of bounded variation; that is,

$$v = \left\{x = (x_k) : \sum_{k=0}^{\infty} \left|x_k - x_{k-1}\right| < \infty, x_{-1} = 0\right\}, \tag{3}$$

where v is a Banach space normed by

$$\|x\| = \sum_{k=0}^{\infty} \left|x_k - x_{k-1}\right|, \tag{4}$$

(see [3]).

Definition 2. Let σ be a mapping of the set of the positive integers into itself having no finite orbits. A continuous linear functional ϕ on l_∞ is said to be an invariant mean or σ-mean if and only if

(i) $\phi(x) \geq 0$ when the sequence $x = (x_k)$ has $x_k \geq 0$ for all k;

(ii) $\phi(e) = 1$, where $e = \{1, 1, 1, \ldots\}$;

(iii) $\phi(x_{\sigma(n)}) = \phi(x)$ for all $x \in l_\infty$.

In case σ is the translation mapping $n \to n+1$, a σ-mean is often called a Banach limit (see [4]), and V_σ, the set of bounded sequences all of whose invariant means are equal, is the set of almost convergent sequences (see [5]).

If $x = (x_k)$, then $Tx = (Tx_k) = (x_{\sigma(k)})$. Then it can be shown that

$$V_\sigma = \left\{x = (x_k) : \sum_{m=1}^{\infty} t_{m,k}(x) = L \text{ uniformly in } k, \right.$$
$$\left. L = \sigma - \lim x\right\}, \tag{5}$$

where $m \geq 0, k > 0$. Consider

$$t_{m,k}(x) = \frac{x_k + x_{\sigma(k)} + \cdots + x_{\sigma^m(k)}}{m+1}, \quad t_{-1,k} = 0, \tag{6}$$

where $\sigma^m(k)$ denote the mth iterate of $\sigma(k)$ at k. The special case of (5) in which $\sigma(n) = n + 1$ was given by Lorentz [5, Theorem 1], and that the general result can be proved in a similar way. It is familiar that a Banach limit extends the limit functional on c.

Theorem 3. *A σ-mean extends the limit functional on c in the sense that $\phi(x) = \lim x$ for all $x \in c$ if and only if σ has no finite orbits; that is to say, if and only if, for all $k \geq 0$, $j \geq 1$, (see [3])*

$$\sigma^j(k) \neq k. \tag{7}$$

Put

$$\phi_{m,k}(x) = t_{m,k}(x) - t_{m-1,k}(x), \tag{8}$$

assuming that $t_{-1,k} = 0$. A straight forward calculation shows (see [6]) that

$$\phi_{m,k}(x) = \begin{cases} \dfrac{1}{m(m+1)} \displaystyle\sum_{j=1}^{m} J\left(x_{\sigma^j(k)} - x_{\sigma^{j-1}(k)}\right) & (m \geq 1), \\ x_k, & (m = 0). \end{cases} \tag{9}$$

For any sequence x, y, and scalar λ, we have

$$\begin{aligned} \phi_{m,k}(x+y) &= \phi_{m,k}(x) + \phi_{m,k}(y), \\ \phi_{m,k}(\lambda x) &= \lambda \phi_{m,k}(x). \end{aligned} \tag{10}$$

Definition 4. A sequence $x \in l_\infty$ is of σ-bounded variation if and only if

(i) $\sum_{k=0}^{\infty} |\phi_{m,k}(x)|$ converges uniformly in n;

(ii) $\lim_{m\to\infty} t_{m,k}(x)$, which must exist, should take the same value for all k.

We denote by BV_σ, the space of all sequences of σ-bounded variation (see [7]):

$$BV_\sigma = \left\{ x \in l_\infty : \sum_m |\phi_{m,k}(x)| < \infty, \text{uniformly in } n \right\}. \tag{11}$$

Theorem 5. BV_σ *is a Banach space normed by*

$$\|x\| = \sup_k \sum_{m=0}^{\infty} |\phi_{m,k}(x)|, \tag{12}$$

(see [8]).

Subsequently, invariant means have been studied by Ahmad and Mursaleen [9], Mursaleen et al. [3, 6, 8, 10–14], Raimi [15], Schaefer [16], Savas and Rhoades [17], Vakeel et al. [18–20], and many others [21–23]. For the first time, I-convergence was studied by Kostyrko et al. [24]. Later on, it was studied by Šalát et al. [25, 26], Tripathy and Hazarika [27], Ebadullah et al. [18–20, 28], and Vakeel et al. [1, 29].

Definition 6 (see [30, 31]). Let X be a nonempty set. Then, a family of sets $I \subseteq 2^X$ (2^X denoting the power set of X) is said to be an ideal in X if

(i) $\emptyset \in I$;

(ii) I is additive; that is, $A, B \in I \Rightarrow A \cup B \in I$;

(iii) I is hereditary that is, $A \in I, B \subseteq A \Rightarrow B \in I$;

An Ideal $I \subseteq 2^X$ is called nontrivial if $I \neq 2^X$. A non-trivial ideal $I \subseteq 2^X$ is called admissible if $\{\{x\} : x \in X\} \subseteq I$.

A non-trivial ideal I is maximal if there cannot exist any non-trivial ideal $J \neq I$ containing I as a subset.

For each ideal I, there is a filter $£(I)$ corresponding to I. That is,

$$£(I) = \{K \subseteq N : K^c \in I\}, \quad \text{where } K^c = N - K. \tag{13}$$

Definition 7 (see [24, 31, 32]). A double sequence $(x_{ij}) \in \omega$ is said to be *I-convergent* to a number L if for every $\epsilon > 0$,

$$\left\{i, j \in \mathbb{N} : \left|x_{ij} - L\right| \geq \epsilon\right\} \in I. \tag{14}$$

In this case, we write $I - \lim x_{ij} = L$.

Definition 8 (see [2]). A double sequence $(x_{ij}) \in \omega$ is said to be I-null if $L = 0$. In this case, we write

$$I - \lim x_{ij} = 0. \tag{15}$$

Definition 9. A double sequence $(x_{ij}) \in \omega$ is said to be I-cauchy if for every $\epsilon > 0$ there exist numbers $m = m(\epsilon)$, $n = n(\epsilon)$ such that

$$\left\{i, j \in \mathbb{N} : \left|x_{ij} - x_{mn}\right| \geq \epsilon\right\} \in I. \tag{16}$$

Definition 10. A double sequence $(x_{ij}) \in \omega$ is said to be I-bounded if there exists $M > 0$ such that

$$\left\{i, j \in \mathbb{N} : \left|x_{ij}\right| > M\right\}. \tag{17}$$

Definition 11. A double-sequence space E is said to be solid or normal if $(x_{ij}) \in E$ implies $(\alpha_{ij} x_{ij}) \in E$ for all sequence of scalars (α_{ij}) with $|\alpha_{ij}| < 1$ for all $i, j \in \mathbb{N}$.

Definition 12 (see [24, 33]). A nonempty family of sets $£(I) \subseteq 2^X$ is said to be filter on X if and only if

(i) $\Phi \notin £(I)$;

(ii) for $A, B \in £(I)$, we have $A \cap B \in £(I)$;

(iii) for each $A \in £(I)$ and $A \subseteq B$ implies $B \in £(I)$.

Definition 13. Let X be a linear space. A function $g : X \to R$ is called a paranorm, if for all $x, y, z \in X$,

(i) $g(x) = 0$ if $x = \theta$;

(ii) $g(-x) = g(x)$;

(iii) $g(x + y) \leq g(x) + g(y)$;

(iv) if (λ_n) is a sequence of scalars with $\lambda_n \to \lambda$ $(n \to \infty)$ and $x_n, a \in X$ with $x_n \to a$ $(n \to \infty)$, in the sense that $g(x_n - a) \to 0$ $(n \to \infty)$, in the sense that $g(\lambda_n x_n - \lambda a) \to 0$ $(n \to \infty)$.

The concept of paranorm is closely related to that of linear metric spaces. It is a generalization of that of absolute value (see [34, 35]).

2. Main Results

In this paper, we introduce the sequence space

$$ {}_2\mathrm{BV}_\sigma^I := \{x = (x_{ij}) \in \omega : \{i, j \in \mathbb{N} : |\phi_{mnij}(x) - L| \geq \epsilon\} \in I \quad \text{for some } L \in \mathbb{C}\}. \tag{18} $$

Theorem 14. ${}_2\mathrm{BV}_\sigma^I$ *is a linear space.*

Proof. Let $(x_{ij}), (y_{ij}) \in {}_2\mathrm{BV}_\sigma^I$ and α, β be two scalars in $\mathbb{C}$. Then for a given $\epsilon > 0$, we have

$$ \begin{gathered} \left\{i, j \in \mathbb{N} : |\phi_{mn,ij}(x) - L_1| \geq \frac{\epsilon}{2}\right\} \in I, \quad \text{for some } L_1 \in \mathbb{C}, \\ \left\{i, j \in \mathbb{N} : |\phi_{mn,ij}(y) - L_2| \geq \frac{\epsilon}{2}\right\} \in I, \quad \text{for some } L_2 \in \mathbb{C}. \end{gathered} \tag{19} $$

Now let,

$$ \begin{gathered} A_1 = \left\{i, j \in \mathbb{N} : |\phi_{mn,ij}(x) - L_1| < \frac{\epsilon}{2}\right\} \in I, \quad \text{for some } L_1 \in \mathbb{C}, \\ A_2 = \left\{i, j \in \mathbb{N} : |\phi_{mn,ij}(y) - L_2| < \frac{\epsilon}{2}\right\} \in I, \quad \text{for some } L_2 \in \mathbb{C} \end{gathered} \tag{20} $$

be such that $A_1^c, A_2^c \in I$. Now consider

$$ \begin{aligned} &|\phi_{mn,ij}(\alpha x + \beta y) - (\alpha L_1 + \beta L_2)| \\ &= |\phi_{mn,ij}(\alpha x) + \phi_{mn,ij}(\beta y) - \alpha L_1 - \beta L_2| \\ &= |\phi_{mn,ij}(\alpha x) - \alpha L_1 + \phi_{mn,ij}(\beta y) - \beta L_2| \\ &\leq |\phi_{mn,ij}(\alpha x) - \alpha L_1| + |\phi_{mn,ij}(\beta y) - \beta L_2| \\ &= |\alpha| |\phi_{mn,ij}(x) - L_1| + |\beta| |\phi_{mn,ij}(y) - L_2| \\ &\leq |\alpha| \frac{\epsilon}{2} + |\beta| \frac{\epsilon}{2} \\ &= (|\alpha| + |\beta|) \frac{\epsilon}{2} \\ &\leq \epsilon' \quad (\text{say}), \end{aligned} \tag{21} $$

this implies that the sequence space

$$ A_3 = \{i, j \in \mathbb{N} : |\phi_{mn,ij}(\alpha x + \beta y) - (\alpha L_1 + \beta L_2)| < \epsilon'\} \in I, \quad \text{for some } L_1, L_2 \in \mathbb{C}. \tag{22} $$

Hence, $(\alpha x + \beta y) \in {}_2\mathrm{BV}_\sigma^I$. Therefore, ${}_2\mathrm{BV}_\sigma^I$ is a linear space. □

Theorem 15. *The space* ${}_2\mathrm{BV}_\sigma^I$ *is a paranormed space, paranormed by*

$$ g(x_{ij}) = \sup_{ij} |\phi_{mn,ij}(x)|. \tag{23} $$

Proof. For $x = (x_{ij}) = 0$, $g(x_{ij}) = 0$ is trivial.
For $x = (x_{ij}) \neq 0$, $g(x_{ij}) \neq 0$, we have

(i) $g(x) = \sup_{ij}|\phi_{mn,ij}(x)| \geq 0$ for all $x \in {}_2\mathrm{BV}_\sigma^I$.

(ii) $g(-x) = \sup_{ij}|\phi_{mn,ij}(-x)| = \sup_{ij}|-\phi_{mn,ij}(x)| = \sup_{ij}|\phi_{mn,ij}(x)| = g(x)$ for all $x \in {}_2\mathrm{BV}_\sigma^I$.

(iii) $g(x + y) = \sup_{ij}|\phi_{mn,ij}(x + y)| \leq \sup_{ij}|\phi_{mn,ij}(x)| + \sup_{ij}|\phi_{mn,ij}(y)| = g(x) + g(y)$.

(iv) Let (λ_{ij}) be a sequence of scalars with $\lambda_{ij} \to \lambda$ $(ij \to \infty)$ and $(x) \in {}_2\mathrm{BV}_\sigma^I$ such that

$$ \phi_{mn,ij}(x) \longrightarrow L \quad (ij \longrightarrow \infty), \tag{24} $$

in the sense that

$$ g(\phi_{mn,ij}(x) - L) \longrightarrow 0 \quad (ij \longrightarrow \infty). \tag{25} $$

Therefore,

$$ \begin{aligned} &g(\lambda_{ij}\phi_{mn,ij}(x) - \lambda L) \\ &\leq g(\lambda_{ij}\phi_{mn,ij}(x)) - g(\lambda L) \\ &= \lambda_{ij} g(\phi_{mn,ij}(x)) - \lambda g(L) \longrightarrow 0 \quad \text{as } (ij \longrightarrow \infty). \end{aligned} \tag{26} $$

Hence, ${}_2\mathrm{BV}_\sigma^I$ is a paranormed space. □

Theorem 16. ${}_2\mathrm{BV}_\sigma^I$ *is a closed subspace of* ${}_2l_\infty^I$.

Proof. Let $(x_{ij}^{(pq)})$ be a cauchy sequence in ${}_2\mathrm{BV}_\sigma^I$ such that $x^{(pq)} \to x$. We show that $x \in {}_2\mathrm{BV}_\sigma^I$. Since $(x_{ij}^{(pq)}) \in {}_2\mathrm{BV}_\sigma^I$, then there exists a_{pq} such that

$$ \{i, j \in \mathbb{N} : |\phi_{mn,ij}(x^{(pq)}) - a_{pq}| \geq \epsilon\} \in I. \tag{27} $$

We need to show that

(i) (a_{pq}) converges to a.

(ii) If $U = \{i, j \in \mathbb{N} : |x_{ij} - a| < \epsilon\}$, then $U^c \in I$.

Since $(x_{ij}^{(pq)})$ is a cauchy sequence in ${}_2\mathrm{BV}_\sigma^I$, then for a given $\epsilon > 0$, there exists $k_0 \in \mathbb{N}$ such that

$$\sup_{ij} \left|\phi_{mn,ij}\left(x_{ij}^{(pq)}\right) - \phi_{mn,ij}\left(x_{ij}^{(rs)}\right)\right| < \frac{\epsilon}{3}, \quad \forall p, q, r, s \geq k_0. \tag{28}$$

For a given $\epsilon > 0$, we have

$$\begin{aligned} B_{pqrs} &= \left\{i, j \in \mathbb{N} : \left|\phi_{mn,ij}\left(x_{ij}^{(pq)}\right) - \phi_{mn,ij}\left(x_{ij}^{(rs)}\right)\right| < \frac{\epsilon}{3}\right\}, \\ B_{pq} &= \left\{i, j \in \mathbb{N} : \left|\phi_{mn,ij}\left(x_{ij}^{(pq)}\right) - a_{pq}\right| < \frac{\epsilon}{3}\right\}, \\ B_{rs} &= \left\{i, j \in \mathbb{N} : \left|\phi_{mn,ij}\left(x_{ij}^{(rs)}\right) - a_{rs}\right| < \frac{\epsilon}{3}\right\}. \end{aligned} \tag{29}$$

Then B_{pqrs}^c, B_{pq}^c, and $B_{rs}^c \in I$. Let $B^c = B_{pqrs}^c \cap B_{pq}^c \cap B_{rs}^c$, where $B = \{i, j \in \mathbb{N} : |a_{pq} - a_{rs}| < \epsilon\}$. Then $B^c \in I$. We choose $k_0 \in B^c$, then for each $p, q, r, s \geq k_0$, we have

$$\begin{aligned} &\left\{i, j \in \mathbb{N} : \left|a_{pq} - a_{rs}\right| < \epsilon\right\} \\ &\quad \supseteq \left\{i, j \in \mathbb{N} : \left|\phi_{mn,ij}\left(x_{ij}^{(pq)}\right) - a_{pq}\right| < \frac{\epsilon}{3}\right\} \\ &\quad \cap \left\{i, j \in \mathbb{N} : \left|\phi_{mn,ij}\left(x_{ij}^{(pq)}\right) - \phi_{mn,ij}\left(x_{ij}^{(rs)}\right)\right| < \frac{\epsilon}{3}\right\} \\ &\quad \cap \left\{i, j \in \mathbb{N} : \left|\phi_{mn,ij}\left(x_{ij}^{(rs)}\right) - a_{rs}\right| < \frac{\epsilon}{3}\right\}. \end{aligned} \tag{30}$$

Then (a_{pq}) is a cauchy sequence of scalars in $\mathbb{C}$, so there exists a scalar $a \in \mathbb{C}$ such that $(a_{pq}) \to a$, as $p, q \to \infty$.

For the next step, let $0 < \delta < 1$ be given. Then, we show that if

$$U = \left\{i, j \in \mathbb{N} : \left|\phi_{mn,ij}(x) - a\right| < \delta\right\}, \tag{31}$$

then $U^c \in I$. Since $\phi_{mn,ij}(x^{(pq)}) \to \phi_{mn,ij}(x)$, then there exists $p_0, q_0 \in \mathbb{N}$ such that

$$P = \left\{i, j \in \mathbb{N} : \left|\phi_{mn,ij}\left(x_{ij}^{(p_0q_0)}\right) - \phi_{mn,ij}(x)\right| < \frac{\delta}{3}\right\} \tag{32}$$

which implies that $P^c \in I$. The number p_0, q_0 can be so chosen that together with (32), we have

$$Q = \left\{i, j \in \mathbb{N} : \left|a_{p_0q_0} - a\right| < \frac{\delta}{3}\right\} \tag{33}$$

such that $Q^c \in I$. Since $\{i, j \in \mathbb{N} : |\phi_{mn,ij}(x_{ij}^{(p_0q_0)}) - a_{p_0q_0}| \geq \delta\} \in I$, then we have a subset S of $\mathbb{N}$ such that $S^c \in I$, where

$$S = \left\{i, j \in \mathbb{N} : \left|\phi_{mn,ij}\left(x_{ij}^{(p_0q_0)}\right) - a_{p_0q_0}\right| < \frac{\delta}{3}\right\}. \tag{34}$$

Let $U^c = P^c \cap Q^c \cap S^c$, where $U = \{i, j \in \mathbb{N} : |\phi_{mn,ij}(x) - a| < \delta\}$.

Therefore, for each $i, j \in U^c$, we have

$$\begin{aligned} &\left\{i, j \in \mathbb{N} : \left|\phi_{mn,ij}(x) - a\right| < \delta\right\} \\ &\quad \supseteq \left\{i, j \in \mathbb{N} : \left|\phi_{mn,ij}\left(x_{ij}^{(p_0q_0)}\right) - \phi_{mn,ij}(x)\right| < \frac{\delta}{3}\right\} \\ &\quad \cap \left\{i, j \in \mathbb{N} : \left|\phi_{mn,ij}\left(x_{ij}^{(p_0q_0)}\right) - a_{p_0q_0}\right| < \frac{\delta}{3}\right\} \\ &\quad \cap \left\{i, j \in \mathbb{N} : \left|a_{p_0q_0} - a\right| < \frac{\delta}{3}\right\}. \end{aligned} \tag{35}$$

Hence, the result ${}_2\mathrm{BV}_\sigma^I \subset {}_2l_\infty^I$ follows. □

Theorem 17. *The space ${}_2\mathrm{BV}_\sigma^I$ is nowhere dense subsets of ${}_2l_\infty^I$.*

Proof. Proof of the result follows from the previous theorem. □

Theorem 18. *The space ${}_2\mathrm{BV}_\sigma^I$ is solid and monotone.*

Proof. Let $(x_{ij}) \in {}_2\mathrm{BV}_\sigma^I$ and α_{ij} be a sequence of scalars with $|\alpha_{ij}| \leq 1$ for all $i, j \in \mathbb{N}$. Then, we have

$$\begin{aligned} \left|\alpha_{ij}\phi_{mn,ij}(x)\right| &\leq \left|\alpha_{ij}\right|\left|\phi_{mn,ij}(x)\right| \\ &\leq \left|\phi_{mn,ij}(x)\right|, \quad \forall i, j \in \mathbb{N}. \end{aligned} \tag{36}$$

The space ${}_2\mathrm{BV}_\sigma^I$ is solid follows from the following inclusion relation:

$$\left\{i, j \in \mathbb{N} : \left|\phi_{mn,ij}(x)\right| \geq \epsilon\right\} \supseteq \left\{i, j \in \mathbb{N} : \left|\alpha_{ij}\phi_{mn,ij}(x)\right| \geq \epsilon\right\}. \tag{37}$$

Also a sequence space is solid implies monotone. Hence, the space ${}_2\mathrm{BV}_\sigma^I$ is monotone. □

Theorem 19. *${}_2c_0^I \subset {}_2\mathrm{BV}_\sigma^I \subset {}_2l_\infty^I$ and the inclusions are proper.*

Proof. Let $x = (x_{ij}) \in {}_2c_0^I$. Then, we have $\{i, j \in \mathbb{N} : |x_{ij}| \geq \epsilon\} \in I$. Since ${}_2c_0 \subset {}_2\mathrm{BV}_\sigma$, $x = (x_{ij}) \in {}_2\mathrm{BV}_\sigma^I$ implies

$$\left\{i, j \in \mathbb{N} : \left|\phi_{mn,ij}(x)\right| \geq \epsilon\right\} \in I. \tag{38}$$

Now let,

$$\begin{aligned} A_1 &= \left\{i, j \in \mathbb{N} : \left|x_{ij}\right| < \epsilon\right\}, \\ A_2 &= \left\{i, j \in \mathbb{N} : \left|\phi_{mn,ij}(x)\right| < \epsilon\right\} \end{aligned} \tag{39}$$

be such that $A_1^c, A_2^c \in I$. As $l_\infty = \{x = (x_{ij}) : \sup_{ij}|x_{ij}| < \infty\}$, taking supremum over i, j we get $A_1^c \subset A_2^c$. Hence, ${}_2c_0^I \subset {}_2\mathrm{BV}_\sigma^I \subset {}_2l_\infty^I$.

Next we show that the inclusion is proper

(i) First for ${}_2c_0^I \subset {}_2\mathrm{BV}_\sigma^I$. Consider $x \in {}_2\mathrm{BV}_\sigma^I$, then by the definition

$$\begin{aligned} {}_2\mathrm{BV}_\sigma^I := \Big\{x = \left(x_{ij}\right) \in \omega : \\ \left\{i, j \in \mathbb{N} : \left|\phi_{mnij}(x) - L\right| \geq \epsilon\right\} \in I \\ \text{for some } L \in \mathbb{C}\Big\}, \end{aligned} \tag{40}$$

we have

$$\phi_{mn,ij}(x) = t_{mn,ij}(x) - t_{(m-1)(n-1),ij}(x), \quad (41)$$

where

$$t_{mn,ij}(x) = \frac{x_{ij} + x_{\sigma(ij)} + x_{\sigma^2(ij)} + \cdots + x_{\sigma^{mn}(ij)}}{mn}. \quad (42)$$

Therefore,

$$\begin{aligned} t_{mn,ij}(x) - t_{(m-1)(n-1),ij}(x) &= \frac{x_{ij} + x_{\sigma(ij)} + x_{\sigma^2(ij)} + \cdots + x_{\sigma^{mn}(ij)}}{mn} - \frac{x_{ij} + x_{\sigma(ij)} + x_{\sigma^2(ij)} + \cdots + x_{\sigma^{(m-1)(n-1)}(ij)}}{(m-1)(n-1)} \\ &= \frac{(m-1)(n-1)\left(x_{ij} + x_{\sigma(ij)} + \cdots + x_{\sigma^{mn}(ij)}\right)}{mn(m-1)(n-1)} - \frac{mn\left(x_{ij} + x_{\sigma(ij)} + \cdots + x_{\sigma^{(m-1)(n-1)}(ij)}\right)}{mn(m-1)(n-1)}. \end{aligned} \quad (43)$$

On solving, we get

$$\phi_{mn,ij}(x) = \frac{mnx_{\sigma^{mn}(ij)}}{mn(m-1)(n-1)} + \frac{(1-m-n)\left(x_{ij} + x_{\sigma(ij)} + x_{\sigma^2(ij)} + \cdots + x_{\sigma^{mn}(ij)}\right)}{mn(m-1)(n-1)}. \quad (44)$$

As σ is a translation map, that is, $\sigma(n) = n+1$, we have

$$\phi_{mn,ij}(x) = \frac{mnx_{(i+m)(j+n)}}{mn(m-1)(n-1)} + \frac{(1-m-n)\left(x_{ij} + x_{(i+1)(j+1)} + \cdots + x_{(i+m)(j+n)}\right)}{mn(m-1)(n-1)}. \quad (45)$$

Taking lim $i, j \to \infty$, we have

$$\begin{aligned} &\lim_{(i,j)\to\infty} \phi_{mn,ij}(x) \\ &= \lim_{(i,j)\to\infty} \left[\left(mnx_{(i+m)(j+n)} + (1-m-n) \times \left(x_{ij} + x_{(i+1)(j+1)} + \cdots + x_{(i+m)(j+n)}\right)\right) \times (mn(m-1)(n-1))^{-1}\right], \\ &L(mn(m-1)(n-1)) \\ &= \lim_{(i,j)\to\infty} \left[mnx_{(i+m)(j+n)} + (1-m-n) \times \left(x_{ij} + x_{(i+1)(j+1)} + \cdots + x_{(i+m)(j+n)}\right)\right]. \end{aligned} \quad (46)$$

Since $m, n, L \neq 0$, therefore $\lim_{i,j\to\infty}\phi_{mn,ij}(x) \neq 0$ which implies that $x \notin ({}_2c_0^I)$. Hence, we get that the inclusion is proper.

(ii) Second for ${}_2\mathrm{BV}_\sigma^I \subset {}_2l_\infty^I$.

The result of this part follows from the proof of Theorem 18. □

Theorem 20. *${}_2c^I \subset {}_2\mathrm{BV}_\sigma^I \subset {}_2l_\infty^I$ and the inclusions are proper.*

Proof. Let $x = (x_{ij}) \in {}_2c^I$. Then, we have

$$\left\{i, j \in \mathbb{N} : \left|x_{ij} - L\right| \geq \epsilon\right\} \in I. \quad (47)$$

Since ${}_2c \subset {}_2\mathrm{BV}_\sigma \subset {}_2l_\infty$, which implies $x = (x_{ij}) \in {}_2\mathrm{BV}_\sigma^I$ implies

$$\left\{i, j \in \mathbb{N} : \left|\phi_{mn,ij}(x) - L\right| \geq \epsilon\right\} \in I. \quad (48)$$

Now let,

$$\begin{aligned} B_1 &= \left\{i, j \in \mathbb{N} : \left|\phi_{i,j} - L\right| < \epsilon\right\}, \\ B_2 &= \left\{i, j \in \mathbb{N} : \left|\phi_{mn,ij}(x) - L\right| < \epsilon\right\} \end{aligned} \quad (49)$$

be such that $B_1^c, B_2^c \in I$. As

$${}_2l_\infty = \left\{x = \left(x_{ij}\right) : \sup_{ij}\left|x_{ij}\right| < \infty\right\}, \quad (50)$$

taking lim sup over i, j, we get $B_1^c \subset B_2^c$. Hence, ${}_2c^I \subset {}_2\mathrm{BV}_\sigma^I \subset {}_2l_\infty^I$.

Next, we show that the inclusion is proper

(i) First for ${}_2c^I \subset {}_2\mathrm{BV}_\sigma^I$. We show that ${}_2c^I \subsetneq {}_2\mathrm{BV}_\sigma^I$.

Let $x = (x_{ij}) \in {}_2\mathrm{BV}_\sigma^I$, then by the definition

$${}_2\mathrm{BV}_\sigma^I := \left\{x = \left(x_{ij}\right) \in \omega : \left\{i, j \in \mathbb{N} : \left|\phi_{mnij}(x) - L\right| \geq \epsilon\right\} \in I \text{ for some } L \in \mathbb{C}\right\}. \quad (51)$$

We have, $|\phi_{mnij}(x) - L| \geq \epsilon$. We say that the $I-\lim(\phi_{mnij}(x)) = L$.

Now considering the case when $|\phi_{mnij}(x) - L| < \epsilon$, then

$$\left|t_{mn,ij}(x) - t_{(m-1)(n-1),ij}(x) - L\right| < \epsilon, \quad (52)$$

when $m, n = 0$, then we have $\phi_{mnij}(x) = t_{ij}(x) = x_{ij}$. Therefore we get,

$$\left|x_{ij} - L\right| < \epsilon \quad \forall i, j \in \mathbb{N}. \quad (53)$$

Hence, $x \notin {}_2c^I = \{i, j \in \mathbb{N} : |x_{ij} - L| \geq \epsilon\} \in I$. Hence, the inclusion is proper.

(ii) Second for ${}_2\mathrm{BV}_\sigma^I \subset {}_2l_\infty^I$.

The result follows from the proof of Theorem 18. □

Acknowledgment

The authors would like to record their gratitude to the reviewer for his careful reading and making some useful corrections which improved the presentation of the paper.

References

[1] A. K. Vakeel and S. Tabassum, "On some new double sequence spaces of invariant means defined by Orlicz functions," *Communications de la Faculté des Sciences de l'Université d'Ankara Séries A*, vol. 60, no. 2, pp. 11–21, 2011.

[2] E. D. Habil, "Double sequences and double series," *The Islamic University Journal, Series of Natural Studies and Engineering*, vol. 14, pp. 1–32, 2006.

[3] M. Mursaleen, "On some new invariant matrix methods of summability," *The Quarterly Journal of Mathematics*, vol. 34, no. 133, pp. 77–86, 1983.

[4] S. Banach, *Theorie des Operations Lineaires*, Warszawa, Poland, 1932.

[5] G. G. Lorentz, "A contribution to the theory of divergent sequences," *Acta Mathematica*, vol. 80, pp. 167–190, 1948.

[6] M. Mursaleen, "Matrix transformations between some new sequence spaces," *Houston Journal of Mathematics*, vol. 9, no. 4, pp. 505–509, 1983.

[7] V. A. Khan, "On a new sequence space defined by Orlicz functions," *Communications de la Faculté des Sciences de l'Université d'Ankara Séries A*, vol. 57, no. 2, pp. 25–33, 2008.

[8] M. Mursaleen and S. A. Mohiuddine, "Some new double sequence spaces of invariant means," *Glasnik Matematički*, vol. 45, no. 1, pp. 139–153, 2010.

[9] Z. U. Ahmad and Mursaleen, "An application of Banach limits," *Proceedings of the American Mathematical Society*, vol. 103, no. 1, pp. 244–246, 1988.

[10] M. Mursaleen and A. Alotaibi, "On I-convergence in random 2-normed spaces," *Mathematica Slovaca*, vol. 61, no. 6, pp. 933–940, 2011.

[11] M. Mursaleen, S. A. Mohiuddine, and O. H. H. Edely, "On the ideal convergence of double sequences in intuitionistic fuzzy normed spaces," *Computers & Mathematics with Applications*, vol. 59, no. 2, pp. 603–611, 2010.

[12] M. Mursaleen and S. A. Mohiuddine, "On ideal convergence of double sequences in probabilistic normed spaces," *Mathematical Reports*, vol. 12, no. 4, pp. 359–371, 2010.

[13] M. Mursaleen and S. A. Mohiuddine, "On ideal convergence in probabilistic normed spaces," *Mathematica Slovaca*, vol. 62, no. 1, pp. 49–62, 2012.

[14] M. Mursaleen, A. Alotaibi, and M. A. Alghamdi, "I-summability and I-approximation through invariant mean," *Journal of Computational Analysis and Applications*, vol. 14, no. 6, pp. 1049–1058, 2012.

[15] R. A. Raimi, "Invariant means and invariant matrix methods of summability," *Duke Mathematical Journal*, vol. 30, pp. 81–94, 1963.

[16] P. Schaefer, "Infinite matrices and invariant means," *Proceedings of the American Mathematical Society*, vol. 36, pp. 104–110, 1972.

[17] E. Savas and B. E. Rhoades, "On some new sequence spaces of invariant means defined by Orlicz functions," *Mathematical Inequalities & Applications*, vol. 5, no. 2, pp. 271–281, 2002.

[18] A. K. Vakeel, K. Ebadullah, and S. Suantai, "On a new I-convergent sequence space," *Analysis*, vol. 32, no. 3, pp. 199–208, 2012.

[19] A. K. Vakeel and K. Ebadullah, "On some I-Convergent sequence spaces defined by a modullus function," *Theory and Applications of Mathematics and Computer Science*, vol. 1, no. 2, pp. 22–30, 2011.

[20] A. K. Vakeel and K. Ebadullah, "I-convergent difference sequence spaces defined by a sequence of moduli," *Journal of Mathematical and Computational Science*, vol. 2, no. 2, pp. 265–273, 2012.

[21] A. Komisarski, "Pointwise I-convergence and I-convergence in measure of sequences of functions," *Journal of Mathematical Analysis and Applications*, vol. 340, no. 2, pp. 770–779, 2008.

[22] V. Kumar, "On I and I^*-convergence of double sequences," *Mathematical Communications*, vol. 12, no. 2, pp. 171–181, 2007.

[23] A. Şahiner, M. Gürdal, S. Saltan, and H. Gunawan, "Ideal convergence in 2-normed spaces," *Taiwanese Journal of Mathematics*, vol. 11, no. 5, pp. 1477–1484, 2007.

[24] P. Kostyrko, T. Šalát, and W. Wilczyński, "I-convergence," *Real Analysis Exchange*, vol. 26, no. 2, pp. 669–685, 2000.

[25] T. Šalát, B. C. Tripathy, and M. Ziman, "On some properties of I-convergence," *Tatra Mountains Mathematical Publications*, vol. 28, pp. 279–286, 2004.

[26] T. Šalát, B. C. Tripathy, and M. Ziman, "On I-convergence field," *Italian Journal of Pure and Applied Mathematics*, no. 17, pp. 45–54, 2005.

[27] B. C. Tripathy and B. Hazarika, "Paranorm I-convergent sequence spaces," *Mathematica Slovaca*, vol. 59, no. 4, pp. 485–494, 2009.

[28] V. A. Khan and K. Ebadullah, "On a new difference sequence space of invariant means defined by Orlicz functions," *Bulletin of the Allahabad Mathematical Society*, vol. 26, no. 2, pp. 259–272, 2011.

[29] A. K. Vakeel and T. Sabiha, "On ideal convergent difference double sequence spaces in 2-normed spaces defined by Orlicz function," *JMI International Journal of Mathematical Sciences*, vol. 1, no. 2, pp. 1–9, 2010.

[30] P. Das, P. Kostyrko, W. Wilczyński, and P. Malik, "I and I^*-convergence of double sequences," *Mathematica Slovaca*, vol. 58, no. 5, pp. 605–620, 2008.

[31] M. Gurdal and M. B. Huban, "On I-convergence of double sequences in the topology induced by random 2-norms," *Matematički Vesnik*, vol. 65, no. 3, pp. 1–13, 2013.

[32] M. Gürdal and A. Şahiner, "Extremal I-limit points of double sequences," *Applied Mathematics E-Notes*, vol. 8, pp. 131–137, 2008.

[33] S. Roy, "Some new type of fuzzy I-convergent double difference sequence spaces," *International Journal of Soft Computing and Engineering*, vol. 1, pp. 429–431, 2012.

[34] I. J. Maddox, *Elements of Functional Analysis*, Cambridge University Press, 1970.

[35] H. Nakano, "Concave modulars," *Journal of the Mathematical Society of Japan*, vol. 5, pp. 29–49, 1953.

On Right Caputo Fractional Ostrowski Inequalities Involving Three Functions

Deepak B. Pachpatte

Department of Mathematics, Dr. B.A.M. University, Aurangabad, Maharashtra 431004, India

Correspondence should be addressed to Deepak B. Pachpatte; pachpatte@gmail.com

Academic Editor: Abdallah El Hamidi

We establish Ostrowski inequalities involving three functions in right Caputo fractional derivative in L_p spaces.

1. Introduction

In 1938, Ostrowski proved the following useful inequality.

Let $f : [a,b] \to R$ be continuous on $[a,b]$ and differentiable on (a,b) whose derivative $f' : [a,b] \to R$ is bounded on (a,b), that is, $\|f'\|_\infty = \sup_{t\in(a,b)}|f'(t)| < \infty$. Then

$$\left| f(x) - \frac{1}{b-a}\int_a^b f(t)\,dt \right| \le \left[\frac{1}{4} + \frac{\left(x - \dfrac{a+b}{2}\right)}{(b-a)^2} \right] \|f'\|_\infty, \tag{1}$$

for any $x \in [a,b]$. The constant $1/4$ is best possible.

In [1, 2] Pachpatte has proved Ostrowski inequality in three independent variables.

In past few years many authors have obtained various generalisation and variant of the above type of inequality and other on fractional as well as time scale calculus see [3–6].

Here we give some basic definition from fractional calculus used in [7–9].

Definition 1. Let $f \in L_1[a,b]$, $\alpha > 0$. The right and left Riemann-Liouville integrals $I_{a+}^{\alpha} f(x)$ and $I_{b-}^{\alpha} f(x)$ of order $\alpha > 0$ with $a \ge 0$ are defined by

$$I_{a+}^{\alpha} f(x) = \frac{1}{\Gamma(\alpha)} \int_a^x (x-t)^{(\alpha-1)} f(t)\,dt, \quad x > a,$$

$$I_{b-}^{\alpha} f(x) = \frac{1}{\Gamma(\alpha)} \int_x^b (t-x)^{(\alpha-1)} f(t)\,dt, \quad x < b, \tag{2}$$

respectively, where $\Gamma(\alpha) = \int_0^\infty e^{-t} t^{\alpha-1}\,dt$ and $I_{a+}^{0} f(x) = I_{b-}^{0} f(x) = f(x)$.

Definition 2 (see [10, page 2]). Let $f \in AC^m([a,b])$ ($f^{(m-1)}$ be in $AC([a,b])$), $m \in N$, $m = \lceil\alpha\rceil$, $\alpha > 0$ ($\lceil\cdot\rceil$ the ceiling of the number). We define the right Caputo fractional derivative of order $\alpha > 0$ by

$$D_{b-}^{\alpha} f(x) = \frac{(-1)^m}{\Gamma(m-\alpha)} \int_x^b (t-x)^{m-\alpha-1} f^{(m)}(t)\,dt, \quad x \le b. \tag{3}$$

If $\alpha = m \in N$, then

$$D_{b-}^{m} f(x) = (-1)^m f^{(m)}(x), \quad \forall x \in [a,b]. \tag{4}$$

If $x > b$, we define $D_{b-}^{m} f(x) = 0$.

Definition 3 (see [9, page 74]). Let $f, g \in AC^m([a,b])$, $\alpha > 0$ with $a \ge 0$, and then the fractional integral $I_{b-}^{\alpha} f(x) g(x)$ is defined by

$$I_{b-}^{\alpha} f(x)\,g(x) = \frac{1}{\Gamma(\alpha)} \int_x^b (t-x)^{(\alpha-1)} f(t)\,g(t)\,dt, \quad x < b. \tag{5}$$

We give here the theorems proved in [10].

Theorem 4. *Let $\alpha > 0$, $m = [\alpha]$, $f \in AC^m([a,b])$. Assume that $f^k(b) = 0$, $k = 1, \ldots, m-1$ and $D_{b-}^{a} f \in L_\infty([a,b])$. Then*

$$\left|\frac{1}{b-a}\int_a^b f(x)\,dx - f(b)\right| \le \frac{\|D_{b-}^{\alpha} f\|_{\infty,[a,b]}}{\Gamma(\alpha+2)}(b-a)^{\alpha}. \quad (6)$$

Theorem 5. *Let $\alpha > 0$, $m = [\alpha]$, $f \in AC^m([a,b])$. Assume that $f^k(b) = 0$, $k = 1, \ldots, m-1$ and $D_{b-}^{a} f \in L_1[a,b]$. Then*

$$\left|\frac{1}{b-a}\int_a^b f(x)\,dx - f(b)\right| \le \frac{\|D_{b-}^{\alpha} f\|_{L_1([a,b])}}{\Gamma(\alpha+1)}(b-a)^{\alpha-1}. \quad (7)$$

Theorem 6. *Let $p, q > 1$; $1/p + 1/q = 1$, $\alpha > 1 - 1/p$, $m = [\alpha]$, $f \in AC^m([a,b])$. Assume that $f^k(b) = 0$, $k = 1, \ldots, m-1$, and $D_{b-}^{\alpha} f, h \in L_q[a,b]$. Then*

$$\left|\frac{1}{b-a}\int_a^b f(x)\,dx - f(b)\right| \le \frac{\|D_{b-}^{\alpha} f\|_{L_q([a,b])}}{\Gamma(\alpha)(p(\alpha-1)+1)^{1/p}(\alpha+1/p)}(b-a)^{\alpha-1+1/p}. \quad (8)$$

2. Main Results

Our main results are given in the following theorems.

Theorem 7. *Let $\alpha > 0$, $m = [\alpha]$, $f, g, h \in AC^m([a,b])$. Assume that $f^k(b) = g^k(b) = h^k(b) = 0$, $k = 1, \ldots, m-1$ and $D_{b-}^{a} f, D_{b-}^{a} g, D_{b-}^{a} h \in L_\infty([a,b])$. Then*

$$\begin{aligned}\Bigg|3\int_a^b f(x)g(x)h(x)\,dx &- \int_a^b [f(b)g(x)h(x) + f(x)g(b)h(x) + f(x)g(x)h(b)]\,dx\Bigg| \\ &\le \|D_{b-}^{\alpha} f\|_\infty I_\alpha^{\alpha+1}|g(b)h(b)| + \|D_{b-}^{\alpha} g\|_\infty I_\alpha^{\alpha+1}|f(b)h(b)| \\ &\quad + \|D_{b-}^{\alpha} h\|_\infty I_\alpha^{\alpha+1}|f(b)g(b)|.\end{aligned} \quad (9)$$

Proof. Let $x \in [a,b]$ we have

$$f(x) - f(b) = \frac{1}{\gamma\alpha}\int_x^b (t-x)^{(\alpha-1)} D_{b-}^{\alpha} f(t)\,dt, \quad (10)$$

$$g(x) - g(b) = \frac{1}{\gamma\alpha}\int_x^b (t-x)^{(\alpha-1)} D_{b-}^{\alpha} g(t)\,dt, \quad (11)$$

$$h(x) - h(b) = \frac{1}{\gamma\alpha}\int_x^b (t-x)^{(\alpha-1)} D_{b-}^{\alpha} h(t)\,dt. \quad (12)$$

Multiplying (10), (11), and (12) by $g(x)h(x)$, $h(x)f(x)$, and $f(x)g(x)$, respectively, and adding them, we have

$$\begin{aligned}3f(x)g(x)h(x) &- f(b)g(x)h(x) - f(x)g(b)h(x) - f(x)g(x)h(b) \\ &= \frac{g(x)h(x)}{\Gamma(\gamma)}\int_x^b (t-x)^{\alpha-1} D_{b-}^{\alpha} f(t)\,dt \\ &\quad + \frac{h(x)f(x)}{\Gamma(\gamma)}\int_x^b (t-x)^{\alpha-1} D_{b-}^{\alpha} g(t)\,dt \\ &\quad + \frac{f(x)g(x)}{\Gamma(\gamma)}\int_x^b (t-x)^{\alpha-1} D_{b-}^{\alpha} h(t)\,dt.\end{aligned} \quad (13)$$

Integrating both sides of (13) with respect to x and rewriting above equation we have

$$\begin{aligned}3\int_a^b f(x)g(x)h(x)\,dx &- \int_a^b [f(b)g(x)h(x) + f(x)g(b)h(x) + f(x)g(x)h(b)]\,dx \\ &= \int_a^b \frac{g(x)h(x)}{\Gamma(\gamma)}\left(\int_x^b (t-x)^{\alpha-1} D_{b-}^{\alpha} f(t)\,dt\right) \\ &\quad + \int_a^b \frac{h(x)f(x)}{\Gamma(\gamma)}\left(\int_x^b (t-x)^{\alpha-1} D_{b-}^{\alpha} g(t)\,dt\right) \\ &\quad + \int_a^b \frac{f(x)g(x)}{\Gamma(\gamma)}\left(\int_x^b (t-x)^{\alpha-1} D_{b-}^{\alpha} h(t)\,dt\right).\end{aligned} \quad (14)$$

From (14) and using the properties of modulus we have

$$\begin{aligned}\Bigg|3\int_a^b f(x)g(x)h(x)\,dx &- \int_a^b [f(b)g(x)h(x) + f(x)g(b)h(x) + f(x)g(x)h(b)]\,dx\Bigg| \\ &= \int_a^b \frac{|g(x)h(x)|}{\Gamma(\gamma)}\left(\int_x^b (t-x)^{\alpha-1}|D_{b-}^{\alpha} f(t)\,dt|\right) \\ &\quad + \int_a^b \frac{|h(x)f(x)|}{\Gamma(\gamma)}\left(\int_x^b (t-x)^{\alpha-1}|D_{b-}^{\alpha} g(t)\,dt|\right) \\ &\quad + \int_a^b \frac{|f(x)g(x)|}{\Gamma(\gamma)}\left(\int_x^b (t-x)^{\alpha-1}|D_{b-}^{\alpha} h(t)\,dt|\right).\end{aligned} \quad (15)$$

It is easy to observe that

$$\left|3\int_a^b f(x)\,g(x)\,h(x)\,dx - \int_a^b [f(b)\,g(x)\,h(x) + f(x)\,g(b)\,h(x) + f(x)\,g(x)\,h(b)]\,dx\right|$$
$$\leq \frac{\|D_{b-}^{\alpha}f\|_{\infty}}{\gamma\alpha}\int_a^b |g(x)\,h(x)|\left(\int_x^b (t-x)^{\alpha-1}dt\right)dx + \frac{\|D_{b-}^{\alpha}g\|_{\infty}}{\gamma\alpha}\int_a^b |h(x)\,f(x)|\left(\int_x^b (t-x)^{\alpha-1}dt\right)dx + \frac{\|D_{b-}^{\alpha}h\|_{\infty}}{\gamma\alpha}\int_a^b |f(x)\,g(x)|\left(\int_x^b (t-x)^{\alpha-1}dt\right)dx$$
$$= \frac{\|D_{b-}^{\alpha}f\|_{\infty}}{\Gamma(\alpha+1)}\int_a^b (b-x)^{\alpha}|g(x)\,h(x)|\,dx + \frac{\|D_{b-}^{\alpha}g\|_{\infty}}{\Gamma(\alpha+1)}\int_a^b (b-x)^{\alpha}|h(x)\,f(x)|\,dx + \frac{\|D_{b-}^{\alpha}h\|_{\infty}}{\Gamma(\alpha+1)}\int_a^b (b-x)^{\alpha}|f(x)\,g(x)|\,dx$$
$$= \|D_{b-}^{\alpha}f\|_{\infty}I_{a+}^{\alpha+1}|g(b)\,h(b)| + \|D_{b-}^{\alpha}g\|_{\infty}I_{a+}^{\alpha+1}|f(b)\,h(b)| + \|D_{b-}^{\alpha}h\|_{\infty}I_{a+}^{\alpha+1}|f(b)\,g(b)|. \tag{16}$$

The proof of the theorem is complete.

Remark 8. If we take $h(t) = 1$, $g(t) = 1$ and hence $\|D_{b-}^{a}h\|_{\infty} = 0$, $\|D_{b-}^{a}g\|_{\infty} = 0$ in Theorem 7, then we get Theorem 4.

Theorem 9. *Let $\alpha \geq 0$, $m = [\alpha]$, $f, g, h \in AC^m([a,b])$. Assume that $f^k(b) = g^k(b) = h^k(b) = 0$, $k = 1, \ldots, m-1$, and $D_{b-}^{\alpha}f, D_{b-}^{\alpha}g, D_{b-}^{\alpha}h \in L_1[a,b]$. Then*

$$\left|3\int_a^b f(x)\,g(x)\,h(x)\,dx - \int_a^b [f(b)\,g(x)\,h(x) + f(x)\,g(b)\,h(x) + f(x)\,g(x)\,h(b)]\,dx\right| \tag{17}$$
$$\leq \|D_{b-}^{\alpha}f\|_{L_1([a,b])}I_{a+}^{\alpha}|g(b)\,h(b)| + \|D_{b-}^{\alpha}g\|_{L_1([a,b])}I_{a+}^{\alpha}|f(b)\,h(b)| + \|D_{b-}^{\alpha}h\|_{L_1([a,b])}I_{a+}^{\alpha}|f(b)\,g(b)|.$$

Proof. From (15) we have

$$\left|3\int_a^b f(x)\,g(x)\,h(x)\,dx - \int_a^b [f(b)\,g(x)\,h(x) + f(x)\,g(b)\,h(x) + f(x)\,g(x)\,h(b)]\,dx\right|$$
$$\leq \int_a^b \frac{|g(x)\,h(x)|}{\Gamma(\alpha)}\left(\int_x^b (t-x)^{\alpha-1}|D_{b-}^{\alpha}f(t)\,dt|\right)dx + \int_a^b \frac{|h(x)\,f(x)|}{\Gamma(\alpha)}\left(\int_x^b (t-x)^{\alpha-1}|D_{b-}^{\alpha}g(t)\,dt|\right)dx + \int_a^b \frac{|f(x)\,g(x)|}{\Gamma(\alpha)}\left(\int_x^b (t-x)^{\alpha-1}|D_{b-}^{\alpha}h(t)\,dt|\right)dx$$
$$\leq \int_a^b \frac{|g(x)\,h(x)|}{\Gamma(\alpha)}(b-x)^{\alpha-1}\left(\int_x^b |D_{b-}^{\alpha}f(t)\,dt|\right)dx + \int_a^b \frac{|h(x)\,f(x)|}{\Gamma(\alpha)}(b-x)^{\alpha-1}\left(\int_x^b |D_{b-}^{\alpha}g(t)\,dt|\right)dx + \int_a^b \frac{|f(x)\,g(x)|}{\Gamma(\alpha)}(b-x)^{\alpha-1}\left(\int_x^b |D_{b-}^{\alpha}h(t)\,dt|\right)dx$$
$$= \|D_{b-}^{\alpha}f\|_{L_1([a,b])}\int_a^b \frac{|g(x)\,h(x)|}{\Gamma(\alpha)}(b-x)^{\alpha-1}dx + \|D_{b-}^{\alpha}g\|_{L_1([a,b])}\int_a^b \frac{|h(x)\,f(x)|}{\Gamma(\alpha)}(b-x)^{\alpha-1}dx + \|D_{b-}^{\alpha}g\|_{L_1([a,b])}\int_a^b \frac{|f(x)\,g(x)|}{\Gamma(\alpha)}(b-x)^{\alpha-1}dx$$
$$\leq \|D_{b-}^{\alpha}f\|_{L_1([a,b])}I_{a+}^{\alpha}|g(b)\,h(b)| + \|D_{b-}^{\alpha}g\|_{L_1([a,b])}I_{a+}^{\alpha}|f(b)\,h(b)| + \|D_{b-}^{\alpha}h\|_{L_1([a,b])}I_{a+}^{\alpha}|f(b)\,g(b)|. \tag{18}$$

This proves the theorem.

Remark 10. If we take $h(t) = 1$, $g(t) = 1$, and hence $\|D_{b-}^{a}h\|_{\infty} = 0$, $\|D_{b-}^{a}g\|_{\infty} = 0$ in Theorem 9, then we get Theorem 5.

Theorem 11. *Let $p, q > 1$; $1/p + 1/q = 1$, $\alpha > 1 - 1/p$, $m = [\alpha]$, $f, g \in AC^m([a,b])$. Assume that $f^k(b) = g^k(b) = h^k(b) = 0$,*

$k = 1, \dots, m-1$, *and* $D^{\alpha}_{b-}f, D^{\alpha}_{b-}g, D^{\alpha}_{b-}h \in L_q[a,b]$. *Then*

$$\begin{aligned}&\Big|3\int_a^b f(x)g(x)h(x)\,dx - \int_a^b [f(b)g(x)h(x) + f(x)g(b)h(x) + f(x)g(x)h(b)]\,dx\Big| \\ &\leq \Gamma\left(\alpha+\frac{1}{p}\right)\Big(\|D^{\alpha}_{b-}f\|_{L_q([a,b])} I^{\alpha}_{a+}|g(b)h(b)| + \|D^{\alpha}_{b-}g\|_{L_q([a,b])} I^{\alpha}_{a+}|f(b)h(b)| + \|D^{\alpha}_{b-}h\|_{L_q([a,b])} I^{\alpha}_{a+}|f(b)g(b)|\Big).\end{aligned} \tag{19}$$

Proof. From (15) we have

$$\begin{aligned}&\Big|3\int_a^b f(x)g(x)h(x)\,dx - \int_a^b [f(b)g(x)h(x) + f(x)g(b)h(x) + f(x)g(x)h(b)]\,dx\Big| \\ &\leq \int_a^b \frac{|g(x)h(x)|}{\Gamma(\alpha)}\left(\int_x^b (t-x)^{\alpha-1}|D^{\alpha}_{b-}f(t)\,dt|\right)dx \\ &+ \int_a^b \frac{|h(x)f(x)|}{\Gamma(\alpha)}\left(\int_x^b (t-x)^{\alpha-1}|D^{\alpha}_{b-}g(t)\,dt|\right)dx \\ &+ \int_a^b \frac{|f(x)g(x)|}{\Gamma(\alpha)}\left(\int_x^b (t-x)^{\alpha-1}|D^{\alpha}_{b-}h(t)\,dt|\right)dx.\end{aligned} \tag{20}$$

Applying Holder's inequality to (21), we get

$$\begin{aligned}&\Big|3\int_a^b f(x)g(x)h(x)\,dx - \int_a^b [f(b)g(x)h(x) + f(x)g(b)h(x) + f(x)g(x)h(b)]\,dx\Big| \\ &\leq \int_a^b \frac{|g(x)h(x)|}{\Gamma(\alpha)}\left(\int_x^b (t-x)^{(\alpha-1)p}dt\right)^{1/p} \times \left(\int_x^b |D^{\alpha}_{b-}f(t)\,dt|^q dt\right)^{1/q} dx \\ &+ \int_a^b \frac{|h(x)f(x)|}{\Gamma(\alpha)}\left(\int_x^b (t-x)^{(\alpha-1)p}dt\right)^{1/p} \times \left(\int_x^b |D^{\alpha}_{b-}g(t)\,dt|^q dt\right)^{1/q} dx \\ &+ \int_a^b \frac{|f(x)g(x)|}{\Gamma(\alpha)}\left(\int_x^b (t-x)^{(\alpha-1)p}dt\right)^{1/p} \times \left(\int_x^b |D^{\alpha}_{b-}h(t)\,dt|^q dt\right)^{1/q} dx \\ &= \frac{\|D^{\alpha}_{b-}f\|}{\Gamma(\alpha)} L_q([a,b]) \int_a^b |g(x)h(x)| \times \left(\int_x^b (t-x)^{(\alpha-1)p}dt\right)^{1/p} \\ &+ \frac{\|D^{\alpha}_{b-}g\|}{\Gamma(\alpha)} L_q([a,b]) \int_a^b |h(x)f(x)| \times \left(\int_x^b (t-x)^{(\alpha-1)p}dt\right)^{1/p} \\ &+ \frac{\|D^{\alpha}_{b-}h\|}{\Gamma(\alpha)} L_q([a,b]) \int_a^b |f(x)g(x)| \times \left(\int_x^b (t-x)^{(\alpha-1)p}dt\right)^{1/p}.\end{aligned} \tag{21}$$

We have

$$\left(\int_x^b (t-x)^{(\alpha-1)p}dt\right)^{1/p} = \frac{(b-x)^{\alpha-1+1/p}}{(p(\alpha-1)+1)^{1/p}}. \tag{22}$$

Substituting (22) into (21), we get the required inequality.

Remark 12. If we take $h(t) = 1, g(t) = 1$ and hence $\|D^{a}_{b-}h\|_{\infty} = 0$, $\|D^{a}_{b-}g\|_{\infty} = 0$ in Theorem 11, then we get Theorem 6.

References

[1] B. G. Pachpatte, "On an inequality of Ostrowski type in three independent variables," *Journal of Mathematical Analysis and Applications*, vol. 249, no. 2, pp. 583–591, 2000.

[2] B. G. Pachpatte, "New Ostrowski and Grüss type inequalities," *Analele Stintifice Ale Universitath Al.I Cuza Iasi, Tomul LI. s. I Mathematica*, vol. 51, no. 2, pp. 377–386, 2005.

[3] G. A. Anastassiou, *Intelligent Mathematics: Computational Analysis*, Springer, Berlin, Germany, 2011.

[4] G. A. Anastassiou, "On right fractional calculus," *Chaos, Solitons and Fractals*, vol. 42, no. 1, pp. 365–376, 2009.

[5] E. A. Bohner, M. Bohner, and T. Matthews, "Time scales ostrowski and gruss type inequalities involving three functions," *Dynamics and Systems Theory*, vol. 12, no. 2, pp. 119–135, 2012.

[6] M. Bohner and T. Matthews, "Ostrowski inequalities on time scales," *Journal of Inequalities in Pure and Applied Mathematics*, vol. 9, no. 1, article 6, 8 pages, 2008.

[7] G. A. Anastassiou, *Fractional Differentiation Inequalities*, Springer, Dordrecht, The Netherlands, 2010.

[8] I. Podlubny, *Fractional Differential Equations*, vol. 198, Academic Press, San Diego, Calif, USA, 1999.

[9] K. S. Miller and B. Ross, *An Introduction to the Fractional Calculus and Fractional Differential Equations*, John Wiley & Sons, New York, NY, USA, 1993.

[10] G. A. Anastassiou, *Advances in Fractional Inequalities*, Springer, New York, NY, USA, 2011.

4

A New Class of Analytic Functions Defined by Using Salagean Operator

R. M. El-Ashwah,[1] M. K. Aouf,[2] A. A. M. Hassan,[3] and A. H. Hassan[3]

[1] *Department of Mathematics, Faculty of Science, Damietta University, New Damietta 34517, Egypt*
[2] *Department of Mathematics, Faculty of Science, Mansoura University, Mansoura 33516, Egypt*
[3] *Department of Mathematics, Faculty of Science, Zagazig University, Zagazig 44519, Egypt*

Correspondence should be addressed to A. H. Hassan; alaahassan1986@yahoo.com

Academic Editor: Yaozhong Hu

We derive some results for a new class of analytic functions defined by using Salagean operator. We give some properties of functions in this class and obtain numerous sharp results including for example, coefficient estimates, distortion theorem, radii of star-likeness, convexity, close-to-convexity, extreme points, integral means inequalities, and partial sums of functions belonging to this class. Finally, we give an application involving certain fractional calculus operators that are also considered.

1. Introduction

Let $\mathcal{A}$ denote the class of functions of the form

$$f(z) = z + \sum_{k=2}^{\infty} a_k z^k \tag{1}$$

that are analytic and univalent in the open unit disc $U = \{z \in \mathbb{C} : |z| < 1\}$.

For $f(z) \in \mathcal{A}$, Salagean [1] introduced the following differential operator: $D^0 f(z) = f(z)$, $D^1 f(z) = zf'(z), \ldots,$ $D^n f(z) = D(D^{n-1} f(z))$ $(n \in N = \{1, 2, \ldots\})$.

We note that

$$D^n f(z) = z + \sum_{k=2}^{\infty} k^n a_k z^k \quad (n \in \mathbb{N}_0 = \mathbb{N} \cup \{0\}). \tag{2}$$

Definition 1 (subordination principle). For two functions f and g, analytic in U, we say that the function $f(z)$ is subordinate to $g(z)$ in U and write $f(z) \prec g(z)$, if there exists a Schwarz function $w(z)$, which (by definition) is analytic in U with $w(0) = 0$ and $|w(z)| < 1$, such that $f(z) = g(w(z))$ $(z \in U)$. Indeed it is known that

$$f(z) \prec g(z) \Longrightarrow f(0) = g(0), \qquad f(U) \subset g(U). \tag{3}$$

Furthermore, if the function g is univalent in U, then we have the following equivalence [2, page 4]:

$$f(z) \prec g(z) \Longleftrightarrow f(0) = g(0), \qquad f(U) \subset g(U). \tag{4}$$

Definition 2 (see [3]). Let $U_{m,n}(\beta, A, B)$ denote the subclass of $\mathcal{A}$ consisting of functions $f(z)$ of the form (1) and satisfy the following subordination:

$$\frac{D^m f(z)}{D^n f(z)} - \beta \left| \frac{D^m f(z)}{D^n f(z)} - 1 \right| \prec \frac{1 + Az}{1 + Bz},$$
$$(-1 \le B < A \le 1;\ -1 \le B < 0;\ \beta \ge 0;$$
$$m \in \mathbb{N};\ n \in \mathbb{N}_0, m > n;\ z \in U). \tag{5}$$

Specializing the parameters $A, B, \beta, m,$ and n, we obtain the following subclasses studied by various authors:

(i)

$$U_{m,n}(\beta,1-2\alpha,-1) = N_{m,n}(\alpha,\beta) = \left\{ f \in \mathcal{A} : \operatorname{Re}\left\{\frac{D^m f(z)}{D^n f(z)} - \alpha\right\} > \beta\left|\frac{D^m f(z)}{D^n f(z)} - 1\right| \quad (0 \le \alpha < 1;\ \beta \ge 0;\ m \in \mathbb{N};\ n \in \mathbb{N}_0;\ m > n;\ z \in U) \right\}, \tag{6}$$

(see Eker and Owa [4]);

(ii)

$$U_{1,0}(\beta,1-2\alpha,-1) = US(\alpha,\beta) = \left\{ f \in \mathcal{A} : \operatorname{Re}\left\{\frac{zf'(z)}{f(z)} - \alpha\right\} > \beta\left|\frac{zf'(z)}{f(z)} - 1\right| \quad (0 \le \alpha < 1;\ \beta \ge 0;\ z \in U) \right\},$$

$$U_{2,1}(\beta,1-2\alpha,-1) = UK(\alpha,\beta) = \left\{ f \in \mathcal{A} : \operatorname{Re}\left\{1 + \frac{zf''(z)}{f'(z)} - \alpha\right\} > \beta\left|\frac{zf''(z)}{f'(z)}\right| \quad (0 \le \alpha < 1;\ \beta \ge 0;\ z \in U) \right\}, \tag{7}$$

(see Shams et al. [5, 6]);

(iii)

$$U_{1,0}(0,A,B) = S^*(A,B) = \left\{ f \in \mathcal{A} : \frac{zf'(z)}{f(z)} \prec \frac{1+Az}{1+Bz} \quad (-1 \le B < A \le 1;\ z \in U) \right\},$$

$$U_{2,1}(0,A,B) = K(A,B) = \left\{ f \in \mathcal{A} : 1 + \frac{zf''(z)}{f'(z)} \prec \frac{1+Az}{1+Bz} \quad (-1 \le B < A \le 1;\ z \in U) \right\}, \tag{8}$$

(see Janowski [7] and Padmanabhan and Ganesan [8]).

Also we note that

$$U_{m,n}(0,A,B) = U(m,n;A,B) = \left\{ f(z) \in \mathcal{A} : \frac{D^m f(z)}{D^n f(z)} \prec \frac{1+Az}{1+Bz} \quad (-1 \le B < A \le 1;\ m \in \mathbb{N};\ n \in \mathbb{N}_0;\ m > n;\ z \in U) \right\}. \tag{9}$$

Let T denote the subclass of functions of $\mathcal{A}$ of the form

$$f(z) = z - \sum_{k=2}^{\infty} a_k z^k, \quad a_k \ge 0. \tag{10}$$

Further, we define the class $TS_\gamma(f,g;\alpha,\beta)$ by

$$TU_{m,n}(\beta,A,B) = U_{m,n}(\beta,A,B) \cap T. \tag{11}$$

For suitable choices of the parameters A, B, β, m, and n, we can get various known or new subclasses of T. For example, we have the following:

(i) $TU_{n+1,n}(\beta,1-2\alpha,-1) = TS(n,\alpha,\beta)$ $(0 \le \alpha < 1, \beta \ge 0, n \in \mathbb{N}_0)$ (see Rosy and Murugusundaramoorthy [9] and Aouf [10]);

(ii) $TU_{1,0}(1,1-2\alpha,-1) = S_pT(\alpha)$ and $TU_{2,1}(1,1-2\alpha,-1) = UCT(\alpha)$ $(0 \le \alpha < 1)$ (see Bharati et al. [11]);

(iii) $TU_{1,0}(0,1-2\alpha,-1) = T^*(\alpha)$ and $TU_{2,1}(0,1-2\alpha,-1) = C(\alpha)$ $(0 \le \alpha < 1)$ (see Silverman [12]).

2. Coefficient Estimates

Unless otherwise mentioned, we assume in the reminder of this paper that $-1 \le B < A \le 1$, $-1 \le B < 0$, $\beta \ge 0$, $m \in \mathbb{N}$, $n \in \mathbb{N}_0$, $m > n$ and $z \in U$.

Now, we will need the following lemma which gives a sufficient condition for functions belonging to the class $U_{m,n}(\beta,A,B)$.

Lemma 3 (see [13]). *A function $f(z)$ of the form (1) is in the class $U_{m,n}(\beta,A,B)$ if*

$$\sum_{k=2}^{\infty} \left[(1+\beta(1+|B|))(k^m - k^n) + |Bk^m - Ak^n|\right]|a_k| \le A - B. \tag{12}$$

In Theorem 4, it is shown that the condition (12) is also necessary for functions $f(z)$ of the form (10) to be in the class $TU_{m,n}(\beta,A,B)$.

Theorem 4. *Let $f(z) \in T$. Then $f(z) \in TU_{m,n}(\beta,A,B)$ if and only if*

$$\sum_{k=2}^{\infty} \left[(1+\beta(1+|B|))(k^m - k^n) + |Bk^m - Ak^n|\right] a_k \le A - B. \tag{13}$$

Proof. In view of Lemma 3, we only need to prove the only if part of Theorem 4. Since $TU_{m,n}(\beta, A, B) \subset U_{m,n}(\beta, A, B)$, for functions $f(z) \in TU_{m,n}(\beta, A, B)$, we can write

$$\left|\frac{p(z)-1}{A-Bp(z)}\right| < 1, \quad \text{where } p(z) = \frac{D^m f(z)}{D^n f(z)} - \beta\left|\frac{D^m f(z)}{D^n f(z)} - 1\right|. \tag{14}$$

then

$$\left|\left(\sum_{k=2}^{\infty}(k^m-k^n)a_k z^k + \beta e^{i\theta}\left|\sum_{k=2}^{\infty}(k^m-k^n)a_k z^k\right|\right) \times \left((A-B)z + \sum_{k=2}^{\infty}(Bk^m - Ak^n)a_k z^k + B\beta e^{i\theta}\left|\sum_{k=2}^{\infty}(k^m-k^n)a_k z^k\right|\right)^{-1}\right| < 1. \tag{15}$$

Since $\mathrm{Re}(z) \le |z|$ $(z \in U)$, then we obtain

$$\mathrm{Re}\left\{\left(\sum_{k=2}^{\infty}(k^m-k^n)a_k z^k + \beta e^{i\theta}\left|\sum_{k=2}^{\infty}(k^m-k^n)a_k z^k\right|\right) \times \left((A-B)z + \sum_{k=2}^{\infty}(Bk^m - Ak^n)a_k z^k + B\beta e^{i\theta}\left|\sum_{k=2}^{\infty}(k^m-k^n)a_k z^k\right|\right)^{-1}\right\} < 1. \tag{16}$$

Now choosing z to be real and letting $z \to 1^-$, we obtain

$$\sum_{k=2}^{\infty}\left[(1+\beta(1-B))(k^m-k^n) - (Bk^m - Ak^n)\right]a_k \le A - B. \tag{17}$$

Or, equivalently

$$\sum_{k=2}^{\infty}\left[(1+\beta(1+|B|))(k^m-k^n) + |Bk^m - Ak^n|\right]a_k \le A - B. \tag{18}$$

This completes the proof of Theorem 4.

Remark 5. (i) The result obtained by Theorem 4 corrects the result obtained by Li and Tang [3, Theorem 1].

(ii) Putting $A = 1 - 2\alpha$ $(0 \le \alpha < 1)$ and $B = -1$ in Theorem 4, we correct the result obtained by Eker and Owa [4, Theorem 2.1].

(iii) Putting $A = 1 - 2\alpha$ $(0 \le \alpha < 1)$, $B = -1$, and $m = n + 1$ $(n \in \mathbb{N}_0)$ in Theorem 4, we obtain the result obtained by Rosy and Murugusudaramoorthy [9, Theorem 2].

Corollary 6. *Let the function $f(z)$ be defined by (10) and let it be in the class $TU_{m,n}(\beta, A, B)$. Then*

$$a_k \le \frac{A-B}{(1+\beta(1+|B|))(k^m-k^n) + |Bk^m - Ak^n|}, \quad (k \ge 2). \tag{19}$$

The result is sharp for the function

$$f(z) = z - \frac{A-B}{(1+\beta(1+|B|))(k^m-k^n) + |Bk^m - Ak^n|}z^k, \quad (k \ge 2). \tag{20}$$

3. Distortion Theorems

Theorem 7. *Let the function $f(z)$ defined by (10) be in the class $TU_{m,n}(\beta, A, B)$. Then*

$$|f(z)| \ge |z| - \frac{A-B}{(1+\beta(1+|B|))(2^m-2^n) + |B2^m - A2^n|}|z|^2 \le |z| + \frac{A-B}{(1+\beta(1+|B|))(2^m-2^n) + |B2^m - A2^n|}|z|^2. \tag{21}$$

The result is sharp.

Proof. In view of Theorem 4, since

$$\Phi(k) = (1+\beta(1+|B|))(k^m-k^n) + |Bk^m - Ak^n| \tag{22}$$

is an increasing function of k $(k \ge 2)$, we have

$$\Phi(2)\sum_{k=2}^{\infty}|a_k| \le \sum_{k=2}^{\infty}\Phi(k)|a_k| \le A - B, \tag{23}$$

that is

$$\sum_{k=2}^{\infty}|a_k| \le \frac{A-B}{\Phi(2)}. \tag{24}$$

Thus we have

$$|f(z)| \le |z| + |z|^2\sum_{k=2}^{\infty}|a_k|,$$

$$|f(z)| \le |z| + \frac{A-B}{(1+\beta(1+|B|))(2^m-2^n) + |B2^m - A2^n|}|z|^2. \tag{25}$$

Similarly, we get

$$|f(z)| \ge |z| - \sum_{k=2}^{\infty}|a_k||z|^k \ge |z| - |z|^2\sum_{k=2}^{\infty}|a_k| \ge |z| - \frac{A-B}{(1+\beta(1+|B|))(2^m-2^n) + |B2^m - A2^n|}|z|^2. \tag{26}$$

Finally the result is sharp for the function

$$f(z) = z - \frac{A - B}{(1 + \beta(1 + |B|))(2^m - 2^n) + |B2^m - A2^n|} z^2, \tag{27}$$

at $z = r$ and $z = re^{i(2k+1)\pi}$ $(k \in \mathbb{Z} = \{\dots, -2, -1, 0, 1, 2, \dots\})$. This completes the proof of Theorem 7.

Theorem 8. *Let the function $f(z)$ defined by (10) be in the class $TU_{m,n}(\beta, A, B)$. Then*

$$\begin{aligned} |f'(z)| &\geq 1 - \frac{2(A - B)}{(1 + \beta(1 + |B|))(2^m - 2^n) + |B2^m - A2^n|}|z| \\ &\leq 1 + \frac{2(A - B)}{(1 + \beta(1 + |B|))(2^m - 2^n) + |B2^m - A2^n|}|z|. \end{aligned} \tag{28}$$

The result is sharp.

Proof. Similarly $\Phi(k)/k$ is an increasing function of k $(k \geq 2)$, in view of Theorem 4, we have

$$\frac{\Phi(2)}{2}\sum_{k=2}^{\infty} k|a_k| \leq \sum_{k=2}^{\infty}\frac{\Phi(k)}{k}k|a_k| = \sum_{k=2}^{\infty}\Phi(k)|a_k| \leq A - B, \tag{29}$$

that is

$$\sum_{k=2}^{\infty} k|a_k| \leq \frac{2(A - B)}{\Phi(2)}. \tag{30}$$

Thus we have

$$\begin{aligned} |f'(z)| &\leq 1 + |z|\sum_{k=2}^{\infty} k|a_k| \\ &\leq 1 + \frac{2(A - B)}{(1 + \beta(1 + |B|))(2^m - 2^n) + |B2^m - A2^n|}|z|. \end{aligned} \tag{31}$$

Similarly

$$\begin{aligned} |f'(z)| &\geq 1 - |z|\sum_{k=2}^{\infty} k|a_k| \\ &\geq 1 - \frac{2(A - B)}{(1 + \beta(1 + |B|))(2^m - 2^n) + |B2^m - A2^n|}|z|. \end{aligned} \tag{32}$$

Finally, we can see that the assertions of Theorem 8 are sharp for the function $f(z)$ defined by (27). This completes the proof of Theorem 8. ∎

4. Radii of Starlikeness, Convexity, and Close-to-Convexity

In this section radii of close-to-convexity, starlikeness, and convexity for functions belonging to the class $TU_{m,n}(\beta, A, B)$ are obtained.

Theorem 9. *Let the function $f(z)$ defined by (10) be in the class $TU_{m,n}(\beta, A, B)$; then*

(i) *$f(z)$ is starlike of order φ $(0 \leq \varphi < 1)$ in $|z| < r_1$, where*

$$r_1 = \inf_{k \geq 2}\left\{\frac{(1 + \beta(1 + |B|))(k^m - k^n) + |Bk^m - Ak^n|}{A - B} \times \left(\frac{1 - \varphi}{k - \varphi}\right)\right\}^{1/(k-1)}, \tag{33}$$

(ii) *$f(z)$ is convex of order φ $(0 \leq \varphi < 1)$ in $|z| < r_2$, where*

$$r_2 = \inf_{k \geq 2}\left\{\frac{(1 + \beta(1 + |B|))(k^m - k^n) + |Bk^m - Ak^n|}{A - B} \times \frac{(1 - \varphi)}{k(k - \varphi)}\right\}^{1/(k-1)}, \tag{34}$$

(iii) *$f(z)$ is close-to-convex of order φ $(0 \leq \varphi < 1)$ in $|z| < r_3$, where*

$$r_3 = \inf_{k \geq 2}\left\{\frac{(1 + \beta(1 + |B|))(k^m - k^n) + |Bk^m - Ak^n|}{A - B} \times \left(\frac{1 - \varphi}{k}\right)\right\}^{1/(k-1)}. \tag{35}$$

Each of these results is sharp for the function $f(z)$ given by (20).

Proof. It is sufficient to show that

$$\left|\frac{zf'(z)}{f(z)} - 1\right| \leq 1 - \varphi, \quad \text{for } |z| < r_1, \tag{36}$$

where r_1 is given by (33). Indeed we find from (10) that

$$\left|\frac{zf'(z)}{f(z)} - 1\right| \leq \frac{\sum_{k=2}^{\infty}(k - 1)a_k|z|^{k-1}}{1 - \sum_{k=2}^{\infty}a_k|z|^{k-1}}. \tag{37}$$

Thus we have

$$\left|\frac{zf'(z)}{f(z)} - 1\right| \leq 1 - \varphi, \tag{38}$$

if and only if

$$\sum_{k=2}^{\infty}\frac{(k - \varphi)a_k|z|^{k-1}}{(1 - \varphi)} \leq 1. \tag{39}$$

But, by Theorem 4, (39) will be true if

$$\frac{(k - \varphi)|z|^{k-1}}{(1 - \varphi)} \leq \frac{(1 + \beta(1 + |B|))(k^m - k^n) + |Bk^m - Ak^n|}{A - B}, \tag{40}$$

that is, if

$$|z| \leq \left\{ \frac{(1+\beta(1+|B|))(k^m - k^n) + |Bk^m - Ak^n|}{A-B} \times \left(\frac{1-\varphi}{k-\varphi} \right) \right\}^{1/(k-1)} \quad (k \geq 2). \tag{41}$$

Or

$$r_1 = \inf_{k\geq 2} \left\{ \frac{(1+\beta(1+|B|))(k^m - k^n) + |Bk^m - Ak^n|}{A-B} \times \left(\frac{1-\varphi}{k-\varphi} \right) \right\}^{1/(k-1)}. \tag{42}$$

This completes the proof of (33).

To prove (34) and (35) it is sufficient to show that

$$\left| 1 + \frac{zf''(z)}{f'(z)} - 1 \right| \leq 1 - \varphi \quad (|z| < r_2;\ 0 \leq \varphi < 1), \tag{43}$$

$$\left| f'(z) - 1 \right| \leq 1 - \varphi \quad (|z| < r_3;\ 0 \leq \varphi < 1), \tag{44}$$

respectively.

5. Extreme Points

Theorem 10. *Let*

$$f_1(z) = z,$$

$$f_k(z) = z - \frac{A-B}{(1+\beta(1+|B|))(k^m - k^n) + |Bk^m - Ak^n|} z^k, \quad (k = 2, 3, \ldots). \tag{45}$$

Then $f(z) \in TU_{m,n}(\beta, A, B)$ if and only if it can be expressed in the following form:

$$f(z) = \sum_{k=1}^{\infty} \eta_k f_k(z), \tag{46}$$

where

$$\eta_k \geq 0, \qquad \sum_{k=1}^{\infty} \eta_k = 1. \tag{47}$$

Proof. Suppose that

$$f(z) = \sum_{k=1}^{\infty} \eta_k f_k(z) = z - \sum_{k=2}^{\infty} \eta_k \frac{A-B}{(1+\beta(1+|B|))(k^m - k^n) + |Bk^m - Ak^n|} z^k. \tag{48}$$

Then, from Theorem 4, we have

$$\sum_{k=2}^{\infty} \left[\{(1+\beta(1+|B|))(k^m - k^n) + |Bk^m - Ak^n|\} \cdot \frac{A-B}{(1+\beta(1+|B|))(k^m - k^n) + |Bk^m - Ak^n|} \eta_k \right] \tag{49}$$

$$= (A-B) \sum_{k=2}^{\infty} \eta_k = (A-B)(1-\eta_1) \leq A - B.$$

Thus, in view of Theorem 4, we find that $f(z) \in TU_{m,n}(\beta, A, B)$.

Conversely, let us suppose that $f(z) \in TU_{m,n}(\beta, A, B)$, then, since

$$a_k \leq \frac{A-B}{(1+\beta(1+|B|))(k^m - k^n) + |Bk^m - Ak^n|}, \quad (k = 2, 3, \ldots). \tag{50}$$

Set

$$\eta_k = \frac{(1+\beta(1+|B|))(k^m - k^n) + |Bk^m - Ak^n|}{A-B} a_k, \quad (k = 2, 3, \ldots), \tag{51}$$

$$\eta_1 = 1 - \sum_{k=2}^{\infty} \eta_k.$$

Thus clearly, we have

$$f(z) = \sum_{k=1}^{\infty} \eta_k f_k(z). \tag{52}$$

This completes the proof of Theorem 10.

Corollary 11. *The extreme points of the class $TU_{m,n}(\beta, A, B)$ are given by*

$$f_1(z) = z,$$

$$f_k(z) = z - \frac{A-B}{(1+\beta(1+|B|))(k^m - k^n) + |Bk^m - Ak^n|} z^k, \quad (k = 2, 3, \ldots). \tag{53}$$

6. Integral Means Inequalities

In 1925, Littlewood [14] proved the following subordination lemma.

Lemma 12. *If the functions f and g are analytic in U with*

$$f(z) \prec g(z) \quad (z \in U), \tag{54}$$

then for $p > 0$ and $z = re^{i\theta}$ $(0 < r < 1)$,

$$\int_0^{2\pi} |f(z)|^p d\theta \leq \int_0^{2\pi} |g(z)|^p d\theta. \tag{55}$$

We now make use of Lemma 12 to prove Theorem 13.

Theorem 13. *Suppose that* $f(z) \in TU_{m,n}(\beta, A, B)$, $p > 0$, $-1 \leq B < A \leq 1$, $\beta > 0$, $m \in \mathbb{N}$, $n \in \mathbb{N}_0$, $m > n$, *and* $f_2(z)$ *is defined by*

$$f_2(z) = z - \frac{A - B}{(1 + \beta(1 + |B|))(2^m - 2^n) + |B2^m - A2^n|} z^2. \tag{56}$$

Then for $z = re^{i\theta}$ $(0 < r < 1)$, *we have*

$$\int_0^{2\pi} |f(z)|^p d\theta \leq \int_0^{2\pi} |f_2(z)|^p d\theta. \tag{57}$$

Proof. For $f(z) = z - \sum_{k=2}^{\infty} a_k z^k$ $(a_k \geq 0)$, (55) is equivalent to prove that

$$\begin{aligned} &\int_0^{2\pi} \left| 1 - \sum_{k=2}^{\infty} a_k z^{k-1} \right|^p d\theta \\ &\leq \int_0^{2\pi} \left| 1 - \frac{A - B}{(1+\beta(1+|B|))(2^m - 2^n) + |B2^m - A2^n|} z \right|^p d\theta. \end{aligned} \tag{58}$$

By applying Littlewood's subordination lemma (Lemma 12), it would suffice to show that

$$\begin{aligned} &1 - \sum_{k=2}^{\infty} a_k z^{k-1} \\ &\prec 1 - \frac{A - B}{(1 + \beta(1 + |B|))(2^m - 2^n) + |B2^m - A2^n|} z. \end{aligned} \tag{59}$$

By setting

$$\begin{aligned} &1 - \sum_{k=2}^{\infty} a_k z^{k-1} \\ &= 1 - \frac{A - B}{(1 + \beta(1 + |B|))(2^m - 2^n) + |B2^m - A2^n|} w(z) \end{aligned} \tag{60}$$

and using (13), we obtain

$$\begin{aligned} &|w(z)| \\ &= \left| \sum_{k=2}^{\infty} \frac{(1 + \beta(1 + |B|))(2^m - 2^n) + |B2^m - A2^n|}{A - B} a_k z^{k-1} \right| \\ &\leq |z| \sum_{k=2}^{\infty} \frac{(1 + \beta(1 + |B|))(2^m - 2^n) + |B2^m - A2^n|}{A - B} a_k \\ &\leq |z| \sum_{k=2}^{\infty} \frac{(1 + \beta(1 + |B|))(k^m - k^n) + |Bk^m - Ak^n|}{A - B} a_k \\ &\leq |z| < 1. \end{aligned} \tag{61}$$

This completes the proof of Theorem 13.

7. Partial Sums

In this section partial sums of functions in the class $U_{m,n}(\beta, A, B)$ are obtained, also we will obtain sharp lower bounds for the ratios of real part of $f(z)$ to $f_n(z)$.

Theorem 14. *Define the partial sums* $f_1(z)$ *and* $f_n(z)$ *by*

$$f_1(z) = z, \qquad f_n(z) = z + \sum_{k=2}^{n} a_k z^k \quad (n \in \mathbb{N} \setminus \{1\}). \tag{62}$$

Let the function $f(z) \in U_{m,n}(\beta, A, B)$ *be given by (1) and let it satisfy the condition (12) and*

$$c_k \geq \begin{cases} 1, & k = 2, 3, \ldots, n, \\ c_{n+1}, & k = n+1, n+2, \ldots, \end{cases} \tag{63}$$

where, for convenience,

$$c_k = \frac{(1 + \beta(1 + |B|))(k^m - k^n) + |Bk^m - Ak^n|}{A - B}, \tag{64}$$

Then

$$\operatorname{Re}\left\{\frac{f(z)}{f_n(z)}\right\} > 1 - \frac{1}{c_{n+1}} \quad (z \in U;\ n \in \mathbb{N}), \tag{65}$$

$$\operatorname{Re}\left\{\frac{f_n(z)}{f(z)}\right\} > \frac{c_{n+1}}{1 + c_{n+1}}. \tag{66}$$

Proof. For the coefficients c_k given by (64) it is not difficult to verify that

$$c_{k+1} > c_k > 1. \tag{67}$$

Therefore we have

$$\sum_{k=2}^{n} |a_k| + c_{n+1} \sum_{k=n+1}^{\infty} |a_k| \leq \sum_{k=2}^{\infty} c_k |a_k| \leq 1. \tag{68}$$

By setting

$$\begin{aligned} g_1(z) &= c_{n+1} \left\{ \frac{f(z)}{f_n(z)} - \left(1 - \frac{1}{c_{n+1}}\right) \right\} \\ &= 1 + \frac{c_{n+1} \sum_{k=n+1}^{\infty} a_k z^{k-1}}{1 + \sum_{k=2}^{n} a_k z^{k-1}} \end{aligned} \tag{69}$$

and applying (68), we find that

$$\left| \frac{g_1(z) - 1}{g_1(z) + 1} \right| < \frac{c_{n+1} \sum_{k=n+1}^{\infty} |a_k|}{2 - 2\sum_{k=2}^{n} |a_k| - c_{n+1} \sum_{k=n+1}^{\infty} |a_k|}. \tag{70}$$

Now

$$\left| \frac{g_1(z) - 1}{g_1(z) + 1} \right| < 1, \tag{71}$$

if

$$\sum_{k=2}^{n} |a_k| + c_{n+1} \sum_{k=n+1}^{\infty} |a_k| \leq 1. \tag{72}$$

From the condition (12), it is sufficient to show that

$$\sum_{k=2}^{n} |a_k| + c_{n+1} \sum_{k=n+1}^{\infty} |a_k| \le \sum_{k=2}^{\infty} c_k |a_k|, \quad (73)$$

which is equivalent to

$$\sum_{k=2}^{n} (c_k - 1) |a_k| + \sum_{k=n+1}^{\infty} (c_k - c_{n+1}) |a_k| \ge 0, \quad (74)$$

which readily yields the assertion (65) of Theorem 14. In order to see that

$$f(z) = z + \frac{z^{n+1}}{c_{n+1}} \quad (75)$$

gives sharp result, we observe that for $z = re^{i\pi/n}$ that $f(z)/f_n(z) = 1 + z^n/c_{n+1} \to 1 - 1/c_{n+1}$ as $z \to 1^-$. Similarly, if we take

$$\begin{aligned} g_2(z) &= (1 + c_{n+1}) \left\{ \frac{f_n(z)}{f(z)} - \frac{c_{n+1}}{1 + c_{n+1}} \right\} \\ &= 1 - \frac{(1 + c_{n+1}) \sum_{k=n+1}^{\infty} a_k z^{k-1}}{1 + \sum_{k=2}^{\infty} a_k z^{k-1}} \end{aligned} \quad (76)$$

and making use of (68), we can deduce that

$$\left| \frac{g_2(z) - 1}{g_2(z) + 1} \right| < \frac{(1 + c_{n+1}) \sum_{k=n+1}^{\infty} |a_k|}{2 - 2\sum_{k=2}^{n} |a_k| - (1 - c_{n+1}) \sum_{k=n+1}^{\infty} |a_k|}, \quad (77)$$

which leads us immediately to the assertion (66) of Theorem 14.

The bound in (66) is sharp for each $n \in \mathbb{N}$ with the extremal function $f(z)$ given by (75). Then the proof of Theorem 14 is completed.

8. Distortion Theorems Involving Fractional Calculus

In this section, we will prove several distortion theorems for functions belonging to the class $TU_{m,n}(\beta, A, B)$. Each of these theorems would involve certain operators of fractional calculus (i.e., fractional integrals and fractional derivatives), which are defined as follows (see, for details, [15–18]). For our present investigation, we recall the following definitions.

Definition 15. The fractional integral of order δ is defined, for a function $f(z)$, by

$$D_z^{-\delta} f(z) = \frac{1}{\Gamma(\delta)} \int_0^z \frac{f(\zeta)}{(z - \zeta)^{1-\delta}} d\zeta \quad (\delta > 0), \quad (78)$$

where the function $f(z)$ is analytic in a simply connected domain of the complex z-plane containing the origin, and the multiplicity of $(z - \zeta)^{\delta-1}$ is removed by requiring $\log(z - \zeta)$ to be real when $z - \zeta > 0$.

Definition 16. The fractional derivative of order δ is defined, for a function $f(z)$, by

$$D_z^{\delta} f(z) = \frac{1}{\Gamma(1-\delta)} \frac{d}{dz} \int_0^z \frac{f(\zeta)}{(z - \zeta)^{\delta}} d\zeta \quad (0 \le \delta < 1), \quad (79)$$

where the function $f(z)$ is constrained, and the multiplicity of $(z - \zeta)^{-\delta}$ is removed as in Definition 15.

Definition 17. Under the hypotheses of Definition 16, the fractional derivative of order δ is defined, for a function $f(z)$, by

$$D_z^{n+\delta} f(z) = \frac{d^n}{dz^n} \left\{ D_z^{\delta} f(z) \right\} \quad (0 \le \delta < 1;\ n \in \mathbb{N}_0). \quad (80)$$

Using Definitions 15, 16, and 17, we obtain

$$\begin{aligned} D_z^{\delta} \{z^k\} &= \frac{\Gamma(k+1)}{\Gamma(k+1-\delta)} z^{k-\delta} \quad (k \in \mathbb{N};\ 0 \le \delta < 1), \\ D_z^{-\delta} \{z^k\} &= \frac{\Gamma(k+1)}{\Gamma(k+1+\delta)} z^{k+\delta} \quad (k \in \mathbb{N};\ \delta > 0), \end{aligned} \quad (81)$$

in terms of Gamma functions.

Theorem 18. *Let the function $f(z)$ defined by (10) be in the class $TU_{m,n}(\beta, A, B)$. Then*

$$\begin{aligned} &\left| D_z^{-\delta} f(z) \right| \\ &\ge \frac{|z|^{1+\delta}}{\Gamma(2+\delta)} \\ &\times \left(1 - \frac{(A-B)}{(2+\delta)\left[(1+\beta(1+|B|))(2^{m-1}-2^{n-1}) + |B2^{m-1} - A2^{n-1}|\right]} |z| \right), \\ &\qquad (\delta > 0;\ z \in U), \\ &\left| D_z^{-\delta} f(z) \right| \\ &\le \frac{|z|^{1+\delta}}{\Gamma(2+\delta)} \\ &\times \left(1 + \frac{(A-B)}{(2+\delta)\left[(1+\beta(1+|B|))(2^{m-1}-2^{n-1}) + |B2^{m-1} - A2^{n-1}|\right]} |z| \right), \\ &\qquad (\delta > 0;\ z \in U). \end{aligned} \quad (82)$$

The results are sharp.

Proof. Let

$$\begin{aligned} F(z) &= \Gamma(2+\delta) z^{-\delta} D_z^{-\delta} f(z) \\ &= z - \sum_{k=2}^{\infty} \frac{\Gamma(k+1)\Gamma(2+\delta)}{\Gamma(k+1+\delta)} a_k z^k \\ &= z - \sum_{k=2}^{\infty} \Lambda(k) a_k z^k, \end{aligned} \quad (83)$$

where

$$\Lambda(k) = \frac{\Gamma(k+1)\Gamma(2+\delta)}{\Gamma(k+1+\delta)}, \quad (k = 2, 3, \ldots). \tag{84}$$

Since $\Lambda(k)$ is a decreasing function of k, we can write

$$0 < \Lambda(k) \le \Lambda(2) = \frac{2}{2+\delta}. \tag{85}$$

Furthermore, in view of Theorem 4, we have

$$\begin{aligned}&[(1+\beta(1+|B|))(2^m - 2^n) + |B2^m - A2^n|] \sum_{k=2}^{\infty} a_k \\ &\le \sum_{k=2}^{\infty} [(1+\beta(1+|B|))(k^m - k^n) + |Bk^m - Ak^n|] a_k \\ &\le A - B.\end{aligned} \tag{86}$$

Then

$$\sum_{k=2}^{\infty} a_k \le \frac{A-B}{[(1+\beta(1+|B|))(2^m - 2^n) + |B2^m - A2^n|]}. \tag{87}$$

Therefore, by using (85) and (87), we can see that

$$\begin{aligned}&|F(z)| \\ &\ge |z| - \Lambda(2)|z|^2 \sum_{k=2}^{\infty} a_k \ge |z| \\ &- \frac{(A-B)}{(2+\delta)[(1+\beta(1+|B|))(2^{m-1} - 2^{n-1}) + |B2^{m-1} - A2^{n-1}|]}|z|^2\end{aligned} \tag{88}$$

and similarly

$$\begin{aligned}&|F(z)| \\ &\le |z| + \Lambda(2)|z|^2 \sum_{k=2}^{\infty} a_k \le |z| \\ &+ \frac{(A-B)}{(2+\delta)[(1+\beta(1+|B|))(2^{m-1} - 2^{n-1}) + |B2^{m-1} - A2^{n-1}|]}|z|^2,\end{aligned} \tag{89}$$

which prove Theorem 18.

Finally, the equalities are attained for the function $f(z)$ defined by

$$\begin{aligned}&D_z^{-\delta} f(z) \\ &= \frac{z^{1+\delta}}{\Gamma(2+\delta)} \\ &\times \left(1 - \frac{(A-B)}{(2+\delta)[(1+\beta(1+|B|))(2^{m-1} - 2^{n-1}) + |B2^{m-1} - A2^{n-1}|]} z\right)\end{aligned} \tag{90}$$

or, equivalently, by $f(z)$ given by (27).

Then the results are sharp, and the proof of Theorem 18 is completed.

Corollary 19. *Under the hypothesis of Theorem 20, $D_z^{-\delta} f(z)$ is included in a disk with its center at the origin and radius R_1 given by*

$$\begin{aligned}&R_1 \\ &= \frac{1}{\Gamma(2+\delta)} \\ &\times \left(1 - \frac{(A-B)}{(2+\delta)[(1+\beta(1+|B|))(2^{m-1} - 2^{n-1}) + |B2^{m-1} - A2^{n-1}|]}\right).\end{aligned} \tag{91}$$

Theorem 20. *Let the function $f(z)$ defined by (10) be in the class $TU_{m,n}(\beta, A, B)$. Then*

$$\begin{aligned}&\left|D_z^{\delta} f(z)\right| \\ &\ge \frac{|z|^{1-\delta}}{\Gamma(2-\delta)} \\ &\times \left(1 - \frac{(A-B)}{(2-\delta)[(1+\beta(1+|B|))(2^{m-1}-2^{n-1})+|B2^{m-1}-A2^{n-1}|]}|z|\right), \\ &(0 \le \delta < 1;\ z \in U), \\ &\left|D_z^{\delta} f(z)\right| \\ &\le \frac{|z|^{1-\delta}}{\Gamma(2-\delta)} \\ &\times \left(1 + \frac{(A-B)}{(2-\delta)[(1+\beta(1+|B|))(2^{m-1}-2^{n-1})+|B2^{m-1}-A2^{n-1}|]}|z|\right), \\ &(0 \le \delta < 1;\ z \in U).\end{aligned} \tag{92}$$

Each of these results is sharp.

Proof. Let

$$\begin{aligned}G(z) &= \Gamma(2-\delta) z^{\delta} D_z^{\delta} f(z) \\ &= z - \sum_{k=2}^{\infty} \frac{\Gamma(k)\Gamma(2-\delta)}{\Gamma(k+1-\delta)} k a_k z^k = z - \sum_{k=2}^{\infty} \Omega(k) k a_k z^k,\end{aligned} \tag{93}$$

where

$$\Omega(k) = \frac{\Gamma(k)\Gamma(2-\delta)}{\Gamma(k+1-\delta)} \quad (k = 2, 3, \ldots). \tag{94}$$

Since $\Omega(k)$ is a decreasing function of k, we can write

$$0 < \Omega(k) \le \Omega(2) = \frac{1}{2-\delta}. \tag{95}$$

Furthermore, in view of Theorem 4, we have

$$\left[(1+\beta(1+|B|))\left(2^{m-1}-2^{n-1}\right)+\left|B2^{m-1}-A2^{n-1}\right|\right]\sum_{k=2}^{\infty}ka_k$$
$$\leq \sum_{k=2}^{\infty}\left[(1+\beta(1+|B|))\left(k^m-k^n\right)+\left|Bk^m-Ak^n\right|\right]a_k$$
$$\leq A-B. \tag{96}$$

Then

$$\sum_{k=2}^{\infty}ka_k \leq \frac{A-B}{\left[(1+\beta(1+|B|))\left(2^{m-1}-2^{n-1}\right)+\left|B2^{m-1}-A2^{n-1}\right|\right]}. \tag{97}$$

Therefore, by using (95) and (97), we can see that

$$|G(z)| \geq |z|-\Omega(2)|z|^2\sum_{k=2}^{\infty}ka_k \geq |z| - \frac{(A-B)}{(2-\delta)\left[(1+\beta(1+|B|))\left(2^{m-1}-2^{n-1}\right)+\left|B2^{m-1}-A2^{n-1}\right|\right]}|z|^2 \tag{98}$$

and similarly

$$|G(z)| \leq |z|+\Omega(2)|z|^2\sum_{k=2}^{\infty}ka_k \leq |z| + \frac{(A-B)}{(2-\delta)\left[(1+\beta(1+|B|))\left(2^{m-1}-2^{n-1}\right)+\left|B2^{m-1}-A2^{n-1}\right|\right]}|z|^2, \tag{99}$$

which together prove the two assertions of Theorem 20.

Finally, the equalities are attained for the function $f(z)$ defined by

$$D_z^{\delta}f(z) = \frac{z^{1-\delta}}{\Gamma(2-\delta)} \times\left(1-\frac{(A-B)}{(2-\delta)\left[(1+\beta(1+|B|))\left(2^{m-1}-2^{n-1}\right)+\left|B2^{m-1}-A2^{n-1}\right|\right]}z\right) \tag{100}$$

or, equivalently, by $f(z)$ given by (27).

Then the result is sharp, and the proof of Theorem 20 is completed.

Corollary 21. *Under the hypothesis of Theorem 20, $D_z^{\delta}f(z)$ is included in a disk with its center at the origin and radius R_2 given by*

$$R_2 = \frac{1}{\Gamma(2-\delta)} \times\left(1-\frac{(A-B)}{(2-\delta)\left[(1+\beta(1+|B|))\left(2^{m-1}-2^{n-1}\right)+\left|B2^{m-1}-A2^{n-1}\right|\right]}\right). \tag{101}$$

Acknowledgment

The authors thank the referee for his valuable suggestions which led to improvement of this study.

References

[1] G. S. Salagean, "Subclasses of univalent functions," in *Complex Analysis: Fifth Romanian-Finnish Seminar*, vol. 1013 of *Lecture Notes in Mathematics*, pp. 362–372, Springer, Berlin, UK, 1983.

[2] S. S. Miller and P. T. Mocanu, *Differential Subordinations: Theory and Applications*, vol. 225 of *Monographs and Textbooks in Pure and Applied Mathematics*, Marcel Dekker, New York, NY, USA, 2000.

[3] S.-H. Li and H. Tang, "Certain new classes of analytic functions defined by using the Salagean operator," *Bulletin of Mathematical Analysis and Applications*, vol. 2, no. 4, pp. 62–75, 2010.

[4] S. S. Eker and S. Owa, "Certain classes of analytic functions involving Salagean operator," *Journal of Inequalities in Pure and Applied Mathematics*, vol. 10, no. 1, pp. 12–22, 2009.

[5] S. Shams, S. R. Kulkarni, and J. M. Jahangiri, "On a class of univalent functions defined by Ruschweyh derivatives," *Kyungpook Mathematical Journal*, vol. 43, no. 4, pp. 579–585, 2003.

[6] S. Shams, S. R. Kulkarni, and J. M. Jahangiri, "Classes of uniformly starlike and convex functions," *International Journal of Mathematics and Mathematical Sciences*, no. 53, pp. 2959–2961, 2004.

[7] W. Janowski, "Some extremal problems for certain families of analytic functions," *Annales Polonici Mathematici*, vol. 28, pp. 648–658, 1973.

[8] K. S. Padmanabhan and M. S. Ganesan, "Convolutions of certain classes of univalent functions with negative coefficients," *Indian Journal of Pure and Applied Mathematics*, vol. 19, no. 9, pp. 880–889, 1988.

[9] T. Rosy and G. Murugusundaramoorthy, "Fractional calculus and their applications to certain subclass of uniformly convex functions," *Far East Journal of Mathematical Sciences*, vol. 15, no. 2, pp. 231–242, 2004.

[10] M. K. Aouf, "A subclass of uniformly convex functions with negative coefficients," *Mathematica*, vol. 52, no. 2, pp. 99–111, 2010.

[11] R. Bharati, R. Parvatham, and A. Swaminathan, "On subclasses of uniformly convex functions and corresponding class of starlike functions," *Tamkang Journal of Mathematics*, vol. 28, no. 1, pp. 17–32, 1997.

[12] H. Silverman, "Univalent functions with negative coefficients," *Proceedings of the American Mathematical Society*, vol. 51, pp. 109–116, 1975.

[13] M. K. Aouf, R. M. El-Ashwah, A. A. M. Hassan, and A. H. Hassan, "On subordination results for certain new classes of analytic functions defined by using Salagean operator," *Bulletin of Mathematical Analysis and Applications*, vol. 4, no. 1, pp. 239–246, 2012.

[14] J. E. Littlewood, "On inequalities in the theory of functions," *Proceedings of the London Mathematical Society*, vol. 23, no. 1, pp. 481–519.

[15] H. M. Srivastava and M. K. Aouf, "A certain fractional derivative operator and its applications to a new class of analytic and multivalent functions with negative coefficients I," *Journal of Mathematical Analysis and Applications*, vol. 171, no. 1, pp. 1–13, 1992.

[16] H. M. Srivastava and M. K. Aouf, "A certain fractional derivative operator and its applications to a new class of analytic and multivalent functions with negative coefficients, II," *Journal of Mathematical Analysis and Applications*, vol. 192, no. 3, pp. 673–688, 1995.

[17] H. M. Srivastava and S. Owa, "Some characterization and distortion theorems involving fractional calculus, generalized hypergeometric functions, Hadamard products, linear operators, and certain subclasses of analytic functions," *Nagoya Mathematical Journal*, vol. 106, pp. 1–28, 1987.

[18] H. M. Srivastava and S. Owa, *Univalent Functions, Fractional Calculus, and Their Applications*, Halsted Press, Horwood Limited, Chichester, UK; John Wiley and Sons, New York, NY, USA, 1989.

5

On Common Random Fixed Points of a New Iteration with Errors for Nonself Asymptotically Quasi-Nonexpansive Type Random Mappings

R. A. Rashwan,[1] P. K. Jhade,[2] and Dhekra Mohammed Al-Baqeri[1]

[1] *Department of Mathematics, University of Assiut, Assiut 71516, Egypt*
[2] *Department of Mathematics, NRI Institute of Information Science & Technology, Bhopal, Madhya Pradesh 462021, India*

Correspondence should be addressed to P. K. Jhade; pmathsjhade@gmail.com

Academic Editor: Stefan Kunis

We prove some strong convergence of a new random iterative scheme with errors to common random fixed points for three and then N nonself asymptotically quasi-nonexpansive-type random mappings in a real separable Banach space. Our results extend and improve the recent results in Kiziltunc, 2011, Thianwan, 2008, Deng et al., 2012, and Zhou and Wang, 2007 as well as many others.

1. Introduction and Preliminaries

The theory of random operators is an important branch of probabilistic analysis which plays a key role in many applied areas. The study of random fixed points forms a central topic in this area. Research of this direction was initiated by Prague School of Probabilistic in connection with random operator theory [1–3]. Random fixed point theory has attracted much attention in recent times since the publication of the survey article by Bharucha-Reid [4] in 1976, in which the stochastic versions of some well-known fixed point theorems were proved.

A lot of efforts have been devoted to random fixed point theory and applications (e.g. see [5–10] and many others).

Let (Ω, Σ) be a measurable space, C a nonempty subset of a separable Banach space E. A mapping $\xi : \Omega \to C$ is called measurable if $\xi^{-1}(B \bigcap C) \in \Sigma$ for every Borel subset B of E.

A mapping $T : \Omega \times C \to C$ is said to be random mapping if for each fixed $x \in C$, the mapping $T(\cdot, x) : \Omega \to C$ is measurable.

A measurable mapping $\xi : \Omega \to C$ is called a random fixed point of the random mapping $T : \Omega \times C \to C$ if $T(w, \xi(w)) = \xi(w)$ for each $w \in \Omega$.

Throughout this paper, we denote the set of all random fixed points of random mapping T by $RF(T)$ and by $T^n(w, x)$ for the nth iterate $T(w, T(, \ldots T(w, x)))$ of T.

The class of asymptotically nonexpansive mappings is a natural generalization of the important class of nonexpansive mappings. Goebel and Kirk [11] proved that if C is a nonempty closed and bounded subset of a uniformly convex Banach space, then every asymptotically nonexpansive self-mapping has a fixed point.

Iterative techniques for asymptotically nonexpansive self-mappings in Banach spaces including Mann type and Ishikawa type iteration processes have been studied extensively by various authors (e.g. see [12–15]).

The strong and weak convergences of the sequence of Mann iterates to a fixed point of quasi-nonexpansive mappings were studied by Petryshyn and Williamson [16]. Subsequently, the convergence of Ishikawa iterates of quasi-nonexpansive mappings in Banach spaces were discussed by Ghosh and Debnath [17]. The previous results and some obtained necessary and sufficient conditions for Ishikawa iterative sequence to converge a fixed point for asymptotically quasi-nonexpansive mappings were extended by Liu [18, 19].

In 2000, Noor [20] introduced a three-step iterative scheme and studied the approximate solutions of variational inclusion in Hilbert spaces. Xu and Noor [21] introduced and studied a three-step iterative scheme for asymptotically nonexpansive mappings, and they proved weak and strong convergences theorems for asymptotically nonexpansive mappings in Banach spaces. In 2005, Suantai [22] defined a new three-step iteration, which is an extension of Noor iterations, and gave some weak and strong convergences theorems of such iterations for asymptotically nonexpansive mappings in uniformly convex Banach spaces.

For nonself nonexpansive mappings, some authors (e.g., see [23–27]) have studied the strong and weak convergences theorems in Hilbert space or uniformly convex Banach spaces.

A subset C of E is said to be a retract of E if there exists a continuous map $P : E \to C$ such that $Px = x$ for all $x \in C$. Every closed convex subset of uniformly convex Banach space is a retract. A map $P : E \to E$ is a retraction if $P^2 = P$. It follows that if a map P is a retraction, then $Py = y$ for all y in the range of P.

The concept of nonself asymptotically nonexpansive mappings was introduced by Chidume et al. [28] in 2003 as the generalization of asymptotically nonexpansive self-mappings.

They studied the following iteration process:

$$x_1 \in C, \quad x_{n+1} = P\left(\left(1-\alpha_n\right)x_n + \alpha_n T(PT)^{n-1}x_n\right), \tag{1}$$

where $T : C \to E$ is an asymptotically nonexpansive nonself mapping, $\{\alpha_n\}$ is a real sequence in $(0,1)$, and P is a nonexpansive retraction from E to C.

Wang [29] generalized the result of Chidume et al. [28] and got some new results. He defined and studied the following iteration process:

$$\begin{gathered} x_{n+1} = P\left(\left(1-\alpha_n\right)x_n + \alpha_n T_1(PT_1)^{n-1}y_n\right), \\ y_n = P\left(\left(1-\beta_n\right)x_n + \beta_n T_2(PT_2)^{n-1}x_n\right), \quad x_1 \in C,\ n \geq 1, \end{gathered} \tag{2}$$

where $T_1, T_2 : C \to E$ are asymptotically nonexpansive nonself mappings and $\{\alpha_n\}$, $\{\beta_n\}$ are real sequences in $[0,1)$.

Now, we introduce the following concepts for nonself mappings

Definition 1 (see [28, 30, 31]). Let C be a nonempty subset of a real separable Banach space and $T : \Omega \times C \to E$ a nonself random mapping. Then, T is said to be

(1) nonexpansive random operator if for arbitrary $x, y \in C$, $\|T(w,x) - T(w,y)\| \leq \|x-y\|$, for all $w \in \Omega$;

(2) nonself asymptotically nonexpansive random mapping if there exists a sequence of measurable functions $r_n(w) : \Omega \to [1,\infty)$ with $\lim_{n\to\infty} r_n(w) = 1$ for each $w \in \Omega$ such that for arbitrary $x, y \in C$,

$$\begin{aligned} &\left\|T(PT)^{n-1}(w,x) - T(PT)^{n-1}(w,y)\right\| \\ &\quad \leq r_n(w)\left\|x-y\right\|, \quad \forall w \in \Omega,\ n \geq 1; \end{aligned} \tag{3}$$

(3) nonself asymptotically quasi-nonexpansive random mapping if $RF(T) \neq \phi$ and there exists a sequence of measurable functions $r_n(w) : \Omega \to [1,\infty)$ with $\lim_{n\to\infty} r_n(w) = 1$ for each $w \in \Omega$ such that

$$\begin{aligned} &\left\|T(PT)^{n-1}(w,\eta(w)) - \xi(w)\right\| \\ &\quad \leq r_n(w)\left\|\eta(w) - \xi(w)\right\|, \quad \forall w \in \Omega,\ n \geq 1, \end{aligned} \tag{4}$$

where $\xi(w) : \Omega \to C$ is a random fixed point of T and $\eta(w) : \Omega \to C$ is any measurable mapping;

(4) nonself asymptotically nonexpansive-type random mapping if

$$\begin{aligned} &\limsup_{n\to\infty}\left\{\sup_{x,y\in C}\left\{\left\|T(PT)^{n-1}(w,x) - T(PT)^{n-1}(w,y)\right\|^2\right.\right. \\ &\quad \left.\left. -\left\|x-y\right\|^2\right\}\right\} \leq 0, \quad \forall w \in \Omega,\ n \geq 1; \end{aligned} \tag{5}$$

(5) Nonself asymptotically quasi-nonexpansive-type random mapping if $RF(T) \neq \phi$, and

$$\begin{aligned} &\limsup_{n\to\infty}\left\{\sup_{\xi(w)\in F}\left\{\left\|T(PT)^{n-1}(w,\eta(w)) - \xi(w)\right\|^2\right.\right. \\ &\quad \left.\left. -\left\|\eta(w) - \xi(w)\right\|^2\right\}\right\} \leq 0, \\ &\quad \forall w \in \Omega,\ n \geq 1, \end{aligned} \tag{6}$$

where $\xi(w) : \Omega \to C$ is a random fixed point of T and $\eta(w) : \Omega \to C$ is any measurable mapping.

Remark 2. (1) If $T : \Omega \times C \to E$ is a nonself asymptotically nonexpansive random mapping, then T is a nonself asymptotically nonexpansive-type random mapping.

(2) If $RF(T) \neq \phi$ and $T : \Omega \times C \to E$ is a nonself asymptotically quasi-nonexpansive random mapping, then T is a nonself asymptotically quasi-nonexpansive-type random mapping.

(3) If $RF(T) \neq \phi$ and $T : \Omega \times C \to E$ is a nonself asymptotically nonexpansive-type random mapping, then T is a nonself asymptotically quasi-nonexpansive-type random mapping.

Remark 3. Observe that for any measurable mapping $\eta(w) : \Omega \to C$ and $\xi(w) \in F$, we have

$$\begin{aligned} &\limsup_{n\to\infty}\left\{\sup_{\xi(w)\in F}\left\{\left\|T(PT)^{n-1}(w,\eta(w)) - \xi(w)\right\|^2\right.\right. \\ &\quad \left.\left. -\left\|\eta(w) - \xi(w)\right\|^2\right\}\right\} \leq 0, \end{aligned} \tag{7}$$

which implies

$$\begin{aligned} &\limsup_{n\to\infty}\left\{\sup_{\xi(w)\in F}\left\{\left(\left\|T(PT)^{n-1}(w,\eta(w)) - \xi(w)\right\|\right.\right.\right. \\ &\quad \left. - \left\|\eta(w) - \xi(w)\right\|\right) \\ &\quad \times \left(\left\|T(PT)^{n-1}(w,\eta(w)) - \xi(w)\right\|\right. \\ &\quad \left.\left.\left. + \left\|\eta(w) - \xi(w)\right\|\right)\right\}\right\} \leq 0. \end{aligned} \tag{8}$$

Therefore,

$$\limsup_{n\to\infty}\left\{\sup_{\xi(w)\in F}\left\{\left(\left\|T(PT)^{n-1}(w,\eta(w))-\xi(w)\right\|-\left\|\eta(w)-\xi(w)\right\|\right)\right\}\right\}\leq 0. \tag{9}$$

In [25], Shahzad studied the following iterative sequences:

$$x_{n+1}=P\left(\left(1-\alpha_n\right)x_n+\alpha_n TP\left[\left(1-\beta_n\right)x_n+\beta_n Tx_n\right]\right),\quad x_1\in C,\quad n\geq 1, \tag{10}$$

where $T: C \to E$ is a nonexpansive nonself mapping, C is a nonempty closed convex nonexpansive retract of a real uniformly convex Banach space E with P being a nonexpansive retraction from E to C, and $\{\alpha_n\}$, $\{\beta_n\}$ are real sequences in $[0,1)$.

Recently, Thianwan [32] generalized the iteration process (10) as follows:

$$\begin{aligned}x_{n+1}&=P\left(\left(1-\alpha_n-\gamma_n\right)x_n+\alpha_n TP\left[\left(1-\beta_n\right)y_n+\beta_n Ty_n\right]+\gamma_n u_n\right),\\ y_n&=P\left(\left(1-\alpha_n'-\gamma_n'\right)x_n+\alpha_n' TP\left[\left(1-\beta_n'\right)x_n+\beta_n' Tx_n\right]+\gamma_n' v_n\right),\quad x_1\in C,\ n\geq 1,\end{aligned} \tag{11}$$

where $\{\alpha_n\}$, $\{\beta_n\}$, $\{\gamma_n\}$, $\{\alpha_n'\}$, $\{\beta_n'\}$, $\{\gamma_n'\}$ are appropriate sequences in $[0,1)$ and $\{u_n\}$, $\{v_n\}$ are bounded sequences in C. He proved weak and strong convergences theorems for nonexpansive nonself mappings in uniformly convex Banach spaces.

In 2011, Kiziltunc [33] studied the strong convergence to a common fixed point of a new iterative scheme for two nonself asymptotically quasi-nonexpansive-type mappings in Banach spaces defined as follows:

$$\begin{aligned}x_{n+1}&=P\left(\left(1-a_n\right)x_n+a_n S(PS)^{n-1}\left(\left(1-\alpha_n\right)y_n+\alpha_n S(PS)^{n-1}y_n\right)\right),\\ y_n&=P\left(\left(1-b_n\right)x_n+b_n T(PT)^{n-1}\left(\left(1-\beta_n\right)x_n+\beta_n T(PT)^{n-1}x_n\right)\right),\\ &\qquad x_1\in C,\ n\geq 1,\end{aligned} \tag{12}$$

where $\{a_n\}$, $\{b_n\}$, $\{\alpha_n\}$, $\{\beta_n\}$ are appropriate sequences in $[0,1)$.

More recently, Deng et al. [34] obtained the strong and weak convergences theorems for common fixed points of two nonself asymptotically nonexpansive mappings in Banach spaces.

The iterative scheme is defined as follows:

$$\begin{aligned}x_{n+1}&=\alpha_{n1}x_n+\beta_{n1}(PT_1)^n y_n+\gamma_{n1}(PT_2)^n y_n,\\ y_n&=\alpha_{n2}x_n+\beta_{n2}(PT_1)^n z_n+\gamma_{n2}(PT_2)^n z_n,\\ z_n&=\alpha_{n3}x_n+\beta_{n3}(PT_1)^n x_n+\gamma_{n3}(PT_2)^n x_n,\end{aligned} \tag{13}$$

where $\{\alpha_{ni}\}$, $\{\beta_{ni}\}$, $\{\gamma_{ni}\}$ $(i=1,2,3)$ are appropriate sequences in $[a,1-a]$ for some $a\in(0,1)$ satisfying $\alpha_{ni}+\beta_{ni}+\gamma_{ni}=1$ $(i=1,2,3)$.

For random operators, Beg and Abbas [30] studied the different random iterative algorithms for weakly contractive and asymptotically nonexpansive random operators on arbitrary Banach space. They also established convergence of an implicit random iterative process to a common fixed point for a finite family of asymptotically quasi-nonexpansive operators. Plubtieng et al. [35, 36] studied weak and strong convergences theorems for a modified random Noor iterative scheme with errors for three asymptotically nonexpansive self-mappings in Banach space defined as follows:

$$\begin{aligned}\xi_{n+1}(w)&=\alpha_n T_1^n\left(w,\eta_n(w)\right)+\beta_n\xi_n(w)+\gamma_n f_n(w),\\ \eta_n(w)&=\alpha_n' T_2^n\left(w,\zeta_n(w)\right)+\beta_n'\xi_n(w)+\gamma_n' f_n'(w),\\ \zeta_n(w)&=\alpha_n'' T_3^n\left(w,\xi_n(w)\right)+\beta_n''\xi_n(w)+\gamma_n'' f_n''(w),\\ &\qquad n\geq 1,\ w\in\Omega,\end{aligned} \tag{14}$$

where $T_1, T_2, T_3 : \Omega\times C \to C$ are three asymptotically nonexpansive random self-mappings, $\xi_1 : \Omega \to C$ is an arbitrary given measurable mapping from Ω to C, $\{f_n(w)\}$, $\{f_n'(w)\}$, $\{f_n''(w)\}$ are bounded sequence of measurable functions from Ω to C, and $\{\alpha_n\}$, $\{\alpha_n'\}$, $\{\alpha_n''\}$, $\{\beta_n\}$, $\{\beta_n'\}$, $\{\beta_n''\}$, $\{\gamma_n\}$, $\{\gamma_n'\}$, $\{\gamma_n''\}$ are sequences of real numbers in $[0,1]$ with $\alpha_n+\beta_n+\gamma_n=\alpha_n'+\beta_n'+\gamma_n'=\alpha_n''+\beta_n''+\gamma_n''=1$.

Remark 4. If $T_1=T_2=T_3=T$ and $\gamma_n=\gamma_n'=\gamma_n''=0$, then (14) becomes as follows:

$$\begin{aligned}\xi_{n+1}(w)&=\alpha_n T^n\left(w,\eta_n(w)\right)+\beta_n\xi_n(w),\\ \eta_n(w)&=\alpha_n' T^n\left(w,\zeta_n(w)\right)+\beta_n'\xi_n(w),\\ \zeta_n(w)&=\alpha_n'' T_3^n\left(w,\xi_n(w)\right)+\beta_n''\xi_n(w),\quad n\geq 1,\ w\in\Omega,\end{aligned} \tag{15}$$

which was studied by Beg and Abbas in [30].

For nonself random mappings, Zhou and Wang [37] studied the approximation of the following iteration process:

$$\begin{aligned}\xi_{n+1}(w)&=P\left(\left(1-\alpha_n\right)\xi_n(w)+\alpha_n T(PT)^{n-1}\left(w,\eta_n(w)\right)\right),\\ \eta_n(w)&=P\left(\left(1-\beta_n\right)\xi_n(w)+\beta_n T(PT)^{n-1}\left(w,\xi_n(w)\right)\right),\\ &\qquad n\geq 1,\ w\in\Omega,\end{aligned} \tag{16}$$

where $T : \Omega \times C \to E$ is an asymptotically nonexpansive nonself random mapping, $\xi_1 : \Omega \to C$ is an arbitrary given measurable mapping from Ω to C, $\{\alpha_n\}$, $\{\beta_n\}$ are sequences in $[0, 1]$, and P is a nonexpansive retraction from E to C.

Saluja [38] and many other authors extended the results of Zhou and Wang [37] by studying multistep random iteration scheme with errors for common random fixed point of a finite family of nonself asymptotically nonexpansive random mapping in real uniformly separable Banach spaces.

Inspired and motivated by [32–34, 37] and others, we introduced a new three-step and N-step random iterative scheme with errors for asymptotically quasi-nonexpansive-type nonself random mappings in a separable Banach space. Some strong convergences theorems are established for these new random iterative schemes with errors in separable Banach space. The iterative scheme for three nonself random mappings is defined as follows.

Definition 5. Let $T_1, T_2, T_3 : \Omega \times C \to C$ be three nonself random mappings, where C is a nonempty closed convex subset of a separable Banach space E, and $P : E \to C$ is a nonexpansive retraction of E onto C. Let $\xi_1(w) : \Omega \to C$ be a measurable mapping. Suppose that $\{\xi_n(w)\}$ is generated iteratively by $\xi_1(w) \in C$, having

$$\begin{aligned}
\xi_{n+1}(w) = {} & P\left[\left(1 - a_n - \sigma_n\right)\xi_n(w) + a_n T_1 (PT_1)^{n-1}\right. \\
& \times \left(w, \left(1 - \alpha_n\right)\eta_n(w)\right. \\
& \left.\left. + \alpha_n T_1 (PT_1)^{n-1}\left(w, \eta_n(w)\right)\right) + \sigma_n f_n(w)\right], \\
\eta_n(w) = {} & P\left[\left(1 - b_n - \delta_n\right)\xi_n(w) + b_n T_2 (PT_2)^{n-1}\right. \\
& \times \left(w, \left(1 - \beta_n\right)\zeta_n(w)\right. \\
& \left.\left. + \beta_n T_2 (PT_2)^{n-1}\left(w, \zeta_n(w)\right)\right) + \delta_n g_n(w)\right], \\
\zeta_n(w) = {} & P\left[\left(1 - c_n - \lambda_n\right)\xi_n(w) + c_n T_3 (PT_3)^{n-1}\right. \\
& \times \left(w, \left(1 - \gamma_n\right)\xi_n(w)\right. \\
& \left.\left. + \gamma_n T_3 (PT_3)^{n-1}\left(w, \xi_n(w)\right)\right) + \lambda_n h_n(w)\right],
\end{aligned} \tag{17}$$

for all $n \geq 1$, $w \in \Omega$, where $\{a_n\}$, $\{b_n\}$, $\{c_n\}$, $\{\alpha_n\}$, $\{\beta_n\}$, $\{\gamma_n\}$, $\{\sigma_n\}$, $\{\delta_n\}$, and $\{\lambda_n\}$ are sequences in $[0, 1]$ such that $a_n + \sigma_n \leq 1$, $b_n + \delta_n \leq 1$, $c_n + \lambda_n \leq 1$, and $\{f_n(w)\}$, $\{g_n(w)\}$, $\{h_n(w)\}$ are bounded sequences of measurable functions from Ω to C for all $w \in \Omega$.

Definition 5 can be extended to N nonself random mappings as follows.

Definition 6. Let $T_1, T_2, \ldots, T_N : \Omega \times C \to C$ be N nonself random mappings, where C is a nonempty closed convex subset of a separable Banach space E, and $P : E \to C$ is a nonexpansive retraction of E onto C. Let $\xi_1(w) : \Omega \to C$ be a measurable mapping. Define sequences function $\{\xi_n^{(N)}(w)\}$, $\{\xi_n^{(N-1)}(w)\}, \ldots, \{\xi_n^{(1)}(w)\}$ in C as follows:

$$\begin{aligned}
\xi_{n+1}(w) = {} & \xi_n^{(N)}(w) \\
= {} & P\left[\left(1 - a_n^{(N)} - \sigma_n^{(N)}\right)\xi_n(w) + a_n^{(N)} T_N (PT_N)^{n-1}\right. \\
& \times \left(w, \left(1 - \alpha_n^{(N)}\right)\xi_n^{(N-1)}(w) + \alpha_n^{(N)} T_N\right. \\
& \left.\left. \times (PT_N)^{n-1}\left(w, \xi_n^{(N-1)}(w)\right)\right) + \sigma_n^{(N)} f_n^{(N)}\right], \\
\xi_n^{(N-1)}(w) = {} & P\left[\left(1 - a_n^{(N-1)} - \sigma_n^{(N-1)}\right)\xi_n(w)\right. \\
& + a_n^{(N-1)} T_{N-1} (PT_{N-1})^{n-1} \\
& \times \left(w, \left(1 - \alpha_n^{(N-1)}\right)\xi_n^{(N-2)}(w)\right. \\
& \left. + \alpha_n^{(N-1)} T_{N-1} (PT_{N-1})^{n-1}\left(w, \xi_n^{(N-2)}(w)\right)\right) \\
& \left. + \sigma_n^{(N-1)} f_n^{(N-1)}\right], \\
\cdots = {} & \cdots \\
\cdots = {} & \cdots \\
\xi_n^{(1)}(w) = {} & P\left[\left(1 - a_n^{(1)} - \sigma_n^{(1)}\right)\xi_n(w) + a_n^{(1)} T_1 (PT_1)^{n-1}\right. \\
& \times \left(w, \left(1 - \alpha_n^{(1)}\right)\xi_n(w) + \alpha_n^{(1)} T_1 (PT_1)^{n-1}\right. \\
& \left.\left. \times \left(w, \xi_n(w)\right)\right) + \sigma_n^{(1)} f_n^{(1)}\right], \\
& n \geq 1, \quad w \in \Omega,
\end{aligned} \tag{18}$$

where $\{a_n^{(i)}\}$, $\{\alpha_n^{(i)}\}$, and $\{\sigma_n^{(i)}\}$ $(i = 1, 2, \ldots, N)$ are sequences in $[0, 1]$ such that $a_n^{(i)} + \sigma_n^{(i)} \leq 1$, for all $(i = 1, 2, \ldots, N)$, and $\{f_n^{(i)}\}$ $(i = 1, 2, \ldots, N)$ are bounded sequences of measurable functions from Ω to C for all $w \in \Omega$.

The following lemma is useful for proving our results.

Lemma 7 (see [39]). *Let $\{a_n\}$, $\{b_n\}$ and $\{m_n\}$ be nonnegative real sequences satisfying*

$$a_{n+1} \leq \left(1 + m_n\right) a_n + b_n, \qquad n \geq 1. \tag{19}$$

If $\sum_{n=1}^{\infty} m_n < \infty$ and $\sum_{n=1}^{\infty} b_n < \infty$, then

(1) $\lim_{n \to \infty} a_n$ *exists;*

(2) $\lim_{n \to \infty} a_n = 0$ *whenever* $\liminf_{n \to \infty} a_n = 0$.

2. Main Results

In this section, we will first prove the strong convergence of the iterative scheme (17) to a common random fixed point for three asymptotically quasi-nonexpansive-type nonself random mappings in a separable Banach space. Then, we extend the obtained results to N asymptotically quasi-nonexpansive-type nonself random mappings by using the iterative scheme (18). Finally, we use Theorem 8 and Condition (A) [40] to obtain a convergences theorem for scheme (17).

Theorem 8. *Let E be a real separable Banach space and C a nonempty closed convex subset of E with P being a nonexpansive retraction. Let $T_i : \Omega \times C \to E$, $i = 1,2,3$, be three asymptotically quasi-nonexpansive-type nonself random mappings with $F = \bigcap_{i=1}^{3} RF(T_i) \neq \phi$, for all $w \in \Omega$. Suppose that $\{\xi_n(w)\}$, $\{\eta_n(w)\}$ and $\{\zeta_n(w)\}$ are the sequences defined as in (17) where $\{a_n\}$, $\{b_n\}$, $\{c_n\}$, $\{\alpha_n\}$, $\{\beta_n\}$, $\{\gamma_n\}$, $\{\sigma_n\}$, $\{\delta_n\}$, and $\{\lambda_n\}$ are sequences in $[0,1]$ such that $a_n + \sigma_n \leq 1$, $b_n + \delta_n \leq 1$, $c_n + \lambda_n \leq 1$ and $\{f_n(w)\}$, $\{g_n(w)\}$, $\{h_n(w)\}$ are bounded sequences of measurable functions from Ω to C with the following restrictions: $\sum_{n=1}^{\infty} \sigma_n < \infty$, $\sum_{n=1}^{\infty} \delta_n < \infty$, and $\sum_{n=1}^{\infty} \lambda_n < \infty$. Then, $\{\xi_n(w)\}$ converge to a common random fixed point of T_1, T_2, and T_3 if and only if*

$$\liminf_{n\to\infty} d(\xi_n(w), F) = 0, \qquad w \in \Omega. \tag{20}$$

Proof. The necessity of (20) is obvious.

Next, we prove the sufficiency of (20). Let $\xi(w) \in F = \bigcap_{i=1}^{3} RF(T_i)$; by the boundedness of the sequences of measurable functions $\{f_n(w)\}$, $\{g_n(w)\}$, $\{h_n(w)\}$, we put for each $w \in \Omega$,

$$M(w) = \max\left\{\sup_{n\geq 1,\xi\in F} \|f_n(w) - \xi(w)\| \vee \times \sup_{n\geq 1,\xi\in F} \|g_n(w) - \xi(w)\| \vee \times \sup_{n\geq 1,\xi\in F} \|h_n(w) - \xi(w)\|\right\}. \tag{21}$$

Then, $M(w) < \infty$ for each $w \in \Omega$.

Since $\xi(w) \in F$ and $\eta(w) : \Omega \to C$ is any measurable mapping, we have

$$\limsup_{n\to\infty}\left\{\sup_{\xi(w)\in F}\left\{\left\|T_i(PT_i)^{n-1}(w,\eta(w)) - \xi(w)\right\| - \|\eta(w) - \xi(w)\|\right\}\right\} \leq 0, \qquad i = 1,2,3. \tag{22}$$

It follows that for any given $\epsilon > 0$, there exists a positive integer n_0 such that for $n \geq n_0$ and $\xi(w) \in F$, we have

$$\sup_{\xi(w)\in F}\left\{\left\|T_i(PT_i)^{n-1}(w,\eta(w)) - \xi(w)\right\| - \|\eta(w) - \xi(w)\|\right\} \leq \epsilon, \qquad i = 1,2,3. \tag{23}$$

Since $\{\xi_n(w)\}$, $\{\eta_n(w)\}$, and $\{\zeta_n(w)\} \subset E$, then we have for $w \in \Omega$,

$$\left\{\left\|T_1(PT_1)^{n-1}(w,\eta_n(w)) - \xi(w)\right\| - \|\eta_n(w) - \xi(w)\|\right\} \leq \epsilon, \qquad \forall n \geq n_0, \ \forall \xi \in F, \tag{24}$$

$$\left\{\left\|T_2(PT_2)^{n-1}(w,\zeta_n(w)) - \xi(w)\right\| - \|\zeta_n(w) - \xi(w)\|\right\} \leq \epsilon, \qquad \forall n \geq n_0, \ \forall \xi \in F, \tag{25}$$

$$\left\{\left\|T_3(PT_3)^{n-1}(w,\xi_n(w)) - \xi(w)\right\| - \|\xi_n(w) - \xi(w)\|\right\} \leq \epsilon, \qquad \forall n \geq n_0, \ \forall \xi \in F. \tag{26}$$

Setting for $w \in \Omega$,

$$\begin{aligned}
\mu_n(w) &= (1-\alpha_n)\eta_n(w) + \alpha_n T_1(PT_1)^{n-1}(w,\eta_n(w)), \\
\nu_n(w) &= (1-\beta_n)\zeta_n(w) + \beta_n T_2(PT_2)^{n-1}(w,\zeta_n(w)), \\
\tau_n(w) &= (1-\gamma_n)\xi_n(w) + \gamma_n T_3(PT_3)^{n-1}(w,\xi_n(w)).
\end{aligned} \tag{27}$$

Thus, for $\xi(w) \in F$ and $w \in \Omega$, using (17) and (24), we have

$$\begin{aligned}
&\|\xi_{n+1}(w) - \xi(w)\| \\
&= \left\|P\left[(1-a_n-\sigma_n)\xi_n(w) + a_n T_1(PT_1)^{n-1} \times (w,\mu_n(w)) + \sigma_n f_n(w)\right] - \xi(w)\right\| \\
&\leq \left\|(1-a_n-\sigma_n)\xi_n(w) + a_n T_1(PT_1)^{n-1} \left(w,(1-\alpha_n)\eta_n(w) + \alpha_n T_1(PT_1)^{n-1}(w,\eta_n(w))\right) + \sigma_n f_n(w) - \xi(w)\right\| \\
&= \left\|(1-a_n-\sigma_n)\xi_n(w) + a_n\xi(w) + \sigma_n\xi(w) - \xi(w) + a_n\left(T_1(PT_1)^{n-1}(w,\mu_n(w)) - \xi(w)\right) + \sigma_n(f_n(w) - \xi(w))\right\| \\
&\leq (1-a_n-\sigma_n)\|\xi_n(w) - \xi(w)\| + a_n\left\|T_1(PT_1)^{n-1}(w,\mu_n(w)) - \xi(w)\right\| + \sigma_n\|f_n(w) - \xi(w)\| \\
&\leq (1-a_n-\sigma_n)\|\xi_n(w) - \xi(w)\| + a_n\left[\left\|T_1(PT_1)^{n-1}(w,\mu_n(w)) - \xi(w)\right\| - \|\mu_n(w) - \xi(w)\|\right] \\
&\quad + a_n\|\mu_n(w) - \xi(w)\| + \sigma_n M(w) \\
&\leq (1-a_n-\sigma_n)\|\xi_n(w) - \xi(w)\| + a_n\epsilon + a_n\|\mu_n(w) - \xi(w)\| + \sigma_n M(w).
\end{aligned} \tag{28}$$

In addition, by (24), we obtain

$$\begin{aligned}
&\|\mu_n(w) - \xi(w)\| \\
&\quad = \left\|(1-\alpha_n)\eta_n(w) + \alpha_n T_1(PT_1)^{n-1}(w,\eta_n(w)) - \xi(w)\right\| \\
&\quad \le (1-\alpha_n)\|\eta_n(w) - \xi(w)\| \\
&\qquad + \alpha_n\left\|T_1(PT_1)^{n-1}(w,\eta_n(w)) - \xi(w)\right\| \\
&\quad \le (1-\alpha_n)\|\eta_n(w) - \xi(w)\| + \alpha_n\epsilon + \alpha_n\|\eta_n(w) - \xi(w)\| \\
&\quad = \|\eta_n(w) - \xi(w)\| + \alpha_n\epsilon.
\end{aligned} \tag{29}$$

Again using (17) and (25), we have

$$\begin{aligned}
&\|\eta_n(w) - \xi(w)\| \\
&\quad = \left\|P\left[(1-b_n-\delta_n)\xi_n(w) + b_n T_2(PT_2)^{n-1}(w,\nu_n(w))\right.\right. \\
&\qquad\qquad \left.\left. +\delta_n g_n(w)\right] - \xi(w)\right\| \\
&\quad \le \left\|(1-b_n-\delta_n)\xi_n(w) + b_n T_2(PT_2)^{n-1}\right. \\
&\qquad \times\left(w,(1-\beta_n)\zeta_n(w) + \beta_n T_2(PT_2)^{n-1}(w,\zeta_n(w))\right) \\
&\qquad \left. +\delta_n g_n(w) - \xi(w)\right\| \\
&\quad = \|(1-b_n-\delta_n)\xi_n(w) + b_n\xi(w) + \delta_n\xi(w) - \xi(w) \\
&\qquad + b_n\left(T_2(PT_2)^{n-1}(w,\nu_n(w)) - \xi(w)\right) \\
&\qquad +\delta_n(g_n(w) - \xi(w))\| \\
&\quad \le (1-b_n-\delta_n)\|\xi_n(w) - \xi(w)\| \\
&\qquad + b_n\left\|T_2(PT_2)^{n-1}(w,\nu_n(w)) - \xi(w)\right\| \\
&\qquad + \delta_n\|g_n(w) - \xi(w)\| \\
&\quad \le (1-b_n-\delta_n)\|\xi_n(w) - \xi(w)\| + b_n\epsilon \\
&\qquad + b_n\|\nu_n(w) - \xi(w)\| + \delta_n M(w).
\end{aligned} \tag{30}$$

In addition, by (25), we have

$$\begin{aligned}
&\|\nu_n(w) - \xi(w)\| \\
&\quad = \left\|(1-\beta_n)\zeta_n(w) + \beta_n T_2(PT_2)^{n-1}(w,\zeta_n(w)) - \xi(w)\right\| \\
&\quad \le (1-\beta_n)\|\zeta_n(w) - \xi(w)\| \\
&\qquad + \beta_n\left\|T_2(PT_2)^{n-1}(w,\zeta_n(w)) - \xi(w)\right\| \\
&\quad \le (1-\beta_n)\|\zeta_n(w) - \xi(w)\| + \beta_n\epsilon + \beta_n\|\zeta_n(w) - \xi(w)\| \\
&\quad = \|\zeta_n(w) - \xi(w)\| + \beta_n\epsilon.
\end{aligned} \tag{31}$$

Also, by (17) and (26), we have

$$\begin{aligned}
&\|\zeta_n(w) - \xi(w)\| \\
&\quad = \left\|P\left[(1-c_n-\lambda_n)\xi_n(w) + c_n T_3(PT_3)^{n-1}\right.\right. \\
&\qquad\qquad \left.\left.\times(w,\tau_n(w)) + \lambda_n h_n(w)\right] - \xi(w)\right\| \\
&\quad \le \left\|(1-c_n-\lambda_n)\xi_n(w) + c_n T_3(PT_3)^{n-1}\right. \\
&\qquad \times\left(w,(1-\gamma_n)\xi_n(w) + \gamma_n T_3(PT_3)^{n-1}(w,\xi_n(w))\right) \\
&\qquad \left. +\lambda_n h_n(w) - \xi(w)\right\| \\
&\quad = \|(1-c_n-\lambda_n)\xi_n(w) + c_n\xi(w) + \lambda_n\xi(w) - \xi(w) \\
&\qquad + c_n\left(T_3(PT_3)^{n-1}(w,\tau_n(w)) - \xi(w)\right) \\
&\qquad +\lambda_n(h_n(w) - \xi(w))\| \\
&\quad \le (1-c_n-\lambda_n)\|\xi_n(w) - \xi(w)\| \\
&\qquad + c_n\left\|T_3(PT_3)^{n-1}(w,\tau_n(w)) - \xi(w)\right\| \\
&\qquad + \lambda_n\|h_n(w) - \xi(w)\| \\
&\quad \le (1-c_n-\lambda_n)\|\xi_n(w) - \xi(w)\| + c_n\epsilon \\
&\qquad + c_n\|\tau_n(w) - \xi(w)\| + \lambda_n M(w).
\end{aligned} \tag{32}$$

In addition, by (26), we have

$$\begin{aligned}
&\|\tau_n(w) - \xi(w)\| \\
&\quad = \left\|(1-\gamma_n)\xi_n(w) + \gamma_n T_3(PT_3)^{n-1}(w,\xi_n(w)) - \xi(w)\right\| \\
&\quad \le (1-\gamma_n)\|\xi_n(w) - \xi(w)\| \\
&\qquad + \gamma_n\left\|T_3(PT_3)^{n-1}(w,\xi_n(w)) - \xi(w)\right\| \\
&\quad \le (1-\gamma_n)\|\xi_n(w) - \xi(w)\| + \gamma_n\epsilon + \gamma_n\|\xi_n(w) - \xi(w)\| \\
&\quad = \|\xi_n(w) - \xi(w)\| + \gamma_n\epsilon.
\end{aligned} \tag{33}$$

Substituting (29), (30), (31), (32), and (33) into (28) and simplifying, we obtain

$$\begin{aligned}
&\|\xi_{n+1}(w) - \xi(w)\| \\
&\quad \le (1-a_n-\sigma_n)\|\xi_n(w) - \xi(w)\| + a_n\epsilon \\
&\qquad + a_n\|\mu_n(w) - \xi(w)\| + \sigma_n M(w) \\
&\quad \le (1-a_n-\sigma_n)\|\xi_n(w) - \xi(w)\| + a_n\epsilon \\
&\qquad + a_n\left[\|\eta_n(w) - \xi(w)\| + \alpha_n\epsilon\right] + \sigma_n M(w) \\
&\quad = (1-a_n-\sigma_n)\|\xi_n(w) - \xi(w)\| + a_n\epsilon \\
&\qquad + a_n\|\eta_n(w) - \xi(w)\| + a_n\alpha_n\epsilon + \sigma_n M(w)
\end{aligned}$$

$$
\begin{aligned}
&\le (1-a_n-\sigma_n)\|\xi_n(w)-\xi(w)\| + a_n\epsilon \\
&\quad + a_n\big[(1-b_n-\delta_n)\|\xi_n(w)-\xi(w)\| + b_n\epsilon \\
&\qquad + b_n\|v_n(w)-\xi(w)\| + \delta_n M(w)\big] \\
&\quad + a_n\alpha_n\epsilon + \sigma_n M(w) \\
&= (1-a_n-\sigma_n)\|\xi_n(w)-\xi(w)\| + a_n\epsilon \\
&\quad + a_n(1-b_n-\delta_n)\|\xi_n(w)-\xi(w)\| + a_nb_n\epsilon \\
&\quad + a_nb_n\|v_n(w)-\xi(w)\| \\
&\quad + a_n\delta_n M(w) + a_n\alpha_n\epsilon + \sigma_n M(w) \\
&= (1-\sigma_n-a_nb_n-a_n\delta_n)\|\xi_n(w)-\xi(w)\| \\
&\quad + a_n\epsilon + a_nb_n\epsilon + a_nb_n\|v_n(w)-\xi(w)\| \\
&\quad + a_n\delta_n M(w) + a_n\alpha_n\epsilon + \sigma_n M(w) \\
&\le (1-\sigma_n-a_nb_n-a_n\delta_n)\|\xi_n(w)-\xi(w)\| \\
&\quad + a_n\epsilon + a_nb_n\epsilon + a_nb_n\big[\|\zeta_n(w)-\xi(w)\| + \beta_n\epsilon\big] \\
&\quad + a_n\delta_n M(w) + a_n\alpha_n\epsilon + \sigma_n M(w) \\
&= (1-\sigma_n-a_nb_n-a_n\delta_n)\|\xi_n(w)-\xi(w)\| \\
&\quad + a_n\epsilon + a_nb_n\epsilon + a_nb_n\|\zeta_n(w)-\xi(w)\| \\
&\quad + a_nb_n\beta_n\epsilon + a_n\delta_n M(w) + a_n\alpha_n\epsilon + \sigma_n M(w) \\
&\le (1-\sigma_n-a_nb_n-a_n\delta_n)\|\xi_n(w)-\xi(w)\| + a_n\epsilon \\
&\quad + a_nb_n\epsilon + a_nb_n(1-c_n-\lambda_n)\|\xi_n(w)-\xi(w)\| \\
&\quad + a_nb_nc_n\epsilon + a_nb_nc_n\|\tau_n(w)-\xi(w)\| \\
&\quad + a_nb_n\lambda_n M(w) + a_nb_n\beta_n\epsilon + a_n\delta_n M(w) \\
&\quad + a_n\alpha_n\epsilon + \sigma_n M(w) \\
&= (1-\sigma_n-a_n\delta_n-a_nb_nc_n-a_nb_n\lambda_n) \\
&\quad \times \|\xi_n(w)-\xi(w)\| + a_n\epsilon + a_nb_n\epsilon \\
&\quad + a_nb_nc_n\epsilon + a_nb_nc_n\|\tau_n(w)-\xi(w)\| \\
&\quad + a_nb_n\lambda_n M(w) + a_nb_n\beta_n\epsilon + a_n\delta_n M(w) \\
&\quad + a_n\alpha_n\epsilon + \sigma_n M(w) \\
&\le (1-\sigma_n-a_n\delta_n-a_nb_nc_n-a_nb_n\lambda_n)\|\xi_n(w)-\xi(w)\| \\
&\quad + a_n\epsilon + a_nb_n\epsilon + a_nb_nc_n\epsilon + a_nb_nc_n \\
&\quad \times \|\xi_n(w)-\xi(w)\| + a_nb_nc_n\gamma_n\epsilon + a_nb_n\lambda_n M(w) \\
&\quad + a_nb_n\beta_n\epsilon + a_n\delta_n M(w) + a_n\alpha_n\epsilon + \sigma_n M(w) \\
&= (1-\sigma_n-a_n\delta_n-a_nb_n\lambda_n)\|\xi_n(w)-\xi(w)\| \\
&\quad + a_n\epsilon + a_nb_n\epsilon + a_nb_nc_n\epsilon + a_nb_nc_n\gamma_n\epsilon \\
&\quad + a_nb_n\lambda_n M(w) + a_nb_n\beta_n\epsilon + a_n\delta_n M(w) \\
&\quad + a_n\alpha_n\epsilon + \sigma_n M(w) \\
&= (1-\sigma_n-a_n\delta_n-a_nb_n\lambda_n)\|\xi_n(w)-\xi(w)\| \\
&\quad + \big[a_n + a_nb_n + a_nb_nc_n + a_nb_nc_n\gamma_n + a_nb_n\beta_n + a_n\alpha_n\big]\epsilon \\
&\quad + \big[a_nb_n\lambda_n + a_n\delta_n + \sigma_n\big]M(w) \\
&\le \|\xi_n(w)-\xi(w)\| + 6\epsilon + (\lambda_n+\delta_n+\sigma_n)M(w).
\end{aligned} \tag{34}
$$

Let $R_n(w) = 6\epsilon + (\lambda_n + \delta_n + \sigma_n)M(w)$; then, $\sum_{n=1}^{\infty} R_n(w) < \infty$ for all $w \in \Omega$.

It follows by (34) that

$$
\begin{aligned}
&\inf_{\xi(w)\in F}\|\xi_{n+1}(w)-\xi(w)\| \\
&\quad \le \inf_{\xi(w)\in F}\|\xi_n(w)-\xi(w)\| + R_n(w), \quad \forall n \ge n_0,\ w \in \Omega.
\end{aligned} \tag{35}
$$

From (35) and $\sum_{n=1}^{\infty} R_n(w) < \infty$ for all $w \in \Omega$, we have

$$
d(\xi_{n+1}(w), F) \le d(\xi_n(w), F) + R_n(w), \quad \forall w \in \Omega. \tag{36}
$$

By Lemma 7 and (36), it follows that $\lim_{n\to\infty} d(\xi_n(w), F)$ exists for all $\xi(w) \in F = \bigcap_{i=1}^{3} RF(T_i)$ and $w \in \Omega$.

Since $\liminf_{n\to\infty} d(\xi_n(w), F) = 0$, then we have

$$
\lim_{n\to\infty} d(\xi_n(w), F) = 0, \quad w \in \Omega. \tag{37}
$$

Next, we prove that $\xi_n(w)$ is a Cauchy sequence in E for each $w \in \Omega$.

For $n \ge n_0$, $m \ge n_1$, and $\xi(w) \in F$, we have by (35) that

$$
\begin{aligned}
&\|\xi_{n+m}(w)-\xi(w)\| \\
&\quad \le \|\xi_{n+m-1}(w)-\xi(w)\| + R_{n+m-1}(w) \\
&\quad \le \|\xi_{n+m-2}(w)-\xi(w)\| + R_{n+m-1}(w) + R_{n+m-2}(w) \\
&\quad \vdots \\
&\quad \le \|\xi_n(w)-\xi(w)\| + \sum_{k=n}^{n+m-1} R_k(w).
\end{aligned} \tag{38}
$$

Therefore, by using (38), we have for each $w \in \Omega$,

$$
\begin{aligned}
&\|\xi_{n+m}(w)-\xi_n(w)\| \\
&\quad \le \|\xi_{n+m}(w)-\xi(w)\| + \|\xi_n(w)-\xi(w)\| \\
&\quad \le 2\|\xi_n(w)-\xi(w)\| + \sum_{k=n}^{\infty} R_k(w).
\end{aligned} \tag{39}
$$

Since $\xi(w) \in F$ and by (39), we have for each $w \in \Omega$,

$$\begin{aligned} &\|\xi_{n+m}(w) - \xi_n(w)\| \\ &\quad \leq 2d(\xi_n(w), F) + \sum_{k=n}^{\infty} R_k(w), \quad \forall n \geq n_0. \end{aligned} \tag{40}$$

Since $\lim_{n\to\infty} d(\xi_n(w), F) = 0$ and $\sum_{n=1}^{\infty} R_n(w) < \infty$, for given $\epsilon > 0$, there exists a positive integer $n_1 \geq n_0$ such that $d(\xi_n(w), F) < \epsilon/4$ and $\sum_{n=1}^{\infty} R_n(w) < \epsilon/2$. We have

$$\|\xi_{n+m}(w) - \xi_n(w)\| < \epsilon, \quad \forall w \in \Omega, \tag{41}$$

or

$$\lim_{n\to\infty} \|\xi_{n+m}(w) - \xi_n(w)\| = 0, \quad \forall w \in \Omega; \tag{42}$$

this shows that $\xi_n(w)$ is a Cauchy sequence in C for each $w \in \Omega$.

Since E is complete and C is a closed subset of E and so it is complete, then there exists a $p(w) \in C$ such that $\xi_n(w) \to p(w)$ as $n \to \infty$, for all $w \in \Omega$.

Now, we show that $p(w) \in F$.

By contradiction, we assume that $p(w)$ does not belong to F. Since F is closed set, $d(p(w), F) > 0$. By using the fact that $\lim_{n\to\infty} d(\xi_n(w), F) = 0$, it follows that for all $\xi(w) \in F$,

$$\begin{aligned} &\|p(w) - \xi(w)\| \\ &\quad \leq \|p(w) - \xi_n(w)\| + \|\xi_n(w) - \xi(w)\|. \end{aligned} \tag{43}$$

This implies that

$$\begin{aligned} d(p(w), F) &\leq \|p(w) - \xi_n(w)\| + d(\xi_n(w), F) \\ &\leq 0 \quad (as\ n \to \infty), \end{aligned} \tag{44}$$

which is a contradiction. Hence, $p(w) \in F$. □

Corollary 9. *Suppose that the conditions in Theorem 8 are satisfied. Then the random iterative sequence $\xi_n(w)$ generated by (17) converges to a common random fixed point $\xi(w)$ if and only if for all $w \in \Omega$, there exists a subsequence $\xi_{n_j}(w)$ of $\xi_n(w)$ which converges to $\xi(w)$.*

Theorem 10. *Let E be a real separable Banach space and C a nonempty closed convex subset of E with P as a nonexpansive retraction. Let $T_i : \Omega \times C \to E$, $i = 1, 2, 3$, be three asymptotically quasi-nonexpansive nonself random mappings with $F = \bigcap_{i=1}^{3} RF(T_i) \neq \phi$, for all $w \in \Omega$. Suppose that $\{\xi_n(w)\}$, $\{\eta_n(w)\}$, and $\{\zeta_n(w)\}$ are the sequences defined as in (17) where $\{a_n\}$, $\{b_n\}$, $\{c_n\}$, $\{\alpha_n\}$, $\{\beta_n\}$, $\{\gamma_n\}$, $\{\sigma_n\}$, $\{\delta_n\}$, and $\{\lambda_n\}$ are sequences in $[0, 1]$ such that $a_n + \sigma_n \leq 1$, $b_n + \delta_n \leq 1$, $c_n + \lambda_n \leq 1$, and $\{f_n(w)\}$, $\{g_n(w)\}$, $\{h_n(w)\}$ are bounded sequences of measurable functions from Ω to C with the following restrictions: $\sum_{n=1}^{\infty} \sigma_n < \infty$, $\sum_{n=1}^{\infty} \delta_n < \infty$ and $\sum_{n=1}^{\infty} \lambda_n < \infty$. Then, $\{\xi_n(w)\}$ converge to a common random fixed point of T_1, T_2, and T_3 if and only if*

$$\liminf_{n\to\infty} d(\xi_n(w), F) = 0, \quad w \in \Omega. \tag{45}$$

Proof. Since $T_i : \Omega \times C \to E$, $i = 1, 2, 3$, are three asymptotically quasi-nonexpansive nonself random mappings, by Remark 2, they are asymptotically quasi-nonexpansive-type nonself random mappings the conclusion of Theorem 10 can be proved from Theorem 8 immediately. □

Theorem 11. *Let E be a real separable Banach space and C be a nonempty closed convex subset of E with P as a nonexpansive retraction. Let $T_i : \Omega \times C \to E$, $i = 1, 2, 3$, be three asymptotically nonexpansive nonself random mappings with $F = \bigcap_{i=1}^{3} RF(T_i) \neq \phi$, for all $w \in \Omega$. Suppose that $\{\xi_n(w)\}$, $\{\eta_n(w)\}$ and $\{\zeta_n(w)\}$ are the sequences defined as in (17) where $\{a_n\}$, $\{b_n\}$, $\{c_n\}$, $\{\alpha_n\}$, $\{\beta_n\}$, $\{\gamma_n\}$, $\{\sigma_n\}$, $\{\delta_n\}$, and $\{\lambda_n\}$ are sequences in $[0, 1]$ such that $a_n + \sigma_n \leq 1$, $b_n + \delta_n \leq 1$, $c_n + \lambda_n \leq 1$, and $\{f_n(w)\}$, $\{g_n(w)\}$, $\{h_n(w)\}$ are bounded sequences of measurable functions from Ω to C with the following restrictions: $\sum_{n=1}^{\infty} \sigma_n < \infty$, $\sum_{n=1}^{\infty} \delta_n < \infty$, and $\sum_{n=1}^{\infty} \lambda_n < \infty$. Then, $\{\xi_n(w)\}$ converge to a common random fixed point of T_1, T_2, and T_3 if and only if*

$$\liminf_{n\to\infty} d(\xi_n(w), F) = 0, \quad w \in \Omega. \tag{46}$$

Proof. Since $T_i : \Omega \times C \to E$, $i = 1, 2, 3$, are three asymptotically nonexpansive nonself random mappings, by Remark 2, they are asymptotically nonexpansive-type nonself random mappings, and therefore they are asymptotically quasi-nonexpansive-type nonself random mappings; the conclusion of Theorem 11 can be obtained from Theorem 8 immediately. □

Now, we can extend and generalize Theorems 8, 10, and 11 by using random iterative scheme (18) as follows.

Theorem 12. *Let E be a real separable Banach space and C a nonempty closed convex subset of E with P as a nonexpansive retraction. Let $T_i : \Omega \times C \to E$, $i = 1, 2, \ldots, N$, be N asymptotically quasi-nonexpansive-type nonself random mappings with $F = \bigcap_{i=1}^{N} RF(T_i) \neq \phi$, for all $w \in \Omega$. Suppose that $\{\xi_n(w)\}$ is the sequence defined as in (18) where $\{a_n^{(i)}\}$, $\{\alpha_n^{(i)}\}$, and $\{\sigma_n^{(i)}\}$ $(i = 1, 2, \ldots, N)$ are sequences in $[0, 1]$ such that $a_n^{(i)} + \sigma_n^{(i)} \leq 1$ for all $i = 1, 2, \ldots, N$ and $\{f_n^{(i)}(w)\}$ $(i = 1, 2, \ldots, N)$ are bounded sequences of measurable functions from Ω to C with the following restrictions: $\sum_{n=1}^{\infty} \sigma_n^{(i)} < \infty$, for all $(i = 1, 2, \ldots, N)$. Then $\{\xi_n(w)\}$ converge to a common random fixed point of $T_1, T_2, \ldots, T_N$ if and only if*

$$\liminf_{n\to\infty} d(\xi_n(w), F) = 0, \quad w \in \Omega. \tag{47}$$

Theorem 13. *Let E be a real separable Banach space and C be a nonempty closed convex subset of E with P as a nonexpansive retraction. Let $T_i : \Omega \times C \to E$, $i = 1, 2, \ldots, N$ be N asymptotically quasi-nonexpansive nonself random mappings with $F = \bigcap_{i=1}^{N} RF(T_i) \neq \phi$, for all $w \in \Omega$. Suppose that $\{\xi_n(w)\}$ be the sequence defined as in (18) where $\{a_n^{(i)}\}$, $\{\alpha_n^{(i)}\}$, and $\{\sigma_n^{(i)}\}$, $(i = 1, 2, \ldots, N)$ are sequences in $[0, 1]$ such that $a_n^{(i)} + \sigma_n^{(i)} \leq 1$ for all $i = 1, 2, \ldots, N$ and $\{f_n^{(i)}(w)\}$, $(i = 1, 2, \ldots, N)$ are bounded sequences of measurable functions from Ω to C with the following restrictions: $\sum_{n=1}^{\infty} \sigma_n^{(i)} < \infty$*

for all $(i = 1, 2, \ldots, N)$. *Then* $\{\xi_n(w)\}$ *converge to a common random fixed point of* $T_1, T_2, \ldots, T_N$ *if and only if*

$$\liminf_{n\to\infty} d(\xi_n(w), F) = 0, \quad w \in \Omega. \tag{48}$$

Theorem 14. *Let E be a real separable Banach space and C be a nonempty closed convex subset of E with P as a nonexpansive retraction. Let* $T_i : \Omega \times C \to E$, $i = 1, 2, \ldots, N$ *be N asymptotically nonexpansive nonself random mappings with* $F = \bigcap_{i=1}^{N} RF(T_i) \neq \phi$, *for all* $w \in \Omega$. *Suppose that* $\{\xi_n(w)\}$ *is the sequence defined as in* (18) *where* $\{a_n^{(i)}\}$, $\{\alpha_n^{(i)}\}$, *and* $\{\sigma_n^{(i)}\}$ $(i = 1, 2, \ldots, N)$ *are sequences in* $[0, 1]$ *such that* $a_n^{(i)} + \sigma_n^{(i)} \leq 1$ *for all* $i = 1, 2, \ldots, N$ *and* $\{f_n^{(i)}(w)\}$ $(i = 1, 2, \ldots, N)$ *are bounded sequences of measurable functions from* Ω *to C with the following restrictions* $\sum_{n=1}^{\infty} \sigma_n^{(i)} < \infty$ *for all* $(i = 1, 2, \ldots, N)$. *Then,* $\{\xi_n(w)\}$ *converge to a common random fixed point of* $T_1, T_2, \ldots, T_N$ *if and only if*

$$\liminf_{n\to\infty} d(\xi_n(w), F) = 0, \quad w \in \Omega. \tag{49}$$

Senter and Dotson [40] defined Condition (A) as follows.

Definition 15 (see [40]). A mapping $T : C \to C$ where C is a subset of a Banach space E with $F(T) \neq \phi$ is said to satisfy Condition (A) if there exists a nondecreasing function $f : [0, \infty) \to [0, \infty)$ with $f(0) = 0$, $f(r) > 0$, for all $r \in (0, \infty)$ such that for all $x \in C$,

$$\|x - Tx\| \geq f(d(x, F(T))), \tag{50}$$

where $d(x, F(T)) = \inf\{\| x - p \| : p \in F(T)\}$.

As an application, we can apply Theorem 8 and Condition (A) to obtain a convergences theorem for scheme (17).

Theorem 16. *Let E be a real uniformly separable Banach space and C a nonempty closed convex subset of E with P as a nonexpansive retraction. Let* $T_i : \Omega \times C \to E$, $i = 1, 2, 3$, *be three asymptotically quasi-nonexpansive-type nonself random mappings with* $F = \bigcap_{i=1}^{3} RF(T_i) \neq \phi$, *for all* $w \in \Omega$. *Suppose that* $\{\xi_n(w)\}$, $\{\eta_n(w)\}$ *and* $\{\zeta_n(w)\}$ *are the sequences defined as in* (17) *where* $\{a_n\}$, $\{b_n\}$, $\{c_n\}$, $\{\alpha_n\}$, $\{\beta_n\}$, $\{\gamma_n\}$, $\{\sigma_n\}$, $\{\delta_n\}$, *and* $\{\lambda_n\}$ *are sequences in* $[0, 1]$ *such that* $a_n + \sigma_n \leq 1$, $b_n + \delta_n \leq 1$, $c_n + \lambda_n \leq 1$ *and* $\{f_n(w)\}$, $\{g_n(w)\}$, $\{h_n(w)\}$ *are bounded sequences of measurable functions from* Ω *to C with the following restrictions:* $\sum_{n=1}^{\infty} \sigma_n < \infty$, $\sum_{n=1}^{\infty} \delta_n < \infty$, *and* $\sum_{n=1}^{\infty} \lambda_n < \infty$. *Suppose one of the mappings* T_i, $i = 1, 2, 3$, *satisfying Condition (A) and the following condition:* $\lim_{n\to\infty} \| \xi_n(w) - T(w, \xi_n(w)) \| = 0$, *for all* $w \in \Omega$. *Then,* $\{\xi_n(w)\}$ *converge to a common random fixed point of* T_1, T_2, *and* T_3.

Proof. From Theorem 8, we have $\lim_{n\to\infty} \| \xi_n(w) - \xi(w) \|$, and $\lim_{n\to\infty} d(\xi_n(w), F)$ exists. Let one of the mappings T_i, say T_1 satisfy Condition (A) and $\lim_{n\to\infty} \| \xi_n(w) - T_1(w, \xi_n(w)) \| = 0$; then, we have for all $w \in \Omega$,

$$\lim_{n\to\infty} f(d(\xi_n(w), F)) \leq \lim_{n\to\infty} \|\xi_n(w) - T_1(w, \xi_n(w))\| = 0. \tag{51}$$

By the property of f and since $\lim_{n\to\infty} d(\xi_n(w), F)$ exists, we have that

$$\lim_{n\to\infty} d(\xi_n(w), F) = 0. \tag{52}$$

By Theorem 8, $\{\xi_n(w)\}$ converge to a common random fixed point of T_1, T_2, and T_3. □

Acknowledgment

The authors would like to extend their sincerest thanks to the anonymous referees and editors for the exceptional review of this work. The suggestions and recommendations in the report increased the quality of their paper.

References

[1] O. Hanš, "Reduzierende zufällige Transformationen," *Czechoslovak Mathematical Journal*, vol. 7, pp. 154–158, 1957.

[2] O. Hanš, "Random operator equations," in *Proceeding of the 4th Barkeley Symposium on Mathematical Statistics and Probability*, vol. 2, pp. 185–202, University of California Press, Berkeley, Calif, USA, 1961.

[3] A. Špaček, "Zufällige Gleichungen," *Czechoslovak Mathematical Journal*, vol. 5, pp. 462–466, 1955.

[4] A. T. Bharucha-Reid, "Fixed point theorems in probabilistic analysis," *Bulletin of the American Mathematical Society*, vol. 82, no. 5, pp. 641–657, 1976.

[5] I. Beg, "Approximation of random fixed points in normed spaces," *Nonlinear Analysis: Theory, Methods & Applications*, vol. 51, no. 8, pp. 1363–1372, 2002.

[6] I. Beg, "Minimal displacement of random variables under Lipschitz random maps," *Topological Methods in Nonlinear Analysis*, vol. 19, no. 2, pp. 391–397, 2002.

[7] I. Beg and N. Shahzad, "Random fixed point theorems for nonexpansive and contractive-type random operators on Banach spaces," *Journal of Applied Mathematics and Stochastic Analysis*, vol. 7, no. 4, pp. 569–580, 1994.

[8] S. Itoh, "Random fixed-point theorems with an application to random differential equations in Banach spaces," *Journal of Mathematical Analysis and Applications*, vol. 67, no. 2, pp. 261–273, 1979.

[9] N. S. Papageorgiou, "Random fixed point theorems for measurable multifunctions in Banach spaces," *Proceedings of the American Mathematical Society*, vol. 97, no. 3, pp. 507–514, 1986.

[10] H. K. Xu, "Some random fixed point theorems for condensing and nonexpansive operators," *Proceedings of the American Mathematical Society*, vol. 110, no. 2, pp. 395–400, 1990.

[11] K. Goebel and W. A. Kirk, "A fixed point theorem for asymptotically nonexpansive mappings," *Proceedings of the American Mathematical Society*, vol. 35, pp. 171–174, 1972.

[12] J. Gornicki, "Weak convergence theorems for asymptotically nonexpansive mappings in uniformly convex Banach spaces," *Commentationes Mathematicae Universitatis Carolinae*, vol. 30, no. 2, pp. 249–252, 1989.

[13] M. O. Osilike and S. C. Aniagbosor, "Weak and strong convergence theorems for fixed points of asymptotically nonexpansive mappings," *Mathematical and Computer Modelling*, vol. 32, no. 10, pp. 1181–1191, 2000.

[14] J. Schu, "Iterative construction of fixed points of asymptotically nonexpansive mappings," *Journal of Mathematical Analysis and Applications*, vol. 158, no. 2, pp. 407–413, 1991.

[15] J. Schu, "Weak and strong convergence to fixed points of asymptotically nonexpansive mappings," *Bulletin of the Australian Mathematical Society*, vol. 43, no. 1, pp. 153–159, 1991.

[16] W. V. Petryshyn and T. E. Williamson, Jr., "Strong and weak convergence of the sequence of successive approximations for quasi-nonexpansive mappings," *Journal of Mathematical Analysis and Applications*, vol. 43, pp. 459–497, 1973.

[17] M. K. Ghosh and L. Debnath, "Convergence of Ishikawa iterates of quasi-nonexpansive mappings," *Journal of Mathematical Analysis and Applications*, vol. 207, no. 1, pp. 96–103, 1997.

[18] Q. H. Liu, "Iterative sequences for asymptotically quasi-nonexpansive mappings," *Journal of Mathematical Analysis and Applications*, vol. 259, no. 1, pp. 1–7, 2001.

[19] Q. H. Liu, "Iterative sequences for asymptotically quasi-nonexpansive mappings with error member," *Journal of Mathematical Analysis and Applications*, vol. 259, no. 1, pp. 18–24, 2001.

[20] M. A. Noor, "New approximation schemes for general variational inequalities," *Journal of Mathematical Analysis and Applications*, vol. 251, no. 1, pp. 217–229, 2000.

[21] B. Xu and M. A. Noor, "Fixed-point iterations for asymptotically nonexpansive mappings in Banach spaces," *Journal of Mathematical Analysis and Applications*, vol. 267, no. 2, pp. 444–453, 2002.

[22] S. Suantai, "Weak and strong convergence criteria of Noor iterations for asymptotically nonexpansive mappings," *Journal of Mathematical Analysis and Applications*, vol. 311, no. 2, pp. 506–517, 2005.

[23] J. S. Jung and S. S. Kim, "Strong convergence theorems for nonexpansive nonself-mappings in Banach spaces," *Nonlinear Analysis: Theory, Methods & Applications*, vol. 33, no. 3, pp. 321–329, 1998.

[24] S. Y. Matsushita and D. Kuroiwa, "Strong convergence of averaging iterations of nonexpansive nonself-mappings," *Journal of Mathematical Analysis and Applications*, vol. 294, no. 1, pp. 206–214, 2004.

[25] N. Shahzad, "Approximating fixed points of non-self nonexpansive mappings in Banach spaces," *Nonlinear Analysis: Theory, Methods & Applications*, vol. 61, no. 6, pp. 1031–1039, 2005.

[26] W. Takahashi and G. E. Kim, "Strong convergence of approximants to fixed points of nonexpansive nonself-mappings in Banach spaces," *Nonlinear Analysis: Theory, Methods & Applications*, vol. 32, no. 3, pp. 447–454, 1998.

[27] H. K. Xu and X. M. Yin, "Strong convergence theorems for nonexpansive non-self-mappings," *Nonlinear Analysis: Theory, Methods & Applications*, vol. 24, no. 2, pp. 223–228, 1995.

[28] C. E. Chidume, E. U. Ofoedu, and H. Zegeye, "Strong and weak convergence theorems for asymptotically nonexpansive mappings," *Journal of Mathematical Analysis and Applications*, vol. 280, no. 2, pp. 364–374, 2003.

[29] L. Wang, "Strong and weak convergence theorems for common fixed point of nonself asymptotically nonexpansive mappings," *Journal of Mathematical Analysis and Applications*, vol. 323, no. 1, pp. 550–557, 2006.

[30] I. Beg and M. Abbas, "Iterative procedures for solutions of random operator equations in Banach spaces," *Journal of Mathematical Analysis and Applications*, vol. 315, no. 1, pp. 181–201, 2006.

[31] Y. X. Tian, S. S. Chang, and J. L. Huang, "On the approximation problem of common fixed points for a finite family of non-self asymptotically quasi-nonexpansive-type mappings in Banach spaces," *Computers & Mathematics with Applications*, vol. 53, no. 12, pp. 1847–1853, 2007.

[32] S. Thianwan, "Weak and strong convergence theorems for new iterations with errors for nonexpansive nonself-mapping," *Thai Journal of Mathematics*, vol. 6, no. 3, pp. 27–38, 2008.

[33] H. Kiziltunc, "On common fixed points of a new iteration for two nonself asymptotically quasi-nonexpansive-type mappings in Banach spaces," *Journal of Nonlinear Analysis and Optimization*, vol. 2, no. 2, pp. 259–267, 2011.

[34] W. Q. Deng, L. Wang, and Y. J. Chen, "Strong and weak convergence theorems for common fixed points of two nonself asymptotically nonexpansive mappings in Banach spaces," *International Mathematical Forum*, vol. 7, no. 9–12, pp. 407–417, 2012.

[35] S. Plubtieng, P. Kumam, and R. Wangkeeree, "Random three-step iteration scheme and common random fixed point of three operators," *Journal of Applied Mathematics and Stochastic Analysis*, vol. 2007, Article ID 82517, 10 pages, 2007.

[36] S. Plubtieng, R. Wangkeeree, and R. Punpaeng, "On the convergence of modified Noor iterations with errors for asymptotically nonexpansive mappings," *Journal of Mathematical Analysis and Applications*, vol. 322, no. 2, pp. 1018–1029, 2006.

[37] X. W. Zhou and L. Wang, "Approximation of random fixed points of non-self asymptotically nonexpansive random mappings," *International Mathematical Forum*, vol. 2, no. 37–40, pp. 1859–1868, 2007.

[38] G. S. Saluja, "Approximation of common random fixed point for a finite family of non-self asymptotically nonexpansive random mappings," *Demonstratio Mathematica*, vol. 42, no. 3, pp. 581–598, 2009.

[39] K. K. Tan and H. K. Xu, "Approximating fixed points of nonexpansive mappings by the Ishikawa iteration process," *Journal of Mathematical Analysis and Applications*, vol. 178, no. 2, pp. 301–308, 1993.

[40] H. F. Senter and W. G. Dotson Jr., "Approximating fixed points of nonexpansive mappings," *Proceedings of the American Mathematical Society*, vol. 44, pp. 375–380, 1974.

6

Further Results on the Stability Analysis of Singular Systems with Time-Varying Delay: A Delay Decomposition Approach

Pin-Lin Liu

Department of Automation, Engineering Institute of Mechatronoptic System, Chienkuo Technology University, Changhua 500, Taiwan

Correspondence should be addressed to Pin-Lin Liu; lpl@ctu.edu.tw

Academic Editor: Malte Braack

This paper deals with the problem of stability analysis for singular systems with time-varying delay. By developing a delay decomposition approach, information of the delayed plant states can be taken into full consideration, and new delay-dependent sufficient stability criteria are obtained in terms of linear matrix inequalities (LMIs), which can be easily solved by various optimization algorithms. The merits of the proposed results lie in their less conservatism which is realized by choosing different Lyapunov matrices in the decomposed intervals and taking the information of the delayed plant states into full consideration. It is proved that the newly proposed criteria may introduce less conservatism than some existing ones. Meanwhile, the computational complexity of the presented stability criteria is reduced greatly since fewer decision variables are involved. Numerical examples are included to show that the proposed method is effective and can provide less conservative results.

1. Introduction

Delay phenomena widely exist in many practical engineering systems, such as aircraft, chemical, and process control systems. The study on time delay systems is thus of great significance both in theory and in practice. The time delay is frequently a source of instability and performance deterioration. Therefore, stability analysis and controller synthesis for time-delay system have been one of the most challenging issues [1–23]. On the other hand, singular time-delay systems, which are also referred to as implicit time-delay systems, descriptor time-delay systems, or generalized differential-difference equations, have strong practical relevance in various engineering systems, including aircraft attitude control, flexible arm control of robots, large-scale electric network control, chemical engineering systems, lossless transmission lines, and so forth (see, e.g., [3–11, 13, 15–21]). For this reason, over the past decades, there has been increasing interest in the stability analysis for singular time-delay systems, and many results have been reported in the literature [6, 9, 15–20].

Recently, some improved delay-dependent stability criteria have been obtained without using different model transformations and bounding techniques for cross terms [4–6, 13, 16, 19]. However, some slack variables are introduced apart from the matrix variables appearing in Lyapunov-Krasovskii functionals (LKFs), and the results are still conservative, which can be seen by applying these types of criteria to the nominal singular system with a constant time delay and comparing with analytical delay limit for stability. Therefore how one can further improve the existing stability criteria is of great importance to the further study of singular time delay systems. In order to reduce the conservatism, a delay decomposition approach was also proposed in [9, 12, 20, 21, 23]. Both by theory analysis and by numerical examples, it is pointed out that the results obtained by delay decomposition approach are much less conservative than some existing ones and include some as their special cases. Therefore by this approach, very significant steps have been made towards the analytical delay limit for the stability of time-delay systems.

Motivated by the above discussions, we propose new stability criteria for singular systems with time-varying delays. The main aim is to derive a maximum admissible upper bound (MAUB) of the time delay such that the time-delay system is asymptotically stable for any delay size less than the MAUB. Accordingly, the obtained MAUB becomes a key performance index to measure the conservatism of a delay-dependent stability condition. The merits of the proposed results lie in their less conservatism which is realized by choosing different Lyapunov matrices in the decomposed intervals and taking the information of the delayed plant states into full consideration. The analysis, eventually, culminates into a stability condition in convex linear matrix inequality (LMI) framework and is solved nonconservatively at boundary conditions using standard LMI solvers. Numerical examples are given to illustrate the effectiveness and less conservatism of the proposed method.

2. Main Result

Consider the following singular system with a time-varying state delay:

$$E\dot{x}(t) = Ax(t) + Bx(t - h(t)), \quad t \geq 0, \tag{1a}$$

$$x(t) = \phi(t), \quad t \in [-h, 0], \tag{1b}$$

where $x(t) \in R^n$ is the state vector; $A$and B are constant matrices with appropriate dimensions; the matrix $E \in R^{n\times n}$ may be singular, without loss generality, we suppose rank $E = r \leq n$; $\phi(t)$ is a smooth vector-valued initial function; $h(t)$ is a time-varying delay in the state; h is an upper bound on the delay $h(t)$.

We consider two different cases for time varying delays as follows.

Case I. $h(t)$ is a differentiable function, satisfying for all $t \geq 0$:

$$0 \leq h(t) \leq h, \qquad \dot{h}(t) \leq h_d. \tag{2}$$

Case II. $h(t)$ is not differentiable or the upper bound of the derivative of $h(t)$ is unknown, and $h(t)$ satisfies

$$0 \leq h(t) \leq h, \tag{3}$$

where h and h_d are some positive constants.

The main objective is to find the range of h and guarantee stability for the singular system with a time-varying state delay (1a) and (1b). Here, definitions and fundamental lemmas are reviewed.

Definition 1 (see [3]). The pair (E, A) is said to be regular if det $(sE - A)$ is not identically zero.

Definition 2 (see [3]). The pair (E, A) is said to be impulse-free if deg(det($sE - A$)) = rank E.

Definition 3. For a given scalar $\overline{h} > 0$, the singular delay system (1a) and (1b) is said to be regular and impulse-free for any constant time delay h satisfying $0 \leq h \leq \overline{h}$, if the pairs (E, A) and $(E, A + B)$ are regular and impulse-free.

Remark 4. The regularity and the absence of impulses of the pair (E, A) ensure the system (1a) and (1b) with time delay $h \neq 0$ to be regular and impulse-free, while the fact that the pair $(E, A + B)$ is regular and impulse-free ensures the system (1a) and (1b) with time delay $h = 0$ to be regular and impulse-free.

Lemma 5 (see [8]). *The singular system $E\dot{x}(t) = Ax(t)$ is regular, impulse free, and stable, if and only if there exists a matrix P such that*

$$P^T E = E^T P \geq 0, \tag{4a}$$

$$A^T P + PA < 0. \tag{4b}$$

Lemma 6 (see [11]). *For any positive semidefinite matrices,*

$$X = \begin{bmatrix} X_{11} & X_{12} & X_{13} \\ X_{12}^T & X_{22} & X_{23} \\ X_{13}^T & X_{23}^T & X_{33} \end{bmatrix} \geq 0, \tag{5a}$$

the following integral inequality holds

$$\begin{aligned} &-\int_{t-h(t)}^{t} \dot{x}^T(s) X_{33} \dot{x}(s)\, ds \\ &\leq \int_{t-h(t)}^{t} \left[x^T(t) \;\; x^T(t-h(t)) \;\; \dot{x}^T(s)\right] \\ &\quad \times \begin{bmatrix} X_{11} & X_{12} & X_{13} \\ X_{12}^T & X_{22} & X_{23} \\ X_{13}^T & X_{23}^T & 0 \end{bmatrix} \begin{bmatrix} x(t) \\ x(t-h(t)) \\ \dot{x}(s) \end{bmatrix} ds. \end{aligned} \tag{5b}$$

Secondary, we introduce the following Schur complement which is essential in the proofs of our results.

Lemma 7 (see [2]). *The following matrix inequality:*

$$\begin{bmatrix} Q(x) & S(x) \\ S^T(x) & R(x) \end{bmatrix} < 0, \tag{6a}$$

$Q(x) = Q^T(x), R(x) = R^T(x)S(x)x,$

$$R(x) < 0, \tag{6b}$$

$$Q(x) < 0, \tag{6c}$$

$$Q(x) - S(x) R^{-1}(x) S^T(x) < 0. \tag{6d}$$

This paper finds new stability criteria less conservative than the existing results.

For the system (1a), (1b), and (2), we give stability condition by using a delay decomposition approach as follows.

Theorem 8. *In Case I, if $0 \leq h(t) \leq \alpha h$, for given three scalars h, α, and h_d, then, for any delay $h(t)$ satisfy $0 \leq h(t) \leq h$, $\dot{h}(t) \leq h_d$, and $0 < \alpha < 1$, the system described by (1a) and (1b) with (2) is asymptotically stable if there exist matrices*

$P = P^T > 0$, $Q_i = Q_i^T > 0$, $R_i = R_i^T > 0, (i = 1, 2, 3)$, *matrix S of appropriate dimensionsand positive semidefinite matrices:* $X = \begin{bmatrix} X_{11} & X_{12} & X_{13} \\ X_{12}^T & X_{22} & X_{23} \\ X_{13}^T & X_{23}^T & X_{33} \end{bmatrix} \geq 0$, $Y = \begin{bmatrix} Y_{11} & Y_{12} & Y_{13} \\ Y_{12}^T & Y_{22} & Y_{23} \\ Y_{13}^T & Y_{23}^T & Y_{33} \end{bmatrix} \geq 0$, $Z = \begin{bmatrix} Z_{11} & Z_{12} & Z_{13} \\ Z_{12}^T & Z_{22} & Z_{23} \\ Z_{13}^T & Z_{23}^T & Z_{33} \end{bmatrix} \geq 0$ *such that*

$$P^T E = E^T P \geq 0, \tag{7a}$$

$$\Omega = \begin{bmatrix} \Omega_{11} & \Omega_{12} & 0 & 0 & \Omega_{15} \\ \Omega_{12}^T & \Omega_{22} & \Omega_{23} & 0 & \Omega_{25} \\ 0 & \Omega_{23}^T & \Omega_{33} & \Omega_{34} & 0 \\ 0 & 0 & \Omega_{34}^T & \Omega_{44} & 0 \\ \Omega_{15}^T & \Omega_{25}^T & 0 & 0 & \Omega_{55} \end{bmatrix} < 0, \tag{7b}$$

$$\begin{gathered} E^T (R_1 - X_{33}) E \geq 0, \qquad E^T (R_2 - Y_{33}) E \geq 0, \\ E^T [R_1 + (1 - h_d) R_3 - Z_{33}] E \geq 0, \end{gathered} \tag{7c}$$

where $U \in R^{n \times (n-r)}$ *is any matrix satisfying* $E^T U = 0$ *and*

$$\begin{aligned} \Omega_{11} &= A^T P + PA + Q_1 + Q_3 + A^T U S^T + S U^T A \\ &\quad + E^T \left(\alpha h Z_{11} + Z_{13} + Z_{13}^T\right) E, \\ \Omega_{12} &= PB + SU^T B + E^T \left(\alpha h Z_{12} - Z_{13} + Z_{23}^T\right) E, \\ \Omega_{15} &= A^T \left[\alpha h R_1 + (1 - \alpha) h R_2 + \alpha h R_3\right], \\ \Omega_{22} &= -(1 - h_d) Q_3 + E^T \left(\alpha h X_{11} + X_{13} + X_{13}^T \right. \\ &\qquad \left. + \alpha h Z_{22} - Z_{23} - Z_{23}^T\right) E, \\ \Omega_{23} &= E^T \left(\alpha h X_{12} - X_{13} + X_{23}^T\right) E, \\ \Omega_{25} &= B^T \left[\alpha h R_1 + (1 - \alpha) h R_2 + \alpha h R_3\right], \\ \Omega_{33} &= Q_2 - Q_1 + E^T \left[\alpha h X_{22} - X_{23} - X_{23}^T \right. \\ &\qquad \left. + (1 - \alpha) h Y_{11} + Y_{13} + Y_{13}^T\right] E, \\ \Omega_{34} &= E^T \left[(1 - \alpha) h Y_{12} - Y_{13} + Y_{23}^T\right] E, \\ \Omega_{44} &= -Q_2 + E^T \left[(1 - \alpha) h Y_{22} - Y_{23} - Y_{23}^T\right] E, \\ \Omega_{55} &= -\left[\alpha h R_1 + (1 - \alpha) h R_2 + \alpha h R_3\right]. \end{aligned} \tag{8}$$

Proof. In Case I, a Lyapunov functional can be constructed as follows:

$$V(x_t) = V_1(x_t) + V_2(x_t) + V_3(x_t), \tag{9}$$

where

$$\begin{aligned} V_1(x_t) &= x^T(t) PEx(t), \\ V_2(x_t) &= \int_{t-\alpha h}^{t} x^T(s) Q_1 x(s)\, ds \\ &\quad + \int_{t-h}^{t-\alpha h} x^T(s) Q_2 x(s)\, ds \\ &\quad + \int_{t-h(t)}^{t} x^T(s) Q_3 x(s)\, ds, \\ V_3(x_t) &= \int_{-\alpha h}^{0} \int_{t+\theta}^{t} \dot{x}^T(s) E^T R_1 E \dot{x}(s)\, ds\, d\theta \\ &\quad + \int_{-h}^{-\alpha h} \int_{t+\theta}^{t} \dot{x}^T(s) E^T R_2 E \dot{x}(s)\, ds\, d\theta \\ &\quad + \int_{-h(t)}^{0} \int_{t+\theta}^{t} \dot{x}^T(s) E^T R_3 E \dot{x}(s)\, ds\, d\theta. \end{aligned} \tag{10}$$

Taking time derivative $V(x_t)$ for $t \in [0, \infty)$ along the trajectory (1a) and (1b) yields

$$\dot{V}(x_t) = \dot{V}_1(x_t) + \dot{V}_2(x_t) + \dot{V}_3(x_t), \tag{11}$$

where

$$\begin{aligned} \dot{V}_1(x_t) &= \dot{x}^T(t) E^T P x(t) + x^T(t) PE\dot{x}(t) \\ &= x^T(t)\left(A^T P + PA\right) x(t) + x^T(t) PBx(t - h(t)) \\ &\quad + x^T(t - h(t)) B^T P x(t), \\ \dot{V}_2(x_t) &= x^T(t)(Q_1 + Q_3) x(t) \\ &\quad - x^T(t - h(t))\left(1 - \dot{h}(t)\right) Q_3 x(t - h(t)) \\ &\quad + x^T(t - \alpha h)(Q_2 - Q_1) x(t - \alpha h) \\ &\quad - x^T(t - h) Q_2 x(t - h) \\ &\leq x^T(t)(Q_1 + Q_3) x(t) \\ &\quad - x^T(t - h(t))(1 - h_d) Q_3 x(t - h(t)) \\ &\quad + x^T(t - \alpha h)(Q_2 - Q_1) x(t - \alpha h) \\ &\quad - x^T(t - h) Q_2 x(t - h), \end{aligned} \tag{12}$$

$$\begin{aligned} \dot{V}_3(x_t) &= \dot{x}^T(t) E^T \left[\alpha h R_1 + (1 - \alpha) h R_2 + h(t) R_3\right] E\dot{x}(t) \\ &\quad - \int_{t-\alpha h}^{t} \dot{x}^T(s) E^T R_1 E \dot{x}(s)\, ds \end{aligned}$$

$$
\begin{aligned}
&- \int_{t-h}^{t-\alpha h} \dot{x}^T(s) E^T R_2 E\dot{x}(s)\, ds \\
&- \left(1 - \dot{h}(t)\right) \int_{t-h(t)}^{t} \dot{x}^T(s) E^T R_3 E\dot{x}(s)\, ds \\
&\leq \dot{x}^T(t) E^T \left[\alpha h R_1 + (1-\alpha) h R_2 + \alpha h R_3\right] E\dot{x}(t) \\
&- \int_{t-\alpha h}^{t} \dot{x}^T(s) E^T R_1 E\dot{x}(s)\, ds \\
&- \int_{t-h}^{t-\alpha h} \dot{x}^T(s) E^T R_2 E\dot{x}(s)\, ds \\
&- (1 - h_d) \int_{t-h(t)}^{t} \dot{x}^T(s) E^T R_3 E\dot{x}(s)\, ds.
\end{aligned} \tag{13}
$$

Now, we estimate the upper bound of the last three terms in inequality (13) as

$$
\begin{aligned}
&- \int_{t-\alpha h}^{t} \dot{x}^T(s) E^T R_1 E\dot{x}(s)\, ds \\
&- \int_{t-h}^{t-\alpha h} \dot{x}^T(s) E^T R_2 E\dot{x}(s)\, ds \\
&- (1 - h_d) \int_{t-h(t)}^{t} \dot{x}^T(s) E^T R_3 E\dot{x}(s)\, ds \\
&\quad = - \int_{t-\alpha h}^{t-h(t)} \dot{x}^T(s) E^T R_1 E\dot{x}(s)\, ds \\
&\quad - \int_{t-h}^{t-\alpha h} \dot{x}^T(s) E^T R_2 E\dot{x}(s)\, ds \\
&\quad - \int_{t-h(t)}^{t} \dot{x}^T(s) E^T \left(R_1 + (1-h_d) R_3\right) E^T \dot{x}(s)\, ds \\
&\quad = - \int_{t-\alpha h}^{t-h(t)} \dot{x}^T(s) E^T \left(R_1 - X_{33}\right) E\dot{x}(s)\, ds \\
&\quad - \int_{t-h}^{t-\alpha h} \dot{x}^T(s) E^T \left(R_2 - Y_{33}\right) E\dot{x}(s)\, ds \\
&\quad - \int_{t-h(t)}^{t} \dot{x}^T(s) E^T \left(R_1 + (1-h_d) R_3 - Z_{33}\right) E\dot{x}(s)\, ds \\
&\quad - \int_{t-\alpha h}^{t-h(t)} \dot{x}^T(s) E^T X_{33} E\dot{x}(s)\, ds \\
&\quad - \int_{t-h}^{t-\alpha h} \dot{x}^T(s) E^T Y_{33} E\dot{x}(s)\, ds \\
&\quad - \int_{t-h(t)}^{t} \dot{x}^T(s) E^T Z_{33} E\dot{x}(s)\, ds.
\end{aligned} \tag{14}
$$

From integral inequality [11], noticing that $E^T(R_1 - X_{33})E \geq 0$, $E^T(R_2 - Y_{33})E \geq 0$, and $E^T(R_1 + (1-h_d)R_3 - Z_{33})E \geq 0$ yields the following:

$$
\begin{aligned}
&- \int_{t-\alpha h}^{t-h(t)} \dot{x}^T(s) X_{33} \dot{x}(s)\, ds \\
&\leq \int_{t-\alpha h}^{t-h(t)} \left[x^T(t-h(t)) \;\; x^T(t-\alpha h) \;\; \dot{x}^T(s)\right] \\
&\quad \times \begin{bmatrix} X_{11} & X_{12} & X_{13} \\ X_{12}^T & X_{22} & X_{23} \\ X_{13}^T & X_{23}^T & 0 \end{bmatrix} \begin{bmatrix} x(t-h(t)) \\ x(t-\alpha h) \\ \dot{x}(s) \end{bmatrix} ds \\
&\leq x^T(t-h(t)) (\alpha h - h(t)) X_{11} x(t-h(t)) \\
&\quad + x^T(t-h(t)) (\alpha h - h(t)) X_{12} x(t-\alpha h) \\
&\quad + x^T(t-h(t)) X_{13} \int_{t-\alpha h}^{t-h(t)} \dot{x}(s)\, ds \\
&\quad + x^T(t-\alpha h) (\alpha h - h(t)) X_{12}^T x(t-h(t)) \\
&\quad + x^T(t-\alpha h) (\alpha h - h(t)) X_{22} x(t-\alpha h) \\
&\quad + x^T(t-\alpha h) X_{23} \int_{t-\alpha h}^{t-h(t)} \dot{x}(s)\, ds \\
&\quad + \int_{t-\alpha h}^{t-h(t)} \dot{x}^T(s)\, ds X_{13}^T x(t-h(t)) \\
&\quad + \int_{t-\alpha h}^{t-h(t)} \dot{x}^T(s)\, ds X_{23}^T x(t-\alpha h) \\
&\leq x^T(t-h(t)) \alpha h X_{11} x(t-h(t)) \\
&\quad + x^T(t-h(t)) \alpha h X_{12} x(t-\alpha h) \\
&\quad + x^T(t-h(t)) X_{13} \int_{t-\alpha h}^{t-h(t)} \dot{x}(s)\, ds \\
&\quad + x^T(t-\alpha h) \alpha h X_{12}^T x(t-h(t)) \\
&\quad + x^T(t-\alpha h) \alpha h X_{22} x(t-\alpha h) \\
&\quad + x^T(t-\alpha h) X_{23} \int_{t-\alpha h}^{t-h(t)} \dot{x}(s)\, ds \\
&\quad + \int_{t-\alpha h}^{t-h(t)} \dot{x}^T(s)\, ds X_{13}^T x(t-h(t)) \\
&\quad + \int_{t-\alpha h}^{t-h(t)} \dot{x}^T(s)\, ds X_{23}^T x(t-\alpha h) \\
&= x^T(t-h(t)) \left[\alpha h X_{11} + X_{13}^T + X_{13}\right] x(t-h(t)) \\
&\quad + x^T(t-h(t)) \left[\alpha h X_{12} - X_{13} + X_{23}^T\right] x(t-\alpha h) \\
&\quad + x^T(t-\alpha h) \left[\alpha h X_{12}^T - X_{13}^T + X_{23}\right] x(t-h(t)) \\
&\quad + x^T(t-\alpha h) \left[\alpha h X_{22} - X_{23} - X_{23}^T\right] x(t-\alpha h).
\end{aligned} \tag{15}
$$

Similarly, we obtain

$$
\begin{aligned}
-\int_{t-h}^{t-\alpha h} & \dot{x}^T(s) Y_{33}\dot{x}(s)\,ds \\
&\le x^T(t-\alpha h)\left[(1-\alpha)hY_{11}+Y_{13}^T+Y_{13}\right]x(t-\alpha h) \\
&\quad + x^T(t-\alpha h)\left[(1-\alpha)hY_{12}-Y_{13}+Y_{23}^T\right]x(t-h) \\
&\quad + x^T(t-h)\left[(1-\alpha)hY_{12}^T-Y_{13}^T+Y_{23}\right]x(t-\alpha h) \\
&\quad + x^T(t-h)\left[(1-\alpha)hY_{22}-Y_{23}-Y_{23}^T\right]x(t-h), \\
-\int_{t-h(t)}^{t} & \dot{x}^T(s) Z_{33}\dot{x}(s)\,ds \\
&\le x^T(t)\left[h(t)Z_{11}+Z_{13}^T+Z_{13}\right]x(t) \\
&\quad + x^T(t)\left[h(t)Z_{12}-Z_{13}+Z_{23}^T\right]x(t-h(t)) \\
&\quad + x^T(t-h(t))\left[h(t)hZ_{12}^T-Z_{13}^T+Z_{23}\right]x(t) \\
&\quad + x^T(t-h(t))\left[h(t)Z_{22}-Z_{23}-Z_{23}^T\right]x(t-h(t)) \\
&\le x^T(t)\left[\alpha hZ_{11}+Z_{13}^T+Z_{13}\right]x(t) \\
&\quad + x^T(t)\left[\alpha hZ_{12}-Z_{13}+Z_{23}^T\right]x(t-h(t)) \\
&\quad + x^T(t-h(t))\left[\alpha hZ_{12}^T-Z_{13}^T+Z_{23}\right]x(t) \\
&\quad + x^T(t-h(t))\left[\alpha hZ_{22}-Z_{23}-Z_{23}^T\right]x(t-h(t)).
\end{aligned} \tag{16}
$$

The operator for term $\dot{x}^T(t)E^T[\alpha hR_1 + (1-\alpha)hR_2 + \alpha hR_3]E\dot{x}(t)$ is as follows:

$$
\begin{aligned}
\dot{x}^T(t)&E^T\left[\alpha hR_1+(1-\alpha)hR_2+\alpha hR_3\right]E\dot{x}(t) \\
&= \left[Ax(t)+Bx(t-h(t))\right]^T \\
&\quad \times \left[\alpha hR_1+(1-\alpha)hR_2+\alpha hR_3\right] \\
&\quad \times \left[Ax(t)+Bx(t-h(t))\right] \\
&= x^T(t)A^T\left[\alpha hR_1+(1-\alpha)hR_2+\alpha hR_3\right]Ax(t) \\
&\quad + x^T(t)A^T\left[\alpha hR_1+(1-\alpha)hR_2+\alpha hR_3\right] \\
&\quad \times Bx(t-h(t)) + x^T(t-h(t))B^T \\
&\quad \times \left[\alpha hR_1+(1-\alpha)hR_2+\alpha hR_3\right] \\
&\quad \times Ax(t) + x^T(t-h(t))B^T \\
&\quad \times \left[\alpha hR_1+(1-\alpha)hR_2+\alpha hR_3\right]Bx(t-h(t)).
\end{aligned} \tag{17}
$$

Furthermore, noting $E^TU = 0$ deduce

$$
0 = 2\dot{x}^T(t)E^TUS^Tx(t). \tag{18}
$$

Combining (11)–(18) yields the following:

$$
\begin{aligned}
\dot{V}(x_t) &\le \xi^T(t)\Xi\xi(t) \\
&\quad - \int_{t-\alpha h}^{t-h(t)} \dot{x}^T(s)E^T(R_1-X_{33})E\dot{x}(s)\,ds \\
&\quad - \int_{t-h}^{t-\alpha h} \dot{x}^T(s)E^T(R_2-Y_{33})E\dot{x}(s)\,ds \\
&\quad - \int_{t-h(t)}^{t} \dot{x}^T(s)E^T\left(R_1+(1-h_d)R_3-Z_{33}\right) \\
&\quad \times E\dot{x}(s)\,ds,
\end{aligned} \tag{19}
$$

where

$$
\xi^T(t) = \left[x^T(t)\;\; x^T(t-h(t))\;\; x^T(t-\alpha h)\;\; x^T(t-h)\right],
$$

$$
\Xi = \begin{bmatrix} \Xi_{11} & \Xi_{12} & 0 & 0 \\ \Xi_{12}^T & \Xi_{22} & \Xi_{23} & 0 \\ 0 & \Xi_{23}^T & \Xi_{33} & \Xi_{34} \\ 0 & 0 & \Xi_{34}^T & \Xi_{44} \end{bmatrix}, \tag{20}
$$

with

$$
\begin{aligned}
\Xi_{11} &= A^TP+PA+Q_1+Q_3+A^TUS^T \\
&\quad + SU^TA+E^T\left(\alpha hZ_{11}+Z_{13}+Z_{13}^T\right)E \\
&\quad + A^T\left[\alpha hR_1+(1-\alpha)hR_2+\alpha hR_3\right]A, \\
\Xi_{12} &= PB+SU^TB+E^T\left(\alpha hZ_{12}-Z_{13}+Z_{23}^T\right)E \\
&\quad + A^T\left[\alpha hR_1+(1-\alpha)hR_2+\alpha hR_3\right]B, \\
\Xi_{22} &= -(1-h_d)Q_3 \\
&\quad + E^T\left(h\alpha X_{11}+X_{13}+X_{13}^T+\alpha hZ_{22}-Z_{23}-Z_{23}^T\right)E \\
&\quad + B^T\left[\alpha hR_1+(1-\alpha)hR_2+\alpha hR_3\right]B, \\
\Xi_{23} &= E^T\left(\alpha hX_{12}-X_{13}+X_{23}^T\right)E, \\
\Xi_{33} &= Q_2-Q_1+E^T\left[\alpha hX_{22}-X_{23}-X_{23}^T\right. \\
&\qquad \left. +(1-\alpha)hY_{11}+Y_{13}+Y_{13}^T\right]E, \\
\Xi_{34} &= E^T\left[(1-\alpha)hY_{12}-Y_{13}+Y_{23}^T\right]E, \\
\Xi_{44} &= -Q_2+E^T\left[(1-\alpha)hY_{22}-Y_{23}-Y_{23}^T\right]E.
\end{aligned} \tag{21}
$$

For system (1a) and (1b), when $\Xi < 0$ if $0 \le h(t) \le \alpha h$ the last three terms in (19) are all less than 0. Thus, by Schur complements [2], we have $\dot{V}(x_t) < 0$.

Theorem 9. *In Case I, if $\alpha h \le h(t) \le h$, for given three scalars h, α, and h_d, then, for any delay $h(t)$ satisfy $0 \le h(t) \le h$, $\dot{h}(t) \le h_d$ and $0 < \alpha < 1$, the system described by* (1a) *and (1b) with (2) is asymptotically stable if there exist matrices* $P = P^T > 0$, $Q_i = Q_i^T > 0$, $R_i = R_i^T > 0$, $(i = 1, 2, 3)$, *matrix* S *of appropriate dimensionsand*

$$X = \begin{bmatrix} X_{11} & X_{12} & X_{13} \\ X_{12}^T & X_{22} & X_{23} \\ X_{13}^T & X_{23}^T & X_{33} \end{bmatrix} \geq 0,$$
$$Y = \begin{bmatrix} Y_{11} & Y_{12} & Y_{13} \\ Y_{12}^T & Y_{22} & Y_{23} \\ Y_{13}^T & Y_{23}^T & Y_{33} \end{bmatrix} \geq 0, \quad (22)$$
$$Z = \begin{bmatrix} Z_{11} & Z_{12} & Z_{13} \\ Z_{12}^T & Z_{22} & Z_{23} \\ Z_{13}^T & Z_{23}^T & Z_{33} \end{bmatrix} \geq 0$$

such that

$$P^T E = E^T P \geq 0, \quad (23a)$$

$$\overline{\Omega} = \begin{bmatrix} \overline{\Omega}_{11} & \overline{\Omega}_{12} & \overline{\Omega}_{13} & 0 & \overline{\Omega}_{15} \\ \overline{\Omega}_{12}^T & \overline{\Omega}_{22} & \overline{\Omega}_{23} & \overline{\Omega}_{24} & \overline{\Omega}_{25} \\ \overline{\Omega}_{13}^T & \overline{\Omega}_{23}^T & \overline{\Omega}_{33} & 0 & 0 \\ 0 & \overline{\Omega}_{24}^T & 0 & \overline{\Omega}_{44} & 0 \\ \overline{\Omega}_{15}^T & \overline{\Omega}_{25}^T & 0 & 0 & \overline{\Omega}_{55} \end{bmatrix} < 0, \quad (23b)$$

$$E^T [R_1 + (1 - h_d) R_3 - X_{33}] E \geq 0,$$
$$E^T [R_2 + (1 - h_d) R_3 - Y_{33}] E \geq 0, \quad (23c)$$
$$E^T (R_2 - Z_{33}) E \geq 0,$$

where $U \in R^{nx(n-r)}$ *is any matrix satisfying* $E^T U = 0$ *and*

$$\overline{\Omega}_{11} = A^T P + PA + Q_1 + Q_3 + A^T U S^T + S U^T A + E^T \left(\alpha h X_{11} + X_{13} + X_{13}^T\right) E,$$
$$\overline{\Omega}_{12} = PB + SU^T B,$$
$$\overline{\Omega}_{13} = E^T \left(\alpha h X_{12} - X_{13} + X_{23}^T\right) E,$$
$$\overline{\Omega}_{15} = A^T \left[\alpha h R_1 + (1 - \alpha) h R_2 + h R_3\right],$$
$$\overline{\Omega}_{22} = - (1 - h_d) Q_3 + E^T \left[(1 - \alpha) h Y_{22} - Y_{23} - Y_{23}^T + (1 - \alpha) h Z_{11} + Z_{13} + Z_{13}^T\right] E,$$
$$\overline{\Omega}_{23} = E^T \left[(1 - \alpha) h Y_{12}^T - Y_{13}^T + Y_{23}\right] E,$$
$$\overline{\Omega}_{24} = E^T \left[(1 - \alpha) h Z_{12} - Z_{13} + Z_{23}^T\right] E,$$
$$\overline{\Omega}_{25} = B^T \left[\alpha h R_1 + (1 - \alpha) h R_2 + h R_3\right],$$
$$\overline{\Omega}_{33} = Q_2 - Q_1 + E^T \left[\alpha h X_{22} - X_{23} - X_{23}^T + (1 - \alpha) h Y_{11} + Y_{13} + Y_{13}^T\right] E,$$
$$\overline{\Omega}_{44} = -Q_2 + E^T \left[(1 - \alpha) h Z_{22} - Z_{23} - Z_{23}^T\right] E,$$
$$\overline{\Omega}_{55} = - \left[\alpha h R_1 + (1 - \alpha) h R_2 + h R_3\right]. \quad (24)$$

Proof. If $\alpha h \leq h(t) \leq h$, it gets

$$\begin{aligned} &- \int_{t-\alpha h}^{t} \dot{x}^T(s) E^T R_1 E \dot{x}(s)\, ds \\ &- \int_{t-h}^{t-\alpha h} \dot{x}^T(s) E^T R_2 E \dot{x}(s)\, ds \\ &- (1 - h_d) \int_{t-h(t)}^{t} \dot{x}^T(s) E^T R_3 E \dot{x}(s)\, ds \\ &= - \int_{t-\alpha h}^{t} \dot{x}^T(s) E^T [R_1 + (1 - h_d) R_3] E \dot{x}(s)\, ds \\ &\quad - \int_{t-h(t)}^{t-\alpha h} \dot{x}^T(s) E^T [R_2 + (1 - h_d) R_3] E \dot{x}(s)\, ds \\ &\quad - \int_{t-h}^{t-h(t)} \dot{x}^T(s) E^T R_2 E \dot{x}(s)\, ds \\ &= - \int_{t-\alpha h}^{t} \dot{x}^T(s) E^T [R_1 + (1 - h_d) R_3 - X_{33}] E \dot{x}(s)\, ds \\ &\quad - \int_{t-h(t)}^{t-\alpha h} \dot{x}^T(s) E^T [R_2 + (1 - h_d) R_3 - Y_{33}] E \dot{x}(s)\, ds \\ &\quad - \int_{t-h}^{t-h(t)} \dot{x}^T(s) E^T (R_2 - Z_{33}) E \dot{x}(s)\, ds \\ &\quad - \int_{t-\alpha h}^{t} \dot{x}^T(s) E^T X_{33} E \dot{x}(s)\, ds \\ &\quad - \int_{t-h(t)}^{t-\alpha h} \dot{x}^T(s) E^T Y_{33} E \dot{x}(s)\, ds \\ &\quad - \int_{t-h}^{t-h(t)} \dot{x}^T(s) E^T Z_{33} E \dot{x}(s)\, ds. \end{aligned} \quad (25)$$

From integral inequality [11], notice that $E^T[R_1 + (1 - h_d)R_3 - X_{33}]E \geq 0$, $E^T(R_2 - Z_{33})E \geq 0$, and $E^T[R_2 + (1 - h_d)R_3 - Y_{33}]E \geq 0$ yields

$$\begin{aligned} &- \int_{t-\alpha h}^{t} \dot{x}^T(s) X_{33} \dot{x}(s)\, ds \\ &\leq x^T(t) \left[\alpha h X_{11} + X_{13}^T + X_{13}\right] x(t) \\ &\quad + x^T(t) \left[\alpha h X_{12} - X_{13} + X_{23}^T\right] x(t - \alpha h) \end{aligned}$$

$$+ x^T(t-\alpha h)\left[\alpha h X_{12}^T - X_{13}^T + X_{23}\right]x(t)$$
$$+ x^T(t-\alpha h)\left[\alpha h X_{22} - X_{23} - X_{23}^T\right]x(t-\alpha h), \tag{26}$$

$$\begin{aligned} -\int_{t-h(t)}^{t-\alpha h} \dot{x}^T(s) Y_{33}\dot{x}(s)\,ds &\le x^T(t-\alpha h)\left[(1-\alpha)hY_{11} + Y_{13}^T + Y_{13}\right] \\ &\times x(t-\alpha h) + x^T(t-\alpha h) \\ &\times \left[(1-\alpha)hY_{12} - Y_{13} + Y_{23}^T\right]x(t-h(t)) \\ &+ x^T(t-h(t))\left[(1-\alpha)hY_{12}^T - Y_{13}^T + Y_{23}\right] \\ &\times x(t-\alpha h) + x^T(t-h(t)) \\ &\times \left[(1-\alpha)hY_{22} - Y_{23} - Y_{23}^T\right]x(t-h(t)), \end{aligned} \tag{27}$$

$$\begin{aligned} -\int_{t-h}^{t-h(t)} \dot{x}^T(s) Z_{33}\dot{x}(s)\,ds &\le x^T(t-h(t))\left[(1-\alpha)hZ_{11} + Z_{13}^T + Z_{13}\right] \\ &\times x(t-h(t)) + x^T(t-h(t)) \\ &\times \left[(1-\alpha)hZ_{12} - Z_{13} + Z_{23}^T\right]x(t-h) \\ &+ x^T(t-h)\left[(1-\alpha)hZ_{12}^T - Z_{13}^T + Z_{23}\right] \\ &\times x(t-h(t)) + x^T(t-h) \\ &\times \left[(1-\alpha)hZ_{22} - Z_{23} - Z_{23}^T\right]x(t-h). \end{aligned} \tag{28}$$

Combining (11)–(18) and (26)–(28) yields

$$\begin{aligned} \dot{V}(x_t) \le{}& \xi^T(t)\overline{\Xi}\xi(t) \\ &- \int_{t-\alpha h}^{t} \dot{x}^T(s)E^T\left[R_1 + (1-h_d)R_3 - X_{33}\right] \\ &\times E\dot{x}(s)\,ds - \int_{t-h(t)}^{t-\alpha h} \dot{x}^T(s)E^T \\ &\times \left[R_2 + (1-h_d)R_3 - Y_{33}\right]E\dot{x}(s)\,ds \\ &- \int_{t-h}^{t-h(t)} \dot{x}^T(s)E^T(R_2 - Z_{33})E\dot{x}(s)\,ds, \end{aligned} \tag{29}$$

where

$$\overline{\Xi} = \begin{bmatrix} \overline{\Xi}_{11} & \overline{\Xi}_{12} & \overline{\Xi}_{13} & 0 \\ \overline{\Xi}_{12}^T & \overline{\Xi}_{22} & \overline{\Xi}_{23} & \overline{\Xi}_{24} \\ \overline{\Xi}_{13}^T & \overline{\Xi}_{23}^T & \overline{\Xi}_{33} & 0 \\ 0 & \overline{\Xi}_{24}^T & 0 & \overline{\Xi}_{44} \end{bmatrix},$$

$$\begin{aligned} \overline{\Xi}_{11} ={}& A^TP + PA + Q_1 + Q_3 + A^TUS^T \\ &+ SU^TA + E^T\left(\alpha h X_{11} + X_{13} + X_{13}^T\right)E \\ &+ A^T\left[\alpha hR_1 + (1-\alpha)hR_2 + hR_3\right]A, \\ \overline{\Xi}_{12} ={}& PB + SU^TB + A^T\left[\alpha hR_1 + (1-\alpha)hR_2 + hR_3\right]B, \\ \overline{\Xi}_{13} ={}& E^T\left(\alpha h X_{12} - X_{13} + X_{23}^T\right)E, \\ \overline{\Xi}_{22} ={}& -(1-h_d)Q_3 + E^T \\ &+ E^T\left[(1-\alpha)hY_{22} - Y_{23} - Y_{23}^T\right. \\ &\left. + (1-\alpha)hZ_{11} + Z_{13} + Z_{13}^T\right]E \\ &+ B^T\left[\alpha hR_1 + (1-\alpha)hR_2 + hR_3\right]B, \\ \overline{\Xi}_{23} ={}& E^T\left[(1-\alpha)hY_{12}^T - Y_{13}^T + Y_{23}\right]E, \\ \overline{\Xi}_{24} ={}& E^T\left[(1-\alpha)hZ_{12} - Z_{13} + Z_{23}^T\right]E, \\ \overline{\Xi}_{33} ={}& Q_2 - Q_1 + E^T\left[\alpha h X_{22} - X_{23} - X_{23}^T\right. \\ &\left. + (1-\alpha)hY_{11}^T + Y_{13} + Y_{13}^T\right]E, \\ \overline{\Xi}_{44} ={}& -Q_2 + E^T\left[(1-\alpha)hZ_{22} - Z_{23} - Z_{23}^T\right]E. \end{aligned} \tag{30}$$

For system (1a) and (1b), when $\Xi < 0$ if $\alpha h \le h(t) \le h$; the last three terms in (29) are all less than 0. Thus, by Schur complements [2], we have $\dot{V}(x_t) < 0$.

Theorem 10. *In Case II, if* $0 \le h(t) \le \alpha h$, *for given two scalars* h *and* α, *then, for any delay* $h(t)$ *satisfy* $0 \le h(t) \le h$ *and* $0 < \alpha < 1$, *the system described by* (1a) *and* (1b) *with* (3) *is asymptotically stable if there exist matrices* $P = P^T > 0$, $Q_i = Q_i^T > 0$, $R_i = R_i^T > 0, (i = 1, 2)$, *matrix* S *of appropriate dimensions and positive semidefinite matrices:*

$$X = \begin{bmatrix} X_{11} & X_{12} & X_{13} \\ X_{12}^T & X_{22} & X_{23} \\ X_{13}^T & X_{23}^T & X_{33} \end{bmatrix} \ge 0,$$
$$Y = \begin{bmatrix} Y_{11} & Y_{12} & Y_{13} \\ Y_{12}^T & Y_{22} & Y_{23} \\ Y_{13}^T & Y_{23}^T & Y_{33} \end{bmatrix} \ge 0, \tag{31}$$
$$Z = \begin{bmatrix} Z_{11} & Z_{12} & Z_{13} \\ Z_{12}^T & Z_{22} & Z_{23} \\ Z_{13}^T & Z_{23}^T & Z_{33} \end{bmatrix} \ge 0$$

such that

$$P^TE = E^TP \ge 0, \tag{32a}$$

$$\Psi = \begin{bmatrix} \Psi_{11} & \Psi_{12} & 0 & 0 & \Psi_{15} \\ \Psi_{12}^T & \Psi_{22} & \Psi_{23} & 0 & \Psi_{25} \\ 0 & \Psi_{23}^T & \Psi_{33} & \Psi_{34} & 0 \\ 0 & 0 & \Psi_{34}^T & \Psi_{44} & 0 \\ \Psi_{15}^T & \Psi_{25}^T & 0 & 0 & \Psi_{55} \end{bmatrix} < 0, \quad (32b)$$

$$E^T(R_1 - X_{33})E \ge 0, \quad E^T(R_2 - Y_{33})E \ge 0, \quad E^T(R_1 - Z_{33})E \ge 0, \quad (32c)$$

where $U \in R^{n\times(n-r)}$ *is any matrix satisfying* $E^TU = 0$ *and*

$$\begin{aligned}
\Psi_{11} &= A^TP + PA + Q_1 + A^TUS^T + SU^TA + E^T\left(hZ_{11} + Z_{13} + Z_{13}^T\right)E, \\
\Psi_{12} &= PB + SU^TB + E^T\left(hZ_{12} - Z_{13} + Z_{23}^T\right)E, \\
\Psi_{15} &= A^T\left[\alpha hR_1 + (1-\alpha)hR_2\right], \\
\Psi_{22} &= E^T\left(\alpha hX_{11} + X_{13} + X_{13}^T + hZ_{22} - Z_{23} - Z_{23}^T\right), \\
\Psi_{23} &= E^T\left(\alpha hX_{12} - X_{13} + X_{23}^T\right)E, \\
\Psi_{25} &= B^T\left[\alpha hR_1 + (1-\alpha)hR_2\right], \\
\Psi_{33} &= Q_2 - Q_1 + E^T\left[\alpha hX_{22} - X_{23} - X_{23}^T + (1-\alpha)hY_{11} + Y_{13} + Y_{13}^T\right]E, \\
\Psi_{34} &= E^T\left[(1-\alpha)hY_{12} - Y_{13} + Y_{23}^T\right]E, \\
\Psi_{44} &= -Q_2 + E^T\left[(1-\alpha)hY_{22} - Y_{23} - Y_{23}^T\right]E, \\
\Psi_{55} &= -\left[\alpha hR_1 + (1-\alpha)hR_2\right].
\end{aligned} \quad (33)$$

Theorem 11. *In Case II, if* $\alpha h \le h(t) \le h$*, for given two scalars* h *and* α*, then, for any delay* $h(t)$ *satisfy* $0 \le h(t) \le h$ *and* $0 < \alpha < 1$*, the system described by (1a) and (1b) with (3) is asymptotically stable if there exist matrices* $P = P^T > 0$*,* $Q_i = Q_i^T > 0$*,* $R_i = R_i^T > 0$*,* $(i = 1, 2)$*, matrix* S *of appropriate dimensions and positive semidefinite matrices:*

$$X = \begin{bmatrix} X_{11} & X_{12} & X_{13} \\ X_{12}^T & X_{22} & X_{23} \\ X_{13}^T & X_{23}^T & X_{33} \end{bmatrix} \ge 0, \quad Y = \begin{bmatrix} Y_{11} & Y_{12} & Y_{13} \\ Y_{12}^T & Y_{22} & Y_{23} \\ Y_{13}^T & Y_{23}^T & Y_{33} \end{bmatrix} \ge 0, \quad Z = \begin{bmatrix} Z_{11} & Z_{12} & Z_{13} \\ Z_{12}^T & Z_{22} & Z_{23} \\ Z_{13}^T & Z_{23}^T & Z_{33} \end{bmatrix} \ge 0 \quad (34)$$

such that

$$P^TE = E^TP \ge 0, \quad (35a)$$

$$\overline{\Psi} = \begin{bmatrix} \overline{\Psi}_{11} & \overline{\Psi}_{12} & \overline{\Psi}_{13} & 0 & \overline{\Psi}_{15} \\ \overline{\Psi}_{12}^T & \overline{\Psi}_{22} & \overline{\Psi}_{23} & \overline{\Psi}_{24} & \overline{\Psi}_{25} \\ \overline{\Psi}_{13}^T & \overline{\Psi}_{23}^T & \overline{\Psi}_{33} & 0 & 0 \\ 0 & \overline{\Psi}_{24}^T & 0 & \overline{\Psi}_{44} & 0 \\ \overline{\Psi}_{15}^T & \overline{\Psi}_{25}^T & 0 & 0 & \overline{\Psi}_{55} \end{bmatrix} < 0, \quad (35b)$$

$$E^T[R_1 - X_{33}]E \ge 0, \quad E^T[R_2 - Y_{33}]E \ge 0, \quad E^T(R_2 - Z_{33})E \ge 0, \quad (35c)$$

where $U \in R^{n\times(n-r)}$ *is any matrix satisfying* $E^TU = 0$ *and*

$$\begin{aligned}
\overline{\Psi}_{11} &= A^TP + PA + Q_1 + A^TUS^T + SU^TA + E^T\left(\alpha hX_{11} + X_{13} + X_{13}^T\right)E, \\
\overline{\Psi}_{12} &= PB + SU^TB, \\
\overline{\Psi}_{13} &= E^T\left(\alpha hX_{12} - X_{13} + X_{23}^T\right)E, \\
\overline{\Psi}_{15} &= A^T\left[\alpha hR_1 + (1-\alpha)hR_2\right], \\
\overline{\Psi}_{22} &= E^T\left[(1-\alpha)hY_{22} - Y_{23} - Y_{23}^T + (1-\alpha)hZ_{11} + Z_{13} + Z_{13}^T\right]E, \\
\overline{\Psi}_{23} &= E^T\left[(1-\alpha)hY_{12}^T - Y_{13}^T + Y_{23}\right]E, \\
\overline{\Psi}_{24} &= E^T\left[(1-\alpha)hZ_{12} - Z_{13} + Z_{23}^T\right]E, \\
\overline{\Psi}_{25} &= B^T\left[\alpha hR_1 + (1-\alpha)hR_2\right], \\
\overline{\Psi}_{33} &= Q_2 - Q_1 + E^T\left[\alpha hX_{22} - X_{23} - X_{23}^T + (1-\alpha)hY_{11} + Y_{13} + Y_{13}^T\right]E, \\
\overline{\Psi}_{44} &= -Q_2 + E^T\left[(1-\alpha)hZ_{22} - Z_{23} - Z_{23}^T\right]E, \\
\overline{\Psi}_{55} &= -\left[\alpha hR_1 + (1-\alpha)hR_2\right].
\end{aligned} \quad (36)$$

In Case II, a Lyapunov functional can be chosen as (11) *with* $Q_3 = R_3 = 0$*. Similar to the above analysis, one can get that* $\dot{V}(x_t) < 0$ *holds if* $\Psi < 0$ $(\overline{\Psi} < 0)$*. Thus, the proof is complete.*

Remark 12. In the proof of Theorems 8–11, the interval $[t-h, t]$ is divided into subintervals $[t-h, t-\alpha h]$ and $[t-\alpha h, t]$; information of delayed state $x(t-\alpha h)$ can be taken into account. It is clear that the Lyapunov function defined in Theorems 8–11 is more general than the ones in [6, 9, 15–21].

Remark 13. In the previous works except [4–6, 13, 16, 19], the time-delay term $h(t)$ was usually estimated as h when estimating the upper bound of some cross term this may lead to increasing conservatism inevitably. In Theorems 8–11, the value of the upper bound of some cross term is estimated more exactly than the previous methods since $h(t)$ is confined to the subintervals $0 \le h(t) \le \alpha h$ or $\alpha h \le h(t) \le h$. So, such decomposition method may lead to reduction of conservatism.

Remark 14. In the stability problem, maximum admissible upper bound (MAUB) $\bar{h}$ that ensures singular system with a time-varying state delay (1a) and (1b) is stabilizable for any h can be determined by solving the following quasi-convex optimization problem when the other bound of time-varying delay h is known.

$$\begin{aligned} &\text{Maximize} \quad && \bar{h} \\ &\text{Subject to} \quad && \text{Theorems 8–11.} \end{aligned} \tag{37}$$

Inequality (37) is a convex optimization problem and can be obtained efficiently using the MATLAB LMI Toolbox.

For seeking an appropriate α satisfying $0 < \alpha < 1$, such that the upper bound h of delay $h(t)$ subjecting to (7a), (7b), and (7c) is maximal, we give an algorithm.

Algorithm 15 (maximizing $h > 0$).

- Step 1. For given h_d, choose an upper bound on h satisfying (7a), (7b), and (7c), then select this upper bound as the initial value h_0 of h.
- Step 2. Set appropriate step lengths, $h_{step}$and α_{step}, for h and α, respectively. Set k as a counter and choose $k = 1$. Meanwhile, let $h = h_0 + h_{step}$ and initial value α_0 of α equal to α_{step}.
- Step 3. Let $\alpha = k\alpha_{step}$; if inequality (7a), (7b), and (7c) is feasible, go to Step 4; otherwise, go to step 5.
- Step 4. Let $h_0 = h, \alpha_0 = \alpha, k = 1$ and $h = h_0 + h_{step}$ then go to Step 3.
- Step 5. Let $k = k + 1$. If $k\alpha_{step} < 1$, then go to Step 3; otherwise, stop.

Remark 16. For Algorithm 15, final h_0 is the desired maximum of the upper bound of delay $h(t)$ satisfying (7a), (7b), and (7c), and α_0 is the corresponding value of α.

Remark 17. Similar to Algorithm 15, we can also find an appropriate scalar α, such that the upper bound of delay $0 \le h(t) \le \alpha h$ subjecting to (32a), (32b), and (32c) attains the maximum.

Remark 18. Similar to Algorithm 15, an algorithm for seeking appropriate α such that the upper bound of delay $\alpha h \le h(t) \le h$subjecting to (23a), (23b), and (23c) and (35a), (35b), and (35c) are maximal can be easily obtained.

3. Illustrative Examples

To show usefulness of our result, let us consider the following numerical examples.

TABLE 1: Comparison of delay-dependent stability conditions in Example 19.

Methods	MAUB ($\bar{h}$)	Number of variables
[15]	1.1547	53
[17]	1.1547	33
[6]	1.1547	24
[16]	1.1547	17
[19]	1.2011	13
[9]	1.1898	13
[18]	1.2052	33
[20]	1.2060	21
Theorem 8 ($\alpha = 0.5$)	2.3092	11

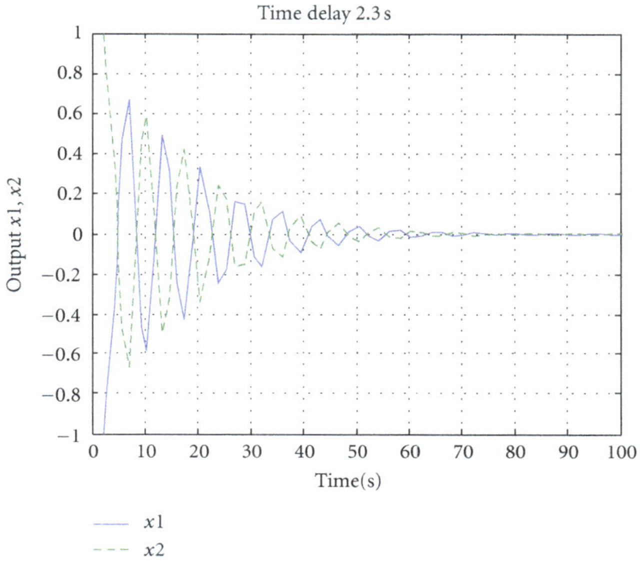

FIGURE 1: The simulation of the Example 19 for $h = 2.3$ sec.

Example 19. Consider the following time-delay singular systems:

$$E\dot{x}(t) = Ax(t) + Bx(t - h(t)), \tag{38}$$

where $E = \begin{bmatrix} 1 & 0 \\ 0 & 0 \end{bmatrix}$, $A = \begin{bmatrix} 0.5 & 0 \\ -1 & -1 \end{bmatrix}$, $B = \begin{bmatrix} -1 & 0 \\ 0 & 0 \end{bmatrix}$.

Now, our problem is to estimate the maximum admissible upper bound (MAUB) $\bar{h}$ to keep the stability of system (38).

Solution 1. Choosing $U = [1 \ \ 0]^T$ and applying the LMI Toolbox in MATLAB (with accuracy 0.01), this above time delay singular system (38) is asymptotically stable for delay time satisfying $h \le 2.3092$. Table 1 lists the results compared with [6, 9, 15–20]. It can be seen from Table 1 that the maximum admissible upper bound (MAUB) $\bar{h}$ by using Theorem 8 is the largest with the fewest variables are computed. Figure 1 shows the simulation of the above system (38) for $h = 2.3$ with the initial state $[-1 \ \ 1]^T$.

Table 2: Comparison of maximum admissible upper bound (MAUB) $\bar{h}$ in Example 20.

Methods	[6]	[13]	[19]	[4]	[10]	Theorem 8 ($\alpha = 0.4$)
$\bar{h}$	1.150	1.1547	1.1547	1.2052	2.3810	2.8863

Table 3: Comparison of maximum admissible upper bound (MAUB) $\bar{h}$ in Example 21.

Methods	[21]	[10]	[11]	Theorem 8 ($\alpha = 0.5$)
$\bar{h}$	1.274	1.281	2.3619	2.9618

Example 20. Consider the following time delay singular systems:

$$E\dot{x}(t) = Ax(t) + Bx(t - h(t)), \tag{39}$$

where $E = \begin{bmatrix} 1 & 0 \\ 0 & 0 \end{bmatrix}$, $A = \begin{bmatrix} 0.5 & 0 \\ 0 & -1 \end{bmatrix}$, $B = \begin{bmatrix} -1 & 1 \\ 0 & 0.5 \end{bmatrix}$.

Now, our problem is to estimate the maximum admissible upper bound (MAUB) $\bar{h}$ to keep the stability of system (39).

Solution 2. Choosing $\alpha = 0.4, U = [1 \ \ 0]^T$ and applying the LMI Toolbox in MATLAB (with accuracy 0.01), this above time delay singular system (39) is asymptotically stable for delay time satisfying $h \le 2.8863$. The results for stability conditions in different methods are compared in Table 2. It can be shown that the delay-dependent stability condition in this paper is the best performance.

Example 21. Consider the following time delay singular systems:

$$E\dot{x}(t) = Ax(t) + Bx(t - h), \tag{40}$$

where $E = \begin{bmatrix} 1 & 0 & 0 \\ 0 & 1 & 0 \\ 0 & 0 & 0 \end{bmatrix}$, $A = \begin{bmatrix} -2 & 0 & 0 \\ 0 & -0.5 & 0 \\ 0 & 0 & 1 \end{bmatrix}$, $B = \begin{bmatrix} -1 & 0 & 1 \\ -1 & -1 & -0.1 \\ -1 & 1 & -0.1 \end{bmatrix}$.

Now, our problem is to estimate the maximum admissible upper bound (MAUB) $\bar{h}$ to keep the stability of system (40).

Solution 3. Choosing $U = [0 \ \ 0 \ \ 1]^T$ and applying the LMI Toolbox in MATLAB (with accuracy 0.01), this above time-delay singular system (40) is asymptotically stable for delay time satisfying $h \le 2.9618$. The maximum admissible upper bounds on the time-delay form Theorem 8 are shown in Table 3. From Table 3, it can be seen that the stability results obtained in the paper are less conservative than those in [10, 11, 21].

4. Conclusion

In this paper, a delay decomposition approach has been developed to investigate the stability of singular systems with a time-varying delay. By developing a delay decomposition approach, the information of the delayed plant states can be taken into full consideration, and new delay-dependent sufficient stability criteria are obtained in terms of linear matrix inequalities (LMIs). Our proposed results are with the form of LMI and can be easily solved by LMI's toolbox in the Matlab without tuning any parameters. It is proved that the obtained results are less conservative than some existing ones. Meanwhile, the computational complexity of the new stability criteria is reduced greatly since fewer decision variables are involved. An algorithm of seeking appropriate tuning parameter is also presented. Numerical examples have illustrated the effectiveness of the proposed methods.

References

[1] M. S. Ali and P. Balasubramaniam, "Exponential stability of time-delay systems with nonlinear uncertainties," *International Journal of Computer Mathematics*, vol. 87, no. 6, pp. 1363–1373, 2010.

[2] S. Boyd, L. El Ghaoui, E. Feron, and V. Balakrishnan, *Linear Matrix Inequalities in System and Control Theory*, SIAM, Philadelphia, Pa, USA, 1994.

[3] L. Dai, *Singular Control Systems*, Springer, New York, NY, USA, 1989.

[4] Y. F. Feng, X. L. Zhu, and Q. L. Zhang, "Delay-dependent stability criteria for singular time-delay systems," *Acta Automatica Sinica*, vol. 36, no. 3, pp. 433–437, 2010.

[5] E. Fridman, "Stability of linear descriptor systems with delay: a Lyapunov-based approach," *Journal of Mathematical Analysis and Applications*, vol. 273, no. 1, pp. 24–44, 2002.

[6] E. Fridman and U. Shaked, "H_∞ control of linear state-delay descriptor systems: an LMI approach," *Linear Algebra and its Applications*, vol. 351, no. 1, pp. 271–302, 2002.

[7] G. Huanli and X. Bugong, "Delay-dependent state feedback robust stabilization for uncertain singular time-delay systems," *Journal of Systems Engineering and Electronics*, vol. 19, no. 4, pp. 758–765, 2008.

[8] F. L. Lewis, "A survey of linear singular systems," *Circuits, Systems, and Signal Processing*, vol. 5, no. 1, pp. 3–36, 1986.

[9] S. Liang, J. Cheng, and M. Huang, "A delay decomposition approach to stability analysis of singular systems with time-varying delay," in *Proceedings of the International Conference on Test and Measurement (ICTM '09)*, pp. 110–114, December 2009.

[10] L. L. Liu, J. G. Peng, and B. W. Wu, "On parameterized Lyapunov-Krasovskii functional techniques for investigating singular time-delay systems," *Applied Mathematics Letters*, vol. 24, no. 5, pp. 703–708, 2011.

[11] P. L. Liu, "Further results on the exponential stability criteria for time delay singular systems with delay-dependence," *International Journal of Innovative Computing, Information and Control*, vol. 8, no. 6, pp. 4015–4024, 2012.

[12] P. L. Liu, "A delay decomposition approach to stability analysis of uncertain systems with time-varying delays," *ISA Transactions*, vol. 51, pp. 694–701, 2012.

[13] H. Su, X. Ji, and J. Chu, "Delay-dependent robust control for uncertain singular time-delay systems," *Asian Journal of Control*, vol. 8, no. 2, pp. 180–189, 2006.

[14] X. Wu, Y. Wang, L. Huang, and Y. Zuo, "Robust stability analysis of delayed Takagi-Sugeno fuzzy Hopfield neural networks with discontinuous activation functions," *Cognitive Neurodynamics*, vol. 4, no. 4, pp. 347–354, 2010.

[15] Z. Wu and W. Zhou, "Delay-dependent robust H_∞ control for uncertain singular time-delay systems," *IET Control Theory and Applications*, vol. 1, no. 5, pp. 1234–1241, 2007.

[16] S. Xu, J. Lam, and Y. Zou, "An improved characterization of bounded realness for singular delay systems and its applications," *International Journal of Robust and Nonlinear Control*, vol. 18, no. 3, pp. 263–277, 2008.

[17] D. Yue and Q. L. Han, "A delay-dependent stability criterion of neutral systems and its application to a partial element equivalent circuit model," *IEEE Transactions on Circuits and Systems I*, vol. 51, no. 12, pp. 685–689, 2004.

[18] X. L. Zhu and G. H. Yang, "New results of stability analysis for singular time-delay systems," in *Proceedings of the American Control Conference*, pp. 4905–4908, St. Louis, Mo, USA, 2009.

[19] S. Zhu, C. Zhang, Z. Cheng, and J. Feng, "Delay-dependent robust stability criteria for two classes of uncertain singular time-delay systems," *IEEE Transactions on Automatic Control*, vol. 52, no. 5, pp. 880–885, 2007.

[20] S. Zhu, Z. Li, and C. Zhang, "Delay decomposition approach to delay-dependent stability for singular time-delay systems," *IET Control Theory and Applications*, vol. 4, no. 11, pp. 2613–2620, 2010.

[21] S. Q. Zhu, Z. L. Cheng, and J. Feng, "Delay-dependent robust stability criterion and robust stabilization for uncertain singular time-delay systems," in *Proceeding of American Control Conference*, pp. 2839–2844, Portland, Ore, USA, June 2005.

[22] X. L. Zhu and G. H. Yang, "Jensen integral inequality approach to stability analysis of continuous-time systems with time-varying delay," *IET Control Theory and Applications*, vol. 2, no. 6, pp. 524–534, 2008.

[23] X. L. Zhu and G. H. Yang, "New results of stability analysis for systems with time-varying delay," *International Journal of Robust and Nonlinear Control*, vol. 20, no. 5, pp. 596–606, 2010.

7

Some Fixed Point Results for Generalized Weak Contraction Mappings in Modular Spaces

Chirasak Mongkolkeha and Poom Kumam

Department of Mathematics, Faculty of Science, King Mongkut's University of Technology Thonburi, Bang Mod, Thung Khru, Bangkok 10140, Thailand

Correspondence should be addressed to Poom Kumam; poom.kum@kmutt.ac.th

Academic Editor: Stefan Kunis

We prove the existence theorem of fixed points for a generalized weak contractive mapping in modular spaces.

1. Introduction

In 1997, Alber and Guerre-Delabriere [1] introduced the concept of weak contraction in Hilbert spaces. Later, Rhoades [2] proved that the result which Alber et al. is also valid in complete metric spaces, the result of Rhoades in the following: A mapping $T : X \to X$ where (X, d) is a metric space, is said to be *weakly contractive* if

$$d(T(x), T(y)) \leq d(x, y) - \phi(d(x, y)), \qquad (1)$$

where $\phi : [0, \infty) \to [0, \infty)$ is continuous and nondecreasing function such that $\phi(t) = 0$ if and only if $t = 0$. In 2008, Dutta and Choudhury [3] introduced a new generalization of contraction in metric spaces and proved the following theorem.

Theorem 1. *Let (X, d) be a complete metric space, and let $T : X \to X$ be a self-mapping satisfying the following inequality:*

$$\psi(d(Tx, Ty)) \leq \psi(d(x, y)) - \phi(d(x, y)), \qquad (2)$$

where $\psi, \phi : [0, \infty) \to [0, \infty)$ are both continuous and monotone nondecreasing function with $\psi(t) = \phi(t) = 0$ if and only if $t = 0$. Then T has a unique fixed point.

We note that, if one takes $\psi(t) = t$, then (2) reduces to (1).

Recall that the theory of modular on linear spaces and the corresponding theory of modular linear spaces were founded by Nakano [4, 5] and redefined by Musielak and Orlicz [6]. Furthermore, the most complete development of these theories is due to Mazur, Luxemburg, and Turpin [7–9]. In the present time, the theory of modular and modular spaces is extensively applied, in particular, in the study of various Orlicz spaces which in their turn have broad applications [10–14]. In many cases, particularly in applications to integral operators, approximation and fixed point theory, modular-type conditions are much more natural as modular-type assumptions can be more easily verified than their metric or norm counterparts. Even though a metric is not defined, many problems in metric fixed point theory can be reformulated in modular spaces. For instance, fixed point theorems are proved in [15, 16] for nonexpansive mappings. The existences for contraction mapping in modular spaces has been studied in [3, 17–24] and the references therein.

From the above mentioned, we will study the existence of fixed point theorems for mappings satisfying generalized weak contraction mappings in modular spaces.

2. Preliminaries

First, we start with a brief recollection of basic concepts and facts in modular spaces.

Definition 2. Let X be a vector space over $\mathbb{R}$ (or $\mathbb{C}$). A functional $\rho : X \to [0,\infty]$ is called a modular if for arbitrary x and y, elements of X, it satisfies the following conditions:

(1) $\rho(x) = 0$ if and only if $x = 0$;

(2) $\rho(\alpha x) = \rho(x)$ for all scalar α with $|\alpha| = 1$;

(3) $\rho(\alpha x + \beta y) \le \rho(x) + \rho(y)$, whenever $\alpha, \beta \ge 0$ and $\alpha + \beta = 1$.
If one replaces (3) by

(4) $\rho(\alpha x+\beta y) \le \alpha^s\rho(x)+\beta^s\rho(y)$, for $\alpha,\beta \ge 0$, $\alpha^s+\beta^s = 1$ with an $s \in (0,1]$, then the modular ρ is called s-convex modular, and if $s = 1$, ρ is called convex modular.

If ρ is modular in X, then the set

$$X_\rho = \{x \in X : \rho(\lambda x) \longrightarrow 0 \text{ as } \lambda \longrightarrow 0\}, \tag{3}$$

is called a *modular space*. X_ρ is a vector subspace of X.

Definition 3. A modular ρ is said to satisfy the Δ_2-condition if $\rho(2x_n) \to 0$, whenever $\rho(x_n) \to 0$ as $n \to \infty$.

Definition 4. Let X_ρ be a modular space.

(1) The sequence $\{x_n\}_{n\in\mathbb{N}}$ in X_ρ is said to be ρ-*convergent* to $x \in X_\rho$ if $\rho(x_n - x) \to 0$, as $n \to \infty$.

(2) The sequence $\{x_n\}_{n\in\mathbb{N}}$ in X_ρ is said to be ρ-*Cauchy* if $\rho(x_n - x_m) \to 0$, as $n, m \to \infty$.

(3) A subset C of X_ρ is said to be ρ-*closed* if the ρ-limit of a ρ-*convergent* sequence of C always belong to C.

(4) A subset C of X_ρ is said to be ρ-*complete* if any ρ-*Cauchy* sequence in C is ρ-*convergent* sequence and its ρ-limit is in C.

(5) A subset C of X_ρ is said to be ρ-*bounded* if any $\delta_\rho(C) = \sup\{\rho(x-y)\ ;\ x, y \in C\} < \infty$.

Definition 5. Let X be a nonempty set and $T : X \to X$. A point $x \in X$ is a fixed point of T if and only if $Tx = x$.

Definition 6. Let C be a subset of a real numbers $\mathbb{R}$. A mapping $T : C \to \mathbb{R}$ is called monotone increasing (or monotone nondecreasing) $x \le y$ if and only if $T(x) \le T(y)$, for all x and y are elements in C. A mapping $T : C \to \mathbb{R}$ is called monotone decreasing (or monotone nonincreasing), $x \ge y$ if and only if $T(x) \ge T(y)$ for all x and y are elements in C.

Definition 7. A sequence $\{a_n\}$ of a real number is said to be monotone increasing (or monotone nondecreasing), if it satisfies $a_1 \le a_2 \le a_3 \le \ldots$. It is also said to be monotone decreasing (or monotone nonincreasing) if it satisfies $a_1 \ge a_2 \ge a_3 \ge \ldots$.

3. A Generalized Weak Contraction in Modular Spaces

In this section, we prove fixed point theorems for mappings satisfying generalized weak contractions in modular spaces.

Proposition 8. *Let ρ be a modular space on X. If $a, b \in \mathbb{R}^+$ with $b \ge a$, then $\rho(ax) \le \rho(bx)$.*

Proof. In case $a = b$, we are done. Suppose $b > a$, and then one has $a/b < 1$ and

$$\begin{aligned}\rho(ax) &= \rho\left(\frac{a}{b}bx\right)\\ &= \rho\left(\frac{a}{b}bx + \left(1-\frac{a}{b}\right)(0)\right)\\ &\le \rho(bx) + \rho(0)\\ &= \rho(bx).\end{aligned} \tag{4}$$

□

Proposition 9. *Let X_ρ be a modular space in which ρ satisfies the Δ_2-condition and let $\{x_n\}_{n\in\mathbb{N}}$ be a sequence in X_ρ. If $\rho(c(x_n - x_{n-1})) \to 0$ as $n \to \infty$, then $\rho(\alpha l(x_n - x_{n-1})) \to 0$ as $n \to \infty$, where $c, l, \alpha \in \mathbb{R}^+$ with $c > l$ and $l/c + 1/\alpha = 1$.*

Proof. Since $\rho(c(x_n - x_{n-1})) \to 0$ as $n \to \infty$, by the Δ_2-condition, we get

$$\rho(2^m c(x_n - x_{n-1})) \longrightarrow 0 \quad \text{as } n \longrightarrow \infty, \tag{5}$$

for $m \in \mathbb{N}$. Using (5), Proposition 8, and the sandwich theorem, we conclude that

$$\rho(2Nc(x_n - x_{n-1})) \longrightarrow 0 \quad \text{as } n \longrightarrow \infty, \tag{6}$$

for $N \in \mathbb{N}$. From the fact that $l/c + 1/\alpha = 1$, we get $\alpha l = (\alpha-1)c \ge c$, then there exist $N_\alpha \in \mathbb{N}$ such that

$$2(N_\alpha - 1)c \le (\alpha-1)c \le 2(N_\alpha)c. \tag{7}$$

By Proposition 8, we get

$$\begin{aligned}&\rho(2(N_\alpha-1)c(x_n - x_{n-1}))\\ &\quad\le \rho((\alpha-1)c(x_n - x_{n-1}))\\ &\quad\le \rho(2(N_\alpha)c(x_n - x_{n-1})).\end{aligned} \tag{8}$$

From (6) and (8), we obtain

$$\begin{aligned}&\lim_{n\to\infty}\rho(\alpha l(x_n - x_{n-1}))\\ &\quad= \lim_{n\to\infty}\rho((\alpha-1)c(x_n - x_{n-1})) = 0.\end{aligned} \tag{9}$$

□

Theorem 10. *Let X_ρ be a ρ-complete modular space, where ρ satisfies the Δ_2-condition. Let $c, l \in \mathbb{R}^+$, $c > l$, and $T : X_\rho \to X_\rho$ be a mapping satisfying the inequality*

$$\begin{aligned}&\psi(\rho(c(Tx - Ty)))\\ &\quad\le \psi(\rho(l(x-y))) - \phi(\rho(l(x-y))),\end{aligned} \tag{10}$$

for all $x, y \in X_\rho$, where $\psi, \phi : [0,\infty) \to [0,\infty)$ are both continuous and monotone nondecreasing functions with $\psi(t) = \phi(t) = 0$ if and only if $t = 0$. Then, T has a unique fixed point.

Proof. Let $x_0 \in X_\rho$, and we construct the sequence $\{x_n\}_{n\in\mathbb{N}}$ by $x_n = Tx_{n-1}$, $n = 1, 2, 3, \ldots$. First, we prove that the sequence $\{\rho(c(Tx_n - Tx_{n-1}))\}$ converges to 0. Indeed

$$\begin{aligned}\psi(\rho(c(x_n - x_{n+1}))) &\le \psi(\rho(l(x_{n-1} - x_n))) \\ &\quad - \phi(\rho(l(x_{n-1} - x_n))) \\ &\le \psi(\rho(l(x_{n-1} - x_n))).\end{aligned} \tag{11}$$

By monotone nondecreasing of ψ and Proposition 8, we have

$$\begin{aligned}\rho(c(x_n - x_{n+1})) &\le \rho(l(x_{n-1} - x_n)) \\ &\le \rho(c(x_{n-1} - x_n)).\end{aligned} \tag{12}$$

This means that the sequence $\{\rho(c(x_n - x_{n-1}))\}$ is monotone decreasing and bounded below. Hence there exists $r \ge 0$ such that

$$\lim_{n\to\infty} \rho(c(x_n - x_{n-1})) = r. \tag{13}$$

If $r > 0$, taking $n \to \infty$ in the inequality (11), we get

$$\begin{aligned}\psi(r) &\le \psi(r) - \phi(r) \\ &< \psi(r),\end{aligned} \tag{14}$$

which is a contradiction, thus $r = 0$. So we have

$$\rho(c(x_n - x_{n-1})) \longrightarrow 0 \quad \text{as } n \longrightarrow \infty. \tag{15}$$

Next, we prove that the sequence $\{cx_n\}_{n\in\mathbb{N}}$ is a ρ-*Cauchy*. Suppose that $\{cx_n\}_{n\in\mathbb{N}}$ is not ρ-*Cauchy*, then there exist $\varepsilon > 0$ and subsequence $\{x_{m_k}\}$, $\{x_{n_k}\}$ with $m_k > n_k \ge k$ such that

$$\rho(c(x_{m_k} - x_{n_k})) \ge \varepsilon, \qquad \rho(c(x_{m_k-1} - x_{n_k})) < \varepsilon. \tag{16}$$

Now, let $\alpha \in \mathrm{R}^+$ such that $l/c + 1/\alpha = 1$, then we get

$$\begin{aligned}\psi(\rho(c(x_{m_k} - x_{n_k}))) &\le \psi(\rho(l(x_{m_k-1} - x_{n_k-1}))) \\ &\quad - \phi(\rho(l(x_{m_k-1} - x_{n_k-1}))) \\ &\le \psi(\rho(l(x_{m_k-1} - x_{n_k-1})))\end{aligned} \tag{17}$$

which implies that

$$\rho(c(x_{m_k} - x_{n_k})) \le \rho(l(x_{m_k-1} - x_{n_k-1})). \tag{18}$$

We have

$$\begin{aligned}\rho(l(x_{m_k-1} - x_{n_k-1})) &= \rho(l(x_{m_k-1} - x_{n_k} + x_{n_k} - x_{n_k-1})) \\ &= \rho\left(\frac{l}{c}c(x_{m_k-1} - x_{n_k}) + \frac{1}{\alpha}\alpha l(x_{n_k} - x_{n_k-1})\right) \\ &\le \rho(c(x_{m_k-1} - x_{n_k})) + \rho(\alpha l(x_{n_k} - x_{n_k-1})) \\ &< \varepsilon + \rho(\alpha l(x_{n_k} - x_{n_k-1})).\end{aligned} \tag{19}$$

By (16), (18), and (19), we get

$$\begin{aligned}\varepsilon &\le \rho(c(x_{m_k} - x_{n_k})) \\ &\le \rho(l(x_{m_k-1} - x_{n_k-1})) \\ &< \varepsilon + \rho(\alpha l(x_{n_k} - x_{n_k-1})).\end{aligned} \tag{20}$$

Using (15) and Proposition 9, we have

$$\lim_{k\to\infty} \rho(\alpha l(x_{n_k} - x_{n_k-1})) = 0. \tag{21}$$

From (20) and (21), we obtain

$$\begin{aligned}&\lim_{k\to\infty} \rho(c(x_{m_k} - x_{n_k})) \\ &= \lim_{k\to\infty} \rho(l(x_{m_k-1} - x_{n_k-1})) = \varepsilon.\end{aligned} \tag{22}$$

Letting $k \to \infty$ in (17), by property of ψ and (22), we get

$$\psi(\varepsilon) \le \psi(\varepsilon) - \phi(\varepsilon) < \psi(\varepsilon) \tag{23}$$

which is a contradiction. Therefore, $\{cx_n\}_{n\in\mathbb{N}}$ is ρ-*Cauchy*. Since X_ρ is ρ-*complete* there exists a point $u \in X_\rho$ such that $\rho(c(x_n - u)) \to 0$ as $n \to \infty$. Consequently, $\rho(l(x_n - u)) \to 0$ as $n \to \infty$. Next, we prove that u is a unique fixed point of T. Putting $x = x_{n-1}$ and $y = u$ in (10), we obtain

$$\begin{aligned}\psi(\rho(c(x_n - Tu))) &\le \psi(\rho(l(x_{n-1} - u))) \\ &\quad - \phi(\rho(l(x_{n-1} - u))).\end{aligned} \tag{24}$$

Taking $n \to \infty$ in the inequality (24), we have

$$\psi(\rho(c(u - Tu))) \le \psi(0) - \phi(0) = 0, \tag{25}$$

which implies that $\rho(c(Tu - u)) = 0$ and $Tu = u$. Suppose that there exists $v \in X_\rho$ such that $Tv = v$ and $v \ne u$, and then we have

$$\begin{aligned}\psi(\rho(c(u - v))) &= \psi(\rho(c(Tu - Tv))) \\ &\le \psi(\rho(l(u - v))) - \phi(\rho(l(u - v))) \\ &< \psi(\rho(l(u - v))) \\ &\le \psi(\rho(c(u - v))),\end{aligned} \tag{26}$$

which is a contradiction. Hence $u = v$ and the proof is complete. □

Corollary 11. *Let X_ρ be a ρ-complete modular space, where ρ satisfies the Δ_2-condition. Let $c, l \in \mathbb{R}^+$, $c > l$, and $T : X_\rho \to X_\rho$ be a mapping satisfying the inequality*

$$\begin{aligned}\rho(c(Tx - Ty)) &\le \rho(l(x - y)) \\ &\quad - \phi(\rho(l(x - y))),\end{aligned} \tag{27}$$

for all $x, y \in X_\rho$, where $\phi : [0, \infty) \to [0, \infty)$ is continuous and monotone nondecreasing function with $\phi(t) = 0$ if and only if $t = 0$. Then, T has a unique fixed point.

Proof. Take $\psi(t) = t$, and then we obtain the Corollary 11. □

Theorem 12. *Let X_ρ be a ρ-complete modular space, where ρ satisfies the Δ_2-condittion and let $T : X_\rho \to X_\rho$ be a mapping satisfying the inequality*

$$\psi(\rho((Tx - Ty))) \le \psi(m(x, y)) - \phi(m(x, y)) \tag{28}$$

for all $x, y \in X_\rho$, where

$$m(x, y) = \max\left\{\rho(x - y), \rho(x - Tx), \rho(y - Ty), \left(\rho\left(\frac{1}{2}(x - Ty)\right) + \rho\left(\frac{1}{2}(y - Tx)\right)\right)(2)^{-1}\right\} \tag{29}$$

and $\psi, \phi : [0, \infty) \to [0, \infty)$ are both continuous and monotone nondecreasing functions with $\psi(t) = \phi(t) = 0$ if and only if $t = 0$. Then, T has a unique fixed point.

Proof. First, we prove that the sequence $\{\rho(c(T^n x - T^{n-1}x))\}$ converges to 0. Since,

$$\begin{aligned}\psi\left(\rho\left(T^n x - T^{n-1}x\right)\right) &\le \psi\left(m\left(T^{n-1}x, T^{n-2}x\right)\right) - \phi\left(m\left(T^{n-1}x, T^{n-2}x\right)\right). \\ &\le \psi\left(m\left(T^{n-1}x, T^{n-2}x\right)\right).\end{aligned} \tag{30}$$

By monotone nondecreasing of ψ, we have

$$\rho\left(T^n x - T^{n-1}x\right) \le m\left(T^{n-1}x, T^{n-2}x\right). \tag{31}$$

From the definition of $m(x, y)$, we get

$$\begin{aligned} m\left(T^{n-1}x, T^{n-2}x\right) &= \max\left\{\rho\left(T^{n-1}x - T^{n-2}x\right), \rho\left(T^n x - T^{n-1}x\right), \frac{\rho\left((1/2)\left(T^n x - T^{n-2}x\right)\right)}{2}\right\} \\ &= \max\left\{\rho\left(T^{n-1}x - T^{n-2}x\right), \rho\left(T^n x - T^{n-1}x\right), \frac{\rho\left(T^n x - T^{n-1}x\right) + \rho\left(T^{n-1}x - T^{n-2}x\right)}{2}\right\} \\ &= \max\left\{\rho\left(T^{n-1}x - T^{n-2}x\right), \rho\left(T^n x - T^{n-1}x\right)\right\}.\end{aligned} \tag{32}$$

If $\rho(T^n x - T^{n-1}x) > \rho(T^{n-1}x - T^{n-2}x) \ge 0$, then $m(T^{n-1}x, T^{n-2}x) = \rho(T^n x - T^{n-1}x)$. Furthermore it is implied that

$$\begin{aligned}\psi\left(\rho\left(T^n x - T^{n-1}x\right)\right) &\le \psi\left(m\left(T^{n-1}x, T^{n-2}x\right)\right) - \phi\left(m\left(T^{n-1}x, T^{n-2}x\right)\right) \\ &\le \psi\left(\rho\left(T^n x - T^{n-1}x\right)\right) - \phi\left(\rho\left(T^n x - T^{n-1}x\right)\right) \\ &< \psi\left(\rho\left(T^n x - T^{n-1}x\right)\right)\end{aligned} \tag{33}$$

which is a contradiction, and, hence,

$$\begin{aligned}\rho\left(T^n x - T^{n-1}x\right) &\le m\left(T^{n-1}x - T^{n-2}x\right) \\ &= \rho\left(T^{n-1}x - T^{n-2}x\right).\end{aligned} \tag{34}$$

So, we have that the sequence $\{\rho(T^n x - T^{n-1}x)\}$ is monotone decreasing and bounded below. Hence there exists $r \ge 0$ such that

$$\lim_{n\to\infty} \rho\left(T^n x - T^{n-1}x\right) = r. \tag{35}$$

If $r > 0$, taking $n \to \infty$ in the inequality (30), we get

$$\begin{aligned}\psi(r) &\le \psi(r) - \phi(r) \\ &< \psi(r)\end{aligned} \tag{36}$$

which is a contradiction, and thus $r = 0$. So, we have

$$\lim_{n\to\infty} \rho\left(T^n x - T^{n-1}x\right) = 0. \tag{37}$$

Next, we prove that the sequence $\{T^n(x)\}_{n\in\mathbb{N}}$ is ρ-*Cauchy*. Suppose $\{T^n(x)\}_{n\in\mathbb{N}}$ is not ρ-*Cauchy*, and there exist $\varepsilon > 0$ and sequence of integers $\{m_k\}$, $\{n_k\}$ with $\{m_k\} > \{n_k\} \ge k$ such that

$$\rho\left(\left(T^{m_k}x - T^{n_k}x\right)\right) \ge \varepsilon, \qquad \rho\left(2\left(T^{m_k-1}x - T^{n_k}x\right)\right) < \varepsilon. \tag{38}$$

Since,

$$\begin{aligned}\psi\left(\rho\left(\left(T^{m_k}x - T^{n_k}x\right)\right)\right) &\le \psi\left(m\left(T^{m_k-1}x, T^{n_k-1}x\right)\right) - \phi\left(m\left(T^{m_k-1}x, T^{n_k-1}x\right)\right) \\ &\le \psi\left(m\left(T^{m_k-1}x, T^{n_k-1}x\right)\right)\end{aligned} \tag{39}$$

which implies that

$$\rho\left(\left(T^{m_k}x - T^{n_k}x\right)\right) \le m\left(T^{m_k-1}x, T^{n_k-1}x\right). \tag{40}$$

On the other hand,

$$\begin{aligned} m&\left(T^{m_k-1}x, T^{n_k-1}x\right) \\ &= \max\left\{\rho\left(T^{m_k-1}x - T^{n_k-1}x\right),\right. \\ &\quad \rho\left(T^{m_k}x - T^{m_k-1}x\right), \\ &\quad \rho\left(T^{n_k}x - T^{n_k-1}x\right), \\ &\quad \left(\rho\left(\frac{1}{2}\left(T^{m_k}x - T^{n_k-1}x\right)\right)\right. \\ &\quad \left.\left.+\rho\left(\frac{1}{2}\left(T^{m_k-1}x - T^{n_k}x\right)\right)\right)(2)^{-1}\right\}, \end{aligned} \tag{41}$$

$$\begin{aligned} \rho&\left(\left(T^{m_k-1}x - T^{n_k-1}x\right)\right) \\ &= \rho\left(\left(T^{m_k-1}x - T^{n_k}x + T^{n_k}x - T^{n_k-1}x\right)\right) \\ &\le \rho\left(2\left(T^{m_k-1}x - T^{n_k}x\right)\right) \\ &\quad + \rho\left(2\left(T^{n_k}x - T^{n_k-1}x\right)\right) \\ &< \varepsilon + \rho\left(2\left(T^{n_k}x - T^{n_k-1}x\right)\right). \end{aligned}$$

For the last term in $m(T^{m_k-1}x, T^{n_k-1}x)$, by Proposition 8, we have

$$\begin{aligned} &\left(\rho\left(\frac{1}{2}\left(T^{m_k}x - T^{n_k-1}x\right)\right) + \rho\left(\frac{1}{2}\left(T^{m_k-1}x - T^{n_k}x\right)\right)\right)(2)^{-1} \\ &= \left(\rho\left(\frac{1}{2}\left(T^{m_k-1}x - T^{n_k}x\right)\right)\right. \\ &\quad + \rho\left(\frac{1}{2}\left(T^{m_k}x - T^{m_k-1}x + T^{m_k-1}x - T^{n_k}x\right)\right. \\ &\quad \left.\left.+\frac{1}{2}\left(T^{n_k}x - T^{n_k-1}x\right)\right)\right)(2)^{-1} \\ &\le \left(\rho\left(\frac{1}{2}\left(T^{m_k-1}x - T^{n_k}x\right)\right)\right. \\ &\quad + \rho\left(T^{m_k}x - T^{m_k-1}x + T^{m_k-1}x - T^{n_k}x\right) \\ &\quad \left.+\rho\left(T^{n_k}x - T^{n_k-1}x\right)\right)(2)^{-1} \\ &\le \left(\rho\left(\frac{1}{2}\left(T^{m_k-1}x - T^{n_k}x\right)\right) + \rho\left(2\left(T^{m_k}x - T^{m_k-1}x\right)\right)\right. \\ &\quad \left.+\rho\left(2\left(T^{m_k-1}x - T^{n_k}x\right)\right) + \rho\left(T^{n_k}x - T^{n_k-1}x\right)\right)(2)^{-1} \\ &< \varepsilon + \frac{\rho\left(T^{n_k}x - T^{n_k-1}x\right) + \rho\left(2\left(T^{m_k}x - T^{m_k-1}x\right)\right)}{2}. \end{aligned} \tag{42}$$

It follow from (41) and (42) that

$$\begin{aligned} m&\left(T^{m_k-1}x, T^{n_k-1}x\right) \\ &= \max\left\{\rho\left(T^{m_k-1}x - T^{n_k-1}x\right),\right. \\ &\quad \rho\left(T^{m_k}x - T^{m_k-1}x\right), \rho\left(T^{n_k}x - T^{n_k-1}x\right), \\ &\quad \left(\rho\left(\frac{1}{2}\left(T^{m_k}x - T^{n_k-1}x\right)\right)\right. \\ &\quad \left.\left.+\rho\left(\frac{1}{2}\left(T^{m_k-1}x - T^{n_k}x\right)\right)\right)(2)^{-1}\right\} \\ &< \max\left\{\varepsilon + \rho\left(2\left(T^{n_k}x - T^{n_k-1}x\right)\right),\right. \\ &\quad \rho\left(T^{m_k}x - T^{m_k-1}x\right), \rho\left(T^{n_k}x - T^{n_k-1}x\right), \varepsilon \\ &\quad \left.+\frac{\rho\left(T^{n_k}x - T^{n_k-1}x\right) + \rho\left(2\left(T^{m_k}x - T^{m_k-1}x\right)\right)}{2}\right\}. \end{aligned} \tag{43}$$

By (37), (38), (40), (43), and the Δ_2-condition of ρ, we have

$$\lim_{k\to\infty} \rho\left(\left(T^{m_k}x - T^{n_k}x\right)\right) = \lim_{k\to\infty} m\left(T^{m_k-1}x, T^{n_k-1}x\right) = \varepsilon. \tag{44}$$

Taking $k \to \infty$ in (39), by (44) and the continuity of ψ, we get

$$\psi(\varepsilon) \le \psi(\varepsilon) - \phi(\varepsilon) < \psi(\varepsilon) \tag{45}$$

which is a contradiction. Hence, $\{T^n(x)\}_{n\in\mathbb{N}}$ is ρ-*Cauchy*. Since X_ρ is ρ-*complete*, there exists a point $u \in X_\rho$ such that $\rho(T^n x - u) \to 0$ as $n \to \infty$. Next, we prove that u is a unique fixed point of T. Suppose that $Tu \ne u$, then $\rho(u - Tu) > 0$.

Since,

$$\begin{aligned} \psi\left(\rho\left(T^n x - Tu\right)\right) \le\ & \psi\left(m\left(T^{n-1}x, u\right)\right) \\ & - \phi\left(m\left(T^{n-1}x, u\right)\right), \end{aligned} \tag{46}$$

$$\begin{aligned} m&\left(T^{n-1}x, u\right) \\ &= \max\left\{\rho\left(T^{n-1}x - u\right), \rho\left(T^{n-1}x - T^n x\right),\right. \\ &\quad \rho(u - Tu), \left(\rho\left(\frac{1}{2}\left(T^{n-1}x - Tu\right)\right)\right. \\ &\quad \left.\left.+\rho\left(\frac{1}{2}\left(u - T^n x\right)\right)\right)(2)^{-1}\right\} \\ &\longrightarrow \max\left\{0, 0, \rho(u - Tu), \frac{\rho((1/2)(u - Tu))}{2}\right\} \\ &= \rho(u - Tu) \end{aligned} \tag{47}$$

as $n \longrightarrow \infty$.

Taking $n \to \infty$ in (46), by using (47), we get

$$\psi(\rho(u - Tu)) \le \psi(\rho(u - Tu)) - \phi(\rho(u - Tu)) < \psi(\rho(u - Tu)) \tag{48}$$

which is a contradiction. Hence, $\rho(u - Tu) = 0$ and $Tu = u$. If there exists point $v \in X_\rho$ such that $Tv = v$ and $u \neq v$, then using an argument similar to the above we get

$$\begin{aligned}\psi(\rho(u - v)) &= \psi(\rho(Tu - Tv)) \\ &\le \psi(m(u, v)) - \phi(m(u, v)) \\ &\le \psi(\rho(u - v)) - \phi(\rho(u - v)) \\ &< \psi(\rho(u - v))\end{aligned} \tag{49}$$

which is a contradiction. Hence, $u = v$ and the proof is complete. □

Corollary 13. *Let X_ρ be a ρ-complete modular space, where ρ satisfies the Δ_2-condition, and let $T : X_\rho \to X_\rho$ be a mapping satisfying the inequality*

$$\rho((Tx - Ty)) \le m(x, y) - \phi(m(x, y)) \tag{50}$$

for all $x, y \in X_\rho$, where $m(x, y) = \max\{\rho(x - y), \rho(x - Tx), \rho(y - Ty), (\rho((1/2)(x - Ty)) + \rho((1/2)(y - Tx)))/2\}$ and $\phi : [0, \infty) \to [0, \infty)$ is continuous and monotone nondecreasing function with $\phi(t) = 0$ if and only if $t = 0$. Then, T has a unique fixed point.

Proof. Taking $\psi(t) = t$, we obtain the Corollary 13. □

Acknowledgments

This work was supported by the Higher Education Research Promotion and National Research University Project of Thailand, Office of the Higher Education Commission. The authors would like to thank the referee for his comments and suggestion. C. Mongkolkeha was supported from the Thailand Research Fund through the the Royal Golden Jubilee Ph.D. Program (Grant no. PHD/0029/2553).

References

[1] Ya. I. Alber and S. Guerre-Delabriere, "Principle of weakly contractive maps in Hilbert spaces," in *New Results in Operator Theory and Its Applications*, I. Gohberg and Yu. Lyubich, Eds., vol. 98 of *Operator Theory: Advances and Applications*, pp. 7–22, Birkhäuser, Basel, Switzerland, 1997.

[2] B. E. Rhoades, "Some theorems on weakly contractive maps," in *Proceedings of the 3rd World Congress of Nonlinear Analysts, Part 4 (Catania, 2000)*, vol. 47, pp. 2683–2693, 2001.

[3] P. N. Dutta and B. S. Choudhury, "A generalisation of contraction principle in metric spaces," *Fixed Point Theory and Applications*, vol. 2008, Article ID 406368, 8 pages, 2008.

[4] H. Nakano, *Modulared Semi-Ordered Linear Spaces*, vol. 1 of *Tokyo Mathematical Book Series*, Maruzen, Tokyo, Japan, 1950.

[5] H. Nakano, *Topology of Linear Topological Spaces*, vol. 3 of *Tokyo Mathematical Book Series*, Maruzen, Tokyo, Japan, 1951.

[6] J. Musielak and W. Orlicz, "On modular spaces," *Studia Mathematica*, vol. 18, pp. 49–65, 1959.

[7] W. A. J. Luxemburg, *Banach function spaces [Ph.D. thesis]*, Delft Institute of Technology, Assen, The Netherlands, 1955.

[8] S. Mazur and W. Orlicz, "On some classes of linear spaces," *Studia Mathematica*, vol. 17, pp. 97–119, 1958.

[9] Ph. Turpin, "Fubini inequalities and bounded multiplier property in generalized modular spaces," *Commentationes Mathematicae*, vol. 1, pp. 331–353, 1978.

[10] R. A. Adams, *Sobolev Spaces*, vol. 65 of *Pure and Applied Mathematics*, Academic Press, New York, NY, USA, 1975.

[11] L. Maligranda, *Orlicz Spaces and Interpolation*, vol. 5 of *Seminars in Mathematics*, Universidade Estadual de Campinas, Departamento de Matemática, Campinas, Brazil, 1989.

[12] J. Musielak, *Orlicz Spaces and Modular Spaces*, vol. 1034 of *Lecture Notes in Mathematics*, Springer, Berlin, Germany, 1983.

[13] S. Rolewicz, *Metric Linear Spaces*, PWN—Polish Scientific Publishers, Warsaw, Poland, 2nd edition, 1984.

[14] M. M. Rao and Z. D. Ren, *Applications of Orlicz Spaces*, vol. 250 of *Monographs and Textbooks in Pure and Applied Mathematics*, Marcel Dekker, New York, NY, USA, 2002.

[15] M. A. Khamsi, W. M. Kozłowski, and S. Reich, "Fixed point theory in modular function spaces," *Nonlinear Analysis. Theory, Methods & Applications*, vol. 14, no. 11, pp. 935–953, 1990.

[16] M. A. Khamsi, "Fixed point theory in modular function spaces," in *Recent Advances on Metric Fixed Point Theory (Seville, 1995)*, vol. 48 of *Ciencias*, pp. 31–57, Universidad de Sevilla, Seville, Spain, 1996.

[17] M. Beygmohammadi and A. Razani, "Two fixed-point theorems for mappings satisfying a general contractive condition of integral type in the modular space," *International Journal of Mathematics and Mathematical Sciences*, vol. 2010, Article ID 317107, 10 pages, 2010.

[18] C. Mongkolkeha and P. Kumam, "Fixed point and common fixed point theorems for generalized weak contraction mappings of integral type in modular spaces," *International Journal of Mathematics and Mathematical Sciences*, vol. 2011, Article ID 705943, 12 pages, 2011.

[19] T. Dominguez Benavides, M. A. Khamsi, and S. Samadi, "Uniformly Lipschitzian mappings in modular function spaces," *Nonlinear Analysis. Theory, Methods & Applications*, vol. 46, no. 2, pp. 267–278, 2001.

[20] M. A. Khamsi, "Uniform noncompact convexity, fixed point property in modular spaces," *Mathematica Japonica*, vol. 40, no. 3, pp. 439–450, 1994.

[21] M. A. Khamsi, "Quasicontraction mappings in modular spaces without Δ_2-condition," *Fixed Point Theory and Applications*, vol. 2008, Article ID 916187, 6 pages, 2008.

[22] P. Kumam, "Fixed point theorems for nonexpansive mappings in modular spaces," *Archivum Mathematicum*, vol. 40, no. 4, pp. 345–353, 2004.

[23] K. Kuaket and P. Kumam, "Fixed points of asymptotic pointwise contractions in modular spaces," *Applied Mathematics Letters*, vol. 24, no. 11, pp. 1795–1798, 2011.

[24] A. Razani and R. Moradi, "Common fixed point theorems of integral type in modular spaces," *Bulletin of the Iranian Mathematical Society*, vol. 35, no. 2, pp. 11–24, 2009.

8

Asymptotic Stability of Solutions to a Nonlinear Urysohn Quadratic Integral Equation

H. H. G. Hashem[1,2] and A. R. Al-Rwaily[2]

[1] *Faculty of Science, Alexandria University, Alexandria, Egypt*
[2] *College of Science & Arts, Qassim University, P.O. Box 6644 Buriadah 81999, Saudi Arabia*

Correspondence should be addressed to H. H. G. Hashem; hendhghashem@yahoo.com

Academic Editor: Seenith Sivasundaram

Here, we prove the existence of L_1-nondecreasing solution to a nonlinear quadratic integral equation of Urysohn type by applying the technique of weak noncompactness. Also, the asymptotic stability of solutions for that quadratic integral equation is studied.

1. Introduction

Integral equations play an important role in many branches of linear and nonlinear functional analysis and their applications in the theory of elasticity, engineering, mathematical physics, and contact mixed problems, and the theory of integral equations is rapidly developing with the help of several tools of functional analysis, topology, and fixed point theory. For details, we refer to [1–23].

Quadratic integral equations often appear in many applications of real world problems, for example, in the theory of radiative transfer, kinetic theory of gases, in the theory of neutron transport, and in the traffic theory (see [12]). The quadratic integral equation can be very often encountered in many applications (see [1, 2, 6–10, 13–26]). However, in most of the previous literature, the main results are realized with the help of the technique associated with the measure of noncompactness. Instead of using the technique of measure of noncompactness, the Tychonoff fixed point theorem is used for some quadratic integral equations [20, 26]. Picard and Adomian decomposition methods are used to compare approximate and exact solutions for quadratic integral equations [13, 19, 22]. Also, nondecreasing solution of a quadratic integral of Urysohn-Stieltjes type is studied in [10].

Let $L_1 = L_1[0,T]$ be the class of Lebesgue integrable functions on $I = [0,T]$ with the standard norm.

Here, we are concerned with the nonlinear quadratic functional integral equation

$$x(t) = f(t, x(\phi_1(t))) + g(t, x(\phi_2(t))) \times \int_0^{\alpha(t)} u(t, s, x(\phi_3(s)))\,ds, \quad t \in I, \tag{1}$$

and we prove the existence of monotonic solutions in L_1 by using the technique of measure of noncompactness. The results of this work generalize those obtained in [18]. Finally, the asymptotic stability of solutions for the quadratic integral equation (1) is studied.

2. Preliminaries

In this section, we collect some definitions and results needed in our further investigations. Assume that the function $f : I \times R \to R$ satisfies Carathèodory condition that is measurable in t for any x and continuous in x for almost all t. Then, to every function $x(t)$ being measurable on the interval I, we may assign the function

$$(Fx)(t) = f(t, x(t)), \quad t \in I. \tag{2}$$

The operator F defined in such a way is called the superposition operator. This operator is one of the simplest and most important operators investigated in the nonlinear functional analysis. For this operator, we have the following theorem due to Krasnosel'skii [3].

Theorem 1. *The superposition operator F maps L_1 into itself if and only if*

$$|f(t,x)| \le c(t) + k|x| \qquad \forall t \in I \tag{3}$$

and $x \in R$, where $c(t)$ is a function from L_1 and k is a nonnegative constant.

Now, let E be a Banach space with zero element θ and X a nonempty bounded subset of E. Moreover denote by $B_r = B(\theta, r)$ the closed ball in E centered at θ and with radius r. In the sequel, we will need some criteria for compactness in measure; the complete description of compactness in measure was given by Banaś [3], but the following sufficient condition will be more convenient for our purposes (see [3]).

Theorem 2. *Let X be a bounded subset of L_1. Assume that there is a family of subsets $(\Omega_c)_{0\le c\le b-a}$ of the interval (a,b) such that meas $\Omega_c = c$ for every $c \in [0, b-a]$, and for every $x \in X$, $x(t_1) \le x(t_2)$, $(t_1 \in \Omega_c, t_2 \notin \Omega_c)$; then, the set X is compact in measure.*

The measure of weak noncompactness defined by De Blasi [11, 27] is given by

$$\beta(X) = \inf(r > 0; \text{there exists a weakly compact subset } Y \text{ of } E \text{ such that } X \subset Y + K_r). \tag{4}$$

The function $\beta(X)$ possesses several useful properties which may be found in [11].

The convenient formula for the function $\beta(X)$ in L_1 was given by Appell and De Pascale (see [27]) as follows:

$$\beta(X) = \lim_{\epsilon\to 0}\left(\sup_{x\in X}\left(\sup\left[\left[\int_D |x(t)|\,dt : D \subset [a,b], \text{meas } D \le \epsilon\right]\right)\right), \tag{5}$$

where the symbol meas D stands for Lebesgue measure of the set D.

Next, we shall also use the notion of the Hausdorff measure of noncompactness χ (see [3]) defined by

$$\chi(X) = \inf(r > 0; \text{there exists a finite subset } Y \text{ of } E \text{ such that } X \subset Y + K_r). \tag{6}$$

In the case when the set X is compact in measure, the Hausdorff and De Blasi measures of noncompactness will be identical. Namely, we have the following (see [11, 27]).

Theorem 3. *Let X be an arbitrary nonempty bounded subset of L_1. If X is compact in measure, then $\beta(X) = \chi(X)$.*

Finally, we will recall the fixed point theorem due to Banaś [5].

Theorem 4. *Let Q be a nonempty, bounded, closed, and convex subset of E, and let $H : Q \to Q$ be a continuous transformation which is a contraction with respect to the Hausdorff measure of noncompactness χ; that is, there exists a constant $\alpha \in [0,1)$ such that $\chi(HX) \le \alpha\chi(X)$ for any nonempty subset X of Q. Then, H has at least one fixed point in the set Q.*

3. Existence Theorem

Let the integral operator H be defined as

$$(Hx)(t) = \int_0^{\alpha(t)} u(t,s,x(s))\,ds,$$
$$(Fx)(t) = f(t,x(t)). \tag{7}$$

Then, (1) may be written in operator form as

$$(Ax)(t) = (Fx(\phi_1))(t) + (Gx(\phi_2))(t)\cdot(Hx(\phi_3))(t), \tag{8}$$

where $(Gx)(t) = g(t,x(t))$.

Consider the following assumptions.

(i) $f, g : I \times R \to R$ are functions such that $f, g : I \times R_+ \to R_+$. Moreover, the functions f, g satisfy Carathèodory condition (i.e., are measurable in t for all $x \in R$ and continuous in x for all $t \in I$), and there exist two functions $a_1, a_2 \in L_1$ and constants $b_1, b_2 > 0$ such that

$$|f(t,x)| \le a_1(t) + b_1|x|,$$
$$|g(t,x)| \le a_2(t) + b_2|x| \quad \forall (t,x) \in I \times R. \tag{9}$$

Apart from this, the functions f and g are nondecreasing in both variables.

(ii) $u : I \times I \times R \to R$ is such that $u(t,s,x) \ge 0$ for $(t,s,x) \in I \times I \times R_+$, and $u(t,s,x)$ satisfies Carathéodory condition (i.e., it is measurable in (t,s) for all $x \in R$ and continuous in x for almost all $(t,s) \in I \times I$).

(iii) There exist a positive constant b_3, a function $a_3 \in L_1$, and a measurable (in both variables) function $k(t,s) = k : I \times I \to R_+$ such that

$$|u(t,s,x)| \le k(t,s)(a_3(t) + b_3|x|) \quad \forall t,s \in I \text{ and for } x \in R, \tag{10}$$

and the integral operator K, generated by the function k and defined by

$$(Kx)(t) = \int_0^t k(t,s)\,x(s)\,ds, \quad t \in I, \tag{11}$$

maps continuously L_1 into L_∞ on I.

(iv) $t \to u(t,s,x)$ is a.e. nondecreasing on I for almost all fixed $s \in I$ and for each $x \in R_+$.

(v) $\alpha : I \to I$ is continuous.

(vi) $\phi_i : I \to I$, $i = 1,2,3$, are increasing, absolutely continuous functions on I, and there exist positive constants B_i, $i = 1,2,3$, such that $\phi_i' \geq B_i$ a.e. on I.

(vii) Let $d > \sqrt{4Mb_2b_3B_1^2B_2B_3(\|a_1\| + M \cdot \|a_2\|\|a_3\|)}$, $M = \|K\|_{L_\infty}$, where $d = B_1B_2B_3 - b_1B_2B_3 - b_2MB_1B_3\|a_3\| - Mb_3B_1B_2\|a_2\|$.

Now, let r be a positive root of the equation

$$b_2b_3B_1Mr^2 - dr + B_1B_2B_3\left(\|a_1\| + M \cdot \|a_2\|\,\|a_3\|\right) = 0, \quad (12)$$

and define the set

$$B_r = \{x \in L_1 : \|x\| \leq r\}. \quad (13)$$

For the existence of at least one L_1-positive solution of the quadratic functional integral equation (1), we have the following theorem.

Theorem 5. *Let the assumptions (i)–(vii) be satisfied.*

If $b_1B_2B_3 + Mb_2B_1B_3\|a_3\| + rb_2b_3M < B_1B_2B_3$, then the quadratic integral equation (1) has at least one solution $x \in L_1$ which is positive and a.e. nondecreasing on I.

Proof. Take an arbitrary $x \in L_1$; then, we get

$$|(Ax)(t)| \leq |a_1(t)| + b_1|x(\phi_1(t))| + (a_2(t) + b_2|x(\phi_2(t))|) \times \int_0^{\alpha(t)} k(t,s)(a_3(t) + b_3|x(\phi_3(s))|)\,ds, \quad (14)$$

which implies that

$$\begin{aligned}
\|(Ax)(t)\| &= \int_0^T |(Ax)(t)|\,dt \\
&\leq \int_0^T |a_1(t)|\,dt + b_1\int_0^T |x(\phi_1(t))|\,dt + \int_0^T (a_2(t)\,dt + b_2|x(\phi_2(t))|) \times \int_0^{\alpha(t)} k(t,s)(a_3(t) + b_3|x(\phi_3(s))|)\,ds\,dt \\
&\leq \|a_1\| + \frac{b_1}{B_1}\int_0^T |x(\phi_1(t))| \cdot \phi_1'(t)\,dt + \int_0^T a_2(t)\int_0^{\alpha(t)} k(t,s)a_3(s)\,ds\,dt \\
&\quad + b_2\int_0^T |x(\phi_2(t))| \int_0^{\alpha(t)} k(t,s)a_3(s)\,ds\,dt + b_3 \times \int_0^T a_2(t)\int_0^{\alpha(t)} k(t,s)|x(\phi_3(s))|\,ds\,dt \\
&\quad + b_2b_3\int_0^T |x(\phi_2(t))| \int_0^{\alpha(t)} k(t,s)|x(\phi_3(s))|\,ds\,dt \\
&\leq \|a_1\| + \frac{b_1}{B_1}\int_0^T |x(\phi_1(t))| \cdot \phi_1'(t)\,dt + \int_0^T a_2(t)\int_0^T k(t,s)a_3(s)\,ds\,dt \\
&\quad + b_2\int_0^T |x(\phi_2(t))| \int_0^T k(t,s)a_3(s)\,ds\,dt + b_3 \times \int_0^T a_2(t)\int_0^T k(t,s)|x(\phi_3(s))|\,ds\,dt \\
&\quad + b_2b_3\int_0^T |x(\phi_2(t))| \int_0^T k(t,s)|x(\phi_3(s))|\,ds\,dt \\
&\leq \|a_1\| + \frac{b_1}{B_1}\int_{\phi_1(0)}^{\phi_1(T)} |x(\phi_1(t))| \cdot \phi_1'(t)\,dt + \int_0^T a_2(t)\int_0^T k(t,s)a_3(s)\,ds\,dt \\
&\quad + \frac{b_2}{B_2}\int_{\phi_2(0)}^{\phi_2(T)} |x(\phi_2(t))| \cdot \phi_2'(t)\int_0^T k(t,s)a_3(s)\,ds\,dt \\
&\quad + \frac{b_3}{B_3}\int_0^T a_2(t)\int_{\phi_3(0)}^{\phi_3(T)} k(t,s)|x(\phi_3(s))| \cdot \phi_3'(s)\,ds\,dt \\
&\quad + \frac{b_2b_3}{B_2B_3}\int_{\phi_2(0)}^{\phi_2(T)} |x(\phi_2(t))| \cdot \phi_2'(t) \times \int_{\phi_3(0)}^{\phi_3(T)} k(t,s)|x(\phi_3(s))| \cdot \phi_3'(s)\,ds\,dt \\
&\leq \|a_1\| + \frac{b_1}{B_1}\int_0^T |x(\theta)|\,d\theta + M\|a_2\|\,\|a_3\| + \frac{Mb_2\|a_3\|}{B_2}\int_0^T |x(\theta)|\,d\theta \\
&\quad + \frac{Mb_3\|a_2\|}{B_3}\int_0^T |x(\theta)|\,d\theta + \frac{Mb_2b_3}{B_2B_3}\int_0^T |x(\theta)|\,d\theta \cdot \int_0^T |x(\theta)|\,d\theta \\
&\leq \|a_1\| + \frac{b_1}{B_1}\|x\| + M\|a_2\|\,\|a_3\| + \frac{Mb_2\|a_3\|}{B_2}\|x\| + \frac{Mb_3\|a_2\|}{B_3}\|x\| + \frac{Mb_2b_3}{B_2B_3}\|x\|^2.
\end{aligned} \quad (15)$$

From this estimate, we show that the operator A maps the ball B_r into itself with

$$r = \frac{d - \sqrt{d^2 - 4Mb_2b_3B_1^2B_2B_3\left(\|a_1\| + M \cdot \|a_2\|\,\|a_3\|\right)}}{2Mb_2b_3B_1}, \quad (16)$$

From assumption (vii) we have

$$0 < d^2 - 4Mb_2b_3B_1^2B_2B_3\left(\|a_1\| + M\cdot\|a_2\|\,\|a_3\|\right) < d^2, \quad (17)$$

which implies that

$$0 < \sqrt{d^2 - 4Mb_2b_3B_1^2B_2B_3\left(\|a_1\| + M\cdot\|a_2\|\,\|a_3\|\right)} < d. \quad (18)$$

Then, d is positive which implies that r is a positive constant.

Now, let Q_r denote the subset of $B_r \in L_1$ consisting of all functions which are positive and a.e. nondecreasing on I.

The set Q_r is nonempty, bounded, convex, and closed (see [3, page 780]). Moreover, this set is compact in measure (see Lemma 2 in [4, page 63]).

From the assumptions, we deduce that the operator A maps Q_r into itself. Since the operator $(Ux)(t) = u(t,s,x)$ is continuous (Theorem 1 in Section 2), then the operator H is continuous, and, hence, the product $G.H$ is continuous. Also, F is continuous. Thus, the operator A is continuous on Q_r.

Let X be a nonempty subset of Q_r. Fix $\epsilon > 0$, and take a measurable subset $D \subset I$ such that meas $D \le \epsilon$. Then, for any $x \in X$, using the same reasoning as in [3, 4], we get

$$\begin{aligned}
&\|Ax\|_{L_1(D)} \\
&= \int_D |(Ax)(t)|\,dt \\
&\le \int_D |a_1(t)|\,dt + b_1\int_D |x(\phi_1(t))|\,dt \\
&\quad + \int_D (a_2(t)\,dt + b_2|x(\phi_2(t))|) \\
&\quad \times \int_0^{\alpha(t)} k(t,s)(a_3(s) + b_3|x(\phi_3(s))|)\,ds\,dt \\
&\le \|a_1\|_{L_1(D)} + \frac{b_1}{B_1}\int_D |x(\phi_1(t))|\cdot\phi_1'(t)\,dt \\
&\quad + \int_D a_2(t)\int_0^T k(t,s)\,a_3(s)\,ds\,dt \\
&\quad + b_2\int_D |x(\phi_2(t))|\int_0^T k(t,s)\,a_3(s)\,ds\,dt + Mb_3 \\
&\quad \times \int_D a_2(t)\int_0^T |x(\phi_3(s))|\,ds\,dt \\
&\quad + Mb_2b_3\int_D |x(\phi_2(t))|\int_0^T |x(\phi_3(s))|\,ds\,dt \\
&\le \|a_1\|_{L_1(D)} + \frac{b_1}{B_1}\int_D |x(\theta)|\,d\theta + M\|a_2\|_{L_1(D)}\|a_3\|_{L_1} \\
&\quad + \frac{Mb_2\|a_3\|_{L_1}}{B_2}\int_D |x(\theta)|\,d\theta \\
&\quad + \frac{Mb_3\|a_2\|_{L_1(D)}}{B_3}\int_0^T |x(\theta)|\,d\theta \\
&\quad + \frac{Mb_2b_3}{B_2B_3}\int_D |x(\theta)|\,d\theta\cdot\int_0^T |x(\theta)|\,d\theta \\
&\le \|a_1\|_{L_1(D)} + \frac{b_1}{B_1}\|x\|_{L_1(D)} + M\|a_2\|_{L_1(D)}\|a_3\|_{L_1} \\
&\quad + \frac{Mb_2\|a_3\|_{L_1}}{B_2}\|x\|_{L_1(D)} \\
&\quad + \frac{Mb_3\|a_2\|_{L_1(D)}}{B_3}\|x\| + \frac{Mb_2b_3}{B_2B_3}\|x\|_{L_1(D)}\cdot\|x\| \\
&\le \|a_1\|_{L_1(D)} + \frac{b_1}{B_1}\|x\|_{L_1(D)} + M\|a_2\|_{L_1(D)}\|a_3\|_{L_1} \\
&\quad + \frac{Mb_2\|a_3\|_{L_1}}{B_2}\|x\|_{L_1(D)} \\
&\quad + \frac{rMb_3\|a_2\|_{L_1(D)}}{B_3} + \frac{rMb_2b_3}{B_2B_3}\|x\|_{L_1(D)}.
\end{aligned} \quad (19)$$

Since

$$\lim_{\epsilon\to 0}\left\{\sup\left\{\int_D |a_i(t)|\,dt : D\subset I,\ \text{meas}\ D < \epsilon\right\}\right\} = 0, \quad i = 1, 2, \quad (20)$$

we obtain

$$\beta(Ax(t)) \le \left[\frac{b_1}{B_1} + \frac{Mb_2\|a_3\|_{L_1}}{B_2} + \frac{Mb_2b_3r}{B_2B_3}\right]\beta(x(t)). \quad (21)$$

This implies that

$$\beta(AX) \le \left[\frac{b_1}{B_1} + \frac{Mb_2\|a_3\|_{L_1}}{B_2} + \frac{Mb_2b_3r}{B_2B_3}\right]\beta(X), \quad (22)$$

where β is the De Blasi measure of weak noncompactness.

Keeping in mind Theorem 3, we can write (22) in the form

$$\chi(AX) \le \left[\frac{b_1}{B_1} + \frac{Mb_2\|a_3\|_{L_1}}{B_2} + \frac{Mb_2b_3r}{B_2B_3}\right]\chi(X), \quad (23)$$

where χ is the Hausdorff measure of noncompactness.

Since $(b_1/B_1) + (Mb_2\|a_3\|_{L_1}/B_2) + (Mb_2b_3r/B_2B_3) < 1$, from Theorem 4 follows that A is contraction with respect to the measure of noncompactness χ. Thus, A has at least one fixed point in Q_r which is a solution of the quadratic functional integral equation. □

4. Asymptotic Stability of the Quadratic Integral Equation

We shall show that the solution of the quadratic integral equation (1) is asymptotically stable on $\mathbb{R}_+$.

Definition 6. The function x is said to be asymptotically stable solution of (1) if for any $\epsilon > 0$ there exists $T' = T'(\epsilon) > 0$ such that for every $t \geq T'$ and for every other solution y of (1),

$$|x(t) - y(t)| \leq \epsilon. \tag{24}$$

Proof. Let r be defined by (16), and consider the following assumptions.

($*$) There exist constants l_1 and l_2 satisfying that

$$\begin{aligned} &|f(t,x) - f(t,y)| \leq l_1 |x-y|, \\ &|g(t,x) - g(t,y)| \leq l_2 |x-y|, \\ &\forall t \in I, x, y \in R_+. \end{aligned} \tag{25}$$

($**$) $2MB_1B_2B_3||a_2||||a_3|| + 2rb_2MB_1B_3||a_2|| + 2rMb_3B_1B_2||a_2|| + 2Mr^2b_2b_3B_1 < \epsilon(B_1B_2B_3 - l_1B_2B_3 - Ml_2B_1B_3||a_3|| - Mrl_2b_3B_1)$.

For solutions $x = x(t)$ and $y = y(t)$ of (1) in B_r, by the assumptions ($*$) and ($**$), we deduce that

$$\begin{aligned} &|x(t) - y(t)| \\ &= |(Ax)(t) - (Ay)(t)| \\ &\leq |f(t, x(\phi_1(t))) - f(t, y(\phi_1(t)))| \\ &\quad + |g(t, x(\phi_2(t))) - g(t, y(\phi_2(t)))| \\ &\quad \times \int_0^{\alpha(t)} |u(t,s,y(\phi_3(s)))| \, ds \\ &\quad + |g(t, y(\phi_2(t)))| \\ &\quad \times \int_0^{\alpha(t)} \left| u(t,s,y(\phi_3(s))) - u(t,s,x(\phi_3(s))) \right| ds \\ &\leq l_1 |x(\phi_1(t)) - y(\phi_1(t))| + l_2 |x(\phi_2(t)) - y(\phi_2(t))| \\ &\quad \times \int_0^T k(t,s) (a_3(s) + b_3 |y(\phi_3(s))|) \, ds \\ &\quad + 2 (a_2(t) + b_2 |y(\phi_2(t))|) \\ &\quad \times \int_0^T k(t,s) (a_3(s) + b_3 |x(\phi_3(s))|) \, ds; \end{aligned} \tag{26}$$

for any $t \in I$, using ($**$), we have

$$\begin{aligned} \|x-y\| &= \int_0^T |x(t) - y(t)| \, dt \\ &\leq \frac{l_1}{B_1} \|x-y\| + \frac{Ml_2}{B_2} \|x-y\| \\ &\quad \times \left[\|a_3\| + \frac{b_3 r}{B_3} \right] + 2M \|a_2\| \|a_3\| \\ &\quad + \frac{2rb_2M \|a_3\|}{B_2} + \frac{2Mr^2b_2b_3}{B_2B_3} + \frac{2Mrb_3 \|a_2\|}{B_3}. \end{aligned} \tag{27}$$

Then,

$$\begin{aligned} &\|x-y\| \\ &\leq \left[2M \|a_2\| \|a_3\| + \frac{2rb_2M \|a_3\|}{B_2} + \frac{2Mr^2b_2b_3}{B_2B_3} + \frac{2Mrb_3 \|a_2\|}{B_3} \right] \\ &\quad \cdot \left[1 - \frac{l_1}{B_1} - \frac{Mrb_3l_2}{B_2B_3} - \frac{l_2M \|a_3\|}{B_2} \right]^{-1} \\ &\leq \epsilon, \quad t \in I. \end{aligned} \tag{28}$$

That is, the solution $x = x(t)$ of (1) is asymptotically stable on R_+. This completes the proof. □

5. Applications

As particular cases of Theorem 5, we can obtain theorems on the existence of positive and a.e. nondecreasing solutions belonging to the space $L_1(I)$ of the following quadratic integral equations.

(1) If $\alpha(t) = 1$, then we obtain the quadratic integral equation

$$\begin{aligned} x(t) &= f(t, x(\phi_1(t))) + g(t, x(\phi_2(t))) \\ &\quad \times \int_0^1 u(t,s,x(\phi_3(s))) \, ds, \quad t \in I. \end{aligned} \tag{29}$$

(2) If $\alpha(t) = 1$ and $f(t,x) = a(t)$, then we obtain the quadratic integral equation

$$x(t) = a(t) + g(t, x(\phi_2(t))) \int_0^1 u(t,s,x(\phi_3(s))) \, ds, \quad t \in I. \tag{30}$$

(3) If $f(t,x) = a(t)$, $u(t,s,x) = h(t,x)$, $\alpha(t) = 1$, and $g(t,s) = 1$, then we obtain the quadratic integral equation

$$x(t) = a(t) + \int_0^1 h(s, x(\phi_3(s))) \, ds, \quad t \in I, \tag{31}$$

which was proved by Banaś in [4].

(4) If $g(t,x) = 0$, then we obtain the functional equation

$$x(t) = f(t, x(\phi_1(t))), \quad t \in I, \tag{32}$$

which is the same results proved by Banaś in [3].

(5) If $f(t,x) = a(t)$, and $u(t,s,x) = k(t,s)h(t,x)$, then we obtain the quadratic integral equation

$$x(t) = a(t) + g(t, x(t)) \int_0^1 k(t,s) h(s, x(\phi_3(s))) \, ds, \quad t \in I, \tag{33}$$

which is the same result proved in [16].

(6) If $f(t,x) = a(t)$, $u(t,s,x) = h(t,x)$, and $\alpha(t) = t$, then we obtain the quadratic integral equation

$$x(t) = a(t) + g(t, x(t)) \int_0^t h(s, x(\phi_3(s))) \, ds, \quad t \in I, \tag{34}$$

which is the same result proved in [18].

Example 7. Let us consider the quadratic integral equation of Urysohn type having the form

$$x(t) = a(t) + x(t) \int_0^1 \frac{t}{t+s} u(t,s,x(s)) \, ds, \quad t \in [0,1]. \tag{35}$$

This equation represents the Hammerstein counterpart of the famous Chandrasekhar quadratic integral equation which has numerous application (cf. [1, 2, 6, 24]). It arose originally in connection with scattering through a homogeneous semi-infinite plane atmosphere [24].

In case $a(t) = 1$ and $u(t,s,x(s)) = \lambda\phi(s)x(s)$, λ is a positive constant. Then, (35) has the form

$$x(t) = 1 + \lambda x(t) \int_0^1 \frac{t\phi(s)}{t+s} x(s) \, ds. \tag{36}$$

In order to apply our results, we have to impose an additional condition that the so-called "characteristic" function ϕ is continuous on I.

In this case, $r = (1 - \sqrt{1 - 4\lambda k_1})/2\lambda k$, and the assumption (vii) may be reduced to $4\lambda k_1 \leq 1$ where $\sup_{s \in I} \phi(s) = k_1$.

Example 8. Consider the following quadratic functional integral equation:

$$x(t) = \frac{1}{6} x(t) + \left[t + \frac{1}{3+t} x\left(\sin\left(t^2 + 3t\right)\right) \right] \times \int_0^t \left[1 + \frac{1}{3+s} x\left(\sin\left(s^2 + 4s\right)\right) \right] ds \quad t \in [0,1]. \tag{37}$$

Taking

$$f(t,x) = \frac{1}{6} x(t), \qquad g(t,x) = 1 + \frac{1}{3+t} x, \qquad u(t,s,x) = t + \frac{1}{3+t} x, \tag{38}$$

then we can easily deduce that

(i) $|u(t,s,x)| \leq 1 + (1/4)|x|$ and $|g(t,x)| \leq t + (1/4)|x|$ (i.e., $\alpha(t) = t$, $a_1(t) = 0$, $a_2(t) = t$, and $a_3(t) = 1$ which implies that $||a_1|| = 0$, $||a_2|| = ||a_3|| = 1$, and $b_1 = 1/6$, $b_2 = b_3 = 1/4$);

(ii) $\phi_1(t) = t$, $\phi_2(t) = \sin(t^2 + 3t)$, and $\phi_3(t) = \sin(t^2 + 4t)$, and then $\phi_1'(t) = 1$, $\phi_2'(t) = (2t+3)\cos(t^2 + 3t) > 2$, and $\phi_3'(t) = (2t+4)\cos(t^2 + 4t) > 3$ (i.e., $B_1 = 1/2$, $B_2 = 2$, $B_3 = 3$, and $M = 1$).

Now, we will calculate r.

Then, $r = 2.30228 > 0$ and $(b_1/B_1) + (Mb_2||a_3||_{L_1}/B_2) + (Mb_2b_3r/B_2B_3) = 0.48231546246258117 < 1$.

Thus, all the assumptions of Theorem 5 are satisfied; so, the quadratic functional integral equation (37) possesses at least one solution being positive, a.e. nondecreasing, and integrable in $[0,1]$.

Acknowledgments

This work is supported by Deanship for Scientific Research, Qassim University. The authors express their gratitude to Deanship for Scientific Research, Qassim University, for their hospitality and their support. The authors are thankful to Professor A. M. A. El-Sayed for his help and encouragement.

References

[1] I. K. Argyros, "Quadratic equations and applications to Chandrasekhar's and related equations," *Bulletin of the Australian Mathematical Society*, vol. 32, no. 2, pp. 275–292, 1985.

[2] I. K. Argyros, "On a class of quadratic integral equations with perturbation," *Functiones et Approximatio Commentarii Mathematici*, vol. 20, pp. 51–63, 1992.

[3] J. Banaś, "On the superposition operator and integrable solutions of some functional equations," *Nonlinear Analysis A*, vol. 12, no. 8, pp. 777–784, 1988.

[4] J. Banaś, "Integrable solutions of Hammerstein and Urysohn integral equations," *Australian Mathematical Society A*, vol. 46, no. 1, pp. 61–68, 1989.

[5] J. Banaś and K. Goebel, *Measures of Noncompactness in Banach Spaces*, vol. 60 of *Lecture Notes in Pure and Applied Mathematics*, Marcel Dekker, New York, NY, USA, 1980.

[6] J. Banaś, M. Lecko, and W. G. El-Sayed, "Eixstence theorems of some quadratic integral equations," *Journal of Mathematical Analysis and Applications*, vol. 227, no. 1, pp. 276–279, 1998.

[7] J. Banaś and A. Martinon, "Monotonic solutions of a quadratic integral equation of Volterra type," *Computers & Mathematics with Applications*, vol. 47, no. 2-3, pp. 271–279, 2004.

[8] J. Banaś, J. Rocha Martin, and K. Sadarangani, "On solutions of a quadratic integral equation of Hammerstein type," *Mathematical and Computer Modelling*, vol. 43, no. 1-2, pp. 97–104, 2006.

[9] J. Banaś and B. Rzepka, "Monotonic solutions of a quadratic integral equation of fractional order," *Journal of Mathematical Analysis and Applications*, vol. 332, no. 2, pp. 1371–1379, 2007.

[10] M. A. Darwish and J. Henderson, "Nondecreasing solutions of a quadratic integral equation of Urysohn-Stieltjes type," *The Rocky Mountain Journal of Mathematics*, vol. 42, no. 2, pp. 545–566, 2012.

[11] F. S. De Blasi, "On a property of the unit sphere in a Banach space," vol. 21, no. 3-4, pp. 259–262, 1977.

[12] K. Deimling, *Nonlinear Functional Analysis*, Springer, Berlin, Germany, 1985.

[13] A. M. A. El-Sayed, M. M. Saleh, and E. A. A. Ziada, "Numerical and analytic solution for nonlinear quadratic integral equations," *Mathematical Sciences Research Journal*, vol. 12, no. 8, pp. 183–191, 2008.

[14] A. M. A. El-Sayed and H. H. G. Hashem, "Carathéodory type theorem for a nonlinear quadratic integral equation," *Mathematical Sciences Research Journal*, vol. 12, no. 4, pp. 71–95, 2008.

[15] A. M. A. El-Sayed and H. H. G. Hashem, "Integrable and continuous solutions of a nonlinear quadratic integral equation," *Electronic Journal of Qualitative Theory of Differential Equations*, vol. 25, pp. 1–10, 2008.

[16] A. M. A. El-Sayed and H. H. G. Hashem, "Integrable solutions for quadratic Hammerstein and quadratic Urysohn functional integral equations," *Commentationes Mathematicae*, vol. 48, no. 2, pp. 199–207, 2008.

[17] A. M. A. El-Sayed and H. H. G. Hashem, "Monotonic solutions of functional integral and differential equations of fractional order," *Electronic Journal of Qualitative Theory of Differential Equations*, no. 7, pp. 1–8, 2009.

[18] A. M. A. El-Sayed and H. H. G. Hashem, "Monotonic positive solution of a nonlinear quadratic functional integral equation," *Applied Mathematics and Computation*, vol. 216, no. 9, pp. 2576–2580, 2010.

[19] A. M. A. El-Sayed, H. H. G. Hashem, and E. A. A. Ziada, "Picard and Adomian methods for quadratic integral equation," *Computational & Applied Mathematics*, vol. 29, no. 3, pp. 447–463, 2010.

[20] A. M. A. El-Sayed, H. H. G. Hashem, and Y. M. Y. Omar, "Positive continuous solution of a quadratic integral equation of fractional orders," *Mathematical Sciences Letters*, vol. 2, no. 1, pp. 19–27, 2013.

[21] H. H. G. Hashem and M. S. Zaki, "Carathèodory theorem for quadratic integral equations of Erdyéli-Kober type," *Journal of Fractional Calculus and Its Applications*, vol. 4, no. 5, pp. 1–8, 2013.

[22] H. O. Bakodah, "The appearance of noise terms in modified adomian decomposition method for quadratic integral equations," *The American Journal of Computational Mathematics*, vol. 2, pp. 125–129, 2012.

[23] W. G. El-Sayed and B. Rzepka, "Nondecreasing solutions of a quadratic integral equation of Urysohn type," *Computers & Mathematics with Applications*, vol. 51, no. 6-7, pp. 1065–1074, 2006.

[24] S. Chandrasekhar, *Radiative Transfar*, Dover, New York, NY, USA, 1960.

[25] M. Cichoń and M. A. Metwali, "On quadratic integral equations in Orlicz spaces," *Journal of Mathematical Analysis and Applications*, vol. 387, no. 1, pp. 419–432, 2012.

[26] H. A. H. Salem, "On the quadratic integral equations and their applications," *Computers & Mathematics with Applications*, vol. 62, no. 8, pp. 2931–2943, 2011.

[27] J. Appell and E. De Pascale, "Su. alcuni parameteri connesi con la misuradi non compacttezza di Hausdorff in spazi di functioni misurablili," *Bollettino Unione Matematica Italiana*, vol. 6, no. 3, pp. 497–515, 1984.

Some Common Fixed Point Results in Rectangular Metric Spaces

Muhammad Arshad,[1] Jamshaid Ahmad,[2] and Erdal Karapınar[3]

[1] *Department of Mathematics, International Islamic University, H-10, Islamabad 44000, Pakistan*
[2] *Department of Mathematics, COMSATS Institute of Information Technology, Chak Shahzad, Islamabad 44000, Pakistan*
[3] *Department of Mathematics, Atilim University, İncek, 06836 Ankara, Turkey*

Correspondence should be addressed to Jamshaid Ahmad; jamshaid_jasim@yahoo.com

Academic Editor: Ahmed Zayed

We obtain sufficient conditions for the existence of unique common fixed point of $(\psi-\phi)$-weakly contractive mappings on complete rectangular metric spaces. In the process, we generalize several fixed point results from the literature. We also give an example to illustrate our work.

1. Introduction

Fixed point theorems are very important tools in nonlinear functional analysis. Banach Contraction Mapping Principle is the most frequently cited fixed point theorem in the literature. It asserts that if X is a complete metric space and $T : X \to X$ is a contraction, that is, there exists $\lambda \in [0, 1)$ such that for all $x, y \in X$,

$$d(Tx, Ty) \le \lambda d(x, y), \tag{1}$$

then T has a unique fixed point. The contraction definition (1) implies that T is uniformly continuous, which is a very strong condition. It is quite natural to ask whether the inequality (1) can be replaced with another inequality which does not force T to be continuous. This question was answered affirmatively by Kannan [1]. A self-mapping $T : X \to X$ has a unique fixed point in a complete metric space (X, d) if there are nonnegative real numbers α, β with $\alpha + \beta < 1$ such that the following inequality is satisfied for all $x, y \in X$:

$$d(Tx, Ty) \le \alpha d(x, Tx) + \beta d(y, Ty). \tag{2}$$

In 2000, Branciari [2] introduced the notion of a generalized (rectangular) metric space where the triangle inequality of a metric space was replaced by another inequality, the so-called rectangular inequality. In this paper [2], the author also extended the celebrated Banach Contraction Mapping Principle in the context of generalized metric spaces. Later, Azam and Arshad [3] obtained sufficient conditions for existence of a unique fixed point of Kannan type mappings in the framework of generalized/rectangular metric spaces. Subsequently, Azam et al. [4] proved an analog of Banach Contraction Principle in the setting of rectangular cone metric spaces. Following this trend, a number of authors focused on rectangular metric spaces and proved the existence and uniqueness of a fixed point for certain type of mappings (see e.g., [5–14]).

Recently, Di Bari and Vetro [15] obtained some common fixed point theorems for mappings satisfying a $(\psi-\phi)$-weakly contractive condition in rectangular metric spaces. In this paper, we prove several fixed point results in rectangular metric spaces that can be considered as a continuation of [15].

We recall some basic definitions and necessary results on the topic in the literature.

Definition 1. Let X be a nonempty set and let $d : X \times X \to [0, +\infty)$ be a mapping such that for all $x, y \in X$ and for all distinct points $u, v \in X$, each of them different from x and y, one has

$$d(x, y) = 0 \quad \text{iff } x = y, \tag{RM1}$$

$$d(x,y) = d(y,x), \tag{RM2}$$

$$d(x,y) \le d(x,u) + d(u,v) + d(v,y) \quad \text{(the rectangular inequality)}. \tag{RM3}$$

Then, the map d is called rectangular (generalized) metric. The pair (X,d) is called a rectangular (generalized) metric space.

To avoid confusion, we prefer to use the term "rectangular metric space" for the spaces under consideration in this paper, because there are some other spaces that are also called generalized metric spaces. We abbreviate a rectangular metric spaces with RMS.

Definition 2 (see [2]). Let (X,d) be a RMS.

(i) A sequence $\{x_n\}$ is called RMS convergent to $x \in X$ if and only if $d(x_n, x) \to 0$ as $n \to +\infty$. In this case, we use the notation $x_n \to x$.

(ii) A sequence $\{x_n\}$ in X is called a RMS Cauchy if and only if for each $\epsilon > 0$, there exists a natural number $N(\epsilon)$ such that $d(x_n, x_m) < \epsilon$ for all $n > m > N(\epsilon)$.

(iii) A RMS (X,d) is called a RMS complete if every RMS Cauchy sequence is RMS convergent in X.

Definition 3. Let F and g be self-mappings of a nonempty set X.

(i) A point $x \in X$ is said to be a common fixed point of F and g if $x = Fx = gx$.

(ii) A point $x \in X$ is called a coincidence point of F and g if $Fx = gx$. And if $w = Fx = gx$, then w is said to be a point of coincidence of F and g.

(iii) The mappings $F, g : X \to X$ are said to be weakly compatible if they commute at their coincidence point that is, $Fgx = gFx$ whenever $gx = Fx$.

Lemma 4. *Let X be a nonempty set. Suppose that the mappings $F, g : X \to X$ have a unique coincidence point z in X. If F and g are weakly compatible, then F and g have a unique common fixed point.*

Proof. Let $z \in X$ be the coincidence point of $F, g : X \to X$, that is,

$$Fz = gz = t. \tag{3}$$

Since F and g are weakly compatible, we observe that

$$\begin{aligned} Fz = gz &\Longrightarrow Fgz = gFz \\ &\Longleftrightarrow Ft = gt. \end{aligned} \tag{4}$$

$F, g : X \to X$ have a unique coincidence point, then $z = t$. Hence, we have $Fz = gz = z$ by (3). □

Let Ψ denote all functions $\phi : [0,\infty) \to [0,\infty)$ such that

(a) ϕ is continuous,

(b) $\phi(t) = 0$ if and only if $t = 0$.

2. Main Result

We start this section with the following theorem.

Theorem 5. *Let (X,d) be a Hausdorff RMS and let $F, g : X \to X$ be self-mappings such that $FX \subset gX$. Assume that (gX, d) is a complete RMS. Suppose that the following condition holds:*

$$\psi(d(Fx,Fy)) \le \psi(M(gx,gy)) - \phi(M(gx,gy)), \tag{5}$$

for all $x, y \in X$ and $\psi, \phi \in \Psi$, where ψ is nondecreasing and

$$M(gx,gy) = \max\{d(gx,gy), d(gx,Fx), d(gy,Fy)\}. \tag{6}$$

Then F and g have a unique coincidence point in X. Moreover, if F and g are weakly compatible, then F and g have a unique common fixed point.

Proof. We first prove that the coincidence point of g and F is unique if it exists. Let z and w be coincidence points of g and F. Thus, there exists some $x, y \in X$ such that $w = Fx = gx$ and $z = Fy = gy$. By (5), we derive that

$$\begin{aligned} \psi(d(w,z)) &= \psi(d(Fx,Fy)) \\ &\le \psi(M(gx,gy)) - \phi(M(gx,gy)), \end{aligned} \tag{7}$$

where

$$\begin{aligned} M(gx,gy) &= \max\{d(gx,gy), d(gx,Fx), d(gy,Fy)\} \\ &= d(gx,gy) = d(w,z). \end{aligned} \tag{8}$$

Thus, we conclude that $z = w$ by (7).

Remember that g and F are weakly compatible. Since z is the unique coincidence point of g and F, the point z is the unique common fixed point of g and F by Lemma 4.

Now, we will prove the existence of a coincidence point of g and F. Let x_0 be an arbitrary point. Since $FX \subset gX$, we define two iterative sequences $\{x_n\}$ and $\{y_n\}$ in X as follows:

$$y_n = gx_{n+1} = Fx_n \tag{9}$$

for all $n = 0, 1, 2, \ldots$. If $y_n = y_{n+1}$ then clearly F and g have a coincidence point in X. Indeed, $y_n = gx_{n+1} = Fx_n = gx_{n+2} = Fx_{n+1} = y_{n+1}$ and x_{n+1} is the desired point. Thus, we assume that $y_n \ne y_{n+1}$, that is, $d(y_n, y_{n+1}) > 0$ for all $n = 0, 1, \ldots$. Moreover, if $y_n = Fx_n = Fx_{n+p} = y_{n+p}$, then we choose $x_{n+p+1} = x_{n+1}$, for all $n \ge 0$.

We assert that

$$\lim_{n\to\infty} d(y_n, y_{n+1}) = 0, \qquad \lim_{n\to\infty} d(y_n, y_{n+2}) = 0. \tag{10}$$

Now from (5), we have

$$\begin{aligned} \psi(d(y_n, y_{n+1})) = \psi(d(Fx_n, Fx_{n+1})) &\le \psi(M(gx_n, gx_{n+1})) \\ &\quad - \phi(M(gx_n, gx_{n+1})), \end{aligned} \tag{11}$$

where

$$\begin{aligned} M(gx_n, gx_{n+1}) &= \max\{d(gx_n, gx_{n+1}), d(gx_n, Fx_n), \\ &\qquad d(gx_{n+1}, Fx_{n+1})\} \\ &= \max\{d(gx_n, gx_{n+1}), d(gx_n, gx_{n+1}), \\ &\qquad d(gx_{n+1}, gx_{n+2})\} \\ &= \max\{d(y_{n-1}, y_n), d(y_{n-1}, y_n), d(y_n, y_{n+1})\} \\ &= \max\{d(y_{n-1}, y_n), d(y_n, y_{n+1})\}. \end{aligned} \tag{12}$$

If $M(gx_n, gx_{n+1}) = d(y_n, y_{n+1})$, then we have

$$\psi(d(y_n, y_{n+1})) \leq \psi(d(y_n, y_{n+1})) - \phi(d(y_n, y_{n+1})), \tag{13}$$

which implies that $\phi(d(y_n, y_{n+1})) = 0$, and hence $d(y_n, y_{n+1}) = 0$. Then $y_n = y_{n+1}$, which contradicts with the initial assumption. Thus, we have $M(gx_n, gx_{n+1}) = d(y_{n-1}, y_n)$, and hence

$$\begin{aligned} \psi(d(y_n, y_{n+1})) &\leq \psi(d(y_{n-1}, y_n)) - \phi(d(y_{n-1}, y_n)) \\ &\leq \psi(d(y_{n-1}, y_n)). \end{aligned} \tag{14}$$

Since ψ is nondecreasing, then $d(y_n, y_{n+1}) \leq d(y_{n-1}, y_n)$ for all $n \geq 0$, that is, the sequence $\{d(y_n, y_{n+1})\}$ is nonincreasing and bounded below. Hence, it converges to a positive number, say $r > 0$. Taking the limit as $n \to \infty$ in (14), we get

$$\psi(r) \leq \psi(r) - \phi(r), \tag{15}$$

which leads to $\phi(r) = 0$, and hence $r = 0$. Thus,

$$\lim_{n\to\infty} d(y_n, y_{n+1}) = 0. \tag{16}$$

In this step, we show that the second limit in (5) is also 0. To prove this claim, we set $x = x_{n-1}$ and $y = x_{n+1}$ in (5) and get that

$$\begin{aligned} &\psi(d(y_n, y_{n+2})) \\ &\quad = \psi(d(Fx_n, Fx_{n+2})) \leq \psi(M(gx_n, gx_{n+2})) \\ &\qquad - \phi(M(gx_n, gx_{n+2})), \end{aligned} \tag{17}$$

where

$$\begin{aligned} &M(gx_n, gx_{n+2}) \\ &\quad = \max\{d(gx_n, gx_{n+2}), d(gx_n, Fx_n), d(gx_{n+2}, Fx_{n+2})\} \\ &\quad = \max\{d(gx_n, gx_{n+2}), d(gx_n, gx_{n+1}), d(gx_{n+2}, gx_{n+3})\} \\ &\quad = \max\{d(y_{n-1}, y_{n+1}), d(y_{n-1}, y_n), d(y_{n+1}, y_{n+2})\}. \end{aligned} \tag{18}$$

We consider all possible cases for $M(gx_n, gx_{n+2})$. If $M(gx_n, gx_{n+2}) = d(y_{n+1}, y_{n+2})$, then we have

$$\psi(d(y_n, y_{n+2})) \leq \psi(d(y_{n+1}, y_{n+2})) - \phi(d(y_{n+1}, y_{n+2})). \tag{19}$$

Letting $n \to \infty$ in (19), the right hand side of (19) tends to 0. Hence, $\lim_{n\to\infty}\psi(d(y_n, y_{n+2})) = 0$. Since, ψ is continuous, we find that $\lim_{n\to\infty} d(y_n, y_{n+2}) = 0$. For the case $M(gx_n, gx_{n+2}) = d(y_{n-1}, y_n)$, we get analogously $\lim_{n\to\infty} d(y_n, y_{n+2}) = 0$.

Let us consider the last case, that is, $M(gx_n, gx_{n+2}) = d(y_{n-1}, y_{n+1})$. The inequality (17) turns into

$$\begin{aligned} 0 &\leq \psi(d(y_n, y_{n+2})) \leq \psi(d(y_{n-1}, y_{n+1})) - \phi(d(y_{n-1}, y_{n+1})) \\ &\leq \psi(d(y_{n-1}, y_{n+1})). \end{aligned} \tag{20}$$

Therefore, the sequence $\{d(y_n, y_{n+2})\}$ is non-increasing and bounded below. Hence, the sequence $\{d(y_n, y_{n+2})\}$ converges to a number, $s \geq 0$. Taking limit as $n \to \infty$ in (20), we get

$$0 \leq \psi(d(s)) \leq \psi(s) - \phi(s), \tag{21}$$

which implies that $\phi(s) = 0$, and hence $s = 0$. In other words,

$$\lim_{n\to\infty} d(y_n, y_{n+2}) = 0. \tag{22}$$

Suppose that $y_n \neq y_m$ for all $m \neq n$ and prove that $\{y_n\}$ is a RMS Cauchy sequence. If possible, let $\{y_n\}$ be not a Cauchy sequence. Then there exists $\epsilon > 0$ for which we can find subsequences $\{y_{m_k}\}$ and $\{y_{n_k}\}$ of $\{y_n\}$ with $n_k > m_k \geq k$ such that

$$d(y_{m_k}, y_{n_k}) \geq \epsilon. \tag{23}$$

Furthermore, corresponding to m_k, we can choose n_k in such a way that it is the smallest integer with $n_k > m_k$ and satisfying (23). Then,

$$d(y_{m_k}, y_{n_k-1}) < \epsilon. \tag{24}$$

Using (23), (24), and the rectangular inequality (RM3), we have

$$\begin{aligned} \epsilon &\leq d(y_{m_k}, y_{n_k}) \\ &\leq d(y_{n_k}, y_{n_k-2}) + d(y_{n_k-2}, y_{n_k-1}) + d(y_{n_k-1}, y_{m_k}) \\ &\leq d(y_{n_k}, y_{n_k-2}) + d(y_{n_k-2}, y_{n_k-1}) + \epsilon. \end{aligned} \tag{25}$$

Taking limit as $k \to \infty$ in (23) and using (16), (22) we get

$$\lim_{k\to\infty} d(y_{m_k}, y_{n_k}) = \epsilon. \tag{26}$$

Again, using the rectangular inequality (RM3), we obtain

$$\begin{aligned} &d(y_{n_k}, y_{m_k}) - d(y_{m_k}, y_{m_k-1}) - d(y_{n_k-1}, y_{n_k}) \\ &\quad \leq d(y_{n_k-1}, y_{m_k-1}) \\ &\quad \leq d(y_{n_k-1}, y_{n_k}) + d(y_{n_k}, y_{m_k}) + d(y_{m_k}, y_{m_k-1}). \end{aligned} \tag{27}$$

Letting $k \to \infty$ in (27), and by using (16) and (22) we get

$$\lim_{k\to\infty} d(y_{n_k-1}, y_{m_k-1}) = \epsilon. \tag{28}$$

Now, we substitute $x = x_{n_k}$ and $y = x_{m_k}$ in (5). Consider

$$\begin{aligned}\psi\left(d\left(y_{n_k}, y_{m_k}\right)\right) &= \psi\left(d\left(Fx_{n_k}, Fx_{m_k}\right)\right) \\ &\leq \psi\left(M\left(gx_{n_k}, gx_{m_k}\right)\right) \\ &\quad - \phi\left(M\left(gx_{n_k}, gx_{m_k}\right)\right)\end{aligned} \tag{29}$$

where

$$\begin{aligned}M\left(gx_{n_k}, gx_{m_k}\right) &= \max\left\{d\left(gx_{n_k}, gx_{m_k}\right), d\left(gx_{n_k}, Fx_{n_k}\right),\right. \\ &\qquad \left. d\left(gx_{m_k}, Fx_{m_k}\right)\right\} \\ &= \max\left\{d\left(y_{n_k-1}, y_{m_k-1}\right), d\left(y_{n_k-1}, y_{n_k}\right),\right. \\ &\qquad \left. d\left(y_{m_k-1}, y_{m_k}\right)\right\}.\end{aligned} \tag{30}$$

Clearly, as $k \to \infty$, we have

$$M\left(gx_{n_k}, gx_{m_k}\right) \longrightarrow \max\{\epsilon, 0, 0\} = \epsilon. \tag{31}$$

Then letting $k \to \infty$ in (29), we have

$$0 \leq \psi(\epsilon) \leq \psi(\epsilon) - \phi(\epsilon). \tag{32}$$

This implies that

$$\phi(\epsilon) = 0, \quad \text{hence } \epsilon = 0, \tag{33}$$

which contradicts the fact that $\epsilon > 0$. Thus, $\{y_n\}$ is a RMS Cauchy sequence. Since (gX, d) is RMS complete, there exists $z \in gX$ such that $y_n \to z$ as $n \to \infty$. Let $y \in X$ such that $gy = z$. Applying the inequality (5), with $x = x_n$, we obtain

$$\psi(d(Fx_n, Fy)) \leq \psi(M(gx_n, gy)) - \phi(M(gx_n, gy)), \tag{34}$$

where

$$\begin{aligned}M(gx_n, gy) &= \max\{d(gx_n, gy), d(gx_n, Fx_n), d(gy, Fy)\} \\ &= \max\{d(gx_n, gy), d(gx_n, gx_{n+1}), \\ &\qquad d(gy, Fy)\}.\end{aligned} \tag{35}$$

Now, if $M(gx_n, gy) = d(gx_n, gy)$ or $M(gx_n, gy) = d(gx_n, gx_{n+1})$, we have

$$\begin{aligned}&d(Fx_n, Fy) \leq d(gx_n, gy) \quad \text{or} \\ &d(Fx_n, Fy) \leq d(gx_n, gx_{n+1}),\end{aligned} \tag{36}$$

since ψ is nondecreasing. In either case, letting $n \to \infty$, we get $gx_{n+1} = Fx_n \to Fy$. Since X is Hausdorff, we deduce that $gy = Fy$. If, on the other hand, $M(gx_n, gy) = d(gy, Fy)$, then taking limit as $n \to \infty$ in

$$\psi(d(Fx_n, Fy)) \leq \psi(d(gy, Fy)) - \phi(d(gy, Fy)), \tag{37}$$

we get $\phi(d(gy, Fy)) = 0$, hence $d(gy, Fy) = 0$, that is, $gy = Fy$. Let $z = gy = Fy$. Then z is a point of coincidence of F and g. Suppose that there exists $n, p \in \mathbb{N}$ such that $y_n = y_{n+p}$, we can choose in such a way that it is the smallest positive integer satisfying $y_n = y_{n+p}$. We aim to prove that $p = 1$, then

$$gx_{n+1} = Fx_n = Fx_{n+1} = y_{n+1}, \tag{38}$$

and so y_{n+1} is a point of coincidence of g and F. Assume that $p > 1$. This implies that $d(y_{n+p-1}, y_{n+p}) > 0$. Now we have

$$\begin{aligned}&\psi(d(y_n, y_{n+1})) \\ &\quad = \psi\left(d\left(y_{n+p}, y_{n+p+1}\right)\right) = \psi\left(d\left(Fx_{n+p}, Fx_{n+p+1}\right)\right) \\ &\quad \leq \psi\left(M\left(gx_{n+p}, gx_{n+p+1}\right)\right) - \phi\left(M\left(gx_{n+p}, gx_{n+p+1}\right)\right),\end{aligned} \tag{39}$$

where

$$\begin{aligned}&M\left(gx_{n+p}, gx_{n+p+1}\right) \\ &\quad = \max\left\{d\left(gx_{n+p}, gx_{n+p+1}\right), d\left(gx_{n+p}, Fx_{n+p}\right),\right. \\ &\qquad \left. d\left(gx_{n+p+1}, Fx_{n+p+1}\right)\right\} \\ &\quad = \max\left\{d\left(gx_{n+p}, gx_{n+p+1}\right), d\left(gx_{n+p}, gx_{n+p+1}\right),\right. \\ &\qquad \left. d\left(gx_{n+p+1}, gx_{n+p+2}\right)\right\} \\ &\quad = \max\left\{d\left(gx_{n+p}, gx_{n+p+1}\right), d\left(gx_{n+p+1}, gx_{n+p+2}\right)\right\} \\ &\quad = \max\left\{d\left(y_{n+p-1}, y_{n+p}\right), d\left(y_{n+p}, y_{n+p+1}\right)\right\} \\ &\quad = \max\{d(y_{n-1}, y_n), d(y_n, y_{n+1})\}.\end{aligned} \tag{40}$$

If $M(gx_{n+p}, gx_{n+p+1}) = d(y_n, y_{n+1})$, then we have

$$\psi(d(y_n, y_{n+1})) = \psi\left(d\left(y_{n+p}, y_{n+p+1}\right)\right) < \psi(d(y_n, y_{n+1})), \tag{41}$$

which is a contradiction. Thus $p = 1$. □

Corollary 6. *Let (X, d) be a Hausdorff and complete RMS and let $F, g : X \to X$ be self-mappings such that $FX \subseteq gX$ satisfying*

$$d(Fx, Fy) \leq k \max\{d(gx, gy), d(gx, Fx), d(gy, Fy)\} \tag{42}$$

for all $x, y \in X$ and $0 \leq k < 1$. Then F and g have a unique common fixed point in X.

Proof. Let $\psi(t) = t$ and $\phi(t) = (1-k)t$. Then by Theorem 5, F and g have a unique common fixed point. □

Corollary 7. *Let (X, d) be a Hausdorff and complete RMS and let $F, g : X \to X$ be self-mappings such that $FX \subseteq gX$ satisfying*

$$d(Fx, Fy) \leq k(d(gx, gy) + d(gx, Fx) + d(gy, Fy)), \tag{43}$$

for all $x, y \in X$ and $0 \leq k < 1$. Then F and g have a unique common fixed point in X.

Proof. It is obvious that

$$d(gx, gy) + d(gx, Fx) + d(gy, Fy) \leq 3\max\{d(gx, gy), d(gx, Fx), d(gy, Fy)\}. \quad (44)$$

Let $\psi(t) = t$ and $\phi(t) = (1 - 3k)t$. Then by Theorem 5, F and g have a unique common fixed point. □

Corollary 8. *Let (X, d) be a Hausdorff and complete RMS and let $F, g : X \to X$ be self-mappings such that $FX \subseteq gX$ satisfying*

$$d(Fx, Fy) \leq M(gx, gy) - \phi(M(gx, gy)), \quad (45)$$

for all $x, y \in X$ where

$$M(gx, gy) = \max\{d(gx, gy), d(gx, Fx), d(gy, Fy)\}. \quad (46)$$

Then F and g have a unique common fixed point.

Proof. Let $\psi(t) = t$. Then by Theorem 5, F and g have a unique common fixed point. □

Theorem 9. *Let (X, d) be a Hausdorff and complete RMS and let $F, g : X \to X$ be self-mappings such that $FX \subseteq gX$ satisfying*

$$\psi(d(Fx, Fy)) \leq \psi(M(gx, gy)) - \phi(M(gx, gy)), \quad (47)$$

for all $x, y \in X$ and $\psi, \phi \in \Psi$, where ψ is nondecreasing and

$$M(gx, gy) = \max\left\{d(gx, gy), d(gy, Fy)\frac{1 + d(gx, Fx)}{1 + d(gy, gy)}\right\}. \quad (48)$$

Then F and g have a unique common fixed point.

Proof. Let $x_0 \in X$ be an arbitrary point. Since $FX \subseteq gX$, we define the sequence $\{x_n\} \subset X$ as follows $gx_n = Fx_{n-1}$ for all $n \geq 1$. Assume that $gx_n \neq gx_{n+1} = Fx_n$ for all $n \geq 0$. Now from (5),we get

$$\psi(d(gx_n, gx_{n+1})) = \psi(d(Fx_{n-1}, Fx_n)) \leq \psi(M(gx_{n-1}, gx_n)) - \phi(M(gx_{n-1}, gx_n)), \quad (49)$$

where

$$M(gx_{n-1}, gx_n) = \max\left\{d(gx_{n-1}, gx_n), d(gx_n, Fx_n)\frac{1 + d(gx_{n-1}, Fx_{n-1})}{1 + d(gx_{n-1}, gx_n)}\right\} \quad (50)$$

$$M(gx_{n-1}, gx_n) = \max\{d(gx_{n-1}, gx_n), d(gx_n, gx_{n+1})\}.$$

The rest of the proof is the same as the proof of Theorem 5. □

By Λ, we denote the class of functions $f : [0, \infty) \to [0, \infty)$ satisfying the following.

(a) f is Lebesgue integrable function on each compact subset of $[0, \infty)$,

(b) $\int_0^{\varepsilon} f(s)ds > 0$ for any $\varepsilon > 0$.

Theorem 10. *Let (X, d) be a Hausdorff RMS and let $F, g : X \to X$ be self-mappings such that $FX \subset gX$. Assume that (gX, d) is a complete RMS and that the following condition holds:*

$$\int_0^{d(Fx,Fy)} f(r)\,dr \leq \int_0^{M(gx,gy)} f(r)\,dr - \int_0^{M(gx,gy)} h(r)\,dr \quad (51)$$

for all $x, y \in X$ and $f, h \in \Lambda$ where

$$M(gx, gy) = \max\{d(gx, gy), d(gx, Fx), d(gy, Fy)\}. \quad (52)$$

Then F and g have a unique common fixed point.

Proof. Let $\psi(r) = \int_0^r f(v)dv$ and $\phi(r) = \int_0^r h(v)dv$. Then ψ and ϕ are function in Ψ. By Theorem 5, F and g have a unique common fixed point. □

Theorem 11. *Let (X, d) be a Hausdorff RMS and let $F, g : X \to X$ be self-mappings such that $FX \subset gX$. Assume that (gX, d) is a complete RMS and that the following condition holds:*

$$\int_0^{d(Fx,Fy)} f(s)\,ds \leq \lambda \int_0^{M(gx,gy)} f(s)\,ds \quad (53)$$

for all $x, y \in X$ and $f \in \Lambda$ and some $0 \leq \lambda < 1$, where

$$M(gx, gy) = \max\{d(gx, gy), d(gx, Fx), d(gy, Fy)\}. \quad (54)$$

Then F and g have a unique common fixed point.

Proof. Let $h(r) = (1 - \lambda)f(r)$. Then by Theorem 10, F and g have a unique common fixed point. □

Theorem 12. *Let (X, d) be a Hausdorff RMS and let $F, g : X \to X$ be a self-mappings such that $FX \subset gX$. Assume that (gX, d) is a complete RMS and that the following condition holds:*

$$\int_0^{d(Fx,Fy)} f(r)\,dr \le \int_0^{M(gx,gy)} f(r)\,dr - \int_0^{M(gx,gy)} h(r)\,dr \tag{55}$$

for all $x, y \in X$ and $f,\ h \in \Lambda$ where

$$M(gx, gy) = \max\left\{d(gx, gy), d(gy, Fy)\frac{1 + d(gx, Fx)}{1 + d(gy, gy)}\right\}. \tag{56}$$

Then F and g have a unique common fixed point.

Proof. Let $\psi(r) = \int_0^r f(v)dv$ and $\phi(r) = \int_0^r h(v)dv$. Then ψ and ϕ are function in Ψ. By Theorem 9, F and g have a unique common fixed point. □

Example 13. Let $X = A \cup B$, where $A = \{1/2, 2/3, 3/4, 4/5\}$ and $B = [1, 3]$. Define the generalized metric d on X as follows:

$$d\left(\frac{1}{2}, \frac{2}{3}\right) = d\left(\frac{3}{4}, \frac{4}{5}\right) = 0.2, \tag{57}$$

$$d\left(\frac{1}{2}, \frac{4}{5}\right) = d\left(\frac{2}{3}, \frac{3}{4}\right) = 0.3, \tag{58}$$

$$d\left(\frac{1}{2}, \frac{3}{4}\right) = d\left(\frac{2}{3}, \frac{4}{5}\right) = 0.6, \tag{59}$$

$$\left(\frac{1}{2}, \frac{1}{2}\right) = d\left(\frac{2}{3}, \frac{2}{3}\right) = d\left(\frac{3}{4}, \frac{3}{4}\right) = d\left(\frac{4}{5}, \frac{4}{5}\right) = 0, \tag{60}$$

$$d(x, y) = |x - y| \quad \text{if } x, y \in B \text{ or } x \in A,\ y \in B \text{ or } x \in B, y \in A. \tag{61}$$

It is easy to show that d does not satisfy the triangle inequality on A. Indeed,

$$0.6 = d\left(\frac{1}{2}, \frac{3}{4}\right) \ge d\left(\frac{1}{2}, \frac{2}{3}\right) + d\left(\frac{2}{3}, \frac{3}{4}\right) = 0.2 + 0.3 = 0.5. \tag{62}$$

Thus (RM3) holds, so d is a rectangular metric. Notice that $(X|_B, d)$ is usual metric space, and hence it is Hausdorff. On the other hand, each singleton is closed and open in $(X|_A, d)$, and hence (X, d) is Hausdorff rectangular metric space.

Let $F, g : X \to X$ be defined as follows:

$$Fx = \begin{cases} \frac{4}{5} & \text{if } x \in [1,3], \\ \frac{3}{4} & \text{if } x \in \left\{\frac{1}{2}, \frac{2}{3}, \frac{3}{4}\right\}, \\ \frac{2}{3} & \text{if } x = \frac{4}{5}, \end{cases}$$
$$gx = \begin{cases} \frac{2}{3} & \text{if } x \in [1,3], \\ \frac{3}{4} & \text{if } x \in \left\{\frac{1}{2}, \frac{3}{4}\right\}, \\ \frac{4}{5} & \text{if } x = \frac{2}{3}, \\ \frac{1}{2} & \text{if } x = \frac{4}{5}. \end{cases} \tag{63}$$

Define $\psi(t) = t$ and $\phi(t) = t/3$. Then F and g satisfy the condition of Theorem 5 and have a unique common fixed point of X, that is, $x = 3/4$.

Acknowledgment

The authors thank the referees for their appreciation, valuable comments, and suggestions.

References

[1] R. Kannan, "Some results on fixed points," *Bulletin of the Calcutta Mathematical Society*, vol. 60, pp. 71–76, 1968.

[2] A. Branciari, "A fixed point theorem of Banach-Caccioppoli type on a class of generalized metric spaces," *Publicationes Mathematicae Debrecen*, vol. 57, no. 1-2, pp. 31–37, 2000.

[3] A. Azam and M. Arshad, "Kannan fixed point theorem on generalized metric spaces," *Journal of Nonlinear Sciences and Its Applications*, vol. 1, no. 1, pp. 45–48, 2008.

[4] A. Azam, M. Arshad, and I. Beg, "Banach contraction principle on cone rectangular metric spaces," *Applicable Analysis and Discrete Mathematics*, vol. 3, no. 2, pp. 236–241, 2009.

[5] C. M. Chen and C. H. Chen, "Periodic points for the weak contraction mappings in complete generalized metric spaces," *Fixed Point Theory and Applications*, vol. 2012, article 79, 2012.

[6] P. Das and L. K. Dey, "A fixed point theorem in a generalized metric space," *Soochow Journal of Mathematics*, vol. 33, no. 1, pp. 33–39, 2007.

[7] P. Das and L. K. Dey, "Fixed point of contractive mappings in generalized metric spaces," *Mathematica Slovaca*, vol. 59, no. 4, pp. 499–504, 2009.

[8] E. Karapinar, "Weak ϕ-contraction on partial metric spaces and existence of fixed points in partially ordered sets," *Mathematica Aeterna*, vol. 1, no. 3-4, pp. 237–244, 2011.

[9] I. M. Erhan, E. Karapinar, and T. Sekulic, "Fixed points of $(\psi-\phi)$ contractions on rectangular metric spaces," *Fixed Point Theory and Applications*, vol. 2012, article 138, 2012.

[10] H. Lakzian and B. Samet, "Fixed points for $(\psi - \phi)$-weakly contractive mappings in generalized metric spaces," *Applied Mathematics Letters*, vol. 25, no. 5, pp. 902–906, 2012.

[11] D. Miheţ, "On Kannan fixed point principle in generalized metric spaces," *Journal of Nonlinear Science and Its Applications*, vol. 2, no. 2, pp. 92–96, 2009.

[12] B. Samet, "A fixed point theorem in a generalized metric space for mappings satisfying a contractive condition of integral type," *International Journal of Mathematical Analysis*, vol. 3, no. 25–28, pp. 1265–1271, 2009.

[13] B. Samet, "Discussion on "A fixed point theorem of Banach-Caccioppoli type on a class of generalized metric spaces" by A. Branciari," *Publicationes Mathematicae Debrecen*, vol. 76, no. 3-4, pp. 493–494, 2010.

[14] I. R. Sarma, J. M. Rao, and S. S. Rao, "Contractions over generalized metric spaces," *Journal of Nonlinear Science and Its Applications*, vol. 2, no. 3, pp. 180–182, 2009.

[15] C. Di Bari and P. Vetro, "Common fixed points in generalized metric spaces," *Applied Mathematics and Computation*, vol. 218, no. 13, pp. 7322–7325, 2012.

10

Common Fixed Points in a Partially Ordered Partial Metric Space

Daniela Paesano and Pasquale Vetro

Dipartimento di Matematica e Informatica, Università Degli Studi di Palermo, Via Archirafi, 34-90123 Palermo, Italy

Correspondence should be addressed to Pasquale Vetro; vetro@math.unipa.it

Academic Editor: Harumi Hattori

In the first part of this paper, we prove some generalized versions of the result of Matthews in (Matthews, 1994) using different types of conditions in partially ordered partial metric spaces for dominated self-mappings or in partial metric spaces for self-mappings. In the second part, using our results, we deduce a characterization of partial metric 0-completeness in terms of fixed point theory. This result extends the Subrahmanyam characterization of metric completeness.

1. Introduction

In the mathematical field of domain theory, attempts were made in order to equip semantics domain with a notion of distance. In particular, Matthews [1] introduced the notion of a partial metric space as a part of the study of denotational semantics of data for networks, showing that the contraction mapping principle can be generalized to the partial metric context for applications in program verification. Moreover, the existence of several connections between partial metrics and topological aspects of domain theory has been lately pointed by other authors as O'Neill [2], Bukatin and Scott [3], Bukatin and Shorina [4], Romaguera and Schellekens [5], and others (see also [6–14] and the references therein).

After the result of Matthews [1], the interest for fixed point theory developments in partial metric spaces has been constantly growing, and many authors presented significant contributions in the directions of establishing partial metric versions of well-known fixed point theorems for the existence of fixed points, common fixed points, and coupled fixed points in classical metric spaces (see e.g., [15, 16]). Obviously, we cannot cite all these papers but we give only a partial list [17–49].

Recently, Romaguera [50] proved that a partial metric space (X, d) is 0-complete if and only if every p^s-Caristi mapping on X has a fixed point. In particular, the result of Romaguera extended Kirk's [51] characterization of metric completeness to a kind of complete partial metric spaces. Successively, Karapinar in [36] extended the result of Caristi and Kirk [52] to partial metric spaces.

In the first part of this paper, following this research direction, we prove some generalized versions of the result of Matthews by using different types of conditions in ordered partial metric spaces for dominated self-mappings or in partial metric spaces for self-mappings. The notion of dominated mapping of economics, finance, trade, and industry is also applied to approximate the unique solution of nonlinear functional equations. In the second part, using the results obtained in the first part, we deduce a characterization of partial metric 0-completeness in terms of fixed point theory. This result extends the Subrahmanyam [53] characterization of metric completeness. For other characterizations of metric completeness in terms of fixed point theory, the reader can see, for example, [54, 55] and for partial metric completeness, [41].

2. Preliminaries

First, we recall some definitions and some properties of partial metric spaces that can be found in [1, 2, 40, 48, 50]. A partial metric on a nonempty set X is a function $p : X \times X \to [0, +\infty)$ such that for all $x, y, z \in X$

$$(p_1)\ x = y \Leftrightarrow p(x,x) = p(x,y) = p(y,y),$$

$$(p_2)\ p(x,x) \le p(x,y),$$

(p_3) $p(x, y) = p(y, x)$,

(p_4) $p(x, y) \leq p(x, z) + p(z, y) - p(z, z)$.

A partial metric space is a pair (X, p) such that X is a nonempty set and p is a partial metric on X. It is clear that if $p(x, y) = 0$, then from (p_1) and (p_2), it follows that $x = y$. But if $x = y$, $p(x, y)$ may not be 0. A basic example of a partial metric space is the pair $([0, +\infty), p)$, where $p(x, y) = \max\{x, y\}$ for all $x, y \in [0, +\infty)$. Other examples of partial metric spaces which are interesting from a computational point of view can be found in [1].

Each partial metric p on X generates a T_0 topology τ_p on X which has as a base the family of open p-balls $\{B_p(x, \varepsilon) : x \in X, \varepsilon > 0\}$, where

$$B_p(x, \varepsilon) = \{y \in X : p(x, y) < p(x, x) + \varepsilon\} \quad (1)$$

for all $x \in X$ and $\varepsilon > 0$.

If p is a partial metric on X, then the function $p^s : X \times X \to [0, +\infty)$ given by

$$p^s(x, y) = 2p(x, y) - p(x, x) - p(y, y) \quad (2)$$

is a metric on X.

Let (X, p) be a partial metric space. A sequence $\{x_n\}$ in (X, p) converges to a point $x \in X$ if and only if $p(x, x) = \lim_{n \to +\infty} p(x, x_n)$.

A sequence $\{x_n\}$ in (X, p) is called a Cauchy sequence if there exists (and is finite) $\lim_{n,m \to +\infty} p(x_n, x_m)$.

A partial metric space (X, p) is said to be complete if every Cauchy sequence $\{x_n\}$ in X converges, with respect to τ_p, to a point $x \in X$ such that $p(x, x) = \lim_{n,m \to +\infty} p(x_n, x_m)$.

A sequence $\{x_n\}$ in (X, p) is called 0-Cauchy if $\lim_{n,m \to +\infty} p(x_n, x_m) = 0$. We say that (X, p) is 0-complete if every 0-Cauchy sequence in X converges, with respect to τ_p, to a point $x \in X$ such that $p(x, x) = 0$.

On the other hand, the partial metric space $(\mathbb{Q} \cap [0, +\infty), p)$, where $\mathbb{Q}$ denotes the set of rational numbers and the partial metric p is given by $p(x, y) = \max\{x, y\}$, provides an example of a 0-complete partial metric space which is not complete.

It is easy to see that every closed subset of a complete partial metric space is complete.

Lemma 1 (see [1, 40]). *Let (X, p) be a partial metric space. Then*

(a) *$\{x_n\}$ is a Cauchy sequence in (X, p) if and only if it is a Cauchy sequence in the metric space (X, p^s).*

(b) *A partial metric space (X, p) is complete if and only if the metric space (X, p^s) is complete. Furthermore, $\lim_{n \to +\infty} p^s(x_n, x) = 0$ if and only if*

$$p(x, x) = \lim_{n \to +\infty} p(x_n, x) = \lim_{n,m \to +\infty} p(x_n, x_m). \quad (3)$$

The following lemma is obvious.

Lemma 2. *Let (X, p) be a partial metric space and $\{x_n\} \subset X$. If $x_n \to x \in X$ and $p(x, x) = 0$, then $\lim_{n \to +\infty} p(x_n, z) = p(x, z)$ for all $z \in X$.*

Define $p(x, A) = \inf\{p(x, a) : a \in A\}$. Then $a \in \overline{A} \Leftrightarrow p(a, A) = p(a, a)$, where $\overline{A}$ denotes the closure of A (for details see [22, Lemma 1]). From

$$p^s(x, a) = 2p(x, a) - p(x, x) - p(a, a) \leq 2p(x, a) \quad (4)$$

for every $a \in A$, we deduce that $p^s(x, A) \leq 2p(x, A)$.

Let X be a nonempty set and $T, f : X \to X$. The mappings T, f are said to be weakly compatible if they commute at their coincidence points (i.e., $Tfx = fTx$ whenever $Tx = fx$). A point $y \in X$ is called a point of coincidence of T and f if there exists a point $x \in X$ such that $y = Tx = fx$.

Lemma 3 (see [56]). *Let X be a nonempty set and the mappings $T, f : X \to X$ have a unique point of coincidence v in X. If T and f are weakly compatible, then T and f have a unique common fixed point.*

Let X be a nonempty set. If (X, p) is a partial metric space and $(X, \preceq)$ is a partially ordered set, then $(X, p, \preceq)$ is called a partially ordered partial metric space. $x, y \in X$ are called comparable if $x \preceq y$ or $y \preceq x$ holds. Let $(X, \preceq)$ be a partially ordered set and $T, f : X \to X$ two mappings. T is called an f-dominated mapping if $Tx \preceq fx$ for every $x \in X$.

3. Main Results

Let (X, p) be a partial metric space and $T, f : X \to X$ be such that $TX \subset fX$. For every $x_0 \in X$ we consider the sequence $\{x_n\} \subset X$ defined by $fx_n = Tx_{n-1}$ for all $n \in \mathbb{N}$ and we say that $\{Tx_n\}$ is a T-f-sequence of the initial point x_0 (see [57]).

Denote with Ψ the family of non-decreasing functions $\psi : [0, +\infty) \to [0, +\infty)$ such that $\psi(t) > 0$ and $\lim_{n \to +\infty} \psi^n(t) = 0$ for each $t > 0$, where ψ^n is the nth iterate of ψ.

Lemma 4. *For every function $\psi \in \Psi$, the following holds, if for each $t > 0$, $\lim_{n \to +\infty} \psi^n(t) = 0$ then $\psi(t) < t$.*

The following theorem is one of our main results, and it ensures the existence of a common fixed point for two self-mappings in the setting of partially ordered partial metric spaces.

Theorem 5. *Let $(X, p, \preceq)$ be a partially ordered partial metric space and $T, f : X \to X$ two mappings such that $TX \subset fX$. Assume that there exists $\psi \in \Psi$ such that*

$$p(Tx, Ty) \leq \max\{\psi(p(fx, fy)), \psi(p(fx, Tx)), \psi(p(fy, Ty))\} \quad (5)$$

for all $x, y \in X$ with fx and fy comparable. If the following conditions hold:

(i) *T is a f-dominated mapping,*

(ii) *either TX or fX is a 0-complete subspace of X,*

(iii) *for a non-increasing sequence* $\{fx_n\} \subset X$ *converging to* $fu \in X$, *we have* $fu \preceq fx_n$ *for all* $n \in \mathbb{N}$ *and* $ffu \preceq fu$,

then T *and* f *have a point of coincidence. Moreover, if* T *and* f *are weakly compatible, then* T *and* f *have a common fixed point.*

Proof. Let $x_0 \in X$ be fixed and $\{Tx_n\}$ be a T-f-sequence of the initial point x_0. As $fx_{n+1} = Tx_n \preceq fx_n$ for all $n \in \mathbb{N}$, then the sequence $\{Tx_n\}$ is non-increasing.

If $Tx_n = Tx_{n-1} = fx_n$ for some $n \in \mathbb{N}$, then $y = Tx_n = fx_n$ is a point of coincidence of T and f. Suppose that $Tx_n \neq Tx_{n-1}$ for all $n \in \mathbb{N}$. Since fx_n and fx_{n+1} are comparable for all $n \in \mathbb{N}$, we have

$$\begin{aligned} p(Tx_{n+1}, Tx_n) &\le \max\{\psi(p(fx_{n+1}, fx_n)), \psi(p(fx_{n+1}, Tx_{n+1})), \\ &\quad \psi(p(fx_n, Tx_n))\} \\ &= \max\{\psi(p(Tx_n, Tx_{n-1})), \psi(p(Tx_{n+1}, Tx_n))\}. \end{aligned} \tag{6}$$

If $\max\{\psi(p(Tx_n, Tx_{n-1})), \psi(p(Tx_{n+1}, Tx_n))\} = \psi(p(Tx_{n+1}, Tx_n))$ from

$$\begin{aligned} p(Tx_{n+1}, Tx_n) &\le \psi(p(Tx_{n+1}, Tx_n)) \\ &< p(Tx_{n+1}, Tx_n), \end{aligned} \tag{7}$$

we obtain a contradiction and so

$$\begin{aligned} &\max\{\psi(p(Tx_n, Tx_{n-1})), \psi(p(Tx_{n+1}, Tx_n))\} \\ &\quad = \psi(p(Tx_n, Tx_{n-1})). \end{aligned} \tag{8}$$

Then, we have

$$p(Tx_{n+1}, Tx_n) \le \psi^n(p(Tx_1, Tx_0)), \quad \forall n \in \mathbb{N} \tag{9}$$

and hence

$$\lim_{n \to +\infty} p(Tx_{n+1}, Tx_n) = 0. \tag{10}$$

Fix $\varepsilon > 0$ and we choose $n(\varepsilon) \in \mathbb{N}$ such that

$$p(Tx_m, Tx_{m+1}) < \varepsilon - \psi(\varepsilon) \tag{11}$$

for all $m \ge n(\varepsilon)$. Let $m \ge n(\varepsilon)$ and we show that

$$p(Tx_m, Tx_{n+1}) < \varepsilon, \quad \forall n \ge m. \tag{12}$$

Clearly, (12) is true for $n = m$. Suppose that (12) holds for some $n \ge m$, as fx_i and fx_j are comparable for all $i, j \in \mathbb{N}$, then

$$\begin{aligned} p(Tx_m, Tx_{n+2}) &\le p(Tx_m, Tx_{m+1}) + p(Tx_{m+1}, Tx_{n+2}) \\ &\le p(Tx_m, Tx_{m+1}) \\ &\quad + \max\{\psi(p(Tx_m, Tx_{n+1})), \psi(p(Tx_m, Tx_{m+1})), \\ &\quad \psi(p(Tx_{n+1}, Tx_{n+2}))\} \\ &< \varepsilon - \psi(\varepsilon) + \max\{\psi(\varepsilon), \psi(\varepsilon - \psi(\varepsilon))\} = \varepsilon. \end{aligned} \tag{13}$$

This implies that (12) holds for $n + 1$ and by induction, it holds for all $n \ge m$. From (12), we deduce that there exists

$$\lim_{n,m \to +\infty} p(Tx_n, Tx_m) = 0 \tag{14}$$

and hence $\{Tx_n\}$ is a 0-Cauchy sequence.

Suppose that TX is a 0-complete subspace of (X, p), then there exists $y \in TX \subset fX$ such that

$$p(y, y) = \lim_{n \to +\infty} p(Tx_n, y) = \lim_{n,m \to +\infty} p(Tx_n, Tx_m) = 0. \tag{15}$$

This holds also if fX is 0-complete with $y \in fX$.

Let $u \in X$ be such that $y = fu$. We show that y is a point of coincidence of T and f. If not, we have $p(fu, Tu) > 0$. This implies that there exists $n_0 \in \mathbb{N}$ such that

$$\begin{aligned} &\max\{\psi(p(Tx_{n-1}, fu)), \psi(p(Tx_{n-1}, Tx_n)), \psi(p(fu, Tu))\} \\ &\quad = \psi(p(fu, Tu)), \end{aligned} \tag{16}$$

for every $n \ge n_0$. By condition (iii), fx_n and fu are comparable for every $n \in \mathbb{N}$ and hence, by condition (5) with $x = x_n$ and $y = u$, we deduce that

$$\begin{aligned} p(Tx_n, Tu) &\le \max\{\psi(p(fx_n, fu)), \psi(p(fx_n, Tx_n)), \\ &\quad \psi(p(fu, Tu))\} \\ &= \max\{\psi(p(Tx_{n-1}, fu)), \psi(p(Tx_{n-1}, Tx_n)), \\ &\quad \psi(p(fu, Tu))\} \\ &= \psi(p(fu, Tu)), \end{aligned} \tag{17}$$

for every $n \ge n_0$. Letting $n \to +\infty$ in the previous inequality and using Lemma 2, we obtain

$$p(fu, Tu) \le \psi(p(fu, Tu)) < p(fu, Tu), \tag{18}$$

which implies $p(fu, Tu) = 0$, that is, $Tu = fu$. Thus, we have shown that $y = fu = Tu$ is a point of coincidence of T and f. If T and f are weakly compatible, then $Ty = Tfu = fTu = fy$. By condition (iii), $fy = ffu \preceq fu$, that is, fy and fu are comparable. Using the contractive condition (5), we get

$$\begin{aligned} p(Tu, Ty) &\le \max\{\psi(p(fu, fy)), \psi(p(fu, Tu)), \psi(p(fy, Ty))\} \\ &= \psi(p(fu, fy)) < p(fu, fy), \end{aligned} \tag{19}$$

which implies $Ty = Tu = y$ and hence y is a common fixed point of T and f.

We shall give a sufficient condition for the uniqueness of the common fixed point in Theorem 5.

Theorem 6. *Let all the conditions of Theorem 5 be satisfied. If the following condition holds:*

(iv) *for all* $x, y \in fX$ *there exists* $v_0 \in X$ *such that* $fv_0 \preceq x$, $fv_0 \preceq y$ *and* $\lim_{n\to+\infty} p(Tv_{n-1}, Tv_n) = 0$, *where* $\{Tv_n\}$ *is the* T-f*-sequence of the initial point* v_0,

then T *and* f *have a unique common fixed point.*

Proof. Let z, w be two common fixed points of T and f with $z \neq w$. If z and w are comparable, then using the contractive condition (5), we deduce that $z = w$. If z and w are not comparable, then there exists $v_0 \in X$ such that $fv_0 \preceq z = fz$, $fv_0 \preceq w = fw$. As T is a f-dominated mapping, we get that

$$fv_1 = Tv_0 \preceq fv_0 \preceq z = fz. \tag{20}$$

To continue, we obtain

$$fv_{n+1} = Tv_n \preceq fv_n \preceq z = fz \quad \forall n \in \mathbb{N} \tag{21}$$

and hence fv_n and fz are comparable.

Using the contractive condition (5) with $x = v_n$ and $y = z$, we get

$$\begin{aligned} p(Tv_n, Tz) &\le \max\{\psi(p(fv_n, fz)), \psi(p(fv_n, Tv_n)), \\ &\quad \psi(p(fz, Tz))\} \qquad (22) \\ &= \max\{\psi(p(Tv_{n-1}, Tz)), \psi(p(Tv_{n-1}, Tv_n)), \\ &\quad \psi(p(Tz, Tz))\} \end{aligned}$$

for all $n \in \mathbb{N}$. Since, the contractive condition (5) ensures that $p(Tz, Tz) = 0$, we have

$$p(Tv_n, Tz) \le \max\{\psi(p(Tv_{n-1}, Tz)), \psi(p(Tv_{n-1}, Tv_n))\}. \tag{23}$$

Now, by condition (iv), $\lim_{n\to+\infty} p(Tv_{n-1}, Tv_n) = 0$ and hence for n sufficiently large, we have

$$p(Tv_n, Tz) \le \psi(p(Tv_{n-1}, Tz)). \tag{24}$$

Without loss of generality, assuming that (24) holds for all $n \in \mathbb{N}$, it follows that

$$p(Tv_n, Tz) \le \psi^n(p(Tv_0, Tz)). \tag{25}$$

Now, letting $n \to +\infty$ in (25), we obtain

$$\lim_{n\to+\infty} p(Tv_n, Tz) = 0. \tag{26}$$

With similar arguments, we deduce that $\lim_{n\to+\infty} p(Tw, Tv_n) = 0$. Hence

$$\begin{aligned} 0 < p(w, z) &= p(Tw, Tz) \\ &\le p(Tw, Tv_n) + p(Tv_n, Tz) \longrightarrow 0 \end{aligned} \tag{27}$$

as $n \to +\infty$, which is a contradiction. Thus T and f have a unique common fixed point. □

As a consequence of Theorem 5, we state the following result.

Theorem 7. *Let* (X, p) *be a partial metric space and* $T, f : X \to X$ *two mappings such that* $TX \subset fX$. *Assume that there exists* $\psi \in \Psi$ *such that*

$$\begin{aligned} &p(Tx, Ty) \\ &\quad \le \max\{\psi(p(fx, fy)), \psi(p(fx, Tx)), \psi(p(fy, Ty))\} \end{aligned} \tag{28}$$

for all $x, y \in X$. *If* TX *or* fX *is a 0-complete subspace of* X, *then* T *and* f *have a unique point of coincidence. Moreover, if* T *and* f *are weakly compatible, then* T *and* f *have a unique common fixed point.*

Proof. Proceeding as in the proof of Theorem 5, we get that T and f have a unique point of coincidence and, by Lemma 3, T and f have a unique common fixed point.

From Theorem 7, we can deduce the following corollaries.

Corollary 8 (Banach type). *Let* (X, p) *be a partial metric space and* $T, f : X \to X$ *two mappings such that* $TX \subset fX$. *Assume that*

$$p(Tx, Ty) \le kp(fx, fy) \tag{29}$$

for all $x, y \in X$, *where* $0 \le k < 1$. *If* TX *or* fX *is a 0-complete subspace of* X, *then* T *and* f *have a unique point of coincidence. Moreover, if* T *and* f *are weakly compatible, then* T *and* f *have a unique common fixed point.*

Corollary 9 (Bianchini type). *Let* (X, p) *be a partial metric space and let* $T, f : X \to X$ *be two mappings such that* $TX \subset fX$. *Assume that*

$$p(Tx, Ty) \le k \max\{p(fx, Tx), p(fy, Ty)\} \tag{30}$$

for all $x, y \in X$, *where* $0 \le k < 1$. *If* TX *or* fX *is a 0-complete subspace of* X, *then* T *and* f *have a unique point of coincidence. Moreover, if* T *and* f *are weakly compatible, then* T *and* f *have a unique common fixed point.*

Corollary 10 (Reich type [58]). *Let* (X, p) *be a partial metric space and let* $T, f : X \to X$ *be two mappings such that* $TX \subset fX$. *Assume that*

$$p(Tx, Ty) \le ap(fx, fy) + bp(fx, Tx) + cp(fy, Ty) \tag{31}$$

for all $x, y \in X$, *where* $a, b, c \ge 0$ *and* $a + b + c < 1$. *If* TX *or* fX *is a 0-complete subspace of* X, *then* T *and* f *have a unique point of coincidence. Moreover, if* T *and* f *are weakly compatible, then* T *and* f *have a unique common fixed point.*

The following example shows that there exist mappings that satisfy the contractive condition (28), but are not quasi-contractions [59].

Example 11. Consider the set $X = \{1, 2, 3\}$ and the function $p : X \times X \to [0, +\infty)$ given by $p(1, 2) = p(2, 3) = 1$, $p(1, 3) = 3/2$, $p(1, 1) = p(3, 3) = 1/2$, $p(2, 2) = 0$, and $p(x, y) = p(y, x)$. Obviously, p is a partial metric on X, but it is not a metric (since $p(x, x) \neq 0$ for $x = 1$ and $x = 3$). Clearly, (X, p) is a 0-complete partial metric space. Let $T, f : X \to X$ be defined by $T1 = 2, T2 = 2, T3 = 1$ and $fx = x$ for every $x \in X$. Take $\psi(t) = (2/3)t$ for every $t \geq 0$.

First, we will check that T and f satisfy the contractive condition (28). If $x, y \in \{1, 2\}$, then $p(Tx, Ty) = p(2, 2) = 0$ and (28) trivially holds. Let, for example, $y = 3$, then we have the following three cases:

$$\begin{aligned} p(T1, T3) &= p(2, 1) = 1 \leq \frac{2}{3} \cdot \frac{3}{2} \\ &= \max\{\psi(p(1, 3)), \psi(p(1, 2)), \psi(p(3, 1))\}; \\ p(T2, T3) &= p(2, 1) = 1 \leq \frac{2}{3} \cdot \frac{3}{2} \\ &= \max\{\psi(p(2, 3)), \psi(p(2, 2)), \psi(p(3, 1))\}; \\ p(T3, T3) &= p(1, 1) = \frac{1}{2} < \frac{2}{3} \cdot \frac{3}{2} \\ &= \max\{\psi(p(3, 3)), \psi(p(3, 1))\}. \end{aligned} \tag{32}$$

Thus, all the conditions of Theorem 7 are satisfied and the existence of a common fixed point of T and f (which is 2) follows. The same conclusion cannot be obtained by the main results from [59]. Indeed, using $p^s(a, b) = 2p(a, b) - p(a, a) - p(b, b)$, and then taking p^s instead of p, $x = 1$, $y = 3$ in (5), we obtain

$$\begin{aligned} L &= p^s(T1, T3) = p^s(2, 1) = \frac{3}{2}, \\ R &= \max\{\psi(p^s(1, 3)), \psi(p^s(1, T1)), \psi(p^s(3, T3)), \\ &\qquad \psi(p^s(1, T3)), \psi(p^s(3, T1))\} \\ &= \frac{2}{3} \max\{p^s(1, 3), p^s(1, 2), p^s(3, 1), p^s(1, 1), p^s(3, 2)\} \\ &= \frac{2}{3} \max\left\{2, \frac{3}{2}, 2, 0, \frac{3}{2}\right\} = \frac{4}{3}. \end{aligned} \tag{33}$$

Since $L > R$, the conclusion follows.

The following example shows that there exist mappings that satisfy the contractive condition (5), but do not satisfy the contractive condition (28).

Example 12. Let $X = [0, 2]$ be endowed with the partial metric

$$p(x, y) = \begin{cases} |x - y| & \text{if } x, y \in [0, 1], \\ \max\{x, y\} & \text{if } \{x, y\} \cap (1, 2] \neq \varnothing. \end{cases} \tag{34}$$

Clearly, (X, p) is a 0-complete partial metric space. Let $T, f : X \to X$ be defined by

$$Tx = \begin{cases} x & \text{if } x \in [0, 1], \\ \dfrac{1 + x}{2} & \text{if } x \in (1, 2] \end{cases} \tag{35}$$

and $fx = x$ for each $x \in X$. As T and f have many common fixed points (each $x \in [0, 1]$ is a common fixed point), then it is immediate to show that T and f do not satisfy the contractive condition (28).

If (X, p) is ordered by

$$x \preceq y \Longleftrightarrow (x = y) \text{ or } (x, y \in (1, 2], x \leq y), \tag{36}$$

then T and f satisfy the contractive condition (5) where $\psi : [0, +\infty) \to [0, +\infty)$ is defined by

$$\psi(t) = \begin{cases} \dfrac{t}{2} & \text{if } t \in [0, 1], \\ \dfrac{1 + t}{2} & \text{if } t \in (1, 2]. \end{cases} \tag{37}$$

Using Theorem 5, we deduce that T and f have a common fixed point.

4. Completeness in Partial Metric Spaces and Fixed Points

In this section, we characterize those partial metric spaces for which every Bianchini mapping has a fixed point in the style of Subrahmanyam characterization of metric completeness. This will be done by means of the notion of 0-completeness which was introduced by Romaguera in [50].

Let (X, p) be a partial metric space and $T : X \to X$ be a mapping. We recall that T is a Bianchini [60] mapping if

$$p(Tx, Ty) \leq k \max\{p(x, Tx), p(y, Ty)\} \tag{38}$$

for all $x, y \in X$, where $0 \leq k < 1$.

Theorem 13. *Let (X, p) be a partial metric space. If every mapping $T : X \to X$ satisfying the following conditions:*

(i) *$p(Tx, Ty) \leq \lambda \max\{p(x, Tx), p(y, Ty)\}$ for all $x, y \in X$, for a fixed $\lambda > 0$,*

(ii) *TX is countable*

has a fixed point, then (X, p) is 0-complete.

Proof. Suppose that there is a 0-Cauchy sequence $\{x_n\}$ of distinct points in (X, p) which is not convergent in (X, p^s). We put $A = \{x_n\}$ and we note that for every $x \notin A$, we have $p^s(x, A) > 0$.

Now, $0 < p^s(x, A) \leq 2p(x, A)$ implies that $p(x, A) > 0$. Since $\{x_n\}$ is a 0-Cauchy sequence in (X, p), there exists a least positive integer $N(x)$ such that

$$\begin{aligned} &p(x_m, x_n) \\ &\quad < \lambda p(x, A) \leq \lambda p(x, x_l), \quad l \in \mathbb{N}, \ \forall m, n \geq N(x). \end{aligned} \tag{39}$$

In particular

$$\begin{aligned} &p\left(x_m, x_{N(x)}\right) \\ &\quad < \lambda p(x, A \setminus \{x\}) \leq \lambda p\left(x, x_{N(x)}\right), \quad \forall m \geq N(x). \end{aligned} \tag{40}$$

For fixed $n \in \mathbb{N}$, since $p^s(x_n, A\backslash\{x_n\}) > 0$ and so $p(x_n, A\backslash \{x_n\}) > 0$, there is $n'(n) > n$ such that

$$p(x_m, x_{n'}) < \lambda p(x_n, A \setminus \{x_n\}) \leq \lambda p(x_n, x_{n'}), \quad \forall m \geq n'. \tag{41}$$

Now, let $T : X \to X$ be defined by

$$Tx = \begin{cases} x_{N(x)} & \text{if } x \notin A, \\ x_{n'} & \text{if } x\,(= x_n) \in A. \end{cases} \tag{42}$$

From the definition of T, we deduce that T satisfies the condition (ii). On the other hand, T satisfies also the condition (i). In fact, (i) is verified by assuming $Tx = x_n$ and $Ty = x_m$, and noting that

$$p(x_m, x_n) < \begin{cases} \lambda p(y, A \setminus \{y\}) \leq \lambda p(y, Ty) & \text{if } n \geq m, \\ \lambda p(x, A \setminus \{x\}) \leq \lambda p(x, Tx) & \text{if } n < m. \end{cases} \tag{43}$$

It is clear that T has not fixed points since $x_{n'} \neq x_n, n = 1, 2, \ldots$. Thus, the assumption that there is a 0-Cauchy sequence $\{x_n\}$ which is not convergent in (X, p^s) leads to a contradiction to Theorem 13 and thereby establishes the same.

If in Theorem 13 we choose $\lambda \in [0, 1)$, by Corollary 9, we obtain the following characterization of 0-completeness for partial metric spaces.

Theorem 14. *A partial metric space (X, p) is 0-complete if and only if every mapping $T : X \to X$ satisfying the following conditions:*

(i) $p(Tx, Ty) \leq \lambda \max\{p(x, Tx), p(y, Ty)\}$ *for all* $x, y \in X$, *for a fixed* $\lambda \in [0, 1)$,

(ii) *TX is countable*

has a fixed point.

In Theorem 14, the class of mappings satisfying (i) and (ii) can be replaced by the class of mappings satisfying (ii) and the following condition:

(i) $p(Tx, Ty) \leq \lambda[p(x, Tx) + p(y, Ty)]$ for all $x, y \in X$, for a fixed $\lambda \in [0, 1/2)$.

Acknowledgment

The authors are grateful to the editor and referees for their valuable suggestions and critical remarks for improving this paper. P. Vetro is supported by Università degli Studi di Palermo (Local University Project R.S. ex 60%).

References

[1] S. G. Matthews, "Partial metric topology," *Annals of the New York Academy of Sciences*, vol. 728, pp. 183–197, 1994, Papers on General Topology and Applications.

[2] S. J. O'Neill, "Partial metrics, valuations and domain theory," *Annals of the New York Academy of Sciences*, vol. 806, pp. 304–315, 1996, Papers on General Topology and Applications.

[3] M. A. Bukatin and J. S. Scott, "Towards computing distances between programs via Scott domains," in *Logical Foundations of Computer Science*, S. Adian, A. Nerode et al., Eds., vol. 1234 of *Lecture Notes in Computer Science*, pp. 33–43, Springer, Berlin, Germany, 1997.

[4] M. A. Bukatin and S. Y. Shorina, "Partial metrics and co-continuous valuations," in *Foundations of Software Science and Computation Structures*, M. Nivat, Ed., vol. 1378 of *Lecture Notes in Computer Science*, pp. 125–139, Springer, Berlin, Germany, 1998.

[5] S. Romaguera and M. Schellekens, "Partial metric monoids and semivaluation spaces," *Topology and Its Applications*, vol. 153, no. 5-6, pp. 948–962, 2005.

[6] R. Heckmann, "Approximation of metric spaces by partial metric spaces," *Applied Categorical Structures*, vol. 7, no. 1-2, pp. 71–83, 1999.

[7] R. D. Kopperman, S. G. Matthews, and H. Pajoohesh, "What do partial metrics represent?" in *Proceedings of the 19th Summer Conference on Topology and Its Applications*, University of CapeTown, 2004.

[8] H. P. A. Künzi, H. Pajoohesh, and M. P. Schellekens, "Partial quasi-metrics," *Theoretical Computer Science*, vol. 365, no. 3, pp. 237–246, 2006.

[9] S. Romaguera and M. Schellekens, "Duality and quasi-normability for complexity spaces," *Applied General Topology*, vol. 3, no. 1, pp. 91–112, 2002.

[10] S. Romaguera and O. Valero, "A quantitative computational model for complete partial metric spaces via formal balls," *Mathematical Structures in Computer Science*, vol. 19, no. 3, pp. 541–563, 2009.

[11] M. P. Schellekens, "A characterization of partial metrizability: domains are quantifiable," *Theoretical Computer Science*, vol. 305, no. 1–3, pp. 409–432, 2003.

[12] M. P. Schellekens, "The correspondence between partial metrics and semivaluations," *Theoretical Computer Science*, vol. 315, no. 1, pp. 135–149, 2004.

[13] P. Waszkiewicz, "Partial metrisability of continuous posets," *Mathematical Structures in Computer Science*, vol. 16, no. 2, pp. 359–372, 2006.

[14] P. Waszkiewicz, "Quantitative continuous domains," *Applied Categorical Structures*, vol. 11, no. 1, pp. 41–67, 2003.

[15] V. Berinde and F. Vetro, "Common fixed points of mappings satisfying implicit contractive conditions," *Fixed Point Theory and Applications*, vol. 2012, article 105, 2012.

[16] L. J. Ćirić, "On contraction type mappings," *Mathematica Balkanica*, vol. 1, pp. 52–57, 1971.

[17] M. Abbas, T. Nazir, and S. Romaguera, "Fixed point results for generalized cyclic contraction mappings in partial metric spaces," *Revista de la Real Academia de Ciencias Exactas*, vol. 106, pp. 287–297, 2012.

[18] T. Abdeljawad, "Fixed points for generalized weakly contractive mappings in partial metric spaces," *Mathematical and Computer Modelling*, vol. 54, pp. 2923–2927, 2011.

[19] T. Abdeljawad, E. Karapinar, and K. Tas, "Existence and uniqueness of common fixed point on partial metric spaces," *Applied Mathematics Letters*, vol. 24, pp. 1894–1899, 2011.

[20] T. Abdeljawad, E. Karapinar, and K. Tas, "A generalized contraction principle with control functions on partial metric spaces," *Computers & Mathematics with Applications*, vol. 63, pp. 716–719, 2012.

[21] I. Altun and A. Erduran, "Fixed point theorems for monotone mappings on partial metric spaces," *Fixed Point Theory and Applications*, vol. 2011, Article ID 508730, 10 pages, 2011.

[22] I. Altun, F. Sola, and H. Simsek, "Generalized contractions on partial metric spaces," *Topology and Its Applications*, vol. 157, no. 18, pp. 2778–2785, 2010.

[23] H. Aydi, "Fixed point theorems for generalized weakly contractive condition in ordered partial metric spaces," *Journal of Nonlinear Analysis and Optimization: Theory and Applications*, vol. 2, pp. 33–48, 2011.

[24] H. Aydi, "Common fixed point results for mappings satisfying (ψ, ϕ)-weak contractions in ordered partial metric spaces," *International Journal of Mathematics and Statistics*, vol. 12, pp. 53–64, 2012.

[25] H. Aydi, E. Karapınar, and W. Shatanawi, "Coupled fixed point results for (ψ, φ)-weakly contractive condition in ordered partial metric spaces," *Computers & Mathematics with Applications*, vol. 62, no. 12, pp. 4449–4460, 2011.

[26] H. Aydi, M. Abbas, and C. Vetro, "Partial Hausdorff metric and Nadler's fixed point theorem on partial metric spaces," *Topology and Its Applications*, vol. 159, no. 14, pp. 3234–3242, 2012.

[27] H. Aydi, "A common fixed point result by altering distances involving a contractive condition of integral type in partial metric spaces," *Demonstratio Mathematica*. In press.

[28] K. P. Chi, E. Karapinar, and T. D. Thanh, "A generalized contraction principle in partial metric spaces," *Mathematical and Computer Modelling*, vol. 55, pp. 1673–1681, 2012.

[29] L. J. Ćirić, B. Samet, H. Aydi, and C. Vetro, "Common fixed points of generalized contractions on partial metric spaces and an application," *Applied Mathematics and Computation*, vol. 218, no. 6, pp. 2398–2406, 2011.

[30] C. Di Bari and P. Vetro, "Fixed points for weak φ-contractions on partial metric spaces," *International Journal of Engineering, Contemporary Mathematics and Sciences*, vol. 1, pp. 5–13, 2011.

[31] C. Di Bari, Z. Kadelburg, H. K. Nashine, and S. Radenović, "Common fixed points of g-quasicontractions and related mappings in 0-complete partial metric spaces," *Fixed Point Theory and Applications*, vol. 2012, article 113, 2012.

[32] D. Ilić, V. Pavlović, and V. Rakočević, "Some new extensions of Banach's contraction principle to partial metric space," *Applied Mathematics Letters*, vol. 24, no. 8, pp. 1326–1330, 2011.

[33] D. Ilić, V. Pavlović, and V. Rakočević, "Extensions of the Zamfirescu theorem to partial metric spaces," *Mathematical and Computer Modelling*, vol. 55, pp. 801–809, 2012.

[34] E. Karapinar, "Weak ϕ-contraction on partial metric spaces," *Journal of Computational Analysis and Applications*, vol. 14, no. 2, pp. 206–210, 2012.

[35] E. Karapinar, "A note on common fixed point theorems in partial metric spaces," *Miskolc Mathematical Notes*, vol. 12, no. 2, pp. 185–191, 2011.

[36] E. Karapinar, "Generalizations of Caristi Kirk's theorem on partial metric spaces," *Fixed Point Theory and Applications*, vol. 2011, article 4, 2011.

[37] E. Karapınar and M. Erhan, "Fixed point theorems for operators on partial metric spaces," *Applied Mathematics Letters*, vol. 24, no. 11, pp. 1894–1899, 2011.

[38] E. Karapınar and U. Yüksel, "Some common fixed point theorems in partial metric spaces," *Journal of Applied Mathematics*, vol. 2011, Article ID 263621, 16 pages, 2011.

[39] H. K. Nashine, Z. Kadelburg, and S. Radenović, "Common fixed point theorems for weakly isotone increasing mappings in ordered partial metric spaces," *Mathematical and Computer Modelling*. In press.

[40] S. Oltra and O. Valero, "Banach's fixed point theorem for partial metric spaces," *Rendiconti dell'Istituto di Matematica dell'Università di Trieste*, vol. 36, no. 1-2, pp. 17–26, 2004.

[41] D. Paesano and P. Vetro, "Suzuki's type characterizations of completeness for partial metric spaces and fixed points for partially ordered metric spaces," *Topology and Its Applications*, vol. 159, no. 3, pp. 911–920, 2012.

[42] S. Romaguera, "Fixed point theorems for generalized contractions on partial metric spaces," *Topology and Its Applications*, vol. 159, no. 1, pp. 194–199, 2012.

[43] S. Romaguera, "Matkowski's type theorems for generalized contractions on (ordered) partial metric spaces," *Applied General Topology*, vol. 12, no. 2, pp. 213–220, 2011.

[44] B. Samet, "Coupled fixed point theorems for a generalized Meir-Keeler contraction in partially ordered metric spaces," *Nonlinear Analysis*, vol. 72, no. 12, pp. 4508–4517, 2010.

[45] B. Samet, M. Rajovic, R. Lazovic, and R. Stoiljkovic, "Common fixed point results for nonlinear contractions in ordered partial metric spaces," *Fixed Point Theory and Applications*, vol. 2011, article 71, 2011.

[46] W. Shatanawi, B. Samet, and M. Abbas, "Coupled fixed point theorems for mixed monotone mappings in ordered partial metric spaces," *Mathematical and Computer Modelling*, vol. 55, pp. 680–687, 2012.

[47] N. Shobkolaei, S. M. Vaezpour, and S. Sedghi, "A common fixed point theorem on ordered partial metric spaces," *Journal of Basic and Applied Scientific Research*, vol. 1, pp. 3433–3439, 2011.

[48] O. Valero, "On Banach fixed point theorems for partial metric spaces," *Applied General Topology*, vol. 6, no. 2, pp. 229–240, 2005.

[49] F. Vetro and S. Radenović, "Nonlinear ψ-quasi-contractions of Ćirić-type in partial metric spaces," *Applied Mathematics and Computation*, vol. 219, no. 4, pp. 1594–1600, 2012.

[50] S. Romaguera, "A Kirk type characterization of completeness for partial metric spaces," *Fixed Point Theory and Applications*, vol. 2010, Article ID 493298, 6 pages, 2010.

[51] W. A. Kirk, "Caristi's fixed point theorem and metric convexity," *Colloquium Mathematicum*, vol. 36, no. 1, pp. 81–86, 1976.

[52] J. Caristi and W. A. Kirk, "Geometric fixed point theory and inwardness conditions," in *The Geometry of Metric and Linear Spaces*, vol. 490 of *Lecture Notes in Mathematics*, pp. 74–83, Springer, Berlin, Germany, 1975.

[53] P. V. Subrahmanyam, "Completeness and fixed-points," *Monatshefte für Mathematik*, vol. 80, no. 4, pp. 325–330, 1975.

[54] S. Park, "Characterizations of metric completeness," *Colloquium Mathematicum*, vol. 49, no. 1, pp. 21–26, 1984.

[55] T. Suzuki, "A generalized Banach contraction principle that characterizes metric completeness," *Proceedings of the American Mathematical Society*, vol. 136, no. 5, pp. 1861–1869, 2008.

[56] M. Abbas and G. Jungck, "Common fixed point results for noncommuting mappings without continuity in cone metric spaces," *Journal of Mathematical Analysis and Applications*, vol. 341, no. 1, pp. 416–420, 2008.

[57] P. Vetro, "Common fixed points in cone metric spaces," *Rendiconti del Circolo Matematico di Palermo*, vol. 56, no. 3, pp. 464–468, 2007.

[58] S. Reich, "Some remarks concerning contraction mappings," *Canadian Mathematical Bulletin*, vol. 14, pp. 121–124, 1971.

[59] L. J. Ćirić, "A generalization of Banach's contraction principle," *Proceedings of the American Mathematical Society*, vol. 45, pp. 267–273, 1974.

[60] R. M. T. Bianchini, "Su un problema di S. Reich riguardante la teoria dei punti fissi," *Bollettino dell'Unione Matematica Italiana*, vol. 5, pp. 103–108, 1972.

General-Appell Polynomials within the Context of Monomiality Principle

Subuhi Khan and Nusrat Raza

Department of Mathematics, Aligarh Muslim University, Aligarh 202002, India

Correspondence should be addressed to Subuhi Khan; subuhi2006@gmail.com

Academic Editor: Jacques Liandrat

A general class of the 2-variable polynomials is considered, and its properties are derived. Further, these polynomials are used to introduce the 2-variable general-Appell polynomials (2VgAP). The generating function for the 2VgAP is derived, and a correspondence between these polynomials and the Appell polynomials is established. The differential equation, recurrence relations, and other properties for the 2VgAP are obtained within the context of the monomiality principle. This paper is the first attempt in the direction of introducing a new family of special polynomials, which includes many other new special polynomial families as its particular cases.

1. Introduction and Preliminaries

The Appell polynomials are very often found in different applications in pure and applied mathematics. The Appell polynomials [1] may be defined by either of the following equivalent conditions: $\{A_n(x)\}(n \in \mathbb{N}_0)$ is an Appell set (A_n being of degree exactly n) if either,

(i) $(d/dx)A_n(x) = nA_{n-1}(x)$ $(n \in \mathbb{N}_0)$ or

(ii) there exists an exponential generating function of the form

$$A(t)\exp(xt) = \sum_{n=0}^{\infty} A_n(x)\frac{t^n}{n!}, \tag{1}$$

where $A(t)$ has (at least the formal) expansion:

$$A(t) = \sum_{n=0}^{\infty} A_n \frac{t^n}{n!} \quad (A_0 \neq 0). \tag{2}$$

Roman [2] characterized Appell sequences in several ways. Properties of Appell sequences are naturally handled within the framework of modern classical umbral calculus by Roman [2]. We recall the following result [2, Theorem 2.5.3], which can be viewed as an alternate definition of Appell sequences.

The sequence $A_n(x)$ is Appell for $g(t)$, if and only if

$$\frac{1}{g(t)}\exp(xt) = \sum_{n=0}^{\infty} A_n(x)\frac{t^n}{n!}, \tag{3}$$

where

$$g(t) = \sum_{n=0}^{\infty} g_n \frac{t^n}{n!} \quad (g_0 \neq 0). \tag{4}$$

In view of (1) and (3), we have

$$A(t) = \frac{1}{g(t)}. \tag{5}$$

The Appell class contains important sequences such as the Bernoulli and Euler polynomials and their generalized forms. Some known Appell polynomials are listed in Table 1.

We recall that, according to the monomiality principle [15, 16], a polynomial set $\{p_n(x)\}_{n\in\mathbb{N}}$ is "quasimonomial", provided there exist two operators $\widehat{M}$ and $\widehat{P}$ playing, respectively, the role of multiplicative and derivative operators, for the family of polynomials. These operators satisfy the following identities, for all $n \in \mathbb{N}$:

$$\widehat{M}\{p_n(x)\} = p_{n+1}(x), \tag{6}$$

$$\widehat{P}\{p_n(x)\} = np_{n-1}(x). \tag{7}$$

Table 1: List of some Appell polynomials.

S. No.	$g(t)$; $A(t)$	Generating functions	Polynomials
(I)	$g(t)=\frac{(e^t-1)}{t}$; $A(t)=\frac{t}{(e^t-1)}$	$\frac{t}{(e^t-1)}e^{xt}=\sum_{n=0}^{\infty}B_n(x)\frac{t^n}{n!}$	The Bernoulli polynomials [3]
(II)	$g(t)=\frac{(e^t+1)}{2}$; $A(t)=\frac{2}{(e^t+1)}$	$\frac{2}{(e^t+1)}e^{xt}=\sum_{n=0}^{\infty}E_n(x)\frac{t^n}{n!}$	The Euler polynomials [3]
(III)	$g(t)=\frac{(e^t-1)^{\alpha}}{t^{\alpha}}$; $A(t)=\frac{t^{\alpha}}{(e^t-1)^{\alpha}}$	$\frac{t^{\alpha}}{(e^t-1)^{\alpha}}e^{xt}=\sum_{n=0}^{\infty}B_n^{(\alpha)}(x)\frac{t^n}{n!}$	The generalized Bernoulli polynomials [4]
(IV)	$g(t)=\frac{(e^t+1)^{\alpha}}{2^{\alpha}}$; $A(t)=\frac{2^{\alpha}}{(e^t+1)^{\alpha}}$	$\frac{2^{\alpha}}{(e^t+1)^{\alpha}}e^{xt}=\sum_{n=0}^{\infty}E_n^{(\alpha)}(x)\frac{t^n}{n!}$	The generalized Euler polynomials [4]
(V)	$g(t)=\frac{(e^{\alpha_1 t}-1)\,(e^{\alpha_2 t}-1)\cdots(e^{\alpha_m t}-1)}{\alpha_1\alpha_2\cdots\alpha_m t^m}$; $A(t)=\frac{\alpha_1\alpha_2\cdots\alpha_m t^m}{(e^{\alpha_1 t}-1)(e^{\alpha_2 t}-1)\cdots(e^{\alpha_m t}-1)}$	$\frac{\alpha_1\alpha_2\cdots\alpha_m t^m}{(e^{\alpha_1 t}-1)(e^{\alpha_2 t}-1)\cdots(e^{\alpha_m t}-1)}e^{xt}=\sum_{n=0}^{\infty}B_n^{(m)}(x\mid\alpha_1,\alpha_2,\ldots,\alpha_m)\frac{t^n}{n!}$	The generalized Bernoulli polynomials of order m [5]
(VI)	$g(t)=\frac{(e^{\alpha_1 t}+1)(e^{\alpha_2 t}+1)\cdots(e^{\alpha_m t}+1)}{2^m}$; $A(t)=\frac{2^m}{(e^{\alpha_1 t}+1)(e^{\alpha_2 t}+1)\cdots(e^{\alpha_m t}+1)}$	$\frac{2^m}{(e^{\alpha_1 t}+1)(e^{\alpha_2 t}+1)\cdots(e^{\alpha_m t}+1)}e^{xt}=\sum_{n=0}^{\infty}E_n^{(m)}(x\mid\alpha_1,\alpha_2,\ldots,\alpha_m)\frac{t^n}{n!}$	The generalized Euler polynomials of order m [5]
(VII)	$g(t)=\frac{e^t-\sum_{h=0}^{m-1}(t^h/h!)}{t^m}$; $A(t)=\frac{t^m}{e^t-\sum_{h=0}^{m-1}(t^h/h!)}$	$\frac{t^m}{e^t-\sum_{h=0}^{m-1}(t^h/h!)}e^{xt}=\sum_{n=0}^{\infty}B_n^{[m-1]}(x)\frac{t^n}{n!}$	The new generalized Bernoulli polynomials [6]

Table 1: Continued.

S. No.	$g(t); A(t)$	Generating functions	Polynomials
(VIII)	$g(t)=\left(\frac{t}{\lambda e^t-1}\right)^{-\alpha}; A(t)=\left(\frac{t}{\lambda e^t-1}\right)^{\alpha}$	$\left(\frac{t}{\lambda e^t-1}\right)^{\alpha} e^{xt}=\sum_{n=0}^{\infty}\mathfrak{B}_n^{(\alpha)}(x;\lambda)\frac{t^n}{n!}$	The Apostol-Bernoulli polynomials of order α [7]
(IX)	$g(t)=\frac{\lambda e^t-1}{t}; A(t)=\frac{t}{\lambda e^t-1}$	$\frac{t}{\lambda e^t-1}e^{xt}=\sum_{n=0}^{\infty}\mathfrak{B}_n(x;\lambda)\frac{t^n}{n!}$	The Apostol-Bernoulli polynomials [7, 8]
(X)	$g(t)=\left(\frac{2}{\lambda e^t+1}\right)^{-\alpha}; A(t)=\left(\frac{2}{\lambda e^t+1}\right)^{\alpha}$	$\left(\frac{2}{\lambda e^t+1}\right)^{\alpha} e^{xt}=\sum_{n=0}^{\infty}\mathscr{E}_n^{(\alpha)}(x;\lambda)\frac{t^n}{n!}, \lvert t+\log\lambda\rvert<\pi;\ 1^{\alpha}:=1$	The Apostol-Euler polynomials of order α [7, 9]
(XI)	$g(t)=\frac{\lambda e^t+1}{2}; A(t)=\frac{2}{\lambda e^t+1}$	$\frac{2}{\lambda e^t+1}e^{xt}=\sum_{n=0}^{\infty}\mathscr{E}_n(x;\lambda)\frac{t^n}{n!}, \lvert t+\log\lambda\rvert<\pi$	The Apostol-Euler polynomials [7–9]
(XII)	$g(t)=e^{-(\xi_0+\xi_1 t+\xi_2 t^2+\cdots+\xi_{r+1}t^{r+1})}$; $A(t)=e^{(\xi_0+\xi_1 t+\xi_2 t^2+\cdots+\xi_{r+1}t^{r+1})}$; $\xi_{r+1}\neq 0$	$e^{(\xi_0+\xi_1 t+\xi_2 t^2+\cdots+\xi_{r+1}t^{r+1})}e^{xt}=\sum_{n=0}^{\infty}A_n(x)\frac{t^n}{n!}$	The generalized Gould-Hopper polynomials [10] (for $r=1$, the Hermite polynomials $H_n(x)$ [11] and for $r=2$, classical 2-orthogonal polynomials)
(XIII)	$g(t)=(1-t)^{m+1}; A(t)=\frac{1}{(1-t)^{m+1}}$	$\frac{1}{(1-t)^{m+1}}e^{xt}=\sum_{n=0}^{\infty}G_n^{(m)}(x)t^n$	The Miller-Lee polynomials [11, 12] (for $m=0$, the truncated exponential polynomials $e_n(x)$ [11] and for $m=\beta-1$, the modified Laguerre polynomials $f_n^{(\beta)}(x)$ [13])
(XIV)	$g(t)=\frac{(e^t+1)}{2t}$; $A(t)=\frac{2t}{(e^t+1)}$	$\frac{2t}{(e^t+1)}e^{xt}=\sum_{n=0}^{\infty}G_n(x)\frac{t^n}{n!}$	The Genocchi polynomials [14]

The operators $\widehat{M}$ and $\widehat{P}$ also satisfy the commutation relation

$$[\widehat{P}, \widehat{M}] = \widehat{P}\widehat{M} - \widehat{M}\widehat{P} = \hat{1} \tag{8}$$

and thus display the Weyl group structure. If the considered polynomial set $\{p_n(x)\}_{n\in\mathbb{N}}$ is quasimonomial, its properties can easily be derived from those of the $\widehat{M}$ and $\widehat{P}$ operators. In fact,

(i) Combining the recurrences (6) and (7), we have

$$\widehat{M}\ \widehat{P}\{p_n(x)\} = np_n(x) \tag{9}$$

which can be interpreted as the differential equation satisfied by $p_n(x)$, if $\widehat{M}$ and $\widehat{P}$ have a differential realization.

(ii) Assuming here and in the sequel $p_0(x) = 1$, then $p_n(x)$ can be explicitly constructed as

$$p_n(x) = \widehat{M}^n\{p_0(x)\} = \widehat{M}^n\{1\} \tag{10}$$

which yields the series definition for $p_n(x)$.

(iii) Identity (10) implies that the exponential generating function of $p_n(x)$ can be given in the form

$$\exp(t\widehat{M})\{1\} = \sum_{n=0}^{\infty} p_n(x)\frac{t^n}{n!} \quad (|t| < \infty). \tag{11}$$

We note that the Appell polynomials $A_n(x)$ are quasimonomial with respect to the following multiplicative and derivative operators:

$$\widehat{M}_A = x + \frac{A'(D_x)}{A(D_x)} \tag{12a}$$

or, equivalently,

$$\widehat{M}_A = x - \frac{g'(D_x)}{g(D_x)}, \tag{12b}$$

$$\widehat{P}_A = D_x, \tag{13}$$

respectively.

The special polynomials of two variables are useful from the point of view of applications in physics. Also, these polynomials allow the derivation of a number of useful identities in a fairly straightforward way and help in introducing new families of special polynomials. For example, Bretti et al. [17] introduced general classes of two variables Appell polynomials by using properties of an iterated isomorphism, related to the Laguerre-type exponentials.

We consider the 2-variable general polynomial (2VgP) family $p_n(x, y)$ defined by the generating function

$$e^{xt}\phi(y,t) = \sum_{n=0}^{\infty} p_n(x,y)\frac{t^n}{n!} \quad (p_0(x,y) = 1), \tag{14}$$

where $\phi(y,t)$ has (at least the formal) series expansion

$$\phi(y,t) = \sum_{n=0}^{\infty}\phi_n(y)\frac{t^n}{n!} \quad (\phi_0(y) \neq 0). \tag{15}$$

We recall that the 2-variable Hermite Kampé de Fériet polynomials (2VHKdFP) $H_n(x, y)$ [18], the Gould-Hopper polynomials (GHP) $H_n^{(m)}(x, y)$ [19], and the Hermite-Appell polynomials (HAP) ${}_HA_n(x, y)$ [20] are defined by the generating functions

$$e^{xt+yt^2} = \sum_{n=0}^{\infty} H_n(x,y)\frac{t^n}{n!}, \tag{16}$$

$$e^{xt+yt^m} = \sum_{n=0}^{\infty} H_n^{(m)}(x,y)\frac{t^n}{n!}, \tag{17}$$

$$A(t)e^{xt+yt^2} = \sum_{n=0}^{\infty} {}_HA_n(x,y)\frac{t^n}{n!}, \tag{18}$$

respectively. Thus, in view of generating functions (14), (16), (17), and (18), we note that the 2VHKdFP $H_n(x, y)$, the GHP $H_n^{(m)}(x, y)$, and the HAP ${}_HA_n(x, y)$ belong to 2VgP family.

In this paper, operational methods are used to introduce certain new families of special polynomials related to the Appell polynomials. In Section 2, some results for the 2-variable general polynomials (2VgP) $p_n(x, y)$ are derived. Further, the 2-variable general-Appell polynomials (2VgAP) ${}_pA_n(x, y)$ are introduced and framed within the context of monomiality principle. In Section 3, the Gould-Hopper-Appell polynomials (GHAP) ${}_HA_n^{(m)}(x, y)$ are considered, and their properties are established. Some members belonging to the Gould-Hopper-Appell polynomial family are given.

2. 2-Variable General-Appell Polynomials

In order to introduce the 2-variable general-Appell polynomials (2VgAP), we need to establish certain results for the 2VgP $p_n(x, y)$. Therefore, first we prove the following results for the 2VgP $p_n(x, y)$.

Lemma 1. *The 2VgP $p_n(x, y)$ defined by generating function (14), where $\phi(y, t)$ is given by (15), are quasimonomial under the action of the following multiplicative and derivative operators:*

$$\widehat{M}_p = x + \frac{\phi'(y, D_x)}{\phi(y, D_x)} \quad \left(\phi'(x,t) := \frac{\partial}{\partial t}\phi(x,t)\right), \tag{19}$$

$$\widehat{P}_p = D_x, \tag{20}$$

respectively.

Proof. Differentiating (14) partially with respect to t, we have

$$\left(x + \frac{\phi'(y,t)}{\phi(y,t)}\right)\{e^{xt}\phi(y,t)\} = \sum_{n=0}^{\infty} p_{n+1}(x,y)\frac{t^n}{n!}. \tag{21}$$

TABLE 2: List of some Gould-Hopper-Appell polynomials ${}_{H}A_n^{(m)}(x, y)$.

S. No.	$g(t); A(t)$	$\widehat{M}_{H^{(m)}A}; \widehat{P}_{H^{(m)}A}$	Generating functions	Polynomials
(I)	$\frac{(e^t - 1)}{t}; \frac{t}{(e^t - 1)}$	$x + myD_x^{m-1} + \frac{((1 - D_x) e^{D_x} - 1)}{D_x (e^{D_x} - 1)}; D_x$	$\frac{t}{(e^t - 1)} e^{xt+yt^m} = \sum_{n=0}^{\infty} {}_{H}B_n^{(m)}(x, y) \frac{t^n}{n!}$	The Gould-Hopper-Bernoulli polynomials
(II)	$\frac{(e^t + 1)}{2}; \frac{2}{(e^t + 1)}$	$x + myD_x^{m-1} - \frac{e^{D_x}}{(e^{D_x} + 1)}; D_x$	$\frac{2}{(e^t + 1)} e^{xt+yt^m} = \sum_{n=0}^{\infty} {}_{H}E_n^{(m)}(x, y) \frac{t^n}{n!}$	The Gould Hopper-Euler polynomials
(III)	$\frac{(e^t - 1)^{\alpha}}{t^{\alpha}}; \frac{t^{\alpha}}{(e^t - 1)^{\alpha}}$	$x + myD_x^{m-1} + \frac{\alpha((1 - D_x) e^{D_x} - 1)}{D_x (e^{D_x} - 1)}; D_x$	$\frac{t^{\alpha}}{(e^t - 1)^{\alpha}} e^{xt+yt^m} = \sum_{n=0}^{\infty} {}_{H}B_n^{(m,\alpha)}(x, y) \frac{t^n}{n!}$	The Gould-Hopper-generalized Bernoulli polynomials
(IV)	$\frac{(e^t + 1)^{\alpha}}{2^{\alpha}}; \frac{2^{\alpha}}{(e^t + 1)^{\alpha}}$	$x + myD_x^{m-1} - \frac{\alpha e^{D_x}}{(e^{D_x} + 1)}; D_x$	$\frac{2^{\alpha}}{(e^t + 1)^{\alpha}} e^{xt+yt^m} = \sum_{n=0}^{\infty} {}_{H}E_n^{(m,\alpha)}(x, y) \frac{t^n}{n!}$	The Gould-Hopper-generalized Euler polynomials
(V)	$\prod_{h=1}^{s} \frac{(e^{\alpha_h t} - 1)}{\alpha_h t^s}$; $\prod_{h=1}^{s} \frac{\alpha_h}{(e^{\alpha_h t} - 1)} t^s$	$x + myD_x^{m-1} + mD_x^{-1} - \sum_{r=1}^{m} \frac{\alpha_r e^{\alpha_r D_x}}{(e^{\alpha_r D_x} - 1)}; D_x$	$\frac{\alpha_1 \cdots \alpha_s t^s}{(e^{\alpha_1 t} - 1) \cdots (e^{\alpha_s t} - 1)} e^{xt+yt^m} = \sum_{n=0}^{\infty} {}_{H}B_n^{(m,s)}(x, y \mid \alpha_1, \ldots, \alpha_s) \frac{t^n}{n!}$	The Gould-Hopper-generalized Bernoulli polynomials of order s
(VI)	$\prod_{h=1}^{s} \frac{(e^{\alpha_h t} + 1)}{2^s}$; $\prod_{h=1}^{s} \frac{2^s}{(e^{\alpha_h t} + 1)}$	$x + myD_x^{m-1} - \sum_{r=1}^{m} \alpha_r e^{\alpha_r D_x}; D_x$	$\frac{2^s}{(e^{\alpha_1 t} + 1) \cdots (e^{\alpha_s t} + 1)} e^{xt+yt^m} = \sum_{n=0}^{\infty} {}_{H}E_n^{(m,s)}(x, y \mid \alpha_1, \ldots, \alpha_s) \frac{t^n}{n!}$	The Gould-Hopper-generalized Euler polynomials of order s
(VII)	$\frac{e^t - \sum_{h=0}^{s-1} (t^h/h!)}{t^s}$; $\frac{t^s}{e^t - \sum_{h=0}^{s-1} (t^h/h!)}$	$x + myD_x^{m-1} + mD_x^{-1} - 1 - \frac{D_x^{m-1}/(m-1)!}{(e^{D_x} - \sum_{h=0}^{m-1} t^h/h!)}; D_x$	$\frac{t^s}{e^t - \sum_{h=0}^{s-1} (t^h/h!)} e^{xt+yt^m} = \sum_{n=0}^{\infty} {}_{H}B_n^{(m,[s-1])}(x, y) \frac{t^n}{n!}$	The Gould-Hopper new generalized Bernoulli polynomials

Table 2: Continued.

S. No.	$g(t)$; $A(t)$	$\widehat{M}_{H^{(m)}A}$; $\widehat{P}_{H^{(m)}A}$	Generating functions	Polynomials
(VIII)	$\left(\frac{t}{\lambda e^t-1}\right)^{-\alpha}$; $\left(\frac{t}{\lambda e^t-1}\right)^{\alpha}$	$x+myD_x^{m-1}+\frac{\lambda e^{D_x}(\alpha-D_x)-\alpha}{D_x(\lambda e^{D_x}-1)}$; D_x	$\left(\frac{t}{\lambda e^t-1}\right)^{\alpha}e^{xt+yt^m}=\sum_{n=0}^{\infty}{}_H\mathfrak{B}_n^{(m,\alpha)}(x,y;\lambda)\frac{t^n}{n!}$	The Gould-Hopper-Apostol-Bernoulli polynomials of order α
(IX)	$\frac{\lambda e^t-1}{t}$; $\frac{t}{\lambda e^t-1}$	$x+myD_x^{m-1}+\frac{\lambda e^{D_x}(1-D_x)-1}{D_x(\lambda e^{D_x}-1)}$; D_x	$\frac{t}{\lambda e^t-1}e^{xt+yt^m}=\sum_{n=0}^{\infty}{}_H\mathfrak{B}_n^{(m)}(x,y;\lambda)\frac{t^n}{n!}$	The Gould-Hopper-Apostol-Bernoulli polynomials
(X)	$\left(\frac{2}{\lambda e^t+1}\right)^{-\alpha}$; $\left(\frac{2}{\lambda e^t+1}\right)^{\alpha}$	$x+myD_x^{m-1}-\frac{\lambda\alpha e^{D_x}}{(\lambda e^{D_x}+1)}$; D_x	$\left(\frac{2}{\lambda e^t+1}\right)^{\alpha}e^{xt+yt^m}=\sum_{n=0}^{\infty}{}_H\mathscr{E}_n^{(m,\alpha)}(x,y;\lambda)\frac{t^n}{n!}, \lvert t+\log\lambda\rvert<\pi$; $1^{\alpha}:=1$	The Gould-Hopper-Apostol-Euler polynomials of order α
(XI)	$\frac{\lambda e^t+1}{2}$; $\frac{2}{\lambda e^t+1}$	$x+myD_x^{m-1}-\frac{\lambda e^{D_x}}{(\lambda e^{D_x}+1)}$; D_x	$\frac{2}{\lambda e^t+1}e^{xt+yt^m}=\sum_{n=0}^{\infty}{}_H\mathscr{E}_n^{(m)}(x,y;\lambda)\frac{t^n}{n!}, \lvert t+\log\lambda\rvert<\pi$	The Gould-Hopper-Apostol-Euler polynomials
(XII)	$e^{-(\sum_{h=0}^{r+1}\xi_h t^h)}$; $e^{(\sum_{h=0}^{r+1}\xi_h t^h)}$; $\xi_{r+1}\neq 0$	$x+myD_x^{m-1}+\sum_{h=1}^{r+1}h\xi_h D_x^{h-1}$; D_x	$e^{(\sum_{h=0}^{r+1}\xi_h t^h+xt+yt^m)}=\sum_{n=0}^{\infty}{}_HA_n^{(m)}(x,y)\frac{t^n}{n!}$	The Gould-Hopper-generalized Gould-Hopper polynomials (for $r=1$, the Gould-Hopper-Hermite polynomials ${}_HH_n^{(m)}(x,y)$ and for $r=2$, the Gould-Hopper classical 2-orthogonal polynomials)
(XIII)	$(1-t)^{s+1}$; $\frac{1}{(1-t)^{s+1}}$	$x+myD_x^{m-1}-\frac{m+1}{1-D_x}$; D_x	$\frac{1}{(1-t)^{s+1}}e^{xt+yt^m}=\sum_{n=0}^{\infty}{}_HG_n^{(m,s)}(x,y)t^n$	The Gould-Hopper-Miller-Lee polynomials (for $s=0$, the Gould-Hopper-truncated exponential polynomials ${}_He_n^{m}(x,y)$ and for $s=\beta-1$ Gould-Hopper-modified Laguerre polynomials ${}_Hf_n^{(m,\beta)}(x,y)$)
(XIV)	$\frac{(e^t+1)}{2t}$; $\frac{2t}{(e^t+1)}$	$x+myD_x^{m-1}+\frac{e^{D_x}(1-D_x)+1}{(e^{D_x}+1)}$; D_x	$\frac{2t}{(e^t+1)}e^{xt+yt^m}=\sum_{n=0}^{\infty}{}_HG_n^{(m)}(x,y)\frac{t^n}{n!}$	The Gould-Hopper-Genocchi polynomials

If $\phi(y,t)$ is an invertible series and $\phi'(y,t)/\phi(y,t)$ has Taylor's series expansion in powers of t, then in view of the identity

$$D_x\left\{e^{xt}\phi(y,t)\right\} = t\left(e^{xt}\phi(y,t)\right) \quad (22)$$

we can write

$$\frac{\phi'(y,D_x)}{\phi(y,D_x)}\left\{e^{xt}\phi(y,t)\right\} = \frac{\phi'(y,t)}{\phi(y,t)}\left(e^{xt}\phi(y,t)\right). \quad (23)$$

Now, using (23) in the l.h.s. of (21), we find

$$\left(x + \frac{\phi'(y,D_x)}{\phi(y,D_x)}\right)\left\{e^{xt}\phi(y,t)\right\} = \sum_{n=0}^{\infty} p_{n+1}(x,y)\frac{t^n}{n!}. \quad (24)$$

Making use of generating function (14) in the l.h.s. of the above equation, we have

$$\left(x + \frac{\phi'(y,D_x)}{\phi(y,D_x)}\right)\left\{\sum_{n=0}^{\infty} p_n(x,y)\frac{t^n}{n!}\right\} = \sum_{n=0}^{\infty} p_{n+1}(x,y)\frac{t^n}{n!} \quad (25)$$

which, on equating the coefficients of like powers of t in both sides, gives

$$\left(x + \frac{\phi'(y,D_x)}{\phi(y,D_x)}\right)\{p_n(x,y)\} = p_{n+1}(x,y). \quad (26)$$

Thus, in view of monomiality principle equation (6), the above equation yields assertion (19) of Lemma 1. Again, using identity (22) in (14), we have

$$D_x\left\{\sum_{n=0}^{\infty} p_n(x,y)\frac{t^n}{n!}\right\} = \sum_{n=1}^{\infty} p_{n-1}(x,y)\frac{t^n}{(n-1)!}. \quad (27)$$

Equating the coefficients of like powers of t in both sides of (27), we find

$$D_x\{p_n(x,y)\} = np_{n-1}(x,y) \quad (n \geq 1) \quad (28)$$

which in view of monomiality principle equation (7) yields assertion (20) of Lemma 1. □

Remark 2. The operators given by (19) and (20) satisfy commutation relation (8). Also, using expressions (19) and (20) in (9), we get the following differential equation satisfied by 2VgP $p_n(x,y)$:

$$\left(xD_x + \frac{\phi'(y,D_x)}{\phi(y,D_x)}D_x - n\right)p_n(x,y) = 0. \quad (29)$$

Remark 3. Since $p_0(x,y) = 1$, therefore, in view of monomiality principle equation (10), we have

$$p_n(x,y) = \left(x + \frac{\phi'(y,D_x)}{\phi(y,D_x)}\right)^n\{1\} \quad (p_0(x,y) = 1). \quad (30)$$

Also, in view of (11), (14), and (19), we have

$$\exp\left(\widehat{M}_p t\right)\{1\} = e^{xt}\phi(y,t) = \sum_{n=0}^{\infty} p_n(x,y)\frac{t^n}{n!}. \quad (31)$$

Now, we proceed to introduce the 2-variable general-Appell polynomials (2VgAP). In order to derive the generating functions for the 2VgAP, we take the 2VgP $p_n(x,y)$ as the base in the Appell polynomials generating function (1). Thus, replacing x by the multiplicative operator $\widehat{M}_p$ of the 2VgP $p_n(x,y)$ in the l.h.s. of (1) and denoting the resultant 2VgAP by ${}_pA_n(x,y)$, we have

$$A(t)e^{(\widehat{M}_p t)} = \sum_{n=0}^{\infty} {}_pA_n(x,y)\frac{t^n}{n!}. \quad (32)$$

Now, using (31) in the exponential term in the l.h.s. of (32), we get the generating function for ${}_pA_n(x,y)$ as

$$A(t)e^{xt}\phi(y,t) = \sum_{n=0}^{\infty} {}_pA_n(x,y)\frac{t^n}{n!}. \quad (33)$$

In view of (5), generating function (33) can be expressed equivalently as

$$\frac{1}{g(t)}e^{xt}\phi(y,t) = \sum_{n=0}^{\infty} {}_pA_n(x,y)\frac{t^n}{n!}. \quad (34)$$

Now, we frame the 2VgAP ${}_pA_n(x,y)$ within the context of monomiality principle formalism. We prove the following results.

Theorem 4. *The 2VgAP ${}_pA_n(x,y)$ are quasimonomial with respect to the following multiplicative and derivative operators:*

$$\widehat{M}_{pA} = x + \frac{\phi'(y,D_x)}{\phi(y,D_x)} + \frac{A'(D_x)}{A(D_x)} \quad (35a)$$

or, equivalently,

$$\widehat{M}_{pA} = x + \frac{\phi'(y,D_x)}{\phi(y,D_x)} - \frac{g'(D_x)}{g(D_x)}, \quad (35b)$$

$$\widehat{P}_{pA} = D_x, \quad (36)$$

respectively.

Proof. Differentiating (33) partially with respect to t, we find

$$\left(x + \frac{\phi'(y,t)}{\phi(y,t)} + \frac{A'(t)}{A(t)}\right)A(t)e^{xt}\phi(y,t) = \sum_{n=0}^{\infty} {}_pA_{n+1}(x,y)\frac{t^n}{n!}. \quad (37)$$

Since $A(t)$ and $\phi(y,t)$ are invertible series of t, therefore, $A'(t)/A(t)$ and $\phi'(y,t)/\phi(y,t)$ possess power series expansions of t. Thus, in view of identity (22), we have

$$\left(x + \frac{\phi'(y,D_x)}{\phi(y,D_x)} + \frac{A'(D_x)}{A(D_x)}\right)\left\{A(t)e^{xt}\phi(y,t)\right\} = \sum_{n=0}^{\infty} {}_pA_{n+1}(x,y)\frac{t^n}{n!} \quad (38)$$

which, on using generating function (33), becomes

$$\left(x+\frac{\phi'(y,D_x)}{\phi(y,D_x)}+\frac{A'(D_x)}{A(D_x)}\right)\left\{\sum_{n=0}^{\infty}{}_pA_n(x,y)\frac{t^n}{n!}\right\}=\sum_{n=0}^{\infty}{}_pA_{n+1}(x,y)\frac{t^n}{n!} \quad (39)$$

or, equivalently,

$$\sum_{n=0}^{\infty}\left(x+\frac{\phi'(y,D_x)}{\phi(y,D_x)}+\frac{A'(D_x)}{A(D_x)}\right)\{{}_pA_n(x,y)\}\frac{t^n}{n!}=\sum_{n=0}^{\infty}{}_pA_{n+1}(x,y)\frac{t^n}{n!}. \quad (40)$$

Now, equating the coefficients of like powers of t in the above equation, we find

$$\left(x+\frac{\phi'(y,D_x)}{\phi(y,D_x)}+\frac{A'(D_x)}{A(D_x)}\right)\{{}_pA_n(x,y)\}={}_pA_{n+1}(x,y) \quad (41)$$

which in view of (6) yields assertion (35a) of Theorem 4. Also, in view of relation (5), assertion (35a) can be expressed equivalently as (35b).

Again, in view of identity (22), we have

$$D_x\{A(t)e^{xt}\phi(y,t)\}=tA(t)e^{xt}\phi(y,t) \quad (42)$$

which on using generating function (33) becomes

$$D_x\left\{\sum_{n=0}^{\infty}{}_pA_n(x,y)\frac{t^n}{n!}\right\}=\sum_{n=1}^{\infty}{}_pA_{n-1}(x,y)\frac{t^n}{(n-1)!}. \quad (43)$$

Equating the coefficients of like powers of t in the above equation, we find

$$D_x\{{}_pA_n(x,y)\}=n\,{}_pA_{n-1}(x,y) \quad (n\ge 1) \quad (44)$$

which, in view of (7), yields assertion (36) of Theorem 4. □

Theorem 5. *The 2VgAP ${}_pA_n(x,y)$ satisfy the following differential equations*

$$\left(xD_x+\frac{\phi'(y,D_x)}{\phi(y,D_x)}D_x+\frac{A'(D_x)}{A(D_x)}D_x-n\right){}_pA_n(x,y)=0 \quad (45a)$$

or, equivalently,

$$\left(xD_x+\frac{\phi'(y,D_x)}{\phi(y,D_x)}D_x-\frac{g'(D_x)}{g(D_x)}D_x-n\right){}_pA_n(x,y)=0. \quad (45b)$$

Proof. Using (35a) and (36) in (9), we get assertion (45a). Further, using (35b) and (36) in (9), we get assertion (45b). □

Note 1. With the help of the results derived above and by taking $A(t)$ (or $g(t)$) of the Appell polynomials listed in Table 1, we can derive the generating function and other results for the members belonging to 2VgAP family.

3. Examples

We consider examples of certain members belonging to the 2VgAP family.

Taking $\phi(y,t)=e^{yt^m}$ (that is, when the 2VgP $p_n(x,y)$ reduces to the GHP $H_n^{(m)}(x,y)$) in generating function (33), we find that the Gould-Hopper-Appell polynomials (GHAP) ${}_HA_n^{(m)}(x,y)$ are defined by the following generating function:

$$A(t)e^{(xt+yt^m)}=\sum_{n=0}^{\infty}{}_HA_n^{(m)}(x,y)\frac{t^n}{n!} \quad (46)$$

or, equivalently,

$$\frac{1}{g(t)}e^{(xt+yt^m)}=\sum_{n=0}^{\infty}{}_HA_n^{(m)}(x,y)\frac{t^n}{n!}. \quad (47)$$

Using (1) in (46) (or (3) in (47)), we get the following series definition for ${}_HA_n^{(m)}(x,y)$ in terms of the Appell polynomials $A_n(x)$:

$${}_HA_n^{(m)}(x,y)=n!\sum_{r=0}^{[n/m]}\frac{A_{n-mr}(x)\,y^r}{(n-mr)!r!}. \quad (48)$$

In view of (35a), (35b), and (36), we note that the GHAP ${}_HA_n^{(m)}(x,y)$ are quasimonomial under the action of the following multiplicative and derivative operators:

$$\widehat{M}_{H^{(m)}A}=x+myD_x^{m-1}+\frac{A'(D_x)}{A(D_x)} \quad (49a)$$

or, equivalently,

$$\widehat{M}_{H^{(m)}A}=x+myD_x^{m-1}-\frac{g'(D_x)}{g(D_x)}, \quad (49b)$$

$$\widehat{P}_{H^{(m)}A}=D_x, \quad (50)$$

respectively. Also, in view of (45a) and (45b), we find that the GHAP ${}_HA_n^{(m)}(x,y)$ satisfy the following differential equation:

$$\left(xD_x+myD_x^m+\frac{A'(D_x)}{A(D_x)}D_x-n\right){}_HA_n^{(m)}(x,y)=0 \quad (51a)$$

or, equivalently,

$$\left(xD_x+myD_x^m-\frac{g'(D_x)}{g(D_x)}D_x-n\right){}_HA_n^{(m)}(x,y)=0. \quad (51b)$$

Remark 6. In view of (16) and (17), we note that, for $m=2$, the GHAP ${}_HA_n^{(m)}(x,y)$ reduce to the Hermite-Appell polynomials (HAP) ${}_HA_n(x,y)$. Therefore, taking $m=2$ in (46), (47), (48), (49a), (49b), (50), (51a), and (51b), we get the corresponding results for the HAP ${}_HA_n(x,y)$.

Remark 7. In view of (16), we note that the 2VHKdFP $H_n(x, y)$ are related to the classical Hermite polynomials $H_n(x)$ [11] or $He_n(x)$ as

$$H_n(2x, -1) = H_n(x), \quad H_n\left(x, -\frac{1}{2}\right) = He_n(x). \tag{52}$$

Therefore, taking $y = -1$ and replacing x by $2x$ (or taking $y = -1/2$) in (46)–(51b), we get the corresponding results for the classical Hermite-Appell polynomials ${}_HA_n(x)$ (or ${}_{He}A_n(x)$).

There are several members belonging to 2VgP family. Thus the results for the corresponding 2VgAP can be obtained by taking other examples. We present the list of some members belonging to the GHAP family in Table 2.

Note 2. Since, for $m = 2$, the GHAP ${}_HA_n^m(x, y)$ reduce to the HAP ${}_HA_n(x, y)$ therefore, for $m = 2$, Table 2 gives the list of the corresponding HAP ${}_HA_n(x, y)$.

The results established in this paper are general and include new families of special polynomials, consequently introducing many new special polynomials.

Appendix

New classes of Bernoulli numbers and polynomials are introduced, which are used to evaluate partial sums involving other special polynomials, see, for example, [21, 22]. Here, we consider the Gould-Hopper-Bernoulli polynomials (GHBP), Gould-Hopper-Euler polynomials (GHEP), Hermite-Bernoulli polynomials (HBP), Hermite-Euler polynomials (HEP), classical Hermite-Bernoulli polynomials (cHBP), and classical Hermite-Euler polynomials (cHEP). We give the surface plots of these polynomials for suitable values of the parameters and indices. Also, we give the graphs of the corresponding single-variable polynomials.

The GHBP ${}_HB_n^m(x, y)$, GHEP ${}_HE_n^m(x, y)$, HBP ${}_HB_n(x, y)$, and HEP ${}_HE_n(x, y)$ are defined by the following series:

$${}_HB_n^{(m)}(x, y) = n! \sum_{r=0}^{[n/m]} \frac{B_{n-mr}(x)\, y^r}{(n - mr)!r!}, \tag{A.1}$$

$${}_HE_n^{(m)}(x, y) = n! \sum_{r=0}^{[n/m]} \frac{E_{n-mr}(x)\, y^r}{(n - mr)!r!}, \tag{A.2}$$

$${}_HB_n(x, y) = n! \sum_{r=0}^{[n/2]} \frac{B_{n-2r}(x)\, y^r}{(n - 2r)!r!}, \tag{A.3}$$

$${}_HE_n(x, y) = n! \sum_{r=0}^{[n/2]} \frac{E_{n-2r}(x)\, y^r}{(n - 2r)!r!}, \tag{A.4}$$

respectively. Taking $y = -1/2$ in (A.3) and (A.4), we get the series definitions for the cHBP ${}_{He}B_n(x)$ and cHEP ${}_{He}E_n(x)$ as

$${}_HB_n(x) = n! \sum_{r=0}^{[n/2]} \left(\frac{-1}{2}\right)^r \frac{B_{n-2r}(x)}{(n - 2r)!r!}, \tag{A.5}$$

$${}_HE_n(x) = n! \sum_{r=0}^{[n/2]} \left(\frac{-1}{2}\right)^r \frac{E_{n-2r}(x)}{(n - 2r)!r!}, \tag{A.6}$$

respectively.

To draw the surface plots of these polynomials, we use the values of the Bernoulli polynomials $B_n(x)$ and the Euler polynomials $E_n(x)$ for $n = 0, 1, 2, 3, 4,$ and 5. We give the list of the first few Bernoulli and the Euler polynomials in Table 3.

Now, we consider the GHBP ${}_HB_n^m(x, y)$, GHEP ${}_HE_n^m(x, y)$, HBP ${}_HB_n(x, y)$, HEP ${}_HE_n(x, y)$, cHBP ${}_{He}B_n(x)$, and cHEP ${}_{He}E_n(x)$, for $m = 3$ and $n = 5$, so that we have

$${}_HB_5^{(3)}(x, y) = B_5(x) + 60B_2(x)\, y, \tag{A.7}$$

$${}_HE_5^{(3)}(x, y) = E_5(x) + 60E_2(x)\, y, \tag{A.8}$$

$${}_HB_5(x, y) = B_5(x) + 20B_3(x)\, y + 60B_1(x)\, y^2, \tag{A.9}$$

$${}_HE_5(x, y) = E_5(x) + 20E_3(x)\, y + 60E_1(x)\, y^2, \tag{A.10}$$

$${}_{He}B_5(x) = B_5(x) - 10B_3(x) + 15B_1(x), \tag{A.11}$$

$${}_{He}E_5(x) = E_5(x) - 10E_3(x) + 15E_1(x), \tag{A.12}$$

respectively. Using the particular values of $B_n(x)$ and $E_n(x)$ given in Table 3, we find

$${}_HB_5^{(3)}(x, y) = x^5 - \frac{5}{2}x^4 + \frac{5}{3}x^3 + 60x^2y - 60xy - \frac{1}{6}x + \frac{1}{10}y, \tag{A.13}$$

$${}_HE_5^{(3)}(x, y) = x^5 - \frac{5}{2}x^4 + 60x^2y + \frac{5}{3}x^2 - 60xy - \frac{1}{2} + \frac{1}{10}y, \tag{A.14}$$

$$\begin{aligned} {}_HB_5(x, y) = {} & x^5 - \frac{5}{2}x^4 + 20x^3y + \frac{5}{3}x^3 + 60xy^2 \\ & - 60x^2y + 10xy - 30y^2 - \frac{1}{6}x, \end{aligned} \tag{A.15}$$

$$\begin{aligned} {}_HE_5(x, y) = {} & x^5 - \frac{5}{2}x^4 + 20x^3y + 60xy^2 \\ & - 30x^2y + \frac{5}{3}x^2 - 30y^2 + \frac{10}{3}y - \frac{1}{2}, \end{aligned} \tag{A.16}$$

$${}_{He}B_5(x) = x^5 - \frac{5}{2}x^4 - \frac{25}{3}x^3 + 15x^2 + \frac{59}{6}x - \frac{15}{2}, \tag{A.17}$$

$${}_{He}E_5(x) = x^5 - \frac{5}{2}x^4 - 10x^3 + \frac{50}{3}x^2 + 15x - \frac{29}{3}, \tag{A.18}$$

respectively.

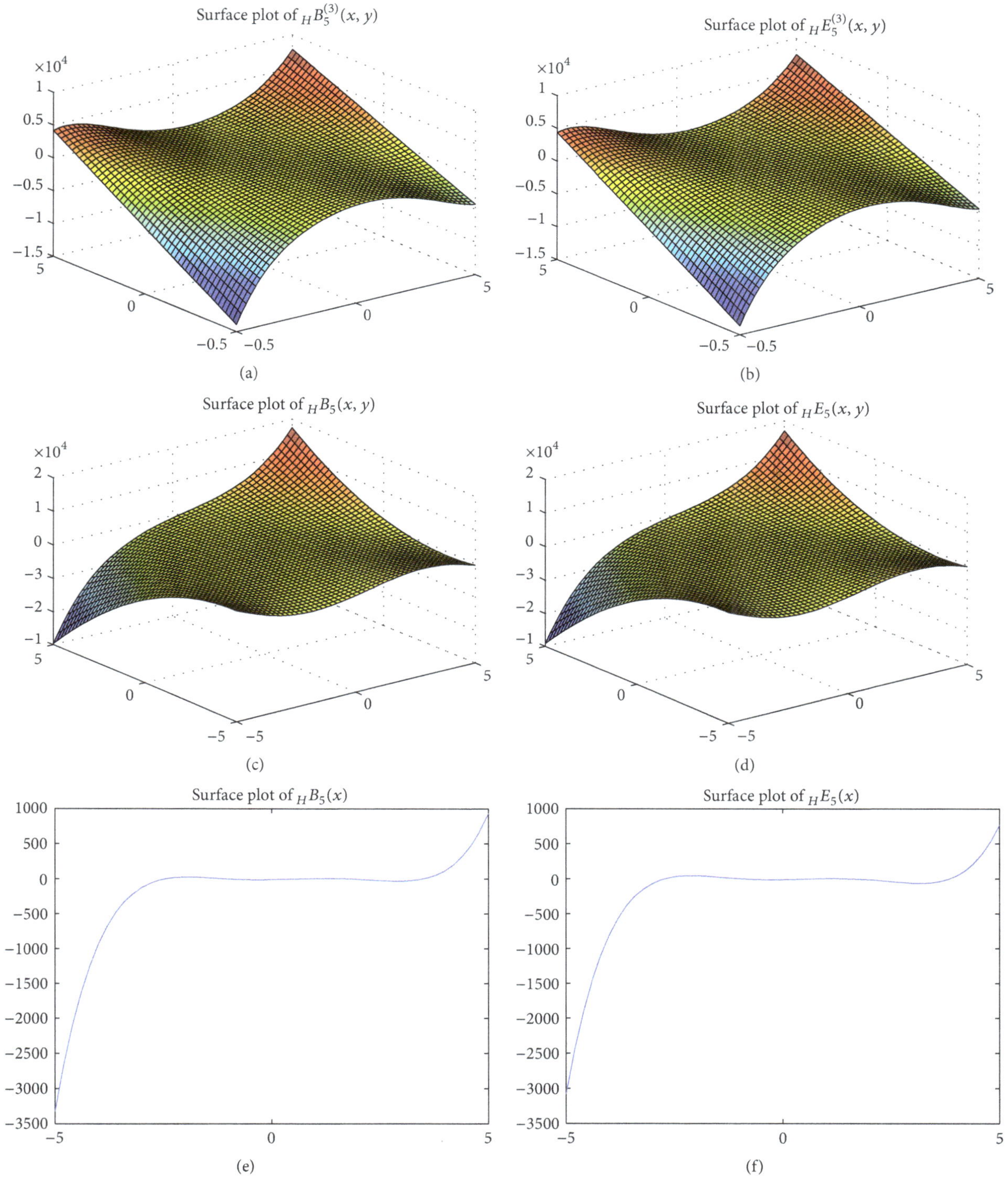

FIGURE 1

TABLE 3: List of the first few Bernoulli and the Euler polynomials.

n	0	1	2	3	4	5
$B_n(x)$	1	$x-\frac{1}{2}$	$x^2-x+\frac{1}{6}$	$x^3-\frac{3}{2}x^2+\frac{x}{2}$	$x^4-2x^3+x^2-\frac{1}{30}$	$x^5-\frac{5}{2}x^4+\frac{5}{3}x^3-\frac{x}{6}$
$E_n(x)$	1	$x-\frac{1}{2}$	x^2-x	$x^3-\frac{3}{2}x^2+\frac{1}{6}$	$x^4-2x^3+\frac{2}{3}x$	$x^5-\frac{5}{2}x^4+\frac{5}{3}x^2-\frac{1}{2}$

In view of equations (A.13)–(A.18), we get Figure 1.

Acknowledgments

The authors are thankful to the anonymous referee for useful suggestions towards the improvement of the paper. This work has been done under the Senior Research Fellowship (Office Memo no. Acad/D-1562/MR) sanctioned to the second author by the University Grants Commission, Government of India, New Delhi.

References

[1] P. Appell, "Sur une classe de polynômes," *Annales Scientifiques de l'École Normale Supérieure*, vol. 9, pp. 119–144, 1880.

[2] S. Roman, *The Umbral Calculus*, vol. 111 of *Pure and Applied Mathematics*, Academic Press, New York, NY, USA, 1984.

[3] E. D. Rainville, *Special Functions*, Macmillan, New York, NY, USA, 1960, reprinted by Chelsea, Bronx, NY, USA, 1971.

[4] A. Erdélyi, W. Magnus, F. Oberhettinger, and F. G. Tricomi, *Higher Transcendental Functions. Vol. III*, McGraw-Hill, New York, NY, USA, 1955.

[5] A. Erdélyi, W. Magnus, F. Oberhettinger, and F. G. Tricomi, *Higher Transcendental Functions. Vol. II*, McGraw-Hill, New York, NY, USA, 1953.

[6] G. Bretti, P. Natalini, and P. E. Ricci, "Generalizations of the Bernoulli and Appell polynomials," *Abstract and Applied Analysis*, vol. 2004, no. 7, pp. 613–623, 2004.

[7] Q.-M. Luo and H. M. Srivastava, "Some generalizations of the Apostol-Bernoulli and Apostol-Euler polynomials," *Journal of Mathematical Analysis and Applications*, vol. 308, no. 1, pp. 290–302, 2005.

[8] T. M. Apostol, "On the Lerch zeta function," *Pacific Journal of Mathematics*, vol. 1, pp. 161–167, 1951.

[9] Q.-M. Luo, "Apostol-Euler polynomials of higher order and Gaussian hypergeometric functions," *Taiwanese Journal of Mathematics*, vol. 10, no. 4, pp. 917–925, 2006.

[10] K. Douak, "The relation of the d-orthogonal polynomials to the Appell polynomials," *Journal of Computational and Applied Mathematics*, vol. 70, no. 2, pp. 279–295, 1996.

[11] L. C. Andrews, *Special Functions for Engineers and Applied Mathematicians*, Macmillan, New York, NY, USA, 1985.

[12] G. Dattoli, S. Lorenzutta, and D. Sacchetti, "Integral representations of new families of polynomials," *Italian Journal of Pure and Applied Mathematics*, no. 15, pp. 19–28, 2004.

[13] W. Magnus, F. Oberhettinger, and R. P. Soni, *Formulas and Theorems for the Special Functions of Mathematical Physics*, vol. 52 of *Die Grundlehren der mathematischen Wissenschaften*, Springer, New York, NY, USA, 3rd edition, 1966.

[14] G. Dattoli, M. Migliorati, and H. M. Srivastava, "Sheffer polynomials, monomiality principle, algebraic methods and the theory of classical polynomials," *Mathematical and Computer Modelling*, vol. 45, no. 9-10, pp. 1033–1041, 2007.

[15] J. F. Steffensen, "The poweroid, an extension of the mathematical notion of power," *Acta Mathematica*, vol. 73, pp. 333–366, 1941.

[16] G. Dattoli, "Hermite-Bessel and Laguerre-Bessel functions: a by-product of the monomiality principle," in *Advanced Special Functions and Applications (Melfi, 1999)*, vol. 1 of *Proc. Melfi Sch. Adv. Top. Math. Phys.*, pp. 147–164, Aracne, Rome, Italy, 2000.

[17] G. Bretti, C. Cesarano, and P. E. Ricci, "Laguerre-type exponentials and generalized Appell polynomials," *Computers & Mathematics with Applications*, vol. 48, no. 5-6, pp. 833–839, 2004.

[18] P. Appell and J. Kampé de Fériet, *Fonctions Hypergéométriques et Hypersphériques: Polynômes d' Hermite*, Gauthier-Villars, Paris, France, 1926.

[19] H. W. Gould and A. T. Hopper, "Operational formulas connected with two generalizations of Hermite polynomials," *Duke Mathematical Journal*, vol. 29, pp. 51–63, 1962.

[20] S. Khan, G. Yasmin, R. Khan, and N. A. M. Hassan, "Hermite-based Appell polynomials: properties and applications," *Journal of Mathematical Analysis and Applications*, vol. 351, no. 2, pp. 756–764, 2009.

[21] G. Dattoli, C. Cesarano, and S. Lorenzutta, "Bernoulli numbers and polynomials from a more general point of view," *Rendiconti di Matematica e delle sue Applicazioni*, vol. 22, pp. 193–202, 2002.

[22] G. Dattoli, S. Lorenzutta, and C. Cesarano, "Finite sums and generalized forms of Bernoulli polynomials," *Rendiconti di Matematica e delle sue Applicazioni*, vol. 19, no. 3, pp. 385–391, 1999.

12

On Certain Classes of Harmonic p-Valent Functions Defined by an Integral Operator

T. M. Seoudy

Department of Mathematics, Faculty of Science, Fayoum University, Fayoum 63514, Egypt

Correspondence should be addressed to T. M. Seoudy; tms00@fayoum.edu.eg

Academic Editor: Frédéric Robert

We obtain coefficient characterization, extreme points, and distortion bounds of certain classes of harmonic p-valent functions defined by an integral operator.

1. Introduction

A continuous complex-valued function $f = u+iv$ defined in a simply connected complex domain D is said to be harmonic in D if both u and v are real harmonic in D. In any simply connected domain, we can write

$$f = h + \overline{g}, \tag{1}$$

where h and g are analytic in D. We call h the analytic part and g the coanalytic part of f. A necessary and sufficient condition for f to be locally univalent and sense preserving in D is that $|h'(z)| > |g'(z)|$ in D (see [1]).

Denote by S_H the class of functions f of the form (1) that are harmonic univalent and sense preserving in the unit disc $U = \{z : |z| < 1\}$ for which $f(0) = f_z(0) - 1 = 0$.

Recently, Jahangiri and Ahuja [2] defined the class $\mathscr{H}_p(p \in \mathbb{N} = \{1, 2, 3, \ldots\})$, consisting of all p-valent harmonic functions $f = h + \overline{g}$ that are sense preserving in U and h, and g are of the form

$$h(z) = z^p + \sum_{k=p+1}^{\infty} a_k z^k, \qquad g(z) = \sum_{k=p}^{\infty} b_k z^k, \quad |b_p| < 1. \tag{2}$$

For $f = h + \overline{g}$ given by (2), we define the modified p-valent Salagean integral operator $I^n_{p,\lambda}$ of f (see [3] and also [4] when $p = 1$) as follows:

$$I^n_{p,\lambda} f(z) = I^n_{p,\lambda} h(z) + (-1)^n \overline{I^n_{p,\lambda} g(z)}, \tag{3}$$

where

$$I^n_{p,\lambda} h(z) = z^p + \sum_{k=p+1}^{\infty} \left(\frac{p}{p+\lambda(k-p)}\right)^n a_k z^k$$
$$(p \in \mathbb{N}; \lambda > 0;\ n \in \mathbb{N}_0 = \mathbb{N} \cup \{0\}),$$
$$I^n_{p,\lambda} g(z) = \sum_{k=p}^{\infty} \left(\frac{p}{p+\lambda(k-p)}\right)^n b_k z^k \tag{4}$$
$$(p \in \mathbb{N}; \lambda > 0;\ n \in \mathbb{N}_0).$$

For $p \in \mathbb{N}$, $\lambda > 0$, $n \in \mathbb{N}_0$, $0 \le \alpha < 1$, and $z \in U$, we let $\mathscr{H}_{p,\lambda}(n;\alpha)$ denote the family of harmonic functions f of the form (2) such that

$$\operatorname{Re}\left\{\frac{I^n_{p,\lambda} f(z)}{I^{n+1}_{p,\lambda} f(z)}\right\} > \alpha, \tag{5}$$

where $I^n_{p,\lambda} f$ is defined by (3).

We let the subclass $\mathscr{H}^-_{p,\lambda}(n;\alpha)$ consists of harmonic functions $f_n = h + \overline{g}_n$ in $\mathscr{H}_{p,\lambda}(n;\alpha)$ so that h and g_n are of the form

$$h(z) = z^p - \sum_{k=p+1}^{\infty} a_k z^k, \qquad g_n(z) = (-1)^n \sum_{k=p}^{\infty} b_k z^k, \tag{6}$$
$$a_k, b_k \ge 0.$$

We note that $\mathscr{H}^{-}_{p,1}(n;\alpha) = \mathscr{H}^{-}_{p}(n;\alpha)$, where the class $\mathscr{H}^{-}_{p}(n;\alpha)$ was defined and studied by Cotirla [5].

In this paper, we obtain coefficient characterization of the classes $\mathscr{H}_{p,\lambda}(n;\alpha)$ and $\mathscr{H}^{-}_{p,\lambda}(n;\alpha)$. We also obtain extreme points and distortion bounds for functions in the class $\mathscr{H}^{-}_{p,\lambda}(n;\alpha)$.

2. Coefficient Characterization

Unless otherwise mentioned, we assume throughout this paper that $p \in \mathbb{N}$, $n \in \mathbb{N}_0$, $0 \le \alpha < 1$, $a_p = 1$, and $\lambda > 0$. We begin with a sufficient condition for functions in $\mathscr{H}_{p,\lambda}(n;\alpha)$.

Theorem 1. *Let $f = h + \overline{g}$ so that h and g are given by (2). Furthermore, let*

$$\sum_{k=p}^{\infty} \left\{ \Psi_{p,\lambda}(n,k,\alpha) |a_k| + \Phi_{p,\lambda}(n,k,\alpha) |b_k| \right\} \le 2, \quad (7)$$

where

$$\Psi_{p,\lambda}(n,k,\alpha) = \left(\left(\frac{p}{p+\lambda(k-p)} \right)^{n} - \alpha \left(\frac{p}{p+\lambda(k-p)} \right)^{n+1} \right) \times (1-\alpha)^{-1}, \quad (8)$$

$$\Phi_{p,\lambda}(n,k,\alpha) = \left(\left(\frac{p}{p+\lambda(k-p)} \right)^{n} + \alpha \left(\frac{p}{p+\lambda(k-p)} \right)^{n+1} \right) \times (1-\alpha)^{-1}. \quad (9)$$

Then, f is sense preserving in U and $f \in \mathscr{H}_{p,\lambda}(n;\alpha)$.

Proof. According to (2) and (3), we only need to show that

$$\operatorname{Re} \left\{ \frac{I^{n}_{p,\lambda} f(z) - \alpha I^{n+1}_{p,\lambda} f(z)}{I^{n+1}_{p,\lambda} f(z)} \right\} \ge 0 \quad (z \in U). \quad (10)$$

It follows that

$$\begin{aligned}
&\operatorname{Re} \left\{ \frac{I^{n}_{p,\lambda} f(z) - \alpha I^{n+1}_{p,\lambda} f(z)}{I^{n+1}_{p,\lambda} f(z)} \right\} \\
&= \operatorname{Re} \Bigg\{ \Bigg((1-\alpha) z^p + \sum_{k=p+1}^{\infty} \times \left[\left(\frac{p}{p+\lambda(k-p)} \right)^{n} - \alpha \left(\frac{p}{p+\lambda(k-p)} \right)^{n+1} \right] a_k z^k \Bigg) \\
&\quad \times \left(z^p + \sum_{k=p+1}^{\infty} \left(\frac{p}{p+\lambda(k-p)} \right)^{n+1} a_k z^k + (-1)^{n+1} \sum_{k=p}^{\infty} \left(\frac{p}{p+\lambda(k-p)} \right)^{n+1} \overline{b}_k \overline{z}^k \right)^{-1} \\
&\quad + \left((-1)^{n} \sum_{k=p}^{\infty} \left[\left(\frac{p}{p+\lambda(k-p)} \right)^{n} + \alpha \left(\frac{p}{p+\lambda(k-p)} \right)^{n+1} \right] \overline{b}_k \overline{z}^k \right) \\
&\quad \times \left(z^p + \sum_{k=p+1}^{\infty} \left(\frac{p}{p+\lambda(k-p)} \right)^{n+1} a_k z^k + (-1)^{n+1} \sum_{k=p}^{\infty} \left(\frac{p}{p+\lambda(k-p)} \right)^{n+1} \overline{b}_k \overline{z}^k \right)^{-1} \Bigg\} \\
&= \operatorname{Re} \Bigg\{ \Bigg((1-\alpha) + \sum_{k=p+1}^{\infty} \times \left[\left(\frac{p}{p+\lambda(k-p)} \right)^{n} - \alpha \left(\frac{p}{p+\lambda(k-p)} \right)^{n+1} \right] a_k z^{k-p} \Bigg) \\
&\quad \times \left(1 + \sum_{k=p+1}^{\infty} \left(\frac{p}{p+\lambda(k-p)} \right)^{n+1} a_k z^{k-p} + (-1)^{n+1} \sum_{k=p}^{\infty} \left(\frac{p}{p+\lambda(k-p)} \right)^{n+1} \overline{b}_k \overline{z}^k z^{-p} \right)^{-1} \\
&\quad + \left((-1)^{n} \sum_{k=p}^{\infty} \left[\left(\frac{p}{p+\lambda(k-p)} \right)^{n} + \alpha \left(\frac{p}{p+\lambda(k-p)} \right)^{n+1} \right] \times \overline{b}_k \overline{z}^k z^{-p} \right)
\end{aligned}$$

$$\times\left(1+\sum_{k=p+1}^{\infty}\left(\frac{p}{p+\lambda(k-p)}\right)^{n+1} \times a_k z^{k-p}+(-1)^{n+1}\sum_{k=p}^{\infty}\left(\frac{p}{p+\lambda(k-p)}\right)^{n+1} \times \overline{b}_k \overline{z}^k z^{-p}\right)^{-1}\Bigg\}$$

$$=\operatorname{Re}\left\{\frac{1-\alpha+A(z)}{1+B(z)}\right\}. \quad (11)$$

For $z=re^{i\theta}$, we have

$$A\left(re^{i\theta}\right)=\sum_{k=p+1}^{\infty}\left[\left(\frac{p}{p+\lambda(k-p)}\right)^{n} -\alpha\left(\frac{p}{p+\lambda(k-p)}\right)^{n+1}\right] a_k r^{k-p} e^{i(k-p)\theta} + (-1)^n\sum_{k=p}^{\infty}\left[\left(\frac{p}{p+\lambda(k-p)}\right)^{n} +\alpha\left(\frac{p}{p+\lambda(k-p)}\right)^{n+1}\right] \times \overline{b}_k r^{k-p} e^{-i(k+p)\theta},$$

$$B\left(re^{i\theta}\right)=\sum_{k=p+1}^{\infty}\left(\frac{p}{p+\lambda(k-p)}\right)^{n+1} a_k r^{k-p} e^{i(k-p)\theta} + (-1)^{n+1}\sum_{k=p}^{\infty}\left(\frac{p}{p+\lambda(k-p)}\right)^{n+1} \overline{b}_k r^{k-p} e^{-i(k+p)\theta}. \quad (12)$$

Setting that

$$\frac{1-\alpha+A(z)}{1+B(z)}=(1-\alpha)\frac{1+w(z)}{1-w(z)}, \quad (13)$$

the proof will be complete if we can show that $|w(z)|<1$. Using the condition (7), we can write

$$|w(z)|=\left|\frac{A(z)-(1-\alpha)B(z)}{A(z)+(1-\alpha)B(z)+2(1-\alpha)}\right|$$

$$=\Bigg|\left(\sum_{k=p+1}^{\infty}\left[\left(\frac{p}{p+\lambda(k-p)}\right)^{n} -\left(\frac{p}{p+\lambda(k-p)}\right)^{n+1}\right] a_k r^{k-p} e^{i(k-p)\theta}\right) \times\left(2(1-\alpha)+\sum_{k=p+1}^{\infty} c_k a_k r^{k-p} e^{i(k-p)\theta} +(-1)^n\sum_{k=p}^{\infty} d_k \overline{b}_k r^{k-p} e^{-i(k+p)\theta}\right)^{-1}$$

$$+\left((-1)^n\sum_{k=p}^{\infty}\left[\left(\frac{p}{p+\lambda(k-p)}\right)^{n} +\alpha\left(\frac{p}{p+\lambda(k-p)}\right)^{n+1}\right] \times \overline{b}_k r^{k-p} e^{-i(k+p)\theta}\right) \times\left(2(1-\alpha)+\sum_{k=p+1}^{\infty} c_k a_k r^{k-p} e^{i(k-p)\theta} +(-1)^n\sum_{k=p}^{\infty} d_k \overline{b}_k r^{k-p} e^{-i(k+p)\theta}\right)^{-1}\Bigg|$$

$$\leq\Bigg|\left(\sum_{k=p+1}^{\infty}\left[\left(\frac{p}{p+\lambda(k-p)}\right)^{n} -\left(\frac{p}{p+\lambda(k-p)}\right)^{n+1}\right]\left|a_k\right| r^{k-p}\right) \times\left(2(1-\alpha)-\sum_{k=p+1}^{\infty} c_k\left|a_k\right| r^{k-p} -\sum_{k=p}^{\infty} d_k\left|b_k\right| r^{k-p}\right)^{-1}$$

$$+\left(\sum_{k=p}^{\infty}\left[\left(\frac{p}{p+\lambda(k-p)}\right)^{n} +\left(\frac{p}{p+\lambda(k-p)}\right)^{n+1}\right]\left|b_k\right| r^{k-p}\right) \times\left(2(1-\alpha)-\sum_{k=p+1}^{\infty} c_k\left|a_k\right| r^{k-p} -\sum_{k=p}^{\infty} d_k\left|b_k\right| r^{k-p}\right)^{-1}\Bigg|$$

$$=\Bigg|\left(\sum_{k=p+1}^{\infty}\left[\left(\frac{p}{p+\lambda(k-p)}\right)^{n}\right.\right.$$

$$-\left(\frac{p}{p+\lambda(k-p)}\right)^{n+1}\Bigg]\left|a_k\right| r^{k-p}\Bigg)$$

$$\times\left(4(1-\alpha)-\sum_{k=p}^{\infty}\left\{c_k\left|a_k\right|+d_k\left|b_k\right|\right\} r^{k-p}\right)^{-1}$$

$$+\left(\sum_{k=p}^{\infty}\left[\left(\frac{p}{p+\lambda(k-p)}\right)^{n}\right.\right.$$

$$\left.\left.+\left(\frac{p}{p+\lambda(k-p)}\right)^{n+1}\right]\left|b_k\right| r^{k-p}\right)$$

$$\left.\times\left(4(1-\alpha)-\sum_{k=p}^{\infty}\left\{c_k\left|a_k\right|+d_k\left|b_k\right|\right\} r^{k-p}\right)^{-1}\right|$$

$$<\left|\left(\sum_{k=p+1}^{\infty}\left[\left(\frac{p}{p+\lambda(k-p)}\right)^{n}\right.\right.\right.$$

$$\left.\left.-\left(\frac{p}{p+\lambda(k-p)}\right)^{n+1}\right]\left|a_k\right|\right)$$

$$\times\left(4(1-\alpha)-\sum_{k=p}^{\infty}\left\{c_k\left|a_k\right|+d_k\left|b_k\right|\right\}\right)^{-1}$$

$$+\left(\sum_{k=p}^{\infty}\left[\left(\frac{p}{p+\lambda(k-p)}\right)^{n}\right.\right.$$

$$\left.\left.+\left(\frac{p}{p+\lambda(k-p)}\right)^{n+1}\right]\left|b_k\right|\right)$$

$$\left.\times\left(4(1-\alpha)-\sum_{k=p}^{\infty}\left\{c_k\left|a_k\right|+d_k\left|b_k\right|\right\}\right)^{-1}\right|$$

$$\leq 1, \tag{14}$$

where

$$c_k=\left(\frac{p}{p+\lambda(k-p)}\right)^{n}+(1-2\alpha)\left(\frac{p}{p+\lambda(k-p)}\right)^{n+1},$$
$$d_k=\left(\frac{p}{p+\lambda(k-p)}\right)^{n}-(1-2\alpha)\left(\frac{p}{p+\lambda(k-p)}\right)^{n+1}. \tag{15}$$

The harmonic functions are as follows:

$$f(z)=z^p+\sum_{k=p+1}^{\infty}\frac{1}{\Psi_{p,\lambda}(n,k,\alpha)}x_k z^k$$
$$+\sum_{k=p}^{\infty}\frac{1}{\Phi_{p,\lambda}(n,k,\alpha)}\overline{y_k z^k}, \tag{16}$$

where $\sum_{k=p+1}^{\infty}|x_k|+\sum_{k=p}^{\infty}|y_k| = 1$ show that the coefficient bound given by (7) is sharp. The functions of the form (8) are in the class $\mathscr{H}_{p,\lambda}(n;\alpha)$ because

$$\sum_{k=p}^{\infty}\left\{\Psi_{p,\lambda}(n,k,\alpha)\left|a_k\right|+\Phi_{p,\lambda}(n,k,\alpha)\left|b_k\right|\right\}$$
$$=1+\sum_{k=p+1}^{\infty}\left|x_k\right|+\sum_{k=p}^{\infty}\left|y_k\right|=2. \tag{17}$$

This completes the proof of Theorem 1. □

In the following theorem, it is shown that the condition (7) is also necessary for functions $f_n = h+\overline{g}_n$, where h and g_n are of the form (6).

Theorem 2. *Let $f_n = h+\overline{g}_n$, where h and g_n are given by (6). Then, $f_n \in \mathscr{H}^{-}_{p,\lambda}(n;\alpha)$ if and only if*

$$\sum_{k=p}^{\infty}\left\{\Psi_{p,\lambda}(n,k,\alpha)\, a_k+\Phi_{p,\lambda}(n,k,\alpha)\, b_k\right\}\leq 2, \tag{18}$$

where $\Psi_{p,\lambda}(n,k,\alpha)$ and $\Phi_{p,\lambda}(n,k,\alpha)$ are given by (8) and (9), respectively.

Proof. Since $\mathscr{H}^{-}_{p,\lambda}(n;\alpha)\subset\mathscr{H}_{p,\lambda}(n;\alpha)$, we only need to prove the "only if" part of the theorem. To this end, for functions $f_n = h+\overline{g}_n$, where h and g_n are given by (6), we notice that the condition $\operatorname{Re}\{I^{n}_{p,\lambda}f(z)/I^{n+1}_{p,\lambda}f(z)\}>\alpha$ is equivalent to

$$\operatorname{Re}\left\{\left((1-\alpha)z^p\right.\right.$$

$$-\sum_{k=p+1}^{\infty}\times\left[\left(\frac{p}{p+\lambda(k-p)}\right)^{n}\right.$$

$$\left.\left.-\alpha\left(\frac{p}{p+\lambda(k-p)}\right)^{n+1}\right]a_k z^k\right)$$

$$\times\left(z^p-\sum_{k=p+1}^{\infty}\left(\frac{p}{p+\lambda(k-p)}\right)^{n+1}a_k z^k\right.$$

$$\left.+(-1)^{2n}\sum_{k=p}^{\infty}\left(\frac{p}{p+\lambda(k-p)}\right)^{n+1}b_k\overline{z}^k\right)^{-1}$$

$$+\left((-1)^{2n-1}\sum_{k=p}^{\infty}\times\left[\left(\frac{p}{p+\lambda(k-p)}\right)^{n}+\alpha\left(\frac{p}{p+\lambda(k-p)}\right)^{n+1}\right]b_k\overline{z}^k\right)\times\left(z^p-\sum_{k=p+1}^{\infty}\left(\frac{p}{p+\lambda(k-p)}\right)^{n+1}a_kz^k+(-1)^{2n}\sum_{k=p}^{\infty}\left(\frac{p}{p+\lambda(k-p)}\right)^{n+1}\overline{b}_k\overline{z}^k\right)^{-1}\Bigg\}\geq 0. \tag{19}$$

The previous required condition (19) must hold for all values of z in U. Upon choosing the values of z on the positive real axis where $0\leq z=r<1$, we must have

$$\left((1-\alpha)-\sum_{k=p+1}^{\infty}\left[\left(\frac{p}{p+\lambda(k-p)}\right)^{n}-\alpha\left(\frac{p}{p+\lambda(k-p)}\right)^{n+1}\right]a_kr^{k-p}\right)\times\left(1-\sum_{k=p+1}^{\infty}\left(\frac{p}{p+\lambda(k-p)}\right)^{n+1}a_kr^{k-p}+\sum_{k=p}^{\infty}\left(\frac{p}{p+\lambda(k-p)}\right)^{n+1}b_kr^{k-p}\right)^{-1}$$
$$+\left(-\sum_{k=p}^{\infty}\left[\left(\frac{p}{p+\lambda(k-p)}\right)^{n}+\alpha\left(\frac{p}{p+\lambda(k-p)}\right)^{n+1}\right]b_kr^{k-p}\right)\times\left(1-\sum_{k=p+1}^{\infty}\left(\frac{p}{p+\lambda(k-p)}\right)^{n+1}a_kr^{k-p}+\sum_{k=p}^{\infty}\left(\frac{p}{p+\lambda(k-p)}\right)^{n+1}b_kr^{k-p}\right)^{-1}\geq 0. \tag{20}$$

If the condition (18) does not hold, then the numerator in (20) is negative for r sufficiently close to 1. Hence there exists $z_0=r_0$ in $(0,1)$ for which the quotient in (20) is negative. This contradicts the required condition for $f_n\in\mathscr{H}^{-}_{p,\lambda}(n;\alpha)$, and so the proof of Theorem 2 is completed. □

3. Extreme Points and Distortion Theorem

Our next theorem is on the extreme points of convex hulls of the class $\mathscr{H}^{-}_{p,\lambda}(n;\alpha)$ denoted by $clco\mathscr{H}^{-}_{p,\lambda}(n;\alpha)$.

Theorem 3. *Let $f_n=h+\overline{g}_n$, where h and g_n are given by (6). Then, $f_n\in\mathscr{H}^{-}_{p,\lambda}(n;\alpha)$ if and only if*

$$f_n(z)=\sum_{k=p}^{\infty}\left[x_kh_k(z)+y_kg_{n_k}(z)\right], \tag{21}$$

where

$$h_1(z)=z^p,\qquad h_k(z)=z^p-\frac{1}{\Psi_{p,\lambda}(n,k,\alpha)}z^k\quad(k=p+1,p+2,p+3,\ldots),$$
$$g_{n_k}(z)=z^p+(-1)^{n-1}\frac{1}{\Phi_{p,\lambda}(n,k,\alpha)}\overline{z}^k\quad(k=p,p+1,p+2,\ldots),$$
$$x_k,y_k\geq 0,\qquad x_p=1-\sum_{k=p+1}^{\infty}x_k-\sum_{k=p}^{\infty}y_k. \tag{22}$$

In particular, the extreme points of the class $\mathscr{H}^{-}_{p,\lambda}(n;\alpha)$ are $\{h_k\}$ and $\{g_{n_k}\}$.

Proof. Suppose that

$$\begin{aligned}f_n(z)&=\sum_{k=p}^{\infty}\left(x_k\,h_k(z)+y_kg_{n_k}(z)\right)\\&=\sum_{k=p}^{\infty}(x_k+y_k)z^p-\sum_{k=p+1}^{\infty}\frac{1}{\Psi_{p,\lambda}(n,k,\alpha)}x_kz^k+(-1)^{n-1}\sum_{k=1}^{\infty}\frac{1}{\Phi_{p,\lambda}(n,k,\alpha)}y_k\overline{z}^k.\end{aligned} \tag{23}$$

Then,

$$\begin{aligned}&\sum_{k=p+1}^{\infty}\Psi_{p,\lambda}(n,k,\alpha)\left(\frac{1}{\Psi_{p,\lambda}(n,k,\alpha)}x_k\right)+\sum_{k=1}^{\infty}\Phi_{p,\lambda}(n,k,\alpha)\left(\frac{1}{\Phi_{p,\lambda}(n,k,\alpha)}y_k\right)\\&\quad=\sum_{k=p+1}^{\infty}x_k+\sum_{k=p}^{\infty}y_k=1-x_p\leq 1,\end{aligned} \tag{24}$$

and so $f_n\in clco\mathscr{H}^{-}_{p,\lambda}(n;\alpha)$.

Conversely, if $f_n \in clco\mathscr{H}^{-}_{p,\lambda}(n;\alpha)$, then

$$a_k \le \frac{1}{\Psi_{p,\lambda}(n,k,\alpha)}, \qquad b_k \le \frac{1}{\Phi_{p,\lambda}(n,k,\alpha)}. \tag{25}$$

Set that

$$x_k = \Psi_{p,\lambda}(n,k,\alpha)\, a_k \quad (k = p+1, p+2, p+3, \ldots),$$
$$y_k = \Phi_{p,\lambda}(n,k,\alpha)\, b_k \quad (k = p, p+1, p+2, \ldots). \tag{26}$$

Then note that by Theorem 2, $0 \le x_k \le 1$, $(k = p+1, p+2, p+3, \ldots)$, and $0 \le y_k \le 1, (k = p, p+1, p+2, \ldots)$. We define that $x_p = 1 - \sum_{k=p+1}^{\infty} x_k - \sum_{k=p}^{\infty} y_k$ and note that by Theorem 2, $x_p \ge 0$. Consequently, we obtain $f_n(z) = \sum_{k=p}^{\infty}\{x_k h_k(z) + y_k g_k(z)\}$ as required. □

The following theorem gives the distortion bounds for functions in the class $\mathscr{H}^{-}_{p,\lambda}(n;\alpha)$ which yields a covering result for this class.

Theorem 4. *Let $f_n(z) \in \mathscr{H}^{-}_{p,\lambda}(n;\alpha)$. Then, for $|z| = r < 1$, we have*

$$\begin{aligned}\left(1 - b_p\right) r^p - \left\{\Gamma_{p,\lambda}(n,\alpha) - \Delta_{p,\lambda}(n,\alpha)\right\} r^{p+1} &\le \left|f_n(z)\right| \\ &\le \left(1 + b_p\right) r^p + \left\{\Gamma_{p,\lambda}(n,\alpha) - \Delta_{p,\lambda}(n,\alpha)\, b_p\right\} r^{p+1},\end{aligned} \tag{27}$$

where

$$\Gamma_{p,\lambda}(n,\alpha) = \frac{1-\alpha}{(p/(p+\lambda))^n - \alpha(p/(p+\lambda))^{n+1}},$$
$$\Delta_{p,\lambda}(n,\alpha) = \frac{1+\alpha}{(p/(p+\lambda))^n - \alpha(p/(p+\lambda))^{n+1}}. \tag{28}$$

The result is sharp.

Proof. We only prove the right-hand inequality. The proof for the left-hand inequality is similar and will be omitted. Let $f_n(z) \in \mathscr{H}^{-}_{p,\lambda}(n;\alpha)$. Taking the absolute value of f_n, we have

$$\begin{aligned}\left|f_n(z)\right| &\le \left(1 + b_p\right) r^p + \sum_{k=p+1}^{\infty} \left(a_k + b_k\right) r^k \\ &\le \left(1 + b_p\right) r^p + \sum_{k=p+1}^{\infty} \left(a_k + b_k\right) r^{p+1} \\ &= \left(1 + b_p\right) r^p + \Gamma_{p,\lambda}(n,\alpha) \\ &\quad \times \sum_{k=p+1}^{\infty} \frac{1}{\Gamma_{p,\lambda}(n,\alpha)} \left(a_k + b_k\right) r^{p+1} \\ &\le \left(1 + b_p\right) r^p + \Gamma_{p,\lambda}(n,\alpha)\, r^{p+1} \\ &\quad \times \sum_{k=p+1}^{\infty} \left\{\Psi_{p,\lambda}(n,k,\alpha)\, a_k + \Phi_{p,\lambda}(n,k,\alpha)\, b_k\right\} \\ &\le \left(1 + b_p\right) r^p \\ &\quad + \left\{\Gamma_{p,\lambda}(n,k,\alpha) - \Delta_{p,\lambda}(n,k,\alpha)\, b_p\right\} r^{p+1}.\end{aligned} \tag{29}$$

The bounds given in Theorem 4 for functions $f_n = h + \overline{g}_n$, where$h$ and g_n of form (6), also hold for functions of form (2) if the coefficient condition (7) is satisfied. The upper bound given for $f \in \mathscr{H}^{-}_{p,\lambda}(n;\alpha)$ is sharp, and the equality occurs for the functions

$$\begin{aligned} f(z) = z^p + b_p\overline{z}^p &- \left(\frac{1-\alpha}{(p/(p+\lambda))^n - \alpha(p/(p+\lambda))^{n+1}}\right. \\ &\left. - \frac{1-\alpha}{(p/(p+\lambda))^n + \alpha(p/(p+\lambda))^{n+1}}\, b_p\right) \overline{z}^{p+1}, \end{aligned} \tag{30}$$

$$\begin{aligned} f(z) = z^p + b_p\overline{z}^p &- \left(\frac{1-\alpha}{(p/(p+\lambda))^n - \alpha(p/(p+\lambda))^{n+1}}\right. \\ &\left. - \frac{1-\alpha}{(p/(p+\lambda))^n + \alpha(p/(p+\lambda))^{n+1}}\, b_p\right) z^{p+1} \end{aligned} \tag{31}$$

showing that the bounds given in Theorem 4 are sharp. □

Remark 5. (i) Putting $\lambda = 1$ in the previous results, we obtain the results of Cotirla [5].

(ii) Putting $\lambda = 1$ in the previous results, we obtain the results of Cotirla [6], when $\beta = 0$.

Acknowledgment

The author is grateful to the referees for their valuable suggestions.

References

[1] J. Clunie and T. Sheil-Small, "Harmonic univalent functions," *Annales Academiæ Scientiarum Fennicæ Mathematica A*, vol. 9, p. 3–25, 1984.

[2] J. M. Jahangiri and O. P. Ahuja, "Multivalent harmonic starlike functions," *Annales Mariae Curie-SkThlodowska A*, vol. 56, pp. 1-13, 2001.

[3] M. K. Aouf, A. O. Mostafa, and R. El-Ashwah, "Sandwich theorems for p-valent functions defined by a certain integral operator," *Mathematical and Computer Modelling*, vol. 53, no. 9-10, pp. 1647–1653, 2011.

[4] F. M. Al-Oboudi, "On univalent functions defined by a generalized Sălăgean operator," *International Journal of Mathematics*

and Mathematical Sciences, vol. 2004, no. 27, pp. 1429–1436, 2004.

[5] L.-I. Cotirla, "Harmonic multivalent functions defined by integral operator," *Studia Universitatis Babeş-Bolyai*, vol. 54, no. 1, pp. 65–74, 2009.

[6] L.-I. Cotirla, "A new class of harmonic multivalent functions defined by an integral operator," *Acta Universitatis Apulensis*, vol. 21, pp. 55–63, 2010.

13

An Expansion Theorem Involving H-Function of Several Complex Variables

Sébastien Gaboury and Richard Tremblay

Department of Mathematics and Computer Science, University of Quebec at Chicoutimi, Chicoutimi, QC, Canada G7H 2B1

Correspondence should be addressed to Sébastien Gaboury; s1gabour@uqac.ca

Academic Editor: Julien Salomon

The aim of this present paper is to obtain a general expansion theorem involving H-functions of several complex variables. This is done by making use of a Taylor-like expansion in terms of a rational function obtained by means of fractional derivatives given recently by the authors. Special cases are also computed.

1. Introduction

In 1971, Osler obtained with the use of Cauchy integral formula for the fractional derivatives the following generalization for the Taylor's series [1]:

$$\sum_{k\in K} a^{-1}\omega^{-\gamma k} f\left(\theta^{-1}\left(\theta(z)\omega^{k}\right)\right) = \sum_{n=-\infty}^{\infty} \frac{D_{z-b}^{an+\gamma}\left[f(z)\theta'(z)\left[(z-z_0)/\theta(z)\right]^{an+\gamma+1}\right]\Big|_{z=z_0}\theta(z)^{an+\gamma}}{\Gamma(an+\gamma+1)}, \tag{1}$$

where a is a positive real number, $b \neq z_0$, $\omega = \exp(2\pi i/a)$, α and γ are arbitrary complex numbers, $f(z)$ is an analytic function in a simply connected region $\mathcal{R}$ and $\theta(z) = (z - z_0)q(z)$ with $q(z)$ is a regular and univalent function without zero in $\mathcal{R}$, and $K = 0, 1, \ldots, [a]$, $[a]$ being the largest integer not greater then a. If $0 < a \leq 1$ and $\theta(z) = (z - z_0)$, then $K = 0$ and the formula (1) reduces to

$$f(z) = \sum_{n=-\infty}^{\infty} \frac{\left[D_{z_0-b}^{an+\gamma} f(z_0)\right](z-z_0)^{an+\gamma+1}}{\Gamma(an+\gamma+1)}. \tag{2}$$

This last formula is usually called the Taylor-Riemann formula and has been studied in several papers [2–6]. But none considered a more general expansion of $f(z)$ in terms of a power series of an arbitrary quadratic, cubic, or higher degrees functions. Recently, the authors [7] obtained the power series of an analytic function $f(z)$ in terms of the rational expression $((z-z_1)/(z-z_2))$ where z_1 and z_2 are two arbitrary points inside the region of analyticity $\mathcal{R}$ of $f(z)$. In particular, we obtain the following expansion:

$$\frac{a^{-1}f(z)(z-z_1)^{\nu}(z-z_2)^{\mu}}{(z_1-z_2)} = \sum_{n=-\infty}^{\infty} \frac{e^{i\pi a(n+1)}\sin((\mu+an+\gamma)\pi)}{\sin((\mu-a+\gamma)\pi)\Gamma(1-\nu+an+\gamma)} \times D_{z-z_2}^{-\nu+an+\gamma}(z-z_2)^{\mu+an+\gamma-1}f(z)\Big|_{z=z_1}\left(\frac{z-z_1}{z-z_2}\right)^{an+\gamma}. \tag{3}$$

Several restrictions are imposed on the functions and parameters in (3). The following list is considered.

(i) ν, μ, and γ are arbitrary complex numbers.

(ii) $0 < a \leq 1$ is a real and n is the integral index of summation.

(iii) z_1, z_2 are fixed points in the z-plane and $\{z \| \theta(z)| = |\theta((z_1+z_2)/2)|\}$, where $\theta(z) = (z-z_1)(z-z_2)$, defines a double-loop curve $C = C_1 \cup C_2$ on which the series (3) converges with $z_1 \neq z_2$.

(iv) z is on the loop C_1 around the point z_1 but $z \neq (z_1 + z_2)/2$ as shown in Figure 1.

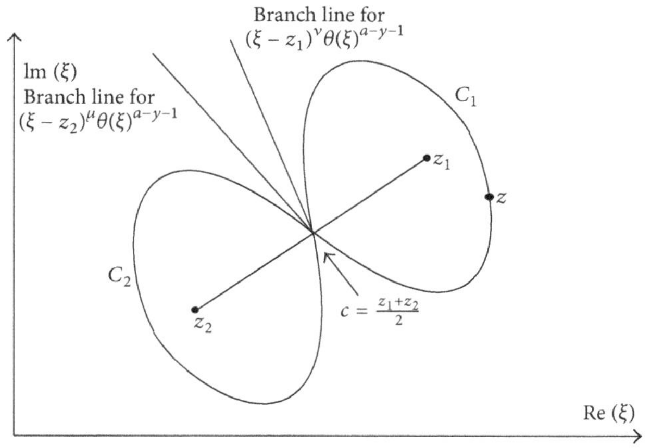

FIGURE 1: Multiloops contour.

The aim of this paper is to obtain a new expansion theorem involving the H-function of r complex variables $z_1, z_2, \dots, z_r$ defined by Srivastava and Panda [8–11]. We will define and represent it in the following form [12, page 251, equation (C.1)]:

$$
\begin{aligned}
H\left[z_1,\dots,z_r\right]
&= H^{0,N:M',N';\dots;M^{(r)},N^{(r)}}_{P,Q:P',Q';\dots;P^{(r)},Q^{(r)}} \\
&\times \left[\begin{array}{c|c} z_1 & \left(a_j;\alpha_j',\dots,\alpha_j^{(r)}\right)_{1,P} : \left(c_j',\gamma_j'\right)_{1,P'};\dots;\left(c_j^{(r)},\gamma_j^{(r)}\right)_{1,P^{(r)}} \\ \vdots & \\ z_r & \left(b_j;\beta_j',\dots,\beta_j^{(r)}\right)_{1,Q} : \left(d_j',\delta_j'\right)_{1,Q'};\dots;\left(d_j^{(r)},\delta_j^{(r)}\right)_{1,Q^{(r)}} \end{array}\right] \\
&= \frac{1}{(2\pi i)^r}\int_{L_1}\cdots\int_{L_r}\phi_1\left(\xi_1\right)\cdots\phi_r\left(\xi_r\right)\psi\left(\xi_1,\dots,\xi_r\right) \\
&\quad \times z_1^{\xi_1}\cdots z_r^{\xi_r}\,d\xi_1\cdots d\xi_r,
\end{aligned}
\tag{4}
$$

where $i=\sqrt{-1}$,

$$
\phi_k\left(\xi_k\right) = \frac{\prod_{j=1}^{M^{(k)}}\Gamma\left(d_j^{(k)}-\delta_j^{(k)}\xi_k\right)\prod_{j=1}^{N^{(k)}}\Gamma\left(1-c_j^{(k)}+\gamma_j^{(k)}\xi_k\right)}{\prod_{j=M^{(k)}+1}^{Q^{(k)}}\Gamma\left(1-d_j^{(k)}+\delta_j^{(k)}\xi_k\right)\prod_{j=N^{(k)}+1}^{P^{(k)}}\Gamma\left(c_j^{(k)}-\gamma_j^{(k)}\xi_k\right)},
\tag{5}
$$

for all $k\in\{1,\dots,r\}$ and

$$
\psi\left(\xi_1,\dots,\xi_r\right) = \frac{\prod_{j=1}^{N}\Gamma\left(1-a_j+\sum_{k=1}^{r}\alpha_j^{(k)}\xi_k\right)}{\prod_{j=N+1}^{P}\Gamma\left(a_j-\sum_{k=1}^{r}\alpha_j^{(k)}\xi_k\right)\prod_{j=1}^{Q}\Gamma\left(1-b_j+\sum_{k=1}^{r}\beta_j^{(k)}\xi_k\right)}.
\tag{6}
$$

Here, for convenience, $(a_j;\alpha_j',\dots,\alpha_j^{(r)})_{1,P}$ abbreviates the p-member array

$$
\left(\left(a_1;\alpha_1',\dots,\alpha_1^{(r)}\right),\dots,\left(a_p;\alpha_p',\dots,\alpha_p^{(r)}\right)\right),
\tag{7}
$$

while $(c_j^{(k)},\gamma_j^{(k)})_{1,P^{(k)}}$ abbreviates the array of $P^{(k)}$ pairs of parameters:

$$
\left(c_1^{(k)},\gamma_1^{(k)}\right),\dots,\left(c_{P^{(k)}}^{(k)},\gamma_{P^{(k)}}^{(k)}\right)\quad (k=1,\dots,r),
\tag{8}
$$

and so on. Suppose, as usual, that the parameters:

$$
\begin{array}{l} a_j,\ j=1,\dots,P;\ c_j^{(k)},\ j=1,\dots,P^{(k)} \\ b_j,\ j=1,\dots,Q;\ d_j^{(k)},\ j=1,\dots,Q^{(k)} \end{array} \quad (\forall k\in\{1,\dots,r\})
\tag{9}
$$

are complex numbers and the associated coefficients

$$
\begin{array}{l} \alpha_j^{(k)},\ j=1,\dots,P;\ \gamma_j^{(k)},\ j=1,\dots,P^{(k)} \\ \beta_j^{(k)},\ j=1,\dots,Q;\ \delta_j^{(k)},\ j=1,\dots,Q^{(k)} \end{array} \quad (\forall k\in\{1,\dots,r\})
\tag{10}
$$

are positive real numbers such that

$$
\Lambda_k := \sum_{j=1}^{P}\alpha_j^{(k)} - \sum_{j=1}^{Q}\beta_j^{(k)} + \sum_{j=1}^{P^{(k)}}\gamma_j^{(k)} - \sum_{j=1}^{Q^{(k)}}\delta_j^{(k)} \le 0,
\tag{11}
$$

$$
\begin{aligned}
\Omega_k := &-\sum_{j=N+1}^{P}\alpha_j^{(k)} - \sum_{j=1}^{Q}\beta_j^{(k)} + \sum_{j=1}^{N^{(k)}}\gamma_j^{(k)} - \sum_{j=N^{(k)}+1}^{P^{(k)}}\gamma_j^{(k)} \\
&+ \sum_{j=1}^{M^{(k)}}\delta_j^{(k)} - \sum_{j=M^{(k)}+1}^{Q^{(k)}}\delta_j^{(k)} > 0,\quad (\forall k\in\{1,\dots,r\}),
\end{aligned}
\tag{12}
$$

where the integers N, P, Q, $M^{(k)}$, $N^{(k)}$, $P^{(k)}$, and $Q^{(k)}$ are constrained by the inequalities $0 \le N \le P$, $Q \ge 0$, $1 \le M^{(k)} \le Q^{(k)}$, and $0 \le N^{(k)} \le P^{(k)}$(for all $k\in\{1,\dots,r\}$) and the equality in (11) holds true for suitably restricted values of the complex variables $z_1,\dots,z_r$.

The multiple Mellin-Barnes contour integral [12, page 251, equation (C.1)] representing the multivariable H-function (4) converges absolutely, under the conditions (12), when

$$
\left|\arg\left(z_k\right)\right| < \frac{1}{2}\Omega_k\pi,\quad (\forall k\in\{1,\dots,r\}),
\tag{13}
$$

the points $z_k = 0$ $(k=1,\dots,r)$ and various exceptional parameter values being tacitly excluded. Furthermore, we have (cf. [9, page 131, equation (1.9)]):

$$
H\left[z_1,\dots,z_r\right] = \begin{cases} O\left(\left|z_1\right|^{\xi_1}\dots\left|z_r\right|^{\xi_r}\right), \\ \quad \left(\max\left\{\left|z_1\right|,\dots,\left|z_r\right|\right\}\longrightarrow 0\right), \\ O\left(\left|z_1\right|^{\eta_1}\dots\left|z_r\right|^{\eta_r}\right), \\ \quad \left(N=0;\min\left\{\left|z_1\right|,\dots,\left|z_r\right|\right\}\longrightarrow\infty\right), \end{cases}
\tag{14}
$$

where $(k = 1, \dots, r)$

$$\xi_k = \min \left\{ \frac{\operatorname{Re}\left(d_j^{(k)}\right)}{\delta_j^{(k)}} \right\}, \quad \left(j = 1, \dots, M^{(k)}\right), \tag{15}$$

$$\eta_k = \max \left\{ \frac{\operatorname{Re}\left(c_j^{(k)} - 1\right)}{\gamma_j^{(k)}} \right\}, \quad \left(j = 1, \dots, N^{(k)}\right), \tag{16}$$

provided that each of the inequalities in (11)–(13) holds true.

Note that throughout this work, we will assume that the convergence and existence conditions corresponding appropriately to the ones detailed above are satisfied by each of the various H-functions involved.

2. Pochhammer Contour Integral Representation for Fractional Derivative

The use of contour of integration in the complex plane provides a very powerful tool in both classical and fractional calculus. The most familiar representation for fractional derivative of order α of $z^p f(z)$ is the Riemann-Liouville integral [13–15], that is,

$$D_z^\alpha z^p f(z) = \frac{1}{\Gamma(-\alpha)} \int_0^z f(\xi) \xi^p (\xi - z)^{-\alpha-1} d\xi, \tag{17}$$

which is valid for $\operatorname{Re}(\alpha) < 0$, $\operatorname{Re}(p) > 1$ and where the integration is done along a straight line from 0 to z in the ξ-plane. By integrating by part m times, we obtain

$$D_z^\alpha z^p f(z) = \frac{d^m}{dz^m} D_z^{\alpha-m} z^p f(z). \tag{18}$$

This allows to modify the restriction $\operatorname{Re}(\alpha) < 0$ to $\operatorname{Re}(\alpha) < m$ [15]. Another used representation for the fractional derivative is the one based on the Cauchy integral formula widely used by Osler [4, 16–18]. These two representations have been used in many interesting research papers. It appears that the less restrictive representation of fractional derivative according to parameters is the Pochhammer's contour definition introduced in [19, 20].

Definition 1. Let $f(z)$ be analytic in a simply connected region $\mathcal{R}$. Let $g(z)$ be regular and univalent on $\mathcal{R}$ and let $g^{-1}(0)$ be an interior point of $\mathcal{R}$ then if α is not a negative integer, p is not an integer, and z is in $\mathcal{R} - \{g^{-1}(0)\}$, we define the fractional derivative of order α of $g(z)^p f(z)$ with respect to $g(z)$ by

$$D_{g(z)}^\alpha g(z)^p f(z) = \frac{e^{-i\pi p} \Gamma(1+\alpha)}{4\pi \sin(\pi p)} \times \int_{C(z+,g^{-1}(0)+,z-,g^{-1}(0)-;F(a),F(a))} \frac{f(\xi) g(\xi)^p g'(\xi)}{(g(\xi) - g(z))^{\alpha+1}} d\xi. \tag{19}$$

For noninteger α and p, the functions $g(\xi)^p$ and $(g(\xi) - g(z))^{-\alpha-1}$ in the integrand have two branch lines which begin, respectively, at $\xi = z$ and $\xi = g^{-1}(0)$, and both pass through the point $\xi = a$ without crossing the Pochhammer contour $P(a) = \{C_1 \cup C_2 \cup C_3 \cup C_4\}$ at any other point as shown in Figure 2. $F(a)$ denotes the principal value of the integrand in (19) at the beginning and ending point of the Pochhammer contour $P(a)$ which is closed on Riemann surface of the multiple-valued function $F(\xi)$.

Remark 2. In Definition 1, the function $f(z)$ must be analytic at $\xi = g^{-1}(0)$. However it is interesting to note here that we could also allow $f(z)$ to have an essential singularity at $\xi = g^{-1}(0)$, and (19) would still be valid.

Remark 3. The Pochhammer contour never crosses the singularities at $\xi = g^{-1}(0)$ and $\xi = z$ in (19), then we know that the integral is analytic for all p and for all α and for z in $\mathcal{R} - \{g^{-1}(0)\}$. Indeed, the only possible singularities of $D_{g(z)}^\alpha g(z)^p f(z)$ are $\alpha = -1, -2, \dots$ and $p = 0, \pm 1, \pm 2, \dots$ which can directly be identified from the coefficient of the integral (19). However, integrating by parts N times the integral in (19) by two different ways, we can show that $\alpha = -1, -2, \dots$, and $p = 0, 1, 2, \dots$ are removable singularities (see [19]).

In 1985, Srivastava and Goyal [21, page 644, equation (17)] obtained the following fractional derivative formula, by using the well-known Riemann-Liouville's definition of the fractional derivative, for the multivariable H-function (4):

$$D_x^\alpha \left\{ x^\beta (x+\xi)^\lambda H\left[z_1 x^{\rho_1} (x+\xi)^{\sigma_1}, \dots, z_r x^{\rho_r} (x+\xi)^{\sigma_r}\right]\right\}$$
$$= \xi^\lambda x^{\beta-\alpha} \sum_{m=0}^{\infty} \frac{(x/\xi)^m}{m!} \times H^{0,N+2:M',N';\dots;M^{(r)},N^{(r)}}_{P+2,Q+2:P',Q';\dots;P^{(r)},Q^{(r)}}$$
$$\times \left[\begin{matrix} z_1 x^{\rho_1} \xi^{\sigma_1} \\ \vdots \\ z_r x^{\rho_r} \xi^{\sigma_r} \end{matrix} \middle| \begin{matrix} (-\lambda; \sigma_1, \dots, \sigma_r), (-\beta - m; \rho_1, \dots, \rho_r), \\ \\ (m - \lambda; \sigma_1, \dots, \sigma_r), (\alpha - \beta - m; \rho_1, \dots, \rho_r), \end{matrix} \right.$$
$$\left. \begin{matrix} \left(a_j; \alpha_j', \dots, \alpha_j^{(r)}\right)_{1,P} : \left(c_j', \gamma_j'\right)_{1,P'}; \dots; \left(c_j^{(r)}, \gamma_j^{(r)}\right)_{1,P^{(r)}} \\ \left(b_j; \beta_j', \dots, \beta_j^{(r)}\right)_{1,Q} : \left(d_j', \delta_j'\right)_{1,Q'}; \dots; \left(d_j^{(r)}, \delta_j^{(r)}\right)_{1,Q^{(r)}} \end{matrix} \right] \tag{20}$$

provided (in addition to the usual convergence and existence conditions) that $\min\{\rho_i, \sigma_i\} > 0$, $i = 1, \dots, r$, $|\arg(x/\xi)| < \pi$, $\operatorname{Re}(\mu) < 0$ and

$$\operatorname{Re}(\beta) + \sum_{i=1}^{r} \rho_i \xi_i > -1, \tag{21}$$

where $\xi_1, \dots, \xi_r$ are given by (15).

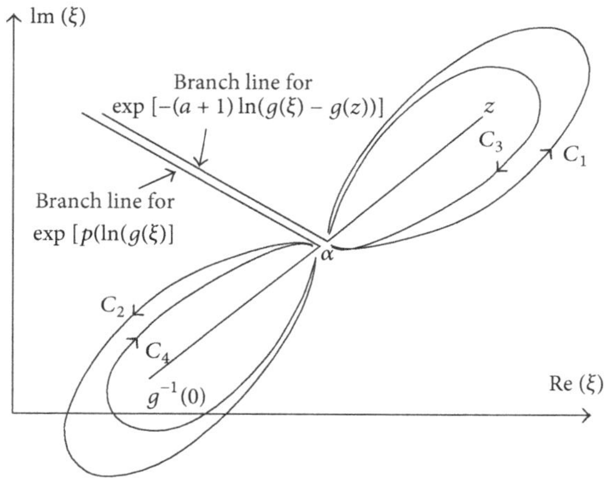

Figure 2: Pochhammer's contour.

Remark 4. Adopting the representation based on the Pochhammer contour for fractional derivative and in view of Remark 3, the restrictions $\operatorname{Re}(\beta) + \sum_{i=1}^{r} \rho_i \xi_i > -1$ can be modified to $\operatorname{Re}(\beta) + \sum_{i=1}^{r} \rho_i \xi_i$ not a negative integer and $\operatorname{Re}(\mu) < 0$ to μ not a negative integer.

Letting $\sigma_i \downarrow 0$, for all $i \in \{1, \ldots, r\}$ and $\lambda = 0$ in (20) yields the following useful particular case:

$$D_x^{\alpha}\left\{x^{\beta} H\left[z_1 x^{\rho_1}, \ldots, z_r x^{\rho_r}\right]\right\} = x^{\beta-\alpha} H^{0,N+1:M',N';\ldots;M^{(r)},N^{(r)}}_{P+1,Q+1:P',Q';\ldots;P^{(r)},Q^{(r)}} \times \left[\begin{array}{c|l} z_1 x^{\rho_1} & (-\beta;\rho_1,\ldots,\rho_r),\left(a_j;\alpha_j',\ldots,\alpha_j^{(r)}\right)_{1,P} : \left(c_j',\gamma_j'\right)_{1,P'};\ldots;\left(c_j^{(r)},\gamma_j^{(r)}\right)_{1,P^{(r)}} \\ \vdots & \\ z_r x^{\rho_r} & (\alpha-\beta;\rho_1,\ldots,\rho_r),\left(b_j;\beta_j',\ldots,\beta_j^{(r)}\right)_{1,Q} : \left(d_j',\delta_j'\right)_{1,Q'};\ldots;\left(d_j^{(r)},\delta_j^{(r)}\right)_{1,Q^{(r)}} \end{array}\right]. \tag{22}$$

3. Expansion Theorem and Special Cases

In this section, we establish the new expansion theorem for the H-functions of several complex variables. Two special cases (presumably new) are also computed to demonstrate the importance of this new expansion theorem to the theory of special functions of mathematical physics.

Theorem 5. *Assume that the convergence and existence conditions corresponding appropriately to the ones detailed in Section 1 are satisfied for each H-functions involved. Assume also that the list of restrictions of (3) is satisfied, then the following expansion*

$$H^{0,N:M',N';\ldots;M^{(r)},N^{(r)}}_{P,Q:P',Q';\ldots;P^{(r)},Q^{(r)}} \times \left[\begin{array}{c|l} y_1 x^{\rho_1} & \left(a_j;\alpha_j',\ldots,\alpha_j^{(r)}\right)_{1,P} : \left(c_j',\gamma_j'\right)_{1,P'};\ldots;\left(c_j^{(r)},\gamma_j^{(r)}\right)_{1,P^{(r)}} \\ \vdots & \\ y_r x^{\rho_r} & \left(b_j;\beta_j',\ldots,\beta_j^{(r)}\right)_{1,Q} : \left(d_j',\delta_j'\right)_{1,Q'};\ldots;\left(d_j^{(r)},\delta_j^{(r)}\right)_{1,Q^{(r)}} \end{array}\right]$$
$$= \frac{(az_1)}{x^{\mu}(x-z_1)^{\nu}} \times \sum_{n=-\infty}^{\infty} \frac{e^{i\pi a(n+1)} \sin((\mu+an+\gamma)\pi)}{\sin((\mu-a+\gamma)\pi)\,\Gamma(1-\nu+an+\gamma)} \times \left(\frac{x-z_1}{x}\right)^{an+\gamma} \times z_1^{\mu+\nu-1} H^{0,N+1:M',N';\ldots;M^{(r)},N^{(r)}}_{P+1,Q+1:P',Q';\ldots;P^{(r)},Q^{(r)}}$$
$$\times \left[\begin{array}{c|l} y_1 z_1^{\rho_1} & (-\mu-an-\gamma+1;\rho_1,\ldots,\rho_r),\left(a_j;\alpha_j',\ldots,\alpha_j^{(r)}\right)_{1,P} : \left(c_j',\gamma_j'\right)_{1,P'};\ldots;\left(c_j^{(r)},\gamma_j^{(r)}\right)_{1,P^{(r)}} \\ \vdots & \\ y_r z_1^{\rho_r} & (1-\mu-\nu;\rho_1,\ldots,\rho_r),\left(b_j;\beta_j',\ldots,\beta_j^{(r)}\right)_{1,Q} : \left(d_j',\delta_j'\right)_{1,Q'};\ldots;\left(d_j^{(r)},\delta_j^{(r)}\right)_{1,Q^{(r)}} \end{array}\right] \tag{23}$$

holds true.

Proof. By setting $f(x) = H[y_1 x^{\rho_1}, \ldots, y_r x^{\rho_r}]$, $z_2 = 0$ in (3) and with the help of (22), we find first that

$$D_x^{-\nu+an+\gamma} x^{\mu+an+\gamma-1} H\left[y_1 x^{\rho_1}, \ldots, y_r x^{\rho_r}\right]\Big|_{x=z_1} = z_1^{\mu+\nu-1} H^{0,N+1:M',N';\ldots;M^{(r)},N^{(r)}}_{P+1,Q+1:P',Q';\ldots;P^{(r)},Q^{(r)}}$$
$$\times \left[\begin{array}{c|l} y_1 z_1^{\rho_1} & (-\mu-an-\gamma+1;\rho_1,\ldots,\rho_r),\left(a_j;\alpha_j',\ldots,\alpha_j^{(r)}\right)_{1,P} : \left(c_j',\gamma_j'\right)_{1,P'};\ldots;\left(c_j^{(r)},\gamma_j^{(r)}\right)_{1,P^{(r)}} \\ \vdots & \\ y_r z_1^{\rho_r} & (1-\mu-\nu;\rho_1,\ldots,\rho_r),\left(b_j;\beta_j',\ldots,\beta_j^{(r)}\right)_{1,Q} : \left(d_j',\delta_j'\right)_{1,Q'};\ldots;\left(d_j^{(r)},\delta_j^{(r)}\right)_{1,Q^{(r)}} \end{array}\right]. \tag{24}$$

Substituting the last result into (3) yields the desired result.

Example 6. For $N = P = Q = 0$ and $\rho_1 = \rho_2 = \cdots = \rho_r = 1$ the multivariable H-function in the L.H.S. of (23) breaks up into product of r H-functions and there holds the following result [22]:

$$H^{0,0:M',N';\ldots;M^{(r)},N^{(r)}}_{0,0:P',Q';\ldots;P^{(r)},Q^{(r)}} \times \left[\begin{array}{c|c} y_1 x & - : \left(c_j', \gamma_j'\right)_{1,P'}; \ldots ; \left(c_j^{(r)}, \gamma_j^{(r)}\right)_{1,P^{(r)}} \\ \vdots & \\ y_r x & - : \left(d_j', \delta_j'\right)_{1,Q'}; \ldots ; \left(d_j^{(r)}, \delta_j^{(r)}\right)_{1,Q^{(r)}} \end{array} \right] = \prod_{i=1}^{r} H^{M^{(i)},N^{(i)}}_{P^{(i)},Q^{(i)}} \left[y_i x \left| \begin{array}{c} \left(c_j^{(i)}, \gamma_j^{(i)}\right)_{1,P^{(i)}} \\ \left(d_j^{(i)}, \delta_j^{(i)}\right)_{1,Q^{(i)}} \end{array} \right. \right]. \tag{25}$$

Now setting $P^{(i)} = Q^{(i)} = M^{(i)} = N^{(i)} = \gamma_j^{(i)} = \delta_j^{(i)} = 1$, $d_j^{(i)} = 0$, $i = 1, \ldots, r$, substituting $c_j^{(i)}$ by $1 - c_j^{(i)}$, $i = 1, \ldots, r$ into (25), replacing x by $-x$, and making use of the following special case [23]

$$H^{1,1}_{1,1}\left[z \left| \begin{array}{c} (1-\nu, 1) \\ (0,1) \end{array} \right. \right] = \Gamma(\nu)(1+z)^{-\nu}, \quad |z| < 1, \tag{26}$$

gives

$$H^{0,0:1,1;\ldots;1,1}_{0,0:1,1;\ldots;1,1} \times \left[\begin{array}{c|c} -y_1 x & - : \left(1 - c_1', 1\right); \ldots ; \left(1 - c_1^{(r)}, 1\right) \\ \vdots & \\ -y_r x & - : (0,1); \ldots ; (0,1) \end{array} \right] = \prod_{i=1}^{r} \Gamma\left(c_1^{(i)}\right)\left(1 - y_i x\right)^{-c_1^{(i)}}. \tag{27}$$

Moreover, we also know from [21, page 649, equation (3.6)] that

$$D_x^{\lambda-\mu} x^{\lambda-1} \prod_{i=1}^{r} \left(1 - x z_i\right)^{-\alpha_i} = \frac{\Gamma(\lambda)}{\Gamma(\mu)} x^{\mu-1} F_D^{(r)}\left[\lambda, \alpha_1, \ldots, \alpha_r; \mu; x z_1, \ldots, x z_r\right], \tag{28}$$

where $F_D^{(r)}$ denotes the Lauricella's hypergeometric function of r variables [24, page 33, equation (4)]. Thus, we have

$$D_x^{-\nu+an+\gamma} x^{\mu+an+\gamma-1} \prod_{i=1}^{r} \left(1 - y_i x\right)^{-c_1^{(i)}} \Big|_{x=z_1} = \frac{\Gamma(\mu + an + \gamma)}{\Gamma(\mu + \nu)} z_1^{\mu+\nu-1} F_D^{(r)} \times \left[\mu + an + \gamma, c_1^{(1)}, \ldots, c_1^{(r)}; \mu + \nu; y_1 z_1, \ldots, y_r z_1\right] \tag{29}$$

with $\max\{|y_1|, \ldots, |y_r|\} < 1$. Combining (27) and (29) with Theorem 5 yields the following new expansion involving the Lauricella's hypergeometric function of r variables

$$\prod_{i=1}^{r} \left(1 - y_i x\right)^{-c_1^{(i)}} = \frac{a}{x^{\mu}(x - z_1)^{\nu}} \times \sum_{n=-\infty}^{\infty} \frac{e^{i\pi a(n+1)} \sin\left((\mu + an + \gamma)\pi\right)\Gamma(\mu + an + \gamma)}{\sin\left((\mu - a + \gamma)\pi\right)\Gamma(1 - \nu + an + \gamma)} \times \frac{z_1^{\mu+\nu}}{\Gamma(\mu+\nu)} F_D^{(r)}\left[\mu + an + \gamma, c_1^{(1)}, \ldots, c_1^{(r)}; \mu + \nu; y_1 z_1, \ldots, y_r z_1\right] \left(\frac{x - z_1}{x}\right)^{an+\gamma} \tag{30}$$

with $\max\{|y_1 z_1|, \ldots, |y_r z_1|\} < 1$.

Example 7. Setting $\rho_1 = \cdots = \rho_r = 1$ and $\alpha_j^{(i)} = \beta_j^{(i)} = \gamma_j^{(i)} = \delta_j^{(i)} = 1$, $i = 1, \ldots, r$ in (23) and with the help of the result given by Srivastava and Panda [10] connecting the generalized Lauricella function and the multivariable H-function:

$$H^{0,P:1,P';\ldots;1,P^{(r)}}_{P,Q:P',Q'+1;\ldots;P^{(r)},Q^{(r)}+1} \times \left[\begin{array}{c|c} x y_1 & \left(a_j; \alpha_j', \ldots, \alpha_j^{(r)}\right)_{1,P} : \left(c_j', \gamma_j'\right)_{1,P'}; \cdots ; \left(c_j^{(r)}, \gamma_j^{(r)}\right)_{1,P^{(r)}} \\ \vdots & \\ x y_r & \left(b_j; \beta_j', \ldots, \beta_j^{(r)}\right)_{1,Q} : (0,1), \left(d_j', \delta_j'\right)_{1,Q'}; \cdots ; (0,1), \left(d_j^{(r)}, \delta_j^{(r)}\right)_{1,Q^{(r)}} \end{array} \right]$$
$$= \frac{\left[\prod_{j=1}^{P} \Gamma(1 - a_j)\right]\left[\prod_{j=1}^{P'} \Gamma\left(1 - c_j'\right)\right] \cdots \left[\prod_{j=1}^{P^{(r)}} \Gamma\left(1 - c_j^{(r)}\right)\right]}{\left[\prod_{j=1}^{Q} \Gamma(1 - b_j)\right]\left[\prod_{j=1}^{Q'} \Gamma\left(1 - d_j'\right)\right] \cdots \left[\prod_{j=1}^{Q^{(r)}} \Gamma\left(1 - d_j^{(r)}\right)\right]} \times F^{P:P';\ldots;P^{(r)}}_{Q:Q';\ldots;Q^{(r)}} \times \left[\begin{array}{c} \left(1 - a_j; \alpha_j', \ldots, \alpha_j^{(r)}\right)_{1,P} : \left(1 - c_j', \gamma_j'\right)_{1,P'}; \ldots ; \left(1 - c_j^{(r)}, \gamma_j^{(r)}\right)_{1,P^{(r)}} \\ \left(1 - b_j; \beta_j', \ldots, \beta_j^{(r)}\right)_{1,Q} : \left(1 - d_j', \delta_j'\right)_{1,Q'}; \ldots ; \left(1 - d_j^{(r)}, \delta_j^{(r)}\right)_{1,Q^{(r)}} \end{array} -x y_1, \ldots, -x y_r \right], \tag{31}$$

we obtain the following expansion involving an extension of the Kampé de Fériet series [24, page 38, equation (24)]

$$F^{P:P';\ldots;P^{(r)}}_{Q:Q';\ldots;Q^{(r)}}\left[\begin{matrix}(1-a_P): & (1-c'_{P'});\ldots; & (1-c^{(r)}_{P^{(r)}}); \\ & & & xy_1,\ldots,xy_r \\ (1-b_Q): & (1-d'_{Q'});\ldots; & (1-d^{(r)}_{Q^{(r)}});\end{matrix}\right]$$

$$= \frac{az_1^{\mu+\nu}}{x^{\mu}(x-z_1)^{\nu}}\sum_{n=-\infty}^{\infty}\frac{e^{i\pi a(n+1)}\sin((\mu+an+\gamma)\pi)\Gamma(\mu+an+\gamma)}{\sin((\mu-a+\gamma)\pi)\Gamma(1-\nu+an+\gamma)\Gamma(\mu+\nu)}$$

$$\times\left(\frac{x-z_1}{x}\right)^{an+\gamma}$$

$$F^{P+1:P';\ldots;P^{(r)}}_{Q+1:Q';\ldots;Q^{(r)}}\left[\begin{matrix}\mu+an+\gamma, & (1-a_P): & (1-c'_{P'});\ldots; \\ \mu+\nu, & (1-b_Q): & (1-d'_{Q'});\ldots;\end{matrix}\right.$$

$$\left.\begin{matrix}(1-c^{(r)}_{P^{(r)}}); & \\ & z_1y_1,\ldots,z_1y_r \\ (1-d^{(r)}_{Q^{(r)}}); & \end{matrix}\right]. \tag{32}$$

If we put $r=1$, $P=1$, $Q=0$, and $z_1=1$ in (32), we have

$${}_1F_0\left[\begin{matrix}1-a_1; & \\ & xy_1 \\ -; & \end{matrix}\right]$$

$$= \frac{a}{x^{\mu}(x-1)^{\nu}}$$

$$\times\sum_{n=-\infty}^{\infty}\frac{e^{i\pi a(n+1)}\sin((\mu+an+\gamma)\pi)\Gamma(\mu+an+\gamma)}{\sin((\mu-a+\gamma)\pi)\Gamma(1-\nu+an+\gamma)\Gamma(\mu+\nu)}$$

$$\times\left(\frac{x-1}{x}\right)^{an+\gamma}\times{}_2F_1\left[\begin{matrix}\mu+an+\gamma, & 1-a_1; & \\ & & y_1 \\ \mu+\nu; & & \end{matrix}\right]. \tag{33}$$

Making the following changes

$$y_1\longmapsto\frac{1-y_1}{2},\qquad a_1\longmapsto 1+\gamma \tag{34}$$

and using the hypergeometric representation for the Jacobi functions [25, page 35]

$$P^{(\theta,\gamma)}_{\mu}(z) = \frac{\Gamma(1+\theta+\mu)}{\Gamma(1+\mu)\Gamma(1+\theta)}{}_2F_1\left[\begin{matrix}\mu+\gamma+\theta+1, & -\mu; & \\ & & \frac{1-z}{2} \\ 1+\theta; & & \end{matrix}\right] \tag{35}$$

yields

$${}_1F_0\left[\begin{matrix}-\gamma; & \\ & \frac{x(1-y_1)}{2} \\ -; & \end{matrix}\right]$$

$$=\left[1-\left(\frac{x(1-y_1)}{2}\right)\right]^{\gamma} = \frac{a}{x^{\mu}(x-1)^{\nu}}$$

$$\times\sum_{n=-\infty}^{\infty}\frac{e^{i\pi a(n+1)}\sin((\mu+an+\gamma)\pi)\Gamma(\mu+an+\gamma)}{\sin((\mu-a+\gamma)\pi)\Gamma(1-\nu+an+\gamma)\Gamma(\mu+\nu)}$$

$$\times\left(\frac{x-1}{x}\right)^{an+\gamma}\times\frac{\Gamma(1+\gamma)\Gamma(\mu+\nu)}{\Gamma(\mu+\nu+\gamma)}P^{(\mu+\nu-1,an-\nu)}_{\gamma}(y_1). \tag{36}$$

References

[1] T. J. Osler, "Taylor's series generalized for fractional derivatives and applications," *SIAM Journal on Mathematical Analysis*, vol. 2, pp. 37–48, 1971.

[2] G. H. Hardy, "Riemann's form of Taylor's series," *Journal of the London Mathematical Society*, vol. 20, pp. 48–57, 1945.

[3] O. Heaviside, *Electromagnetic Theory*, vol. 2, Dover, New York, NY, USA, 1950.

[4] T. J. Osler, *Leibniz rule, the chain rule and Taylor's theorem for fractional derivatives [Ph.D. thesis]*, New York University, 1970.

[5] B. Riemann, *Versuch Einer Allgemeinen Auffasung Der Integration Und Differentiation, the Collection Works of Bernhard Riemann*, Dover, New York, NY, USA, 1953.

[6] Y. Watanabe, "Zum riemanschen binomischen lehrsatz," *Proceedings of the Physico-Mathematical Society of Japan*, vol. 14, pp. 22–35, 1932.

[7] R. Tremblay, S. Gaboury, and B. J. Fuguere, "Taylor-like expansion in terms of a rational function obtained by means of fractional derivatives," *Integral Transforms and Special Functions*. In press.

[8] H. M. Srivastava and R. Panda, "Some expansion theorems and generating relations for the *H* function of several complex variables I," *Commentarii Mathematici Universitatis Sancti Pauli*, vol. 24, no. 2, pp. 119–137, 1975.

[9] H. M. Srivastava and R. Panda, "Expansion theorems for the *H* function of several complex variables," *Journal für die Reine und Angewandte Mathematik*, vol. 288, pp. 129–145, 1976.

[10] H. M. Srivastava and R. Panda, "Some bilateral generating functions for a class of generalized hypergeometric polynomials," *Journal für die Reine und Angewandte Mathematik*, vol. 283/284, pp. 265–274, 1976.

[11] H. M. Srivastava and R. Panda, "Some expansion theorems and generating relations for the *H* function of several complex variables. II," *Commentarii Mathematici Universitatis Sancti Pauli*, vol. 25, no. 2, pp. 167–197, 1976.

[12] H. M. Srivastava, K. C. Gupta, and S. P. Goyal, *The H-functions of one and two variables with Applications*, South Asian, New Delhi, India, 1982.

[13] A. Erdélyi, "An integral equation involving Legendre functions," *SIAM Journal on Applied Mathematics*, vol. 12, pp. 15–30, 1964.

[14] J. Liouville, "Mémoire sur le calcul des différentielles a indices quelconques," *Journal de l'École Polytechnique*, vol. 13, pp. 71–162, 1832.

[15] M. Riesz, "L'intégrale de Riemann-Liouville et le problème de Cauchy," *Acta Mathematica*, vol. 81, pp. 1–223, 1949.

[16] T. J. Osler, "The fractional derivative of a composite function," *SIAM Journal on Mathematical Analysis*, vol. 1, pp. 288–293, 1970.

[17] T. J. Osler, "Leibniz rule for fractional derivatives generalized and an application to infinite series," *SIAM Journal on Applied Mathematics*, vol. 18, pp. 658–674, 1970.

[18] T. J. Osler, "Fractional Derivatives and Leibniz Rule," *The American Mathematical Monthly*, vol. 78, no. 6, pp. 645–649, 1971.

[19] J. -L. Lavoie, T. J. Osler, and R. Tremblay, *Fundamental Properties of Fractional Derivatives via Pochhammer Integrals*, Lecture Notes in Mathematics, Springer, 1976.

[20] R. Tremblay, *Une Contribution à la théorie de la Dérivée Fractionnaire [Ph.D. thesis]*, Laval University, Québec, Canada, 1974.

[21] H. M. Srivastava and S. P. Goyal, "Fractional derivatives of the H-function of several variables," *Journal of Mathematical Analysis and Applications*, vol. 112, no. 2, pp. 641–651, 1985.

[22] R. K. Saxena, "On the H-function of n-variables," *Kyungpook Mathematical Journal*, vol. 17, no. 2, pp. 221–226, 1977.

[23] A. M. Mathai, R. K. Saxena, and H. J. Haubold, *The H-Function: Theory and Applications*, Springer, New York, NY, USA, 2009.

[24] H. M. Srivastava and P. W. Karlsson, *Multiple Gaussian Hypergeometric Series*, Halsted, Chichester, UK; John Wiley and Sons, New York, NY, USA, 1985.

[25] H. M. Srivastava and H. L. Manocha, *A Treatise on Generating Functions*, Ellis Horwood Series: Mathematics and its Applications, Halsted, Chichester, UK; John Wiley and Sons, New York, NY, USA, 1984.

14

Mathematical Analysis and Numerical Simulations for a System Modeling Acid-Mediated Tumor Cell Invasion

Christian Märkl,[1] Gülnihal Meral,[2] and Christina Surulescu[3]

[1] *Institut für Numerische und Angewandte Mathematik, Universität Stuttgart, Pfaffenwaldring 57, 70569 Stuttgart, Germany*
[2] *Department of Mathematics, Faculty of Arts and Sciences, Bülent Ecevit University, 67100 Zonguldak, Turkey*
[3] *Technische Universität Kaiserslautern, Felix Klein Zentrum für Mathematik, Paul Ehrlich Strasse, 67663 Kaiserslautern, Germany*

Correspondence should be addressed to Christina Surulescu; surulescu@mathematik.uni-kl.de

Academic Editor: Liancheng Wang

This work is concerned with the mathematical analysis of a model proposed by Gatenby and Gawlinski (1996) in order to support the hypothesis that tumor-induced alteration of microenvironmental pH may provide a simple but comprehensive mechanism to explain cancer invasion. We give an intuitive proof for the existence of a solution under general initial conditions upon using an iterative approach. Numerical simulations are also performed, which endorse the predictions of the model when compared with experimentally observed qualitative facts.

1. Introduction

Despite major progress in medicine and science there still are incurable diseases which can threaten human lives. Cancer is among the most severe ones and it manifests itself as an uncontrolled growth of cells which are produced by the organism subsequently to mutations. Cancer cells migrate through the surrounding tissue and degrade it on their way toward blood vessels and distal organs where they initiate and develop further tumors, a process known as metastasis.

In the last decades various classes of models have been proposed aiming to provide a quantitative description of tumor growth. They range from the microscopic level of intracellular signaling pathways conditioning the growth of neoplastic tissue by stimulation or inhibition of apoptosis (e.g., by the influence of tumor necrosis factors [1]) or tumor cell motility, for example, by restructuring the cytoskeleton or by producing matrix degrading enzymes [2], through the level of cell-cell or cell-tissue interactions and up to the macroscopic level characterizing the behavior of the entire cell population. Multiscale settings like those in [3, 4] involve several of these scales and offer a systemic approach to the modeling process.

When ignoring the setups relying on mechanical force balance and/or on the theory of mixtures, in the study of tumor invasion and metastasis one can distinguish between the so-called kinetic approach and the direct modeling at the macroscopic level. In the former a mesoscopic model is considered, consisting of an integro-partial differential equation for the evolution of the cell density, possibly coupled with integro-differential and/or reaction-diffusion equations for the fibre density of the extracellular matrix (ECM) and the chemotactic signal (see, e.g., [5] and the references therein [3, 4]). Then with an appropriate scaling the macroscopic limit is deduced, usually leading to a Keller-Segel-type model or some hyperbolic systems; see, for example, [6]. The macroscopic approach involves the largest class of existing models and directly accounts for processes at the level of cell populations, leading to systems of reaction-diffusion (transport) equations like, for example, in [7, 8].

The role of tumor microenvironment in determining cancer malignancy has been put in evidence in several references; see, for example, [9, 10]. For instance hypoxia and acidity are factors that can trigger the progression from benign to malignant growth. In order to survive in the unfavourable environment they create, cancer cells upregulate certain proton extrusion mechanisms [11], the consequence of which is that the extracellular tumor environment has an acidic pH, which boosts apoptosis of normal cells and thus allows the neoplastic tissue to extend in the space becoming available.

Hence the pH level directly influences the metastatic potential of tumor cells [12, 13]. These facts led Gatenby and Gawlinski [8, 14] to propose a model for the acid-mediated tumor invasion, which describes the interaction between the density of normal cells, tumor cells, and the concentration of H^+ protons produced by the latter via reaction-diffusion equations. Starting from this model, travelling waves have been used to explain the aggressive action of cancer cells on their surroundings [15]. Further settings issued from Gatenby and Gawlinski's model involve nutrient dynamics influenced by both vascular and avascular growth of multicellular tumor spheroids [16, 17] assuming rotational symmetry and investigating existence and qualitative properties of the solutions.

In this work we reconsider the model in [8], whereby also explicitly allowing for crowding effects (due to competition with cancer cells) in the growth of normal cells. (This can be also done for the growth term in the tumor cell equation; however it does not change the analysis nor the qualitative behavior of the solutions to the system. Moreover, its biological motivation is not strong enough, since the competition between the two cell types does not really affect the growth, but rather the invasion of the neoplastic tissue: the acidity increase in the peritumoral environment is a byproduct of the enhanced glycolysis of cancer cells and not produced with the purpose of killing the normal cells and eluding concurrence.) For this setting we perform mathematical analysis and numerical simulations in order to verify the model predictions with respect to experimentally observed qualitative facts. In order to prove the existence of a unique (weak) solution we propose an intuitive method relying on an iterative procedure which has also been applied in [18, 19] in a different context (one of the supplementary difficulties here is the diffusion coefficient being nonconstant, but depending on the solution itself) and allows to avoid the use of operator semigroups.

2. Problem Setting

The model by Gatenby and Gawlinski [8] describes the evolution of normal and tumor cell density, respectively, in a domain where these cell types interact on the basis of pH value modifications. The mathematical description of these processes is ensured by the following system of reaction-diffusion equations for the normal cell density $N(t,\mathbf{x})$, the tumor cell density $K(t,\mathbf{x})$ (both in cells/cm^3), and the concentration of excessive $H^+(t,\mathbf{x})$ ion concentration (in Mol):

$$\begin{aligned}
\frac{\partial N}{\partial t} &= w_N N\left(1-\frac{N}{K_N}-\theta\frac{K}{K_K}\right)-d_N HN \quad \text{in } (0,T)\times\Omega,\\
\frac{\partial K}{\partial t} &= w_K K\left(1-\frac{K}{K_K}\right)+\nabla\cdot\left(D_K\left(1-\frac{N}{K_N}\right)\nabla K\right) \quad \text{in } (0,T)\times\Omega,\\
\frac{\partial H}{\partial t} &= w_H K - d_H H + D_H\Delta H \quad \text{in } (0,T)\times\Omega,
\end{aligned} \tag{1}$$

where $\theta \le 1/2$ denotes the strength parameter for the competition between normal cells and neoplastic tissue. Thereby, $\Omega \subset \mathbb{R}^n$ $(n = 1,2,3)$ is a regular-enough and bounded domain and only microscopically small processes are considered at the interface between tumor and healthy tissue. Observe that the diffusion coefficient of the cancer cells depends on the normal cell density: when the healthy tissue is at its carrying capacity the neoplastic tissue cannot diffuse; thus the tumor is confined. It can only spread if the surrounding normal tissue is diminished from its carrying capacity and this is assumed to happen due to lowering the pH level upon secretion of H^+ protons by cancer cells.

The constants w_N and w_K are given in $1/\mathrm{s}$ and represent the growth rates, and the constants K_N and K_K are expressed in cells/cm^3 and provide the carrying capacities of the normal and tumor cells, respectively. The death rate d_N of the normal cells is measured in $1/(\mathrm{Mol}\cdot\mathrm{s})$, and the diffusion coefficient of tumor cells in the absence of normal cells is given in cm^2/s. The production rate w_H of H^+ protons is expressed in $\mathrm{Mol}\cdot\mathrm{cm}^3/(\mathrm{cells}\cdot\mathrm{s})$, the uptake rate d_H (due for instance to proton buffering (see, e.g., [20] and the references therein) and/or various ion exchangers between intracellular and extracellular domains (see e.g. [21])) is measured in $1/\mathrm{s}$, and their diffusion coefficient of D_H in cm^2/s.

We also assume that there is no exchange of cells and H^+ protons through the boundary of the considered domain; thus

$$\frac{\partial K}{\partial \mathbf{n}} = \frac{\partial H}{\partial \mathbf{n}} = 0 \quad \text{in } (0,T)\times\partial\Omega, \tag{2}$$

where $\mathbf{n}$ denotes the outer unit normal vector to $\partial\Omega$.

The initial conditions are given by

$$K(0,\mathbf{x}) = K_0(\mathbf{x}), \qquad N(0,\mathbf{x}) = N_0(\mathbf{x}), \qquad H(0,\mathbf{x}) = H_0(\mathbf{x}) \quad \text{in } \Omega. \tag{3}$$

Thereby $K_0(\mathbf{x})$, $N_0(\mathbf{x})$, and $H_0(\mathbf{x})$ are strictly positive functions, which satisfy the no-flux condition:

$$\frac{\partial K_0}{\partial \mathbf{n}} = \frac{\partial H_0}{\partial \mathbf{n}} = 0 \quad \text{on } \partial\Omega. \tag{4}$$

In order to render the system (1) dimensionless we use the following transformations:

$$\widetilde{N} = \frac{N}{K_N}, \qquad \widetilde{K} = \frac{K}{K_K}, \qquad \widetilde{H} = H\cdot\frac{d_H}{w_H K_K}, \qquad \widetilde{t} = w_N\cdot t, \qquad \widetilde{\mathbf{x}} = \sqrt{\frac{w_N}{D_H}}\cdot\mathbf{x} \tag{5}$$

along with the notations

$$\delta_N = \frac{d_N w_H K_K}{d_H w_N}, \qquad \rho_K = \frac{w_K}{w_N}, \qquad \Delta_K = \frac{D_K}{D_H}, \qquad \delta_H = \frac{d_H}{w_N}. \tag{6}$$

We obtain the system

$$\frac{\partial N}{\partial t} = N(1 - N - \theta K) - \delta_N HN,$$
$$\frac{\partial K}{\partial t} = \rho_K K(1 - K) + \nabla \cdot (\Delta_K (1 - N) \nabla K), \quad (7)$$
$$\frac{\partial H}{\partial t} = \delta_H K - \delta_H H + \Delta H,$$

where for simplicity the tilde notations have been ignored. The stability analysis of this system (with $\theta = 0$) has been performed in [8], leading to biologically significant predictions.

3. Existence and Uniqueness of Solutions

In this section we provide a natural proof for the existence and uniqueness of a weak solution to the system (1), with initial data (3) and boundary conditions (2). We make use of an iterative procedure instead of the classical approach via semigroup theory; this is more intuitive and allows for a separate treatment of the three equations in each step.

Consider the function spaces

$$X := L^{\infty}(0, T; H^1(\Omega)),$$
$$Y := \{u \in L^2(0, T; H^2(\Omega)) : u_t \in L^2(0, T; L^2(\Omega))\}, \quad (8)$$
$$Z := L^{\infty}(0, T; L^2(\Omega)).$$

Definition 1. A weak solution of (1) with boundary conditions (2) and initial data (3) is a triple (H, N, K) of functions in $X \times Y \times Z$, such that for all $\phi \in H^1(\Omega)$ a.e. in $[0, T]$ the following three equations are satisfied:

$$\begin{aligned}
&\int_\Omega w_H K \phi \, d\mathbf{x} \\
&\quad = \int_\Omega H_t \phi \, d\mathbf{x} + \int_\Omega D_H \nabla H \nabla \phi \, d\mathbf{x} + \int_\Omega d_H H \phi \, d\mathbf{x}, \\
&\int_\Omega w_N N \left(1 - \frac{N}{K_N} - \theta \frac{K}{K_K}\right) \phi \, d\mathbf{x} \\
&\quad = \int_\Omega N_t \phi \, d\mathbf{x} + \int_\Omega N d_N H \phi \, d\mathbf{x}, \\
&\int_\Omega w_K K \left(1 - \frac{K}{K_K}\right) \phi \, d\mathbf{x} \\
&\quad = \int_\Omega K_t \phi \, d\mathbf{x} + \int_\Omega D_K \left(1 - \frac{N}{K_N}\right) \nabla K \nabla \phi \, d\mathbf{x}.
\end{aligned} \quad (9)$$

Theorem 2. *There exists $T > 0$, such that the system* (1) *with initial data* (3) *and boundary conditions* (2) *satisfying*

$$H_0 \in H^1(\Omega) \cap C(\Omega), \qquad N_0 \in L^{\infty}(\Omega) \cap H^1(\Omega),$$
$$K_0 \in H^1(\Omega), \quad (10)$$
$$H_0 \geq C_H > 0, \qquad 0 < N_0 \leq \frac{K_N}{2}, \qquad 0 < K_0 \leq K_K$$

has a unique solution $(H, K) \in (X \times X) \cap (Y \times Y)$ and $N \in Z$.

We set

$$T := \prod_{i=1}^{6} T_i \quad (11)$$

with $T_i \leq 1$ to be defined below.

In order to prove Theorem 2 we construct a sequence

$$(H^m, K^m)_{m \in \mathbb{N}_0} \in (X \times X) \cap (Y \times Y),$$
$$(N^m)_{m \in \mathbb{N}_0} \in Z \quad (12)$$

and prove its convergence towards the weak solution of the system.

Let $(H^0, K^0) \in (X \times X) \cap (Y \times Y)$ and $N^0 \in Z$ be the weak solution to the homogeneous system

$$\frac{\partial H^0}{\partial t} - D_H \Delta H^0 + d_H H^0 = 0, \quad (13)$$
$$\frac{\partial N^0}{\partial t} + d_N H^0 N^0 = 0, \quad (14)$$
$$\frac{\partial K^0}{\partial t} - \nabla \cdot \left(D_K \left(1 - \frac{N^0}{K_N}\right) \nabla K^0\right) = 0, \quad (15)$$

while $(H^m, K^m)_{m \in \mathbb{N}_0} \in (X \times X) \cap (Y \times Y)$ and $(N^m)_{m \in \mathbb{N}_0} \in Z$ is the weak solution to

$$\frac{\partial H^{m+1}}{\partial t} - D_H \Delta H^{m+1} + d_H H^{m+1} = w_H K^m, \quad (16)$$
$$\frac{\partial N^{m+1}}{\partial t} + d_N H^{m+1} N^{m+1} = w_N N^m \left(1 - \frac{N^m}{K_N} - \theta \frac{K^m}{K_K}\right), \quad (17)$$
$$\frac{\partial K^{m+1}}{\partial t} - \nabla \cdot \left(D_K \left(1 - \frac{N^{m+1}}{K_N}\right) \nabla K^{m+1}\right) = w_K K^m \left(1 - \frac{K^m}{K_K}\right) \quad (18)$$

with the corresponding initial and boundary conditions (2) and (3).

The existence and uniqueness of the functions $(H^m, K^m, N^m)_{m \in \mathbb{N}_0}$ in the above sequence are ensured by the following.

Lemma 3 (properties of the iteration sequence). *Under assumptions* (10) *there exists $T > 0$ such that*

(i) *there exists a unique weak solution to the systems* (13)–(15) *and* (16)–(18) *with conditions* (3) *and* (2), *and for every $m \in \mathbb{N}_0$ it holds that*

$$N^m, N_t^m \in L^{\infty}((0, T] \times \Omega), \quad (19)$$
$$H^m, K^m \in L^2(0, T; H^2(\Omega)) \cap L^{\infty}(0, T; H^1(\Omega)),$$
$$H_t^m, K_t^m \in L^2(0, T; L^2(\Omega)); \quad (20)$$

(ii) *the functions* H^m, N^m, *and* K^m *are positive for all* $m \in \mathbb{N}_0$. *Moreover, the following inequalities hold:*

$$H^m(t,\mathbf{x}) \geq C_H e^{-d_H t}, \qquad N^m(t,\mathbf{x}) \leq \frac{K_N}{2},$$
$$K^m(t,\mathbf{x}) \leq K_K \quad \textit{for a.e.}\ \mathbf{x} \in \Omega,\ t \in [0,T]; \tag{21}$$

(iii) *the functions* H^m, N^m, *and* K^m *satisfy for adequate constants* $C(\Omega, T)$ *and for all* $m \in \mathbb{N}_0$ *the estimates*

$$\|H^m\|_X + \|H^m\|_{L^2(0,T;H^2(\Omega))} \leq C(\Omega,T)\left(\|K_0\|_{H^1(\Omega)} + \|H_0\|_{H^1(\Omega)}\right), \tag{22}$$

$$\|N^m\|_X^2 \leq C(\Omega,T)\|N_0\|_{H^1(\Omega)}^2, \tag{23}$$

$$\|K^m\|_X + \|K^m\|_{L^2(0,T;H^2(\Omega))} \leq 2C(\Omega,T)\|K_0\|_{H^1(\Omega)}. \tag{24}$$

Remark 4. From (19) it follows that

$$N^m \in L^\infty\left(0,T;L^2(\Omega)\right) \tag{25}$$

for all $m \in \mathbb{N}_0$.

Proof of Lemma 3. We perform mathematical induction with respect to m.

Induction Start. The proof of the claims in Lemma 3 for $m = 0$ is done separately for each of (13)–(15).

(a) With the substitution

$$\widetilde{H}^0(t,\mathbf{x}) = H^0(t,\mathbf{x})e^{d_H t}. \tag{26}$$

Equation (13) becomes the heat equation

$$\widetilde{H}_t^0 - D_H\Delta\widetilde{H}^0 = 0; \tag{27}$$

thus by the theory of linear parabolic differential equations (see, e.g., [22]) and with the assumption $H_0 \in H^1(\Omega)$ it follows that there exists a unique solution H^0 of (13) such that

$$H^0 \in L^2\left(0,T;H^2(\Omega)\right) \cap L^\infty\left(0,T;H^1(\Omega)\right),$$
$$H_t^0 \in L^2\left(0,T;L^2(\Omega)\right). \tag{28}$$

This weak solution also satisfies

$$\|H^0\|_X + \|H^0\|_{L^2(0,T;H^2(\Omega))} \leq C(\Omega,T)\|H_0\|_{H^1(\Omega)}. \tag{29}$$

Further it is known (see, e.g., [23]) that the solution of (13) can be written explicitly with respect to the initial condition H_0 and the heat kernel and it is therefore positive.

(b) Equation (14) is linear and has a positive solution:

$$N^0(t,\mathbf{x}) = N_0 e^{-\int_0^t d_N H^0(s,\mathbf{x})ds} > 0 \tag{30}$$

which depends on $H^0(t,\mathbf{x})$. It follows immediately that

$$\|N^0\|_{L^\infty(0,T;H^1(\Omega))}^2 = \left\|N_0 e^{-\int_0^t d_N H^0 dt}\right\|_{L^\infty(0,T;H^1(\Omega))}^2 \leq \|N_0\|_{H^1(\Omega)}^2, \tag{31}$$

and thus the estimation (23) for $m = 0$ is obtained.

The corresponding statement (19) for N^0 is to be justified below.

(c) In order to prove the claims of Lemma 3 for K^0 we show first that

$$N^0 \in L^\infty((0,T]\times\Omega), \tag{32}$$

$$N_t^0 \in L^\infty((0,T]\times\Omega). \tag{33}$$

The former follows from

$$\|N^0\|_{L^\infty((0,T]\times\Omega)} \overset{(30)}{=} \left\|N_0 e^{-\int_0^t d_N H^0 dt}\right\|_{L^\infty((0,T]\times\Omega)} \leq \|N_0\|_{L^\infty(\Omega)} < \infty. \tag{34}$$

For $t \geq \delta > 0$ it is

$$\|N_t^0\|_{L^\infty((0,T]\times\Omega)} \overset{(30)}{=} d_N\left\|N_0 \cdot e^{-\int_0^t d_N H^0 dt} \cdot H^0\right\|_{L^\infty((0,T]\times\Omega)} < \infty. \tag{35}$$

For $t \to 0$ we can consider (27). For its solution it holds (see, e.g., [22]) that

$$\lim_{(t,\mathbf{x})\to(0,\mathbf{x}^0)} \widetilde{H}^0(t,\mathbf{x}) = H_0\left(\mathbf{x}^0\right) \quad \text{for every } \mathbf{x}^0 \in \Omega. \tag{36}$$

Therefore,

$$\lim_{(t,\mathbf{x})\to(0,\mathbf{x}^0)} H^0(t,\mathbf{x}) = \lim_{(t,\mathbf{x})\to(0,\mathbf{x}^0)} \widetilde{H}^0(t,\mathbf{x})e^{-d_H t} = H_0\left(\mathbf{x}^0\right) \tag{37}$$

and finally (33) follows, thus also (19) for $m = 0$.

The following proof of (20) and (24) upon starting from (15) relies on Theorem 7.1.5 in Evans [22]. However, that result cannot be directly applied to the present case, since the diffusion coefficient $a(t,\mathbf{x}) = D_K(1 - (N^0(t,\mathbf{x})/K_N))$ in (15) depends on time.

Let

$$k_m(t) := \sum_{i=1}^m d_m^i(t)w_i \tag{38}$$

with functions $w_i = w_i(\mathbf{x})$ such that

$$\{w_i\}_{i=1}^\infty \text{ is an orthogonal basis of } H^1(\Omega),$$
$$\{w_i\}_{i=1}^\infty \text{ is an orthonormal basis of } L^2(\Omega). \tag{39}$$

Considering the symmetric bilinear form

$$A[k_m,k_m] := \int_\Omega a(t,\mathbf{x})(\nabla k_m)^2 d\mathbf{x}, \tag{40}$$

the dependence of the coefficient $a(t, \mathbf{x})$ on t leads in its time derivative

$$\frac{d}{dt}A\left[k_m, k_m\right] = \int_\Omega a'(t,\mathbf{x})\left(\nabla k_m\right)^2 d\mathbf{x} + 2\int_\Omega a(t,\mathbf{x})\left(\nabla k_m\right)'\nabla k_m d\mathbf{x} \tag{41}$$

to a supplementary summand

$$\int_\Omega a'(t,\mathbf{x})\left(\nabla k_m\right)^2 d\mathbf{x} = -\int_\Omega \frac{D_K}{K_N}\left(N^0\right)'(t,\mathbf{x})\left(\nabla k_m\right)^2 d\mathbf{x}, \tag{42}$$

where for shortness we denoted by $'$ the derivative with respect to t.

The rest of the proof of Theorem 7.1.5 in [22] can now be adapted to obtain for an arbitrary $\zeta > 0$ the estimate

$$\begin{aligned}\left\|k_m'\right\|^2_{L^2(\Omega)} &+ \frac{d}{dt}\left(\frac{1}{2}A\left[k_m, k_m\right]\right) \\ &\leq \frac{C}{\zeta}\left(\left\|k_m\right\|^2_{H^1(\Omega)} + \|f\|^2_{L^2(\Omega)}\right) \\ &\quad + 2\zeta\left\|k_m'\right\|^2_{L^2(\Omega)} + \frac{1}{2}\int_\Omega \frac{D_K}{K_N}\left(N^0\right)'\left(\nabla k_m\right)^2 d\mathbf{x}.\end{aligned} \tag{43}$$

Now let (recall (33))

$$M_{N^0} := \frac{D_K}{K_N}\left\|N_t^0\right\|_{L^\infty((0,T]\times\Omega)}. \tag{44}$$

Upon integrating with respect to t one can majorize

$$\begin{aligned}\int_0^T \int_\Omega \frac{D_K}{K_N}\left(N^0\right)'\left(\nabla k_m\right)^2 d\mathbf{x}\, dt &\leq M_{N^0}\int_0^T \left\|\nabla k_m\right\|^2_{L^2(\Omega)} dt \\ &\leq M_{N^0}\left\|k_m\right\|^2_{L^2(0,T;H^1(\Omega))} \\ &\leq \gamma(\Omega, T) < \infty,\end{aligned} \tag{45}$$

with $\gamma(\Omega, T)$ an adequate constant. The rest of the proof can be done as in Theorem 7.1.5 in [22], upon taking into account (32) and $K_0 \in H^1(\Omega)$ in order to show that there exists a unique weak solution $K^0(t, \mathbf{x})$ to (15) such that

$$\begin{gathered}K^0 \in L^2\left(0,T;H^2(\Omega)\right) \cap L^\infty\left(0,T;H^1(\Omega)\right), \\ K_t^0 \in L^2\left(0,T;L^2(\Omega)\right), \\ \left\|K^0\right\|_X + \left\|K^0\right\|_{L^2(0,T;H^2(\Omega))} \leq C(\Omega,T)\left\|K_0\right\|_{H^1(\Omega)}.\end{gathered} \tag{46}$$

Since

$$K_0^0(\mathbf{x}) > 0, \tag{47}$$

it follows from the weak maximum principle that $K^0(t, \mathbf{x}) > 0$ and thus also the positivity of $K^0(t, \mathbf{x})$.

The proof of the inequalities (21) for $m = 0$ does not differ from the one for a general $m \in \mathbb{N}$ given below and is therefore omitted here.

With (a)–(c) we proved all statements of Lemma 3 for $m = 0$.

Induction Hypothesis. Assume the assertions of the lemma hold for an arbitrary $m \in \mathbb{N}_0$.

Inductive Step. The proof for $m + 1$ is to be done separately for each of (16)–(18). Since for a corresponding embedding constant $c_1 := c_1(\Omega, T)$

$$\int_0^T \left\|K^m\right\|^2_{L^2(\Omega)} dt \leq c_1 \int_0^T \left\|K^m\right\|^2_{H^1(\Omega)} dt \underset{(24)}{\overset{\text{ind. hyp.}}{\leq}} 4c_1 C^2(\Omega,T)\, T\left\|K_0\right\|^2_{H^1(\Omega)} < \infty \tag{48}$$

and thus

$$K^m \in L^2\left(0,T;L^2(\Omega)\right), \tag{49}$$

the existence of a unique weak solution to (16), (2), and (3) follows from the theory of linear parabolic differential equations. The solution $H^{m+1}(t, \mathbf{x})$ satisfies

$$\begin{aligned}&H^{m+1} \in L^2\left(0,T;H^2(\Omega)\right) \cap L^\infty\left(0,T;H^1(\Omega)\right), \\ &H_t^{m+1} \in L^2\left(0,T;L^2(\Omega)\right), \\ &\left\|H^{m+1}\right\|_X + \left\|H^{m+1}\right\|_{L^2(0,T;H^2(\Omega))} \\ &\quad \leq C_1(\Omega,T)\left(2w_H C(\Omega,T)\sqrt{c_1 T}\left\|K_0\right\|_{H^1(\Omega)} + \left\|H_0\right\|_{H^1(\Omega)}\right) \\ &\quad \leq \mathscr{C}(\Omega,T)\left(\left\|K_0\right\|_{H^1(\Omega)} + \left\|H_0\right\|_{H^1(\Omega)}\right),\end{aligned} \tag{50}$$

with $\mathscr{C}(\Omega, T) = \max\{C_1(\Omega, T), C_1(\Omega, T)2w_H C(\Omega, T)\sqrt{c_1 T}\}$.

In order to establish the lower bound for H^{m+1} define an auxiliary function $\psi^{m+1}(t, \mathbf{x}) := H^{m+1}(t, \mathbf{x}) - C_H e^{-d_H t}$, for which it holds

$$\left\langle \psi_t^{m+1}(t), \phi\right\rangle + D_H\int_\Omega \nabla\psi^{m+1}\nabla\phi\, d\mathbf{x} + d_H\int_\Omega \psi^{m+1}\phi\, d\mathbf{x} = \left\langle w_H K^m, \phi\right\rangle. \tag{51}$$

For every nonnegative $\phi \in H^1(\Omega)$ the right-hand side is positive. Further, $\psi^{m+1}(0, \mathbf{x}) \geq 0$ by construction; thus it follows with the weak maximum principle that $\psi^{m+1} \geq 0$ a.e. which leads to $H^{m+1}(t, \mathbf{x}) \geq C_H e^{-d_H t}$.

Now (17) is a linear, inhomogeneous differential equation, with solution

$$N^{m+1}(t,\mathbf{x}) = e^{-\alpha(t,\mathbf{x})}\left(N_0(\mathbf{x}) + \int_0^t \beta(s,\mathbf{x})\, e^{\alpha(s,\mathbf{x})} ds\right), \tag{52}$$

where

$$\alpha(t,\mathbf{x}) = \int_0^t d_N H^{m+1}(v,\mathbf{x})\,dv,$$

$$\beta(s,\mathbf{x}) = w_N N^m(s,\mathbf{x})\left(1 - \frac{N^m(s,\mathbf{x})}{K_N} - \theta\frac{K^m(s,\mathbf{x})}{K_K}\right). \tag{53}$$

In order to prove (19) for $m+1$ we have to show that

$$N^{m+1} \in L^\infty((0,T]\times\Omega), \tag{54}$$

$$N_t^{m+1} \in L^\infty((0,T]\times\Omega). \tag{55}$$

Obviously, the first assertion (54) holds, due to the induction hypothesis.

Next, estimate

$$\begin{aligned}&\left\|N_t^{m+1}\right\|_{L^\infty((0,T]\times\Omega)}\\ &\quad\le w_N\left\|N^m\right\|_{L^\infty((0,T]\times\Omega)}\left\|1-\frac{N^m}{K_N}-\theta\frac{K^m}{K_K}\right\|_{L^\infty((0,T]\times\Omega)}\\ &\qquad + d_N\left\|N^{m+1}\right\|_{L^\infty((0,T]\times\Omega)}\left\|H^{m+1}\right\|_{L^\infty((0,T]\times\Omega)} < \infty,\end{aligned} \tag{56}$$

due to (19).

Using again the induction hypothesis, the regularity of the initial data, and the properties of the solutions to the heat equations it follows immediately that $\|H^{m+1}\|_{L^\infty((0,T]\times\Omega)} < \infty$, which leads to

$$\left\|N_t^{m+1}\right\|_{L^\infty((0,T]\times\Omega)} < \infty \tag{57}$$

and thus (55) is proved.

Now we prove the positivity of N^{m+1} and the corresponding inequality in (21). To this aim use the induction hypothesis to observe that

$$\begin{aligned}N^{m+1}(t,\mathbf{x}) &\le \frac{K_N}{2}e^{-\alpha(t,\mathbf{x})} + \int_0^t \beta(s,\mathbf{x})\,e^{-(\alpha(t,\mathbf{x})-\alpha(s,\mathbf{x}))}ds\\ &\le \frac{K_N}{2}e^{-\alpha(t,\mathbf{x})} + w_N\frac{K_N}{2}\int_0^t e^{-(\alpha(t,\mathbf{x})-\alpha(s,\mathbf{x}))}ds.\end{aligned} \tag{58}$$

Next notice that there exists a positive constant $\widetilde{C}_H$ such that $H^{m+1}(t,\mathbf{x}) \ge \widetilde{C}_H$ for a.e. $\mathbf{x}\in\Omega$, $t\in[0,T]$. This leads to the estimate

$$\begin{aligned}N^{m+1}(t,\mathbf{x}) &\le \frac{K_N}{2}e^{-d_N\widetilde{C}_H t} + w_N\frac{K_N}{2}\frac{1}{d_N\widetilde{C}_H}\left(1-e^{-d_N\widetilde{C}_H t}\right)\\ &\le \frac{K_N}{2}\left(\left(1-\frac{w_N}{d_N\widetilde{C}_H}\right)e^{-d_N\widetilde{C}_H t} + \frac{w_N}{d_N\widetilde{C}_H}\right) \le \frac{K_N}{2}.\end{aligned} \tag{59}$$

This in turn immediately implies via (52) the positivity of N^{m+1}.

In the next step we prove the estimate (23) for $N^{m+1}(t,\mathbf{x})$.

Due to (52) we get

$$\begin{aligned}&\left\|N^{m+1}(t)\right\|_{H^1(\Omega)}^2\\ &\quad= \left\|e^{-\alpha(t)}N_0 + e^{-\alpha(t)}\int_0^t \beta(s)e^{\alpha(s)}ds\right\|_{H^1(\Omega)}^2\\ &\quad\le 2\left\|N_0\right\|_{H^1(\Omega)}^2\\ &\qquad + 2w_N^2\left\|\int_0^t N^m(s)\left(1-\frac{N^m(s)}{K_N}-\theta\frac{K^m(s)}{K_K}\right)ds\right\|_{H^1(\Omega)}^2\\ &\quad\le 2\left\|N_0\right\|_{H^1(\Omega)}^2\\ &\qquad + 2w_N^2\left\|\int_0^t\left(N^m(s)-\frac{(N^m(s))^2}{K_N^2}\right)ds\right\|_{H^1(\Omega)}^2\\ &\quad\le 2\left\|N_0\right\|_{H^1(\Omega)}^2\\ &\qquad + 4w_N^2\left(\left\|\int_0^t N^m(s)ds\right\|_{H^1(\Omega)}^2\right.\\ &\qquad\left. + \frac{1}{K_N^2}\left\|\int_0^t (N^m(s))^2 ds\right\|_{H^1(\Omega)}^2\right)\\ &\quad\le \left\|N_0\right\|_{H^1(\Omega)}^2\left[2+4w_N^2 C(\Omega,T)T^2\right]\\ &\quad\le \mathscr{C}(\Omega,T)\left\|N_0\right\|_{H^1(\Omega)}^2\end{aligned} \tag{60}$$

by (23) and the induction hypothesis.

In order to prove the assertions of Lemma 3 for $K^{m+1}(t,\mathbf{x})$ one can apply Theorem 7.1.5 in [22], with (54), (55), and the same justification as for the induction start at (c).

With an adequate embedding constant $c_2 := c_2(\Omega,T)$

$$\begin{aligned}&\int_0^T\left\|K^m\left(1-\frac{K^m}{K_K}\right)\right\|_{L^2(\Omega)}^2 dt\\ &\quad\le \int_0^T\left(\left\|K^m\right\|_{L^2(\Omega)} + \left\|\frac{(K^m)^2}{K_K}\right\|_{L^2(\Omega)}\right)^2 dt\\ &\quad\le 2\int_0^T\left\|K^m\right\|_{L^2(\Omega)}^2 dt + 2\int_0^T\left\|\frac{(K^m)^2}{K_K}\right\|_{L^2(\Omega)}^2 dt\\ &\quad\le 2c_1^2\int_0^T\left\|K^m\right\|_{H^1(\Omega)}^2 dt + 2\frac{c_2^4}{K_K^2}\int_0^T\left\|K^m\right\|_{H^1(\Omega)}^4 dt\\ &\quad\le 8c_1^2C^2(\Omega,T)\left\|K_0\right\|_{H^1(\Omega)}^2 T_1T_2\\ &\qquad + 32\frac{c_2^4}{K_K^2}C^4(\Omega,T)\left\|K_0\right\|_{H^1(\Omega)}^4 T_1T_2 < \infty,\end{aligned} \tag{61}$$

by (24) and the induction hypothesis; therefore $K^m(1-(K^m/K_K)) \in L^2(0,T;L^2(\Omega))$ and $K_0(\mathbf{x}) \in H^1(\Omega)$. By applying

Theorem 7.1.5 in [22] it follows that (18) has a unique weak solution $K^{m+1}(t,\mathbf{x})$ with

$$K^{m+1} \in L^2\left(0,T;H^2(\Omega)\right) \cap L^\infty\left(0,T;H^1(\Omega)\right), \quad K_t^{m+1} \in L^2\left(0,T;L^2(\Omega)\right). \tag{62}$$

Now choose T_1 such that $\max\{T_1C^2(\Omega,T), T_1C^4(\Omega,T)\} \le 1$ and

$$T_2 := \min\left\{\frac{1}{2}, \frac{1}{16w_K^2c_1^2\|K_0\|}, \frac{K_K^2}{64w_K^2c_2^4\|K_0\|^3}\right\}. \tag{63}$$

Then

$$\int_0^T \left\|w_K K^m\left(1-\frac{K^m}{K_K}\right)\right\|_{L^2(\Omega)}^2 dt \le \|K_0\|_{H^1(\Omega)} \tag{64}$$

and thus the estimate

$$\left\|K^{m+1}\right\|_X + \left\|K^{m+1}\right\|_{L^2(0,T;H^2(\Omega))} \le 2C(\Omega,T)\|K_0\|_{H^1(\Omega)} \tag{65}$$

holds.

In order to prove the positivity of K^{m+1} we introduce an auxiliary function

$$\xi^{m+1}(t,\mathbf{x}) := -At\exp(1-\eta t) + K_K - K^{m+1}(t,\mathbf{x}), \tag{66}$$

for A positive and large enough and η a positive constant to be correspondingly chosen (see below). With the aid of this function we show that for all $m \in \mathbb{N}_0$

$$K^{m+1} \le K_K, \tag{67}$$

on an adequate time interval.

Proof (of the Statement (67))

Induction Start. The proof of (67) for $m = 0$ is identical to the one for $m+1$.

Induction Hypothesis. Assume assertion (67) holds for an arbitrary $m \in \mathbb{N}_0$.

Inductive Step. Upon using (66) in (18) we get

$$\frac{\partial \xi^{m+1}}{\partial t} - \nabla\cdot\left(D_K\left(1-\frac{N^{m+1}}{K_N}\right)\nabla\xi^{m+1}\right) = A(\eta t-1)\exp(1-\eta t) - w_K K^m\left(1-\frac{K^m}{K_K}\right). \tag{68}$$

Since

$$K^m\left(1-\frac{K^m}{K_K}\right) \le K_K, \tag{69}$$

for the right-hand side of (68) we have that

$$A(\eta t-1)\exp(1-\eta t) - w_K K^m\left(1-\frac{K^m}{K_K}\right) \ge 0 \tag{70}$$

holds for $t < T_3$ with correspondingly chosen T_3 and η such that $\eta t > 1$.

Since by construction $\xi^{m+1}(0,\mathbf{x}) \ge 0$, we can apply the weak maximum principle for $t \le T_3$ to show that

$$A(\eta t-1)\exp(1-\eta t) + K_K - K^{m+1}(t,\mathbf{x}) = \xi^{m+1} \ge 0, \tag{71}$$

from which it also follows that

$$K^{m+1} \le K_K. \tag{72}$$

This completes the proof of the statement (67).

In virtue of (67), for $T \le 1/w_K$ the right-hand side in (18) is positive. Since by hypothesis $K_0^{m+1} > 0$, the weak maximum principle implies the positivity of K^{m+1}. This ends the proof of all statements in Lemma 3 for an arbitrary $m \in \mathbb{N}_0$ and therefore the proof of the lemma itself. □

Now we are able to pass to the following.

Proof (of Theorem 2)

Existence. In order to prove the existence of a weak solution to (1) and (2) we show that the iterative sequence $(N^m, K^m, H^m)_{m\in\mathbb{N}_0}$ is Cauchy.

Due to the completeness of $H^1(\Omega)$ and $L^2(\Omega)$, this will imply the convergence of the sequence to some limit functions N, K, and H, these being solutions to (1) and (2).

Consider an arbitrary $m \in \mathbb{N}_0$. Since $H_0^m, H_0^{m+1} \in H^1(\Omega)$, and $K^m, K^{m+1} \in L^2(0,T;L^2(\Omega))$, it follows that

$$H_0^{m+1} - H_0^m \in H^1(\Omega), \quad K^{m+1} - K^m \in L^2\left(0,T;L^2(\Omega)\right). \tag{73}$$

Next, one can apply Theorem 7.1.5 in [22] to the difference $H^{m+1} - H^m$ to deduce the estimate

$$\left\|H^{m+1} - H^m\right\|_X^2 \le C(\Omega,T)\int_0^T \left\|w_H K^m - w_H K^{m-1}\right\|_{L^2(\Omega)}^2 dt. \tag{74}$$

The right-hand side above can be further estimated and with the embedding constant $c_3 := c_3(\Omega,T)$ it follows that

$$\begin{aligned}\left\|H^{m+1} - H^m\right\|_X^2 &\le C(\Omega,T)\,w_H^2c_3^2\int_0^T \left\|K^m - K^{m-1}\right\|_{H^1(\Omega)}^2 dt \\ &\le C(\Omega,T)\,w_H^2c_3^2T_4\left\|K^m - K^{m-1}\right\|_X^2 \\ &\le \frac{1}{2}\left\|K^m - K^{m-1}\right\|_X^2,\end{aligned} \tag{75}$$

where

$$T_4 = \min\left\{\frac{1}{4}, \frac{1}{4C(\Omega,T)\,w_H^2c_3^2}\right\}. \tag{76}$$

In order to obtain a corresponding estimate for the sequence $(N^m)_{m\in\mathbb{N}}$, consider two consecutive terms in (17) written for N^m and N^{m+1} and substract. This leads to

$$\frac{\partial}{\partial t}\left(N^{m+1}-N^m\right)+d_N\left(H^{m+1}N^{m+1}-H^mN^m\right) = w_N\left(N^m\left(1-\frac{N^m}{K_N}-\frac{K^m}{K_K}\right) -N^{m-1}\left(1-\frac{N^{m-1}}{K_N}-\frac{K^{m-1}}{K_K}\right)\right). \quad (77)$$

Denote $h(N^m, N^{m-1}) := w_N(N^m(1-(N^m/K_N)-(K^m/K_K))-N^{m-1}(1-(N^{m-1}/K_N)-(K^{m-1}/K_K)))$.

Now multiply with $(N^{m+1}-N^m)$ and integrate with respect to $\mathbf{x}$ to infer

$$\frac{1}{2}\int_\Omega \frac{\partial}{\partial t}\left(N^{m+1}-N^m\right)^2 d\mathbf{x} + d_N\int_\Omega\left(N^{m+1}-N^m\right)^2 H^{m+1}d\mathbf{x} = \int_\Omega\left(h\left(N^m,N^{m-1}\right)-d_N N^m\left(H^{m+1}-H^m\right)\right)\times\left(N^{m+1}-N^m\right)d\mathbf{x}. \quad (78)$$

Thus

$$\begin{aligned}\frac{d}{dt}\left\|N^{m+1}-N^m\right\|^2_{L^2(\Omega)} &\le 2w_N\int_\Omega\left|\left(N^m\left(1-\frac{N^m}{K_N}-\frac{K^m}{K_K}\right)-N^{m-1}\left(1-\frac{N^{m-1}}{K_N}-\frac{K^{m-1}}{K_K}\right)\right)\times\left(N^{m+1}-N^m\right)\right|d\mathbf{x}\\ &\quad+2d_N\int_\Omega\left|N^m\left(H^{m+1}-H^m\right)\left(N^{m+1}-N^m\right)\right|d\mathbf{x}\\ &\le\left[2w_N\left\|N^m\left(1-\frac{N^m}{K_N}-\frac{K^m}{K_K}\right)-N^{m-1}\left(1-\frac{N^{m-1}}{K_N}-\frac{K^{m-1}}{K_K}\right)\right\|_{L^2(\Omega)}\right.\\ &\quad\left.+2d_N\left\|N^m\left(H^{m+1}-H^m\right)\right\|_{L^2(\Omega)}\right]\times\left\|N^{m+1}-N^m\right\|_{L^2(\Omega)}.\end{aligned} \quad (79)$$

Next we estimate the above terms.

Table 1: Parameter values used in the model.

Parameters	Range
K_N	$5\times10^7/\text{cm}^3$
K_K	$5\times10^7/\text{cm}^3$
w_N	$1\times10^{-6}/\text{s}$
w_K	$1\times10^{-6}/\text{s}$
D_K	$2\times10^{-10}/\text{cm}^2/\text{s}$
D_H	$5\times10^{-6}/\text{cm}^2/\text{s}$
w_H	$2.2\times10^{-17}\text{M}\cdot\text{cm}^3/\text{s}$
d_H	$1.1\times10^{-4}/\text{s}$
d_N	$0\to10/\text{M}\cdot\text{s}$

Let (recall (19))

$$M_{\max} := \max\left\{M_{N^m} := \left\|N^m\right\|_{L^\infty((0,T]\times\Omega)}, N_{N^{m-1}} := \left\|N^{m-1}\right\|_{L^\infty((0,T]\times\Omega)}\right\}. \quad (80)$$

With the embedding constant $c_4 := c_4(\Omega, T)$ we obtain for the first term on the right-hand side of (79)

$$\begin{aligned}&2w_N\left\|N^m-N^{m-1}-\frac{(N^m)^2}{K_N}+\frac{(N^{m-1})^2}{K_N}-\frac{N^mK^m}{K_K}+\frac{N^{m-1}K^{m-1}}{K_K}\right\|_{L^2(\Omega)}\\ &\le 2w_N\left\|N^m-N^{m-1}\right\|_{L^2(\Omega)}+\frac{4w_NM_{\max}}{K_N}\left\|N^m-N^{m-1}\right\|_{L^2(\Omega)}+2\frac{w_NK_N}{K_K}\left\|K^m-K^{m-1}\right\|_{L^2(\Omega)}+2w_N\left\|N^m-N^{m-1}\right\|_{L^2(\Omega)}\\ &\le C_{\widetilde{N}}\left\|N^m-N^{m-1}\right\|_{L^2(\Omega)}+C_{\widetilde{K}}\left\|K^m-K^{m-1}\right\|_{L^2(\Omega)}\end{aligned} \quad (81)$$

with $C_{\widetilde{N}} := 4w_N(1+(M_{\max}/K_N))$ and $C_{\widetilde{K}} := 2w_NK_N/K_K$.

Now for the second term on the right-hand side of (79)

$$\begin{aligned}2d_N\left\|N^m(H^{m+1}-H^m)\right\|_{L^2(\Omega)} &\le d_Nc_4\frac{K_N}{2}\left\|H^{m+1}-H^m\right\|_{H^1(\Omega)}\\ &= C_{\widetilde{H}}\left\|H^{m+1}-H^m\right\|_{H^1(\Omega)},\end{aligned} \quad (82)$$

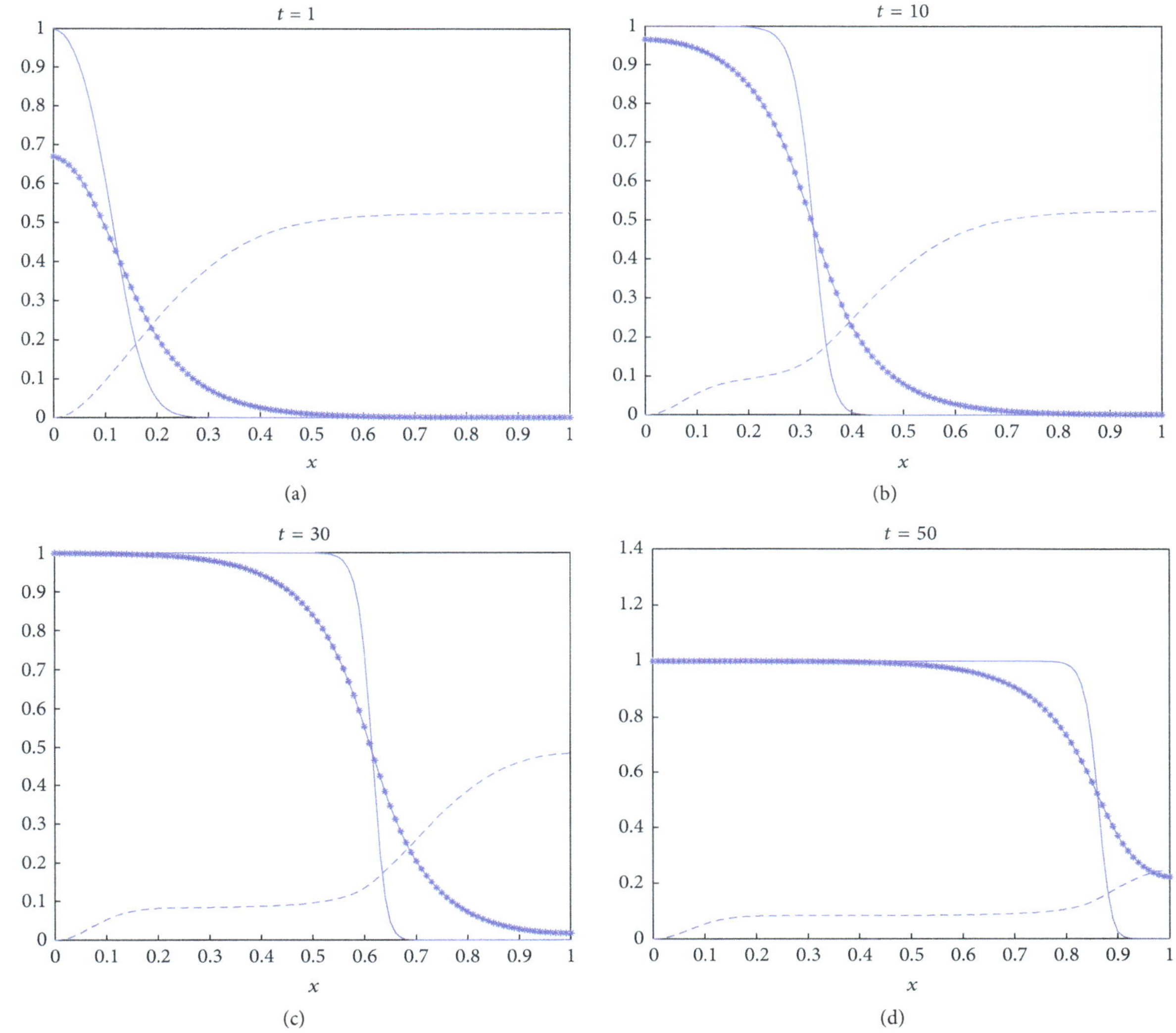

FIGURE 1: Variations of H^+ protons (starred), normal tissue (dashed), and neoplastic tissue (solid) for an aggressive tumor.

with $C_{\widetilde{H}} := d_N K_N c_4$. The two estimates above thus lead to

$$\begin{aligned}
\frac{d}{dt}\left\|N^{m+1}-N^{m}\right\|_{L^2(\Omega)}^{2}
&\le \frac{1}{2}\Big(C_{\widetilde{N}}\left\|N^{m}-N^{m-1}\right\|_{L^2(\Omega)}
+ C_{\widetilde{K}}\left\|K^{m}-K^{m-1}\right\|_{L^2(\Omega)}
+ C_{\widetilde{H}}\left\|H^{m+1}-H^{m}\right\|_{H^1(\Omega)}\Big)^{2}
+ \frac{1}{2}\left\|N^{m+1}-N^{m}\right\|_{L^2(\Omega)}^{2} \\
&\le C_{\widetilde{N}}^{2}\left\|N^{m}-N^{m-1}\right\|_{L^2(\Omega)}^{2}
+ C_{\widetilde{K}}^{2}\left\|K^{m}-K^{m-1}\right\|_{L^2(\Omega)}^{2}
+ C_{\widetilde{H}}^{2}\left\|H^{m+1}-H^{m}\right\|_{H^1(\Omega)}^{2}
+ \frac{1}{2}\left\|N^{m+1}-N^{m}\right\|_{L^2(\Omega)}^{2}.
\end{aligned} \tag{83}$$

Applying Gronwall's inequality we deduce

$$\begin{aligned}
\left\|N^{m+1}-N^{m}\right\|_{L^2(\Omega)}^{2}
\le e^{t/2}\int_0^t \Big(C_{\widetilde{N}}^{2}\left\|N^{m}-N^{m-1}\right\|_{L^2(\Omega)}^{2}
+ C_{\widetilde{K}}^{2}\left\|K^{m}-K^{m-1}\right\|_{L^2(\Omega)}^{2}
+ C_{\widetilde{H}}^{2}\left\|H^{m+1}-H^{m}\right\|_{H^1(\Omega)}^{2}\Big)\,ds,
\end{aligned} \tag{84}$$

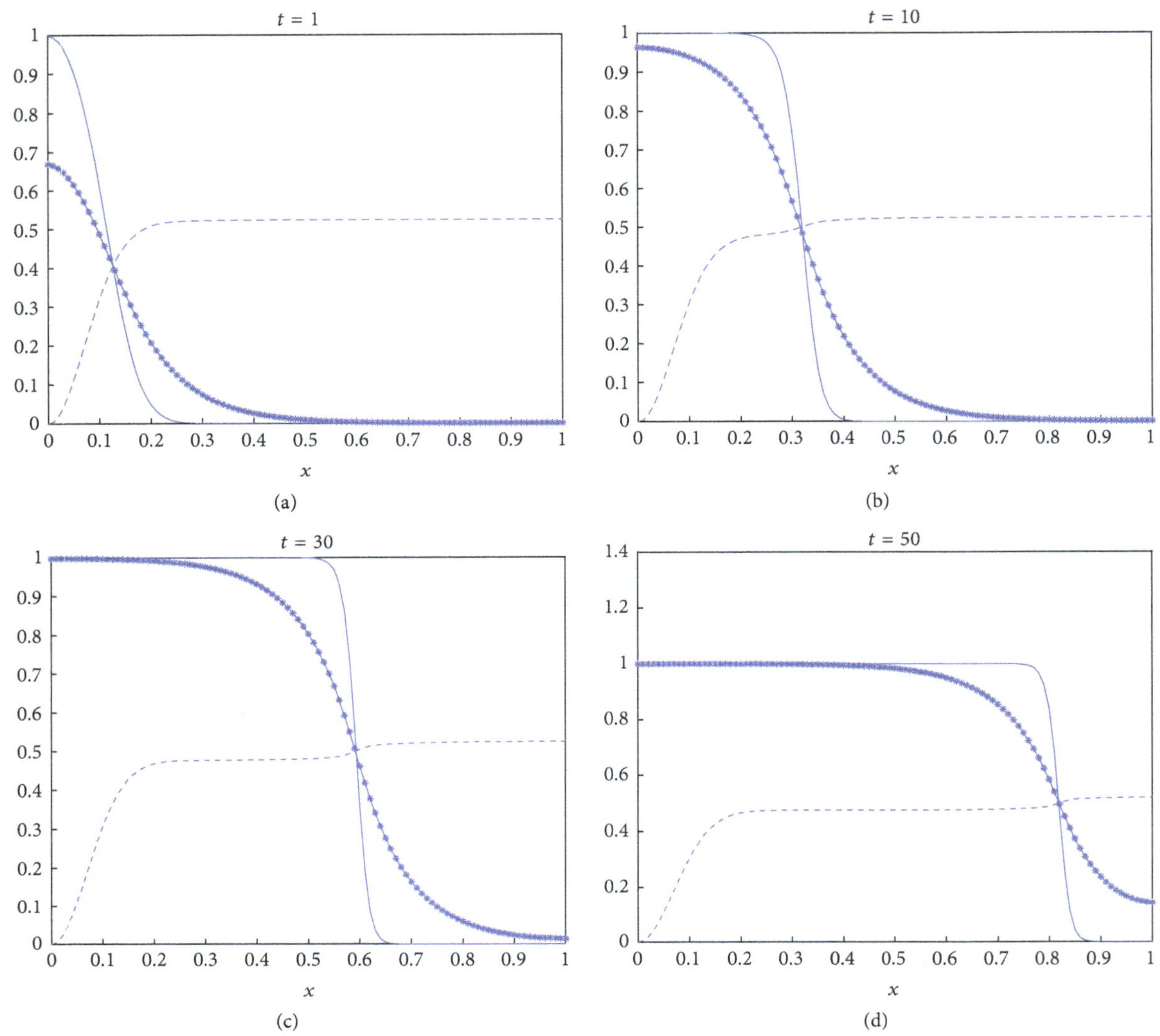

FIGURE 2: Variations of H^+ protons (starred), normal tissue (dashed), and neoplastic tissue (solid) for a less-aggressive tumor comparing to the one in Figure 1.

and finally with $D(\Omega,T) = e^{T/2}\max\{C_{\widetilde{N}}^2, C_{\widetilde{K}}^2, C_{\widetilde{H}}^2\}$ we get

$$\begin{aligned}
&\left\|N^{m+1} - N^{m}\right\|^2_{L^\infty(0,T;L^2(\Omega))} \\
&\quad \le D(\Omega,T)\left(\left\|N^{m} - N^{m-1}\right\|^2_{L^\infty(0,T;L^2(\Omega))}\right. \\
&\qquad + \left\|K^{m} - K^{m-1}\right\|^2_{L^\infty(0,T;L^2(\Omega))} \\
&\qquad \left. + \left\|H^{m+1} - H^{m}\right\|^2_{X}\right) T_5 \\
&\quad \le \frac{1}{4}\left(\left\|N^{m} - N^{m-1}\right\|^2_{L^\infty(0,T;L^2(\Omega))}\right. \\
&\qquad + \left\|H^{m+1} - H^{m}\right\|^2_{X} \\
&\qquad \left. + \left\|K^{m} - K^{m-1}\right\|^2_{L^\infty(0,T;L^2(\Omega))}\right) \\
&\quad \le \frac{1}{4}\left(\left\|N^{m} - N^{m-1}\right\|^2_{L^\infty(0,T;L^2(\Omega))}\right. \\
&\qquad \left. + \frac{5}{4}\left\|K^{m} - K^{m-1}\right\|^2_{X}\right).
\end{aligned} \tag{85}$$

T_5 is chosen such that

$$D(\Omega,T)\,T_5 \le \frac{1}{4}. \tag{86}$$

Now since K_0^m, $K_0^{m+1} \in H^1(\Omega)$ and $K^m(1-(K^m/K_K))$, $K^{m+1}(1-(K^{m+1}/K_K)) \in L^2(0,T;L^2(\Omega))$, we get

$$\begin{gathered}
K_0^{m+1} - K_0^{m} \in H^1(\Omega), \\
\left[w_K K^{m+1}\left(1 - \frac{K^{m+1}}{K_K}\right) - w_K K^{m}\left(1 - \frac{K^{m}}{K_K}\right)\right] \\
\in L^2\left(0,T;L^2(\Omega)\right).
\end{gathered} \tag{87}$$

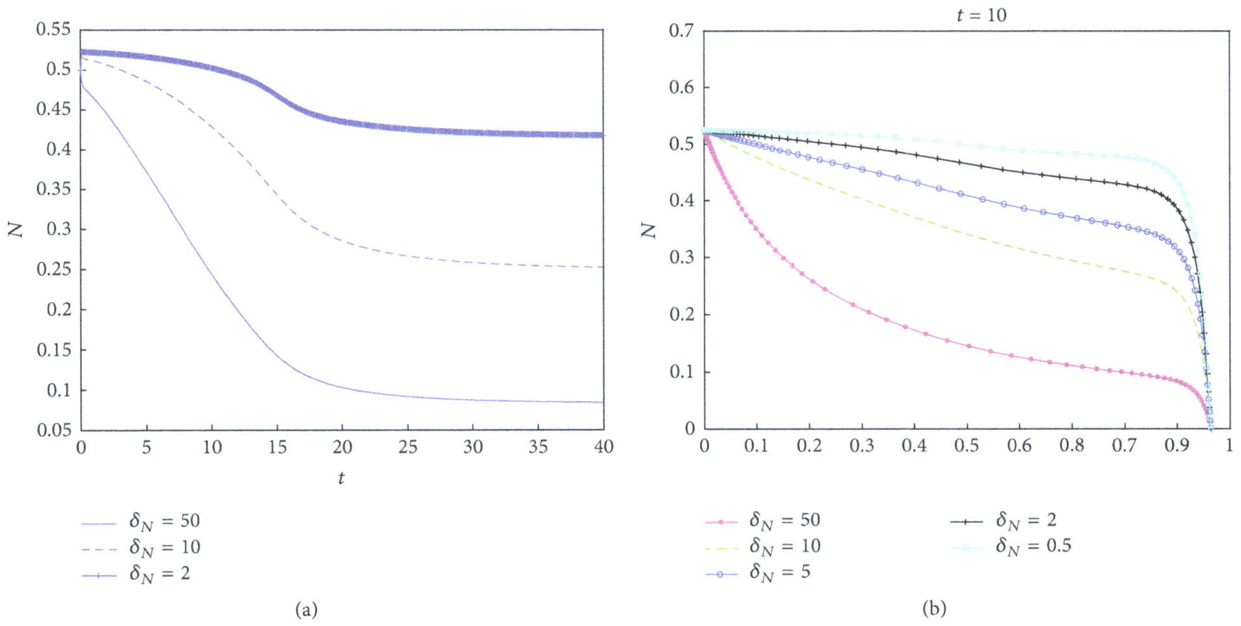

FIGURE 3: (a) Evolution of the normal cell density for several different values of δ_N. (b) Normal cell density with respect to the H^+ proton concentration for several different values of δ_N.

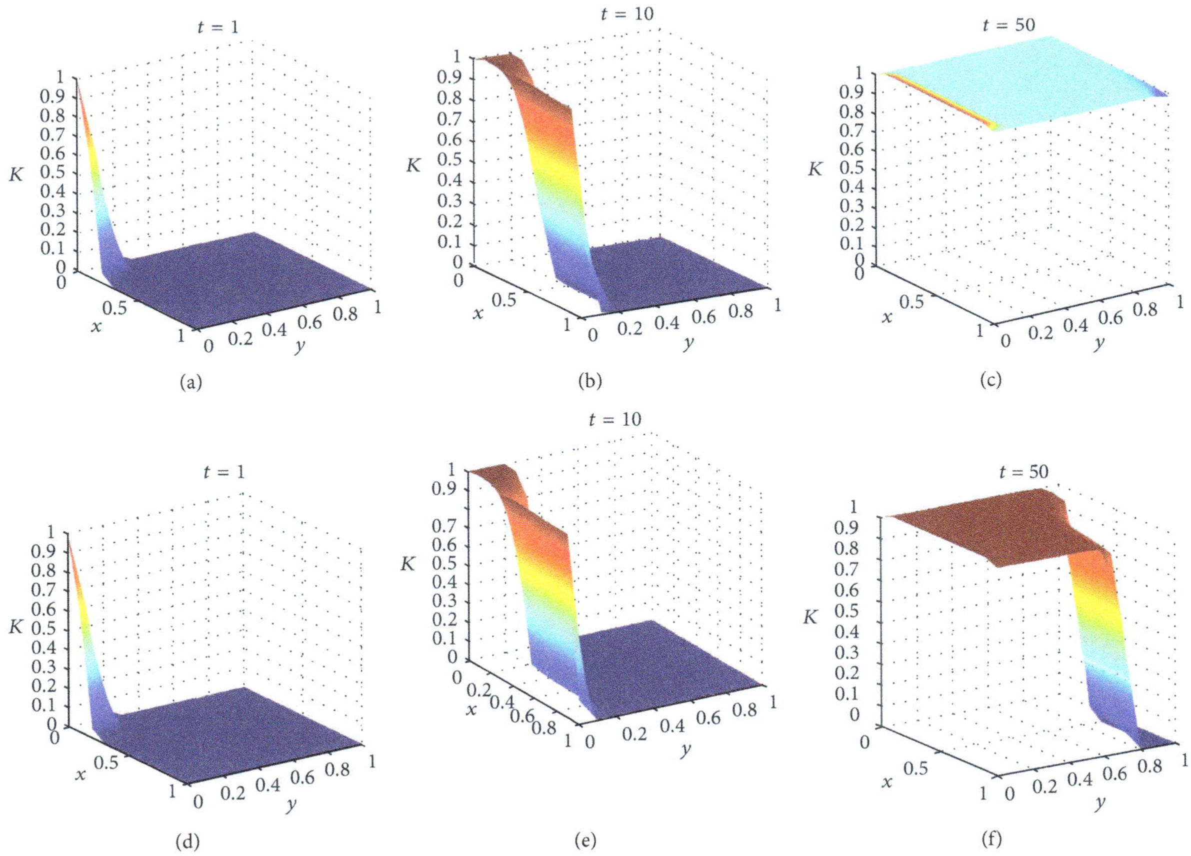

FIGURE 4: Variations of cancer cells for the cases of an aggressive ((a)–(c)) and a less-aggressive ((d)–(f)) tumor.

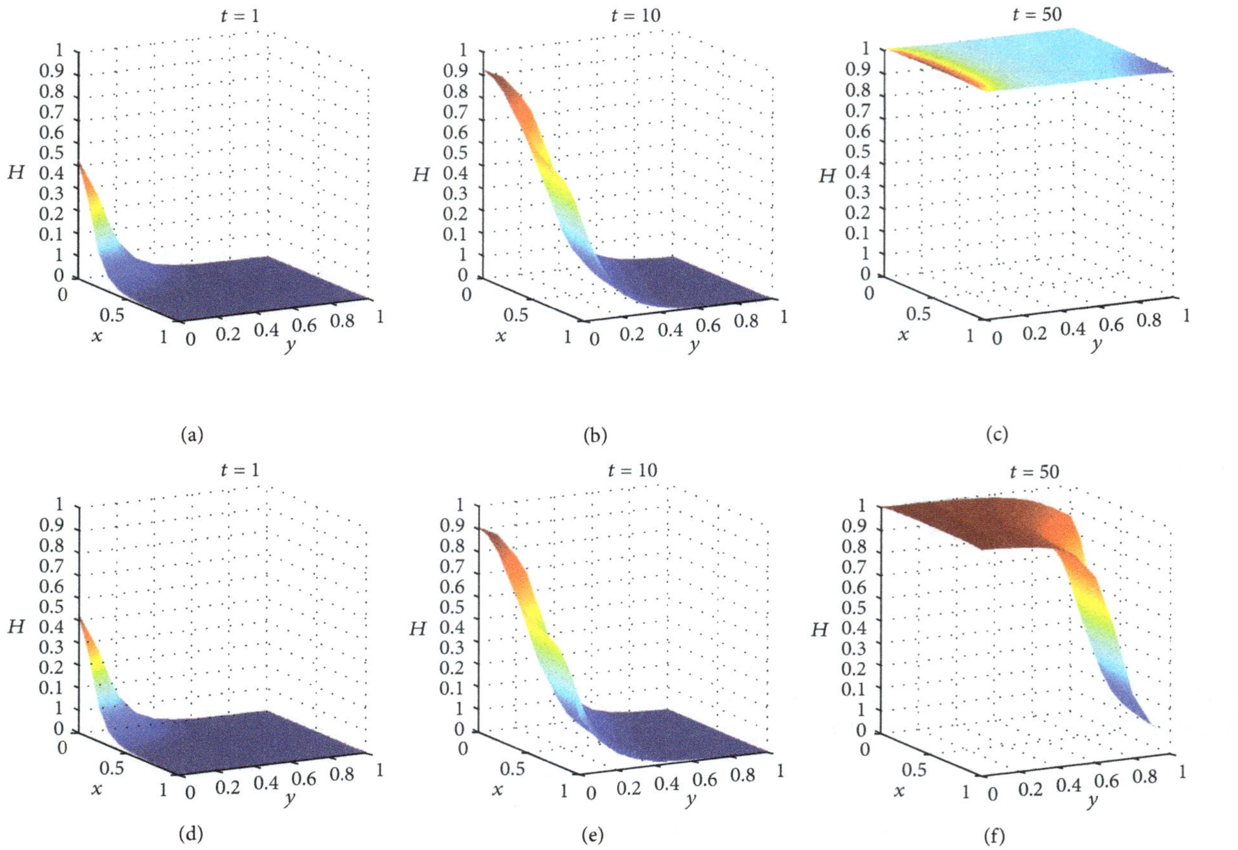

FIGURE 5: Variations of proton concentration for the cases of an aggressive ((a)–(c)) and a less-aggressive ((d)–(f)) tumor.

Theorem 7.1.5 in [22] can be applied to the difference $K^{m+1} - K^m$, leading to

$$\left\|K^{m+1} - K^m\right\|_X^2 \leq C(\Omega, T) \int_0^T \left\| w_K K^m \left(1 - \frac{K^m}{K_K}\right) - w_K K^{m-1} \left(1 - \frac{K^{m-1}}{K_K}\right)\right\|_{L^2(\Omega)}^2 dt. \quad (88)$$

The right hand side of this inequality can further be majorized and with the embedding constants $c_5 := c_5(\Omega, T)$ and $c_6 := c_6(\Omega, T)$ it follows that

$$\begin{aligned}
&\int_0^T \left\| w_K K^m \left(1 - \frac{K^m}{K_K}\right) - w_K K^{m-1} \left(1 - \frac{K^{m-1}}{K_K}\right)\right\|_{L^2(\Omega)}^2 dt \\
&\leq 2 \int_0^T \frac{w_K^2}{K_K^2} \left\|(K^m)^2 - (K^{m-1})^2\right\|_{L^2(\Omega)}^2 + w_K^2 \left\|K^m - K^{m-1}\right\|_{L^2(\Omega)}^2 dt \\
&\leq 2 \int_0^T c_5^4 \frac{w_K^2}{K_K^2} \left\|K^m - K^{m-1}\right\|_{H^1(\Omega)}^2 \left\|K^m + K^{m-1}\right\|_{H^1(\Omega)}^2 + w_K^2 c_6^2 \left\|K^m - K^{m-1}\right\|_{H^1(\Omega)}^2 dt \\
&\leq \left(\int_0^T \left(4c_5^4 \frac{w_K^2}{K_K^2} \left[\left\|K^m\right\|_{H^1(\Omega)}^2 + \left\|K^{m-1}\right\|_{H^1(\Omega)}^2\right] + 2w_K^2 c_6^2\right) dt\right) \left\|K^m - K^{m-1}\right\|_X^2 \\
&\leq \left(32C^2(\Omega, T)\, c_5^4 \frac{w_K^2}{K_K^2} T_6 \left\|K_0\right\|_{H^1(\Omega)}^2 + 2w_K^2 c_6^2 T_6\right) \times \left\|K^m - K^{m-1}\right\|_X^2 \\
&\leq \frac{1}{4} \left\|K^m - K^{m-1}\right\|_X^2,
\end{aligned} \quad (89)$$

where

$$T_6 := \min\left\{\frac{1}{8}, \frac{1}{8\kappa}, \frac{1}{8\lambda}\right\},$$
$$k := 32C^2(\Omega, T)\, c_5^4 \frac{w_K^2}{K_K^2} \left\|K_0\right\|_{H^1(\Omega)}^2, \qquad \lambda := 2w_K^2 c_6^2. \quad (90)$$

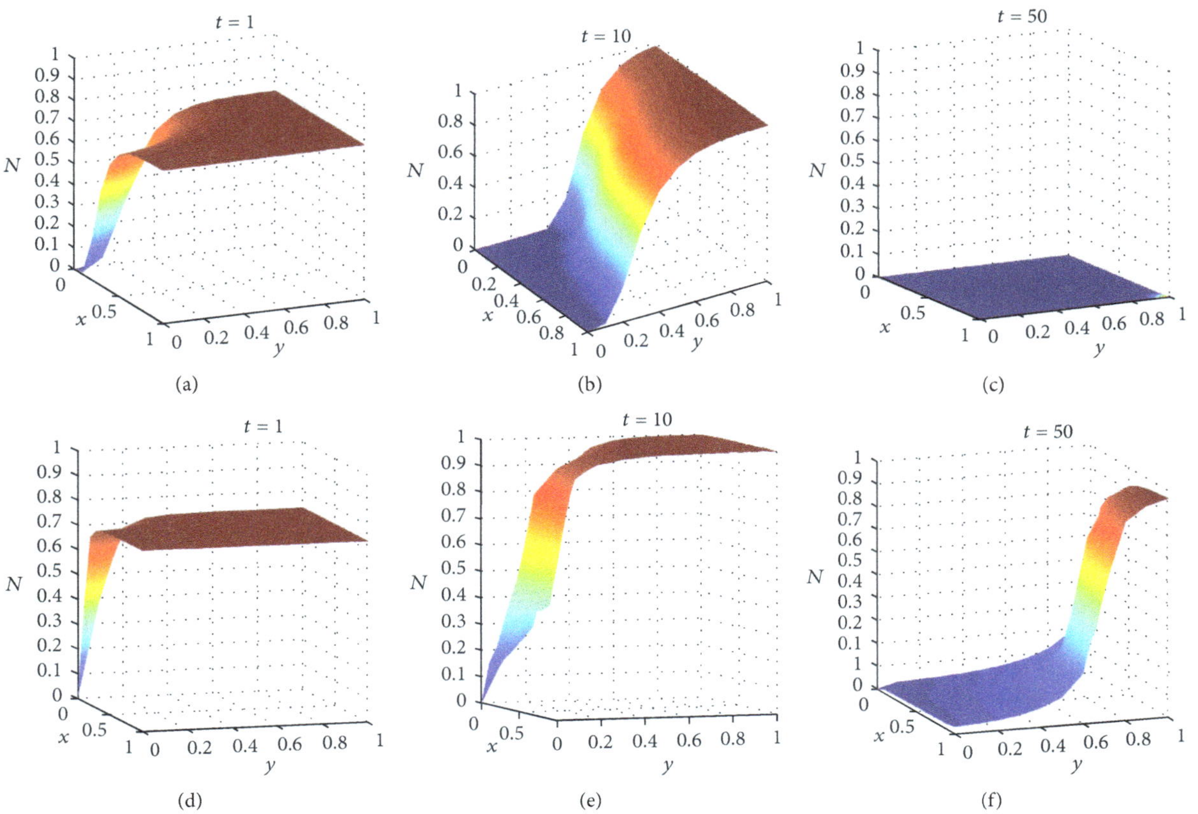

FIGURE 6: Variations of healthy tissue for the cases of an aggressive ((a)–(c)) and a less-aggressive ((d)–(f)) tumor.

Thus, putting all together,

$$\left\|K^{m+1}-K^{m}\right\|_X^2+\left\|N^{m+1}-N^{m}\right\|_{L^\infty(0,T;L^2(\Omega))}^2+\left\|H^{m+1}-H^{m}\right\|_X^2 \le \frac{1}{4}\left(3\left\|K^{m}-K^{m-1}\right\|_X^2+\left\|N^{m}-N^{m-1}\right\|_{L^\infty(0,T;L^2(\Omega))}^2\right). \tag{91}$$

Therefore, (H^m, N^m, K^m) is a Cauchy sequence in $X \times L^\infty(0,T;L^2(\Omega)) \times X$, from which the existence of a weak solution follows.

Uniqueness. Let (K_1, N_1, H_1) and (K_2, N_2, H_2) be two solutions to (1)–(3). Due to the previous estimates

$$\|K_1-K_2\|_X^2 \le \frac{1}{4}\|K_1-K_2\|_X^2, \tag{92}$$

$$\|H_1-H_2\|_X^2 \le \frac{1}{4}\|K_1-K_2\|_X^2 \tag{93}$$

$$\|N_1-N_2\|_{L^\infty(0,T;L^2(\Omega))}^2 \le \frac{1}{4}\|N_1-N_2\|_{L^\infty(0,T;L^2(\Omega))}^2+\frac{1}{4}\|H_1-H_2\|_X^2 + \frac{1}{4}\|K_1-K_2\|_X^2; \tag{94}$$

thus $K_1 = K_2$ (92) and with (93) it follows that $H_1 = H_2$. Finally, (94) implies that $N_1 = N_2$. This completes the proof of the uniqueness.

Regularity of the Solution. From (20) it follows that (K^m, H^m) is uniformly bounded with respect to m in $Y \times Y$; therefore Y is compactly embedded in $L^2(0,T;H^1(\Omega))$. This implies that for $m \to \infty$ we have $(K, H) \in Y \times Y$.

Theorem 5. *(The local solution from Theorem 2 exists globally).*

The proof follows upon sequentially extending the time interval on which the solution exists: the previously deduced estimates allow for a bootstrap of the local existence proof in a subsequent step on the time interval $[T, 2T]$, then on $[2T, T]$, and so forth. Eventually the existence of a unique solution in shown on $[0, \mathbb{T}]$ for any bounded $\mathbb{T}$.

4. Numerical Simulations

In this section we perform the numerical simulation of the system (7). The boundary conditions for K and H are the no-flux boundary conditions given by (2). We assume that initially the normal cells are at half of their carrying capacity, while the tumor cells can be close to theirs and thus prone to invade the surrounding tissue. Since the pH level is lowered by the cancer cells the concentration of protons is taken proportional to the density of the latter. Thus using these assumptions we choose for the initial conditions of the system (7)

$$K(0,x) = \exp\left(-\frac{|x|^2}{\epsilon}\right), \quad \epsilon > 0,$$
$$N(0,x) = 0.5 * \left(1 - \exp\left(-\frac{|x|^2}{\epsilon}\right)\right), \quad \epsilon > 0, \tag{95}$$
$$H(0,x) = \zeta K(0,x), \quad \zeta \in [0,1),$$

where $x \in [0,1]$ and $x \in [0,1] \times [0,1]$ in one- and two-dimensional cases, respectively. In our computations we have taken both ζ and the strength parameter θ to be 0.5.

For the discretization of the model the finite difference method is employed. Thereby in the one-dimensional case the interval $[0,1]$ is divided into m parts with $m+1$ nodes, whereas in two dimensions each of the axes is divided into m parts, thus obtaining a $(m+1) \times (m+1)$ mesh in two dimensions. Moreover the nodes are reordered in a row-wise manner, leading to a total of $(m+1)^2$ nodes. Thereby the subindex i in the discretized equations denotes the spatial node x_i, where $i = 1, 2, \ldots, m+1$ and $i = 1, 2, \ldots, (m+1)^2$ in one and two dimensions, respectively. We use forward differences for the time derivatives in the system. The central difference is used for the diffusion term in the equation characterizing the ion concentration:

$$\frac{H_i^{n+1} - H_i^n}{\Delta t} = \delta_H K_i^n - \delta_H H_i^{n+1} + \left(\frac{H_{i-1}^{n+1} - 2H_i^{n+1} + H_{i+1}^{n+1}}{\Delta x^2}\right), \tag{96}$$

where n denotes the time level and Δt and Δx are the time and space increments, respectively. The discretized equations (96) for each i leads to the following system of equations of the form

$$\mathbf{A_H H}^{n+1} = \mathbf{H}^n + \vartheta_{\mathbf{H}}^n \tag{97}$$

with $\mathbf{H}^{n+1}$ and $\vartheta_{\mathbf{H}}^n$ standing for the vectors containing the values of H and $\Delta t \delta_H K$ at the $(n+1)$st and nth time levels at the discretized space points, respectively; $\mathbf{A_H}$ being the tridiagonal and block tridiagonal matrix for the one- and two-dimensional cases, respectively. The updated values $\mathbf{H}^{n+1}$ are used to find the values of the normal cell density at the time level $n+1$ by solving

$$N_i^{n+1} = \frac{1}{1 + \Delta t \delta_N H_i^{n+1}} \left[N_i^n + \Delta t N_i^n \left(1 - \theta K_i^n - N_i^n\right)\right] \tag{98}$$

for each space point i.

In order to write the term characterizing the dispersion of the neoplastic tissue into the healthy tissue we make use of the nonstandard finite difference scheme [24], that is,

$$\nabla (D(N) \nabla K)|_{x_i} = \frac{1}{2(\Delta x)^2} \sum_{k \in N_i} \left(D\left(N_k^{n+1}\right) + D\left(N_i^{n+1}\right)\right) \times \left(K_k^{n+1} - K_i^{n+1}\right), \tag{99}$$

where $D(N) = \delta_k(1-N)$ and $N_i \subset I$ is the index set pointing at the direct neighbours of x_i on the grid. Thus N_i has two elements for the one-dimensional case and four elements for the two-dimensional case and the matrix-vector form of the scheme reads

$$\mathbf{A_K K}^{n+1} = \mathbf{K}^n + \vartheta_{\mathbf{K}}^{\tilde{n}}, \tag{100}$$

where $\mathbf{A_K}$ is the $(m+1) \times (m+1)$ tridiagonal matrix or the block tridiagonal matrix of size $(m+1)^2 \times (m+1)^2$ for the one- and two-dimensional cases, respectively, coming from the nonstandard finite difference discretization, $\vartheta_{\mathbf{K}}^{\tilde{n}}$ is the vector coming from the proliferation term with the entries $\rho_k K_i^n(1 - K_i^n - N_i^{n+1})$, and $\mathbf{K}^{n+1}$ and $\mathbf{K}^n$ are the vectors containing the K values at the discretized points for the time levels $n+1$ and n, respectively.

Throughout our simulations we use the biological parameter values from Table 1 which is reproduced from [24].

According to a linear stability analysis performed in [24], the parameter δ_N plays a crucial role in characterizing the aggressivity of the tumor. There, $\delta_N = 1$ was shown to be the crossover value; for $\delta_N < 1$ the tumor is less aggressive, whereas for $\delta_N > 1$ it becomes highly aggressive. This will be an important factor in our computations, too.

4.1. The One-Dimensional Case. We consider the space interval $[0,1]$ and choose for the time and space increments $\Delta t = 0.1$ and $\Delta x = 0.01$, respectively.

In Figures 1 and 2 we present the simulations with $\delta_N = 50$ (an aggressive tumor) and $\delta_N = 0.5$ (a less-aggressive one), respectively. The rest of parameters are taken from Table 1. As time progresses, the difference between these two cases is more visible. In the aggressive case at later times the cancer cells invade a larger region and destroy the healthy tissue much more than in the case of a less-aggressive tumor (Figure 2).

In the next set of graphs we consider the negative effect of the aggressivity parameter δ_N on the normal cell density. We know from the nondimensionalization addressed in Section 2 that the (nondimensionalized) parameter δ_N is proportional to the death rate d_N and inversely proportional to the proton reabsorption rate d_H.

Figure 3(a) shows the evolution of the normal cell density with respect to δ_N for the times up to $t = 40$ at a fixed space point $x = 0.4$. One can notice that a more aggressive tumor (larger δ_N or equivalently larger d_N) leads to a faster decay in the density of the normal cells, as expected. On the other hand, Figure 3(a) also supports the intuitive fact that in an organism which can poorly buffer the issuing excessive protons (smaller d_H and larger δ_N, resp.) the pH value will decay faster as well, thus triggering the decay of normal cell density, which in such an acid environment can no longer be sustained at a physiologically convenient level.

Figure 3(b) illustrates the normal cell density at $t = 10$ depending on the concentration of H^+ protons for several different values of δ_N. In an organism whose normal cells are more sensitive to pH variations (larger d_N) the density of these cells will decay faster for the same concentration of H^+ protons. It can also be seen that a smaller reabsorption rate of excessive protons leads to a faster decay of normal cell density.

4.2. The Two-Dimensional Case. We perform 2D simulations in the unit square $[0,1] \times [0,1]$ using $\delta_H = 70$ and still with the parameter values in Table 1. Analogously to our computations in 1D we consider two different cases: $\delta_N = 12.5$ (an aggressive tumor) and $\delta_N = 0.5 < 1$ (a less-aggressive one). We use $\Delta t = 0.01$ as the time increment and for the spatial discretization we take 11 nodes on each axis, leading to 121 nodes in the computational domain.

The evolution of cancer cells is plotted in Figure 4. At an earlier time (e.g., $t = 1$, see Figures 4(a) and 4(d)) the difference between the less- and the higher-aggressive tumors is not relevant. However, as time progresses the difference starts to be visible (e.g., at $t = 10$, see Figures 4(b) and 4(e)), whereas by $t = 50$ the more-aggressive tumor (Figure 4(c)) invaded almost the whole domain and the less-aggressive one was not able to penetrate that far.

Also observe that the proton concentration varies proportionally to the tumor cell density and is inversely proportional to the normal cell density (Figures 5 and 6, resp.). Moreover, the healthy tissue is completely destroyed in the case of an aggressive tumor (Figure 6(c)) by $t = 50$.

Acknowledgments

C. Surulescu acknowledges the support of the Baden-Württemberg Foundation. G. Meral acknowledges the support of LLP Erasmus Staff Mobility Programme during her visit to the University of Kaiserslautern.

References

[1] A. Ashkenazi and V. M. Dixit, "Death receptors: signaling and modulation," *Science*, vol. 281, no. 5381, pp. 1305–1308, 1998.

[2] H. Yamaguchi, F. Pixley, and J. Condeelis, "Invadopodia and podosomes in tumor invasion," *European Journal of Cell Biology*, vol. 85, no. 3-4, pp. 213–218, 2006.

[3] J. Kelkel and C. Surulescu, "On some models for cancer cell migration through tissue networks," *Mathematical Biosciences and Engineering*, vol. 8, no. 2, pp. 575–589, 2011.

[4] J. Kelkel and C. Surulescu, "A multiscale approach to cell migration in tissue networks," *Mathematical Models and Methods in Applied Sciences*, vol. 22, no. 3, Article ID 1150017, 25 pages, 2012.

[5] T. Hillen, "M^5 mesoscopic and macroscopic models for mesenchymal motion," *Journal of Mathematical Biology*, vol. 53, no. 4, pp. 585–616, 2006.

[6] A. Chauviere, T. Hillen, and L. Preziosi, "Modeling cell movement in anisotropic and heterogeneous network tissues," *Networks and Heterogeneous Media*, vol. 2, no. 2, pp. 333–357, 2007.

[7] A. R. A. Anderson, M. A. J. Chaplain, E. L. Newman, R. J. C. Steele, and A. M. Thompson, "Mathematical modeling of tumor invasion and metastasis," *Journal of Theoretical Medicine*, vol. 2, pp. 129–154, 2000.

[8] R. A. Gatenby and E. T. Gawlinski, "A reaction-diffusion model of cancer invasion," *Cancer Research*, vol. 56, no. 24, pp. 5745–5753, 1996.

[9] R. A. Gatenby and R. J. Gillies, "Glycolysis in cancer: a potential target for therapy," *International Journal of Biochemistry and Cell Biology*, vol. 39, no. 7-8, pp. 1358–1366, 2007.

[10] D. Hanahan and R. A. Weinberg, "Hallmarks of cancer: the next generation," *Cell*, vol. 144, no. 5, pp. 646–674, 2011.

[11] H. Izumi, T. Torigoe, H. Ishiguchi et al., "Cellular pH regulators: potentially promising molecular targets for cancer chemotherapy," *Cancer Treatment Reviews*, vol. 29, no. 6, pp. 541–549, 2003.

[12] O. R. Abakarova, "The metastatic potential of tumors depends on the pH of host tissues," *Bulletin of Experimental Biology and Medicine*, vol. 120, no. 6, pp. 1227–1229, 1995.

[13] R. Martínez-Zaguilán, E. A. Seftor, R. E. B. Seftor, Y. W. Chu, R. J. Gillies, and M. J. C. Hendrix, "Acidic pH enhances the invasive behavior of human melanoma cells," *Clinical and Experimental Metastasis*, vol. 14, no. 2, pp. 176–186, 1996.

[14] R. A. Gatenby and E. T. Gawlinski, "The glycolytic phenotype in carcinogenesis and tumor invasion: insights through mathematical models," *Cancer Research*, vol. 63, no. 14, pp. 3847–3854, 2003.

[15] A. Fasano, M. A. Herrero, and M. R. Rodrigo, "Slow and fast invasion waves in a model of acid-mediated tumour growth," *Mathematical Biosciences*, vol. 220, no. 1, pp. 45–56, 2009.

[16] K. Smallbone, D. J. Gavaghan, R. A. Gatenby, and P. K. Maini, "The role of acidity in solid tumour growth and invasion," *Journal of Theoretical Biology*, vol. 235, no. 4, pp. 476–484, 2005.

[17] L. Bianchini and A. Fasano, "A model combining acid-mediated tumour invasion and nutrient dynamics," *Nonlinear Analysis: Real World Applications*, vol. 10, no. 4, pp. 1955–1975, 2009.

[18] J. Kelkel and C. Surulescu, "A weak solution approach to a reaction-diffusion system modeling pattern formation on seashells," *Mathematical Methods in the Applied Sciences*, vol. 32, no. 17, pp. 2267–2286, 2009.

[19] J. Kelkel and C. Surulescu, "On a stochastic reaction-diffusion system modeling pattern formation on seashells," *Journal of Mathematical Biology*, vol. 60, no. 6, pp. 765–796, 2010.

[20] A. S. Silva, J. A. Yunes, R. J. Gillies, and R. A. Gatenby, "The potential role of systemic buffers in reducing intratumoral extracellular pH and acid-mediated invasion," *Cancer Research*, vol. 69, no. 6, pp. 2677–2684, 2009.

[21] I. F. Tannock and D. Rotin, "Acid pH in tumors and its potential for therpeutic exploitation," *Cancer Research*, vol. 49, no. 16, pp. 4373–4384, 1989.

[22] L. C. Evans, *Partial Differential Equations*, vol. 19, American Mathematical Society, Providence, RI, USA, 1998.

[23] A. D. Polyanin, *Handbook of Linear Partial Differential Equations for Engineers and Scientists*, Chapman & Hall, New York, NY, USA, 2002.

[24] H. J. Eberl and L. Demaret, "A finite difference scheme for a degenerated diffusion equation arising in microbial ecology," *Electronic Journal of Differential Equations*, vol. 15, pp. 77–95, 2007.

15

Generalized Abel Inversion Using Extended Hat Functions Operational Matrix

Manoj P. Tripathi,[1,2] Ram K. Pandey,[1] Vipul K. Baranwal,[1] and Om P. Singh[1]

[1] *Department of Applied Mathematics, Indian Institute of Technology, Banaras Hindu University, Varanasi 221005, India*
[2] *Department of Mathematics, Udai Pratap Autonomous College, Varanasi 221002, India*

Correspondence should be addressed to Om P. Singh; singhom@gmail.com

Academic Editor: Frédéric Robert

Abel type integral equations play a vital role in the study of compressible flows around axially symmetric bodies. The relationship between emissivity and the measured intensity, as measured from the outside cylindrically symmetric, optically thin extended radiation source, is given by this equation as well. The aim of the present paper is to propose a stable algorithm for the numerical inversion of the following generalized Abel integral equation: $I(y) = a(y)\int_{\alpha}^{y}((r^{\mu-1}\varepsilon(r))/(y^{\mu}-r^{\mu})^{\gamma})dr + b(y)\int_{y}^{\beta}((r^{\mu-1}\varepsilon(r))/(r^{\mu}-y^{\mu})^{\gamma})dr$, $\alpha \le y \le \beta$, $0 < \gamma < 1$, using our newly constructed extended hat functions operational matrix of integration, and give an error analysis of the algorithm. The earlier numerical inversions available for the above equation assumed either $a(y) = 0$ or $b(y) = 0$.

1. Introduction

Abel integral equation [1] occurs in many branches of science and technology, such as plasma diagnostics and flame studies, where the most common problem of deduction of radial distributions of some important physical quantity from measurement of line-of-sight projected values is encountered. For a cylindrically symmetric, optically thin plasma source, the relation between radial distribution of the emission coefficient and the intensity measured from outside of the radial source is described by Abel transform. The challenging task of reconstruction of emission coefficient from its projection is known as Abel inversion. The earliest application, due to Mach [2], arose in the study of compressible flows around axially symmetric bodies.

The Abel integral equation is given by

$$I(y) = 2\int_{y}^{1}\frac{\varepsilon(r)\,r}{\sqrt{r^2-y^2}}\,dr, \quad 0 \le y \le 1, \tag{1}$$

where $\varepsilon(r)$ and $I(y)$ represent, respectively, the emissivity and measured intensity, as measured from outside the source [3].

The analytical inversion formula for (1) is given as [4]

$$\varepsilon(r) = -\frac{1}{\pi}\int_{r}^{1}\frac{1}{\sqrt{y^2-r^2}}\frac{dI(y)}{d(y)}dy, \quad 0 \le r \le 1. \tag{2}$$

There are several analytic and numerical inversion formulae available in the literature [1, 5–20].

Singh et al. [19] constructed an operational matrix of integration based on orthonormal Bernstein polynomials and used it to propose a stable algorithm to invert the following form of Abel integral equation:

$$I(y) = 2\int_{0}^{y}\frac{\varepsilon(r)\,r}{\sqrt{r^2-y^2}}dr, \quad 0 \le y \le 1. \tag{3}$$

In 2010, Singh et al. [20] constructed yet another operational matrix of integration based on orthonormal Bernstein polynomials and used it to propose an algorithm to invert the Abel integral equation (1).

In 2008, Chakrabarti [21] employed a direct function theoretic method to determine the closed form solution of the following generalized Abel integral equation:

$$I(y) = a(y) \int_{\alpha}^{y} \frac{r^{\mu-1}\varepsilon(r)}{(y^{\mu} - r^{\mu})^{\gamma}} dr + b(y) \int_{y}^{\beta} \frac{r^{\mu-1}\varepsilon(r)}{(r^{\mu} - y^{\mu})^{\gamma}} dr,$$
$$\alpha \le y \le \beta, \quad 0 < \gamma < 1, \tag{4}$$

where the coefficients $a(y)$ and $b(y)$ do not vanish simultaneously. But the numerical inversion is still needed for its application in physical models since the experimental data for the intensity $I(y)$ is available only at a discrete set of points, and it may also be distorted by the noise.

This motivated us to look for a stable algorithm which can be used for numerical inversion of the Abel integral equation (4) obtained by joining the two integrals (1) and (3). In this paper, we construct extended hat functions operational matrix of integration to invert the generalized Abel integral equation (4). Using hat functions for approximation of emissivity and intensity profiles has an edge over the earlier works of Singh et al. [19, 20], where they have used orthonormal Bernstein polynomials to approximate those physical quantities in the sense that a general formula for $n \times n$ operational matrix of integration is obtained in the earlier case whereas no such formula is available for the latter case. In Sections 3 and 4, we give the error estimate and the stability analysis followed by numerical examples to illustrate the efficiency and stability of the proposed algorithm.

The above two forms (1) and (3) of Abel integral equations are obtained by taking $\gamma = 1/2$ and

(i) $a(y) = 0$, $b(y) = 2$, $\beta = 1$, $\mu = 2$;

(ii) $a(y) = 2$, $b(y) = 0$, $\alpha = 0$, $\mu = 1$, respectively, in (4).

Mostly for $\mu = 1, 2$ and $\gamma = 1/2$ the generalized Abel integral equation models the physical problems but the integral equation for $\mu = 2$ can be reduced to the case $\mu = 1$, by change of variables. So we restrict ourselves to $\mu = 1$ only.

2. Extended Hat Functions and Their Operational Matrices for Abel Inversion

Hat functions are defined on the domain $[0, 1]$. These are continuous functions with shape of hats, when plotted on two-dimensional planes. The interval $[0, 1]$ is divided into n subintervals $[ih, (i+1)h]$, $i = 0, 1, 2, \ldots, n-1$, of equal lengths h where $h = 1/n$. The hat function's family of first $(n + 1)$ hat functions is defined as follows:

$$\psi_0(t) = \begin{cases} \dfrac{h-t}{h}, & 0 \le t < h, \\ 0, & \text{otherwise}, \end{cases} \tag{5}$$

$$\psi_i(t) = \begin{cases} \dfrac{t-(i-1)h}{h}, & (i-1)h \le t < ih, \\ \dfrac{(i+1)h-t}{h}, & ih \le t < (i+1)h, \\ & i = 1, 2, \ldots, n-1, \\ 0, & \text{otherwise}, \end{cases} \tag{6}$$

$$\psi_n(t) = \begin{cases} \dfrac{t-(1-h)}{h}, & 1-h \le t \le 1, \\ 0, & \text{otherwise}. \end{cases} \tag{7}$$

We modify these functions by adding characteristics functions $\chi_{[-\tau,0)}$ and $\chi_{(1,1+\tau]}$ to ψ_0 and ψ_n, respectively, to yield a new class of extended hat functions $\overline{\psi}_i(t)$ defined over $[-\tau, 1+\tau]$ for $\tau > 0$, and these are given by

$$\overline{\psi}_0(t) = \chi_{[-\tau,0)}(t) + \psi_0(t), \tag{8}$$

$$\overline{\psi}_i(t) = \psi_i(t), \quad \text{for } 1 \le i \le n-1, \tag{9}$$

$$\overline{\psi}_n(t) = \psi_n(t) + \chi_{(1,1+\tau]}(t). \tag{10}$$

Thus, the supports of $\overline{\psi}_0(t)$ and $\overline{\psi}_n(t)$ are extended to $[-\tau, h]$ and $[1-h, 1+\tau]$, respectively. These extended hat functions $\overline{\psi}_j(t)$ are continuous, linearly independent and are in $L^2[-\tau, 1+\tau]$. As $\tau \to 0$, obviously, the extended hat functions will converge to the traditional hat functions.

A function $f \in L^2[0, 1]$ may be approximated in vector form as

$$f(t) \simeq \sum_{i=0}^{i=n} f_i \overline{\psi}_i(t) = F_{n+1}^T \overline{\Psi}_{n+1}(t) = \overline{\Psi}_{n+1}^T(t) F_{n+1}, \tag{11}$$

where

$$F_{n+1} \triangleq [f_0, f_1, f_2, \ldots, f_n]^T,$$
$$\overline{\Psi}_{n+1}(t) \triangleq [\overline{\psi}_0(t), \overline{\psi}_1(t), \overline{\psi}_2(t), \ldots, \overline{\psi}_n(t)]^T. \tag{12}$$

The important aspect of using extended hat functions in the approximation of function $f(t)$ lies in the fact that the coefficients f_i in (11) are given by

$$f_i = f(ih), \quad i = 0, 1, 2 \ldots, n. \tag{13}$$

Taking $\alpha = 0$, $\beta = 1$, and $\mu = 2$ and by change of variables, the Abel integral equation (4) reduces to

$$I(\sqrt{y}) = \frac{a(\sqrt{y})}{2} \int_0^y \frac{\varepsilon(\sqrt{r})}{(y-r)^{\gamma}} dr + \frac{b(\sqrt{y})}{2} \int_y^1 \frac{\varepsilon(\sqrt{r})}{(r-y)^{\gamma}} dr,$$
$$0 \le y \le 1, \tag{14}$$

which may be written as

$$I_1(y) = a_1(y) \int_0^y \frac{\eta(r)}{(y-r)^{\gamma}} dr + b_1(y) \int_y^1 \frac{\eta(r)}{(r-y)^{\gamma}} dr,$$
$$0 \le y \le 1, \tag{15}$$

where $I_1(y) = I(\sqrt{y})$, $\eta(r) = \varepsilon(\sqrt{r})$, $a_1(y) = a(\sqrt{y})/2$, and $b_1(y) = b(\sqrt{y})/2$.

Instead of considering (15), we consider the more general equation of the form:

$$I_1(y) = a_1(y)\int_{-\tau}^{y}\frac{\eta(r)}{(y-r)^\gamma}dr + b_1(y)\int_{y}^{1+\tau}\frac{\eta(r)}{(r-y)^\gamma}dr, \quad 0 \le y \le 1. \tag{16}$$

Using (11), the functions $I_1(y)$ and $\eta(r)$ may be approximated as

$$I_1(y) \simeq F_{n+1}^T\overline{\Psi}_{n+1}(y), \qquad \eta(r) \simeq C_{n+1}^T\overline{\Psi}_{n+1}(r). \tag{17}$$

Thus the problem of Abel inversion is reduced to finding the unknown matrix C_{n+1}. Substituting (17) into (16), we get

$$F_{n+1}^T\overline{\Psi}_{n+1}(y) = C_{n+1}^T\left[a_1(y)\int_{-\tau}^{y}\frac{\overline{\Psi}_{n+1}(r)}{(y-r)^\gamma}dr + b_1(y)\int_{y}^{1+\tau}\frac{\overline{\Psi}_{n+1}(r)}{(r-y)^\gamma}dr\right], \quad 0 \le y \le 1. \tag{18}$$

The integrals in (18) involve, evaluating integrals of the type $\int_y^{1+\tau}((\overline{\Psi}_{n+1}(r))/(r-y)^\gamma)dr$ and $\int_{-\tau}^{y}((\overline{\psi}_i(r))/(y-r)^\gamma)dr$. Let

$$\Phi_i^L(y) = \int_y^{1+\tau}\frac{\overline{\psi}_i(r)}{(r-y)^\gamma}dr, \qquad \Phi_i^U(y) = \int_{-\tau}^{y}\frac{\overline{\psi}_i(r)}{(y-r)^\gamma}dr \quad \text{for } i = 0, 1, 2\dots, n, \tag{19}$$

and compute the two operational matrices of integration to evaluate these integrals. The scheme of derivation of these two operational matrices is based on the following theorems.

Theorem 1. *The functions $\Phi_i^L(y) \in L^2[-\tau, 1+\tau]$ for $i = 0, 1, 2, \dots, n$.*

Proof. We prove the theorem for $i = 1, 2\dots, n-1$. The proofs for $i = 0$ and $i = n$ are skipped as they may be proved on the same pattern. Based on subdivision of interval $[-\tau, 1+\tau]$, we calculate $\Phi_i^L(y)$ by considering the following cases.

(i) When $-\tau \le y < (i-1)h$, then

$$\begin{aligned}\Phi_i^L(y) &= \int_y^{1+\tau}\frac{\overline{\psi}_i(r)}{(r-y)^\gamma}dr \\ &= \int_y^{(i-1)h}\frac{\overline{\psi}_i(r)}{(r-y)^\gamma}dr + \int_{(i-1)h}^{ih}\frac{\overline{\psi}_i(r)}{(r-y)^\gamma}dr + \int_{ih}^{(i+1)h}\frac{\overline{\psi}_i(r)}{(r-y)^\gamma}dr + \int_{(i+1)h}^{1+\tau}\frac{\overline{\psi}_i(r)}{(r-y)^\gamma}dr \\ &= \int_{(i-1)h}^{ih}\frac{\overline{\psi}_i(r)}{(r-y)^\gamma}dr + \int_{ih}^{(i+1)h}\frac{\overline{\psi}_i(r)}{(r-y)^\gamma}dr\end{aligned}$$

(as the support of $\overline{\psi}_i(t)$ lies in $[(i-1)h, (i+1)h]$)

$$= \int_{(i-1)h}^{ih}\frac{r-(i-1)h}{h(r-y)^\gamma}dr + \int_{ih}^{(i+1)h}\frac{(i+1)h-r}{(r-y)^\gamma}dr. \tag{20}$$

Changing the variable, $r - y = t$, we get

$$\Phi_i^L(y) = \frac{1}{h(1-\gamma)(2-\gamma)}\left[((i-1)h-y)^{2-\gamma} - 2(ih-y)^{2-\gamma} + ((i+1)h-y)^{2-\gamma}\right]. \tag{21}$$

(ii) When $(i-1)h \le y < ih$, then

$$\Phi_i^L(y) = \int_y^{ih}\frac{\overline{\psi}_i(r)}{(r-y)^\gamma}dr + \int_{ih}^{(i+1)h}\frac{\overline{\psi}_i(r)}{(r-y)^\gamma}dr. \tag{22}$$

Adopting the same procedure as in (i), we get

$$\Phi_i^L(y) = \frac{1}{h(1-\gamma)(2-\gamma)} \times \left[((i+1)h-y)^{2-\gamma} - 2(ih-y)^{2-\gamma}\right]. \tag{23}$$

(iii) When $ih \le y < (i+1)h$, then

$$\begin{aligned}\Phi_i^L(y) &= \int_y^{(i+1)h}\frac{\overline{\psi}_i(r)}{(r-y)^\gamma}dr = \int_y^{(i+1)h}\frac{(i+1)h-r}{h(r-y)^\gamma}dr \\ &= \frac{1}{h(1-\gamma)(2-\gamma)}\left[((i+1)h-y)^{2-\gamma}\right].\end{aligned} \tag{24}$$

(iv) When $(i+1)h \le y \le 1+\tau$, then $\Phi_i^L(y) = 0$. Hence

$$\Phi_i^L(y) = \frac{1}{h(1-\gamma)(2-\gamma)}\begin{cases}((i-1)h-y)^{2-\gamma} - 2(ih-y)^{2-\gamma} + ((i+1)h-y)^{2-\gamma}, & -\tau \le y < (i-1)h,\\ ((i+1)h-y)^{2-\gamma} - 2(ih-y)^{2-\gamma}, & (i-1)h \le y < ih,\\ ((i+1)h-y)^{2-\gamma}, & ih \le y < (i+1)h,\\ 0, & \text{otherwise.}\end{cases} \tag{25}$$

Thus, from (25), we see that $\|\Phi_i^L(y)\|_2 < \infty$ and hence $\Phi_i^L(y) \in L^2[-\tau, 1+\tau]$ for $\tau > 0$ and bounded. This completes the proof. □

Therefore, from (11), we get

$$\Phi_i^L(y) \simeq \sum_{j=0}^{n} c_{ij}\overline{\psi}_j(y), \quad \text{for } i = 0, 1, 2 \ldots, n. \tag{26}$$

Theorem 2. *The coefficients c_{ij} in (26) are given by*

(i) *for $i = 0, 1, 2, \ldots, n-1$, $j = 0, 1, 2 \ldots, n$,*

$$c_{ij} = \begin{cases} \beta, & \text{for } j = i, \\ \beta\left[(i-j+1)^{2-\gamma} - 2(i-j)^{2-\gamma} + (i-j-1)^{2-\gamma}\right], & \text{for } j < i, \\ 0, & \text{for } j > i, \end{cases} \tag{27}$$

(ii) *for $i = n$, $j = 0, 1, 2, \ldots, n$,*

$$c_{nj} = \begin{cases} \beta\left[(n-j-1)^{2-\gamma} - (n-j)^{2-\gamma} + (2-\gamma)(n-j)^{1-\gamma}\right] + \dfrac{1}{(1-\gamma)}\left[(1+\tau-jh)^{1-\gamma} - (1-jh)^{1-\gamma}\right], & \text{for } j < n, \\ \dfrac{1}{(1-\gamma)}\tau^{1-\gamma}, & \text{for } j = n, \end{cases} \tag{28}$$

where $\beta = h^{1-\gamma}/((1-\gamma)(2-\gamma))$.

Proof. (i) When $j = i$, $c_{ii} = \Phi_i^L(ih) = \beta$ follows from (25).
When $j < i$, which is equivalent to $j \le i-1$, we get

$$\begin{aligned} c_{ij} = \Phi_i^L(jh) &= \int_{jh}^{1+\tau} \frac{\overline{\psi}_i(r)}{(r-jh)^{\gamma}}dr \\ &= \left(\int_{(i-1)h}^{ih} + \int_{ih}^{(i+1)h}\right)\frac{\overline{\psi}_i(r)}{(r-jh)^{\gamma}}dr, \end{aligned} \tag{29}$$

since the support of $\overline{\psi}_i(t)$ lies in $[(i-1)h, (i+1)h]$. Using (6) and (9), we get from change of variable

$$c_{ij} = \beta\left[(i-j+1)^{2-\gamma} - 2(i-j)^{2-\gamma} + (i-j-1)^{2-\gamma}\right]. \tag{30}$$

Similarly, for $j > i$, that is, $j \ge i+1$, $c_{ij} = 0$ (it follows trivially from (19)).

(ii) For $j < n$,

$$c_{nj} = \Phi_n^L(jh) = \left(\int_{jh}^{1} + \int_{1}^{1+\tau}\right)\frac{\overline{\psi}_n(r)}{(r-jh)^{\gamma}}dr. \tag{31}$$

From (10), we have

$$c_{nj} = \int_{(n-1)h}^{nh} \frac{(r-(n-1)h)}{h(r-jh)^{\gamma}}dr + \int_{1}^{1+\tau}\frac{1}{(r-jh)^{\gamma}}dr, \tag{32}$$

so

$$c_{nj} = \beta\left[(n-j-1)^{2-\gamma} - 2(n-j)^{2-\gamma} + (2-\gamma)(i-j)^{2-\gamma}\right] + \frac{1}{(1-\gamma)}\left[(1+\tau-jh)^{1-\gamma} - (1-jh)^{1-\gamma}\right]. \tag{33}$$

Similarly, $c_{nn} = \Phi_n^L(1) = \int_1^{1+\tau}(\overline{\psi}_n(r)/(r-1)^{\gamma}) = (1/(1-\gamma))\tau^{1-\gamma}$, thus, proving the theorem. □

Similar arguments prove the following theorem.

Theorem 3. *The functions $\Phi_i^U(y) \in L^2[-\tau, 1+\tau]$ for $i = 0, 1, 2, \ldots, n$.*

From Theorem 3 and (11), we have

$$\Phi_i^U(y) \simeq \sum_{j=0}^{n} d_{ij}\overline{\psi}_j(y). \tag{34}$$

The coefficients d_{ij}'s are given as follows.

(i) For $i = 0, 1, 2, \ldots, n-1$; $j = 0, 1, 2, \ldots, n$,

$$d_{ij} = \begin{cases} \beta, & \text{for } j = i, \\ \beta\left[(j-i+1)^{2-\gamma} - 2(j-i)^{2-\gamma} + (j-i-1)^{2-\gamma}\right], & \text{for } j > i, \\ 0, & \text{for } j < i, \end{cases} \tag{35}$$

and

(ii) for $i = n$; $j = 0, 1, 2 \ldots, n$,

$$d_{nj} = \begin{cases} \dfrac{1}{(1-\gamma)}\tau^{1-\gamma}, & \text{for } j = 0, \\ \beta\left[(j-1)^{2-\gamma} - j^{2-\gamma} + (2-\gamma)\,j^{1-\gamma}\right] + \dfrac{1}{(1-\gamma)}\left[(\tau+jh)^{1-\gamma} - (jh)^{1-\gamma}\right], & \text{for } 1 \le j \le n. \end{cases} \tag{36}$$

Using (12) and (26), the following integral may be written as

$$\int_{y}^{1+\tau} \frac{\overline{\Psi}_{n+1}(r)}{(r-y)^{\gamma}}dr = \left[\Phi_0^L(y), \Phi_1^L(y), \ldots, \Phi_n^L(y)\right]^T. \tag{37}$$

Substituting the approximation of $\Phi_i^L(y)$ from (26) in (37), we get

$$\int_y^{1+\tau} \frac{\overline{\Psi}_{n+1}(r)}{(r-y)^\gamma} dr = P_{n+1}^L \overline{\Psi}_{n+1}(y), \tag{38}$$

where P_{n+1}^L is a $(n+1)\times(n+1)$ matrix whose $(i+1, j+1)$th entry is c_{ij}, given by (27) and (28) for $i, j = 0, 1, 2, \ldots, n$. The matrix P_{n+1}^L is given as

$$P_{n+1}^L = \frac{h^{1-\gamma}}{(1-\gamma)(2-\gamma)} \times \begin{bmatrix} 1 & 0 & 0 & 0 & 0 & 0 & 0 & 0 \\ \xi_1 & 1 & 0 & 0 & 0 & 0 & 0 & 0 \\ \xi_2 & \xi_1 & 1 & 0 & 0 & 0 & 0 & 0 \\ \xi_3 & \xi_2 & \xi_1 & 1 & 0 & 0 & 0 & 0 \\ \xi_4 & \xi_3 & \xi_2 & \xi_1 & 1 & 0 & 0 & 0 \\ \cdots & \cdots & \cdots & \cdots & \cdots & 1 & 0 & 0 \\ \xi_{n-1} & \xi_{n-2} & \cdots & \xi_3 & \xi_2 & \xi_1 & 1 & 0 \\ \zeta_n & \zeta_{n-1} & \cdots & \zeta_4 & \zeta_3 & \zeta_2 & \zeta_1 & \zeta_0 \end{bmatrix}_{(n+1)\times(n+1)}, \tag{39}$$

where

$$\zeta_0 = \frac{(2-\gamma)}{h^{1-\gamma}} \tau^{1-\gamma}, \tag{40}$$

$$\zeta_s = \left[(s-1)^{2-\gamma} - s^{2-\gamma} + (2-\gamma) s^{1-\gamma}\right] + \frac{(2-\gamma)}{h^{1-\gamma}} \left[(\tau + sh)^{1-\gamma} - (sh)^{1-\gamma}\right], \quad s = 1, 2, 3, \ldots, n, \tag{41}$$

$$\xi_k = (k+1)^{2-\gamma} - 2k^{2-\gamma} + (k-1)^{2-\gamma}, \quad k = 1, 2, 3, \ldots, n-1. \tag{42}$$

The matrix P_{n+1}^L is called extended hat functions lower operational matrix for Abel's inversion.

Remark 4. It is evident from (40) that when $\tau = 0$, then $\zeta_0 = 0$, and so the lower triangular matrix P_{n+1}^L becomes a singular matrix. In this case, the singularity of the matrix P_{n+1}^L makes it redundant for numerical computation since the invertibility of the matrix is required to obtain the solution. To make the matrix P_{n+1}^L invertible, we introduced a positive parameter τ and extend the traditional hat function to the interval $[-\tau, 1+\tau]$.

Similarly, using (19) and (34)–(36), we construct the extended hat functions upper operational matrix for Abel's inversion, P_{n+1}^U, such that

$$\int_{-\tau}^{y} \frac{\overline{\Psi}_{n+1}(r)}{(y-r)^\gamma} dr = P_{n+1}^U \overline{\Psi}_{n+1}(y), \tag{43}$$

where

$$P_{n+1}^U = \frac{h^{1-\gamma}}{(1-\gamma)(2-\gamma)} \times \begin{bmatrix} \zeta_0 & \zeta_1 & \zeta_2 & \zeta_3 & \zeta_4 & \cdots & \zeta_{n-1} & \zeta_n \\ 0 & 1 & \xi_1 & \xi_2 & \xi_3 & \cdots & \xi_{n-2} & \xi_{n-1} \\ 0 & 0 & 1 & \xi_1 & \xi_2 & \xi_3 & \cdots & \xi_{n-2} \\ 0 & 0 & 0 & 1 & \xi_1 & \xi_2 & \cdots & \xi_{n-3} \\ 0 & 0 & 0 & 0 & 1 & \xi_1 & \cdots & \xi_{n-4} \\ \cdots & \cdots & \cdots & \cdots & \cdots & \cdots & \cdots & \cdots \\ 0 & \cdots & \cdots & \cdots & \cdots & 0 & 1 & \xi_1 \\ 0 & \cdots & \cdots & \cdots & \cdots & \cdots & 0 & 1 \end{bmatrix}_{(n+1)\times(n+1)}. \tag{44}$$

The various entries ζ_0, ζ_s, and ξ_k are given by (40)–(42), for $k = 1, 2, 3, \ldots, n-1$ and $s = 1, 2, 3, \ldots, n$.

If we partition the matrix P_{n+1}^L in four blocks as $P_{n+1}^L = \begin{bmatrix} A & O \\ B & \zeta_0 \end{bmatrix}$, then $P_{n+1}^U = \begin{bmatrix} \zeta_0 & B' \\ O & A^T \end{bmatrix}$, where $B = [\,\zeta_n \ \cdots \ \zeta_2 \ \zeta_1\,]_{1\times n}$, $B' = [\,\zeta_1 \ \zeta_2 \ \cdots \ \zeta_n\,]_{1\times n}$, O is a $n\times 1$ null vector, and

$$A = \begin{bmatrix} 1 & 0 & 0 & 0 & 0 & 0 \\ \xi_1 & 1 & 0 & 0 & 0 & 0 \\ \xi_2 & \xi_1 & 1 & 0 & 0 & 0 \\ \xi_3 & \xi_2 & \xi_1 & 1 & 0 & 0 \\ \xi_{n-2} & \cdots & \cdots & \xi_1 & 1 & 0 \\ \xi_{n-1} & \xi_{n-2} & \cdots & \cdots & \xi_1 & 1 \end{bmatrix}_{n\times n}. \tag{45}$$

Using (38) and (43), (18) may be written as

$$F_{n+1}^T \overline{\Psi}_{n+1}(y) = C_{n+1}^T \left[a(y) P_{n+1}^U + b(y) P_{n+1}^L\right] \overline{\Psi}_{n+1}(y). \tag{46}$$

Solving the above system of linear equations, we obtain

$$C_{n+1}^T = F_{n+1}^T \left[a(y) P_{n+1}^U + b(y) P_{n+1}^L\right]^{-1}. \tag{47}$$

Substituting the value of C_{n+1}^T from (47) into (17), the approximate emissivity $\eta(r)$ is given by

$$\eta(r) = F_{n+1}^T \left[a(y) P_{n+1}^U + b(y) P_{n+1}^L\right]^{-1} \overline{\Psi}_{n+1}(r). \tag{48}$$

3. Error Analysis

In this section, an error analysis of our proposed algorithm is given. Let $f \in L^2[0,1]$, and then, using (11), it is approximated as

$$f(t) \simeq \sum_{i=0}^{i=n} f_i \,\overline{\psi}_i(t) = \sum_{i=0}^{i=n} f(ih)\,\psi_i(t). \tag{49}$$

The above approximation gives exact values at nodal points. We denote the right-hand side of (49) by $\tilde{f}(t)$, and then, for $jh \le t < (j+1)h$, $j = 0, 1, \ldots, n-1$, we have

$$\begin{aligned}\tilde{f}(t) &= \sum_{i=0}^{i=n} f(ih)\,\psi_i(t) = f(jh)\,\psi_j(t) + f((j+1)h)\,\psi_{j+1}(t) \\ &= f(jh)\left(\frac{(j+1)h - t}{h}\right) + f(jh+h)\left(\frac{t - jh}{h}\right) \\ &\qquad\qquad \text{(using (6))} \\ &= (j+1) f(jh) + t\frac{(f(jh+h) - f(jh))}{h} \\ &\quad - jf(jh+h).\end{aligned} \tag{50}$$

Expanding $f(t)$ in Taylor series at $t = jh$, we obtain

$$f(t) = \sum_{r=0}^{\infty} \frac{(t - jh)^r}{r!} f^{(r)}(jh). \tag{51}$$

Thus, from (50) and (51), we get

$$\begin{aligned}\left|f(t) - \tilde{f}(t)\right| = \Bigg| & jh\frac{(f(jh+h) - f(jh))}{h} \\ & - t\frac{(f(jh+h) - f(jh))}{h} \\ & + \sum_{r=1}^{\infty} \frac{(t - jh)^r}{r!} f^{(r)}(jh)\Bigg|\end{aligned} \tag{52}$$

as $h \to 0$, and we have

$$\begin{aligned}\left|f(t) - \tilde{f}(t)\right| &= \left|jhf'(jh) - tf'(jh) + \sum_{r=1}^{\infty} \frac{(t - jh)^r}{r!} f^{(r)}(jh)\right| \\ &= \left|\sum_{r=2}^{\infty} \frac{(t - jh)^r}{r!} f^{(r)}(jh)\right| \\ &\le \frac{(t - jh)^2}{2!} + O\left|(t - jh)^3\right| \le \frac{n^{-2}}{2} + O\left(n^{-3}\right),\end{aligned} \tag{53}$$

as $(t - jh) < n^{-1}$, thus proving the following theorem.

Theorem 5. *The absolute error* $|f(t) - \tilde{f}(t)|$ *associated with the approximation* (49) *is of the order* $O(n^{-2})$.

4. Numerical Results and Stability Analysis

In this section, we discuss the implementation of our proposed algorithm and investigate its accuracy and stability by applying it on test functions with known analytical Abel inverse.

For, it is always desirable to test the behaviour of a numerical inversion method using simulated data for which the exact results are known, and thus making a comparison between inverted results and theoretical data is possible. We have tested our algorithm on several well-known test profiles that are commonly encountered in experimental data and widely used by researchers [7, 15, 20]. The accuracy of the proposed algorithm is demonstrated by calculating the parameters of absolute error $\Delta\eta(r_i)$, average deviation σ also known as root mean square error (RMS). They are calculated using the following equations:

$$\Delta\eta(r_i) = \left|\eta_a(r_i) - \eta_c(r_i)\right|, \tag{54}$$

$$\begin{aligned}\sigma_{n+1} &= \left\{\frac{1}{(n+1)}\sum_{i=0}^{n}\left[\eta_a(r_i) - \eta_c(r_i)\right]^2\right\}^{1/2} \\ &= \left\{\frac{1}{(n+1)}\sum_{i=0}^{n}\left[\Delta\eta(r_i)\right]^2\right\}^{1/2} = \|\Delta\eta\|_2,\end{aligned} \tag{55}$$

where $\eta_c(r_i)$ is the emission coefficient calculated at point r_i using (48) and $\eta_a(r_i)$ is the exact analytical emissivity at the corresponding point. Note that σ, henceforth, denoted by σ_{n+1} (for computational convenience) is the discrete l^2 norm of the absolute error $\Delta\eta$ denoted by $\|\Delta\eta\|_2$. Note that the calculation of σ_{n+1} in (55) is performed by taking different values of n. In all the test profiles, the exact and noisy intensity profiles are denoted by $I_1(y)$ and $I_1^\delta(y)$, respectively, where $I_1^\delta(y)$ is obtained by adding a noise δ to $I_1(y)$ such that $I_1^\delta(y_i) = I_1(y_i) + \delta\theta_i$, where $y_i = ih$, $i = 0, 1, 2, \ldots, n$, $nh = 1$, and θ_i is the uniform random variable with values in $[-1, 1]$ such that $\mathrm{Max}_{0\le i\le n}|I_1^\delta(y_i) - I_1(y_i)| \le \delta$.

The following test problems are solved with and without noise to illustrate the efficiency and stability of our method by choosing three different values of the noises δ_k as $\delta_0 = 0$, $\delta_1 = \sigma_{n+1}$, and $\delta_2 = p\%$ of μ_{n+1}, where we mean $\mu_{n+1} = (1/(n+1))\sum_{i=0}^{n} I_1(y)$. In each of the test problems given in this section we have taken positive parameter $\tau = 0.0001$ and $p = 0.1$, except for Example 10, where $\tau = 0$ has been used. The absolute errors between exact and approximate emissivities, corresponding to different noises δ_k, $k = 0, 1, 2$ have been denoted by $E_0(r)$, $E_1(r)$, and $E_2(r)$, respectively. In the text boxes of the figures, the notations $E0(r)$, $E1(r)$, and $E2(r)$ have been used for $E_0(r)$, $E_1(r)$, and $E_2(r)$, respectively.

Though the stability of the algorithm is illustrated by various numerical experiments performed in this section, we analyze it also mathematically as follows.

The reconstructed emissivities $\eta_{c_\delta}(r)$ (with δ noise) and $\eta_{c_0}(r)$ (without noise) are obtained with and without noise term in the intensity profile $I_1(y)$, and using (48) these are given by

$$\begin{aligned}\eta_{c_\delta}(r) &= F_{n+1}^{\delta T}\left[a(y)\,P_{n+1}^U + b(y)\,P_{n+1}^L\right]^{-1}\overline{\Psi}_{n+1}(r), \\ \eta_{c_0}(r) &= F_{n+1}^{T}\left[a(y)\,P_{n+1}^U + b(y)\,P_{n+1}^L\right]^{-1}\overline{\Psi}_{n+1}(r),\end{aligned} \tag{56}$$

TABLE 1: Noise reduction $h(r)$ for $n = 100$ at different values of δ, $\tau = 0$, $\gamma = 1/2$.

r	$h(r)$ for $\delta = 0.01$	$h(r)$ for $\delta = 0.001$	$h(r)$ for $\delta = 0.0001$
0	0.019718	0.0019718	0.00019718
0.1	0.0040696	0.00040696	0.000040696
0.2	0.0035401	0.00035401	0.000035401
0.3	0.0033118	0.00033118	0.000033118
0.4	0.0032047	0.00032047	0.000032047
0.5	0.0031726	0.00031726	0.000031726
0.6	0.0032047	0.00032047	0.000032047
0.7	0.0033118	0.00033118	0.000033118
0.8	0.0035401	0.00035401	0.000035401
0.9	0.0040696	0.00040696	0.000040696
1	0.019718	0.0019718	0.00019718

where F^{δ}_{n+1} and F_{n+1} are known matrices, and they are obtained from the following equations:

$$I_1^{\delta}(y) = I_1(y) + \delta\theta_i \simeq F^{\delta T}_{n+1}\overline{\Psi}_{n+1}(y),$$
$$I_1(y) \simeq F^{T}_{n+1}\overline{\Psi}_{n+1}(y). \quad (57)$$

Hence

$$\eta_{c_\delta}(r) - \eta_{c_0}(r) = \left(F^{\delta T}_{n+1} - F^{T}_{n+1}\right)\left[a(y)P^{U}_{n+1} + b(y)P^{L}_{n+1}\right]^{-1}\overline{\Psi}_{n+1}(r). \quad (58)$$

Writing $H^{T}_{n+1} = F^{\delta T}_{n+1} - F^{T}_{n+1}$ and replacing random noise $\delta\theta_i$ by its maximum value δ, we get

$$\left\|\eta_{c_\delta}(r) - \eta_{c_0}(r)\right\| = \left\|H^{T}_{n+1}\right\| \left\|\left[a(y)P^{U}_{n+1} + b(y)P^{L}_{n+1}\right]^{-1}\right\| \left\|\overline{\Psi}_{n+1}(r)\right\|. \quad (59)$$

Let $h(r) = \eta_{c_\delta}(r) - \eta_{c_0}(r) = (F^{\delta T}_{n+1} - F^{T}_{n+1})[a(y)P^{U}_{n+1} + b(y)P^{L}_{n+1}]^{-1}\overline{\Psi}_{n+1}(r)$, then $h(r)$ reflects the noise reduction capability of the algorithm and its values at various points, and its graph is shown in Table 1 and Figure 1, respectively.

Table 1 demonstrates the noise filtering capability of the algorithm for three different noise outputs. From Table 1 and Figure 1 we see that noise reduction is symmetric about the point $r = 0.5$, and the maximum reduction in noise is achieved at $r = 0.5$ for all the three levels of noises $\delta = 0.01$, 0.001, and 0.0001 introduced in $I_1(y)$. The general behaviour of the noise reduction is the same irrespective of the value of δ. In the interval $[0.02, 0.98]$ the algorithm is stable, whereas the noise filtering capability decreases continuously and then jumps symmetrically in $[0, 0.02) \cup (0.98, 1]$.

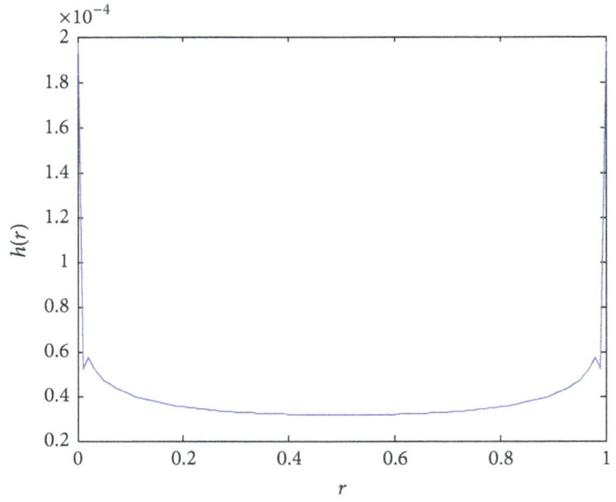

FIGURE 1: Noise reduction $h(r)$ for $n = 100$, $\delta = 0.0001$, $\tau = 0$, and $\gamma = 1/2$.

Example 6. Consider the generalized Abel integral equation:

$$\frac{4}{3}y^{3/2}(1+y) + \frac{2}{3}y^{3/2}\left[2\sqrt{(1-y)y^3} + \sqrt{y-y^2}\right] = a(y)\int_0^y \frac{\eta(r)}{\sqrt{(y-r)}}dr + b(y)\int_y^1 \frac{\eta(r)}{\sqrt{(r-y)}}dr, \quad (60)$$
$$0 \le y \le 1,$$

where $a(y) = y + 1$, $b(y) = y^2$, with the exact analytical solution $\eta(r) = r$.

The absolute errors $E_k(r)$ have been calculated for $n = 1000$ and are given in Table 2. The value of δ_1 is 2.3829×10^{-15}, for $n = 1000$. As $\delta_2 = 5.3347 \times 10^{-4} > 10^{10}\delta_1$, the absolute error $E_2(r)$ is appreciably higher than $E_0(r)$ and $E_1(r)$. The Figure 2 compares the absolute errors $E_0(r)$ and $E_1(r)$ for noise $\delta_1 = 5.9827 \times 10^{-16}$, $n = 100$.

Example 7. In this example, we consider the following Abel's integral equation [7, 20]:

$$\int_0^y \frac{\eta(r)}{\sqrt{(y-r)}}dr = y^{11/2}, \quad 0 \le y \le 1,$$
$$\text{with solution } \eta(r) = \frac{2^{10}11[\Gamma(11/2)]^2}{2\pi\Gamma(11)}r^5. \quad (61)$$

The absolute errors corresponding to different noises are given in Table 3. The values of various parameters are given as:

$\sigma_{3001} = 9.4002 \times 10^{-8} (= \delta_1, n = 3000)$, $\sigma_{2001} = 2.1116 \times 10^{-7} (= \delta_1, n = 2000)$, $\sigma_{1001} = 8.4162 \times 10^{-7} (= \delta_1, n = 1000)$, $\mu_{3001} = 0.1540$, $\mu_{2001} = 0.1540$, and $\mu_{1001} = 0.1542$. Taking $p = 0.1$, the various values of respective δ_2 are given in order as 1.5396×10^{-4}, 1.5402×10^{-4}, and 1.5419×10^{-4}.

TABLE 2: The absolute errors $E_k(r)$, at different nodal points r, for $n = 1000$, in Example 6.

r	0.0	0.2	0.4	0.6	0.8	1.0
$E_0(r)$	0	7.494×10^{-16}	1.2212×10^{-15}	1.8874×10^{-15}	8.1046×10^{-15}	1.5543×10^{-15}
$E_1(r)$	1.132×10^{-13}	2.4591×10^{-14}	8.3822×10^{-15}	1.7764×10^{-14}	2.931×10^{-14}	3.5194×10^{-14}
$E_2(r)$	0.011106	0.0064263	0.0025148	0.01039	0.00029049	0.00014606

TABLE 3: The absolute errors $E_k(r)$, at different nodal points r, for $n = 3000, 2000$, and 1000.

	n	$r = 0.0$	$r = 0.2$	$r = 0.4$	$r = 0.6$	$r = 0.8$	$r = 1.0$
	3000	0	1.9712×10^{-9}	1.5849×10^{-8}	5.361×10^{-8}	1.2724×10^{-7}	2.4875×10^{-7}
$E_0(r)$	2000	0	4.4178×10^{-7}	3.5563×10^{-8}	1.2035×10^{-7}	2.8575×10^{-7}	5.5872×10^{-7}
	1000	0	1.7515×10^{-8}	1.4137×10^{-7}	4.7899×10^{-7}	1.138×10^{-6}	2.2262×10^{-6}
	3000	4.6073×10^{-6}	1.5029×10^{-6}	2.5454×10^{-6}	5.3295×10^{-7}	7.3491×10^{-7}	2.0451×10^{-6}
$E_1(r)$	2000	1.0031×10^{-5}	1.1159×10^{-6}	4.5957×10^{-6}	2.3615×10^{-7}	4.4563×10^{-6}	9.2247×10^{-7}
	1000	1.7061×10^{-5}	7.7631×10^{-6}	7.528×10^{-6}	4.4903×10^{-6}	1.5261×10^{-5}	4.2806×10^{-6}
	3000	0.0061012	0.0011742	0.0011595	0.0039968	0.00064593	0.0029774
$E_2(r)$	2000	0.0068177	0.0068177	0.00078732	0.00067568	0.0016351	4.2642×10^{-5}
	1000	0.0001341	0.00067286	0.0010734	0.00040813	0.00098541	0.00074723

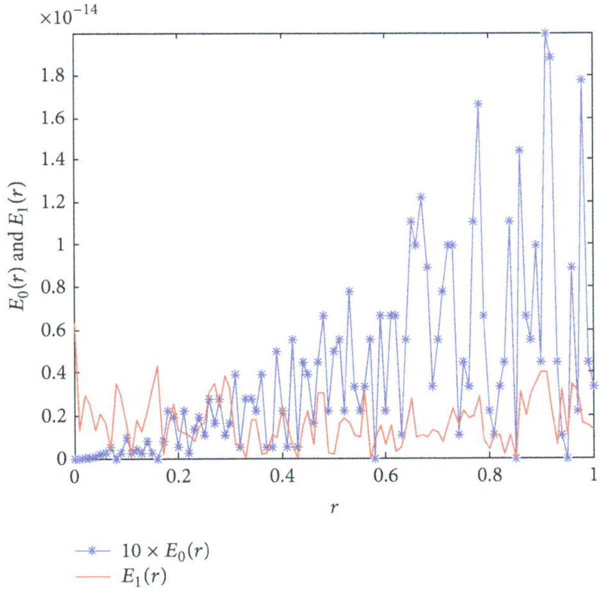

FIGURE 2: Absolute errors $E_0(r)$ and $E_1(r)$ for $n = 100$, $\delta_1 = 5.9827 \times 10^{-16}$.

In Figure 3, the exact and reconstructed emissivities (with δ_2 noise) have been shown for $n = 50$, and the two emissivities match very well even for higher noise δ_2 introduced in the intensity profile. For $n = 100$, Figures 4 and 5 show the absolute errors $E_0(r)$, $E_1(r)$ and $E_0(r)$, $E_2(r)$, respectively.

Example 8. In this example we consider the following Abel's integral equation [22]:

$$\int_0^y \frac{\eta(r)}{\sqrt{(y-r)}} dr = I_1(y), \quad 0 \le y \le 1, \tag{62}$$

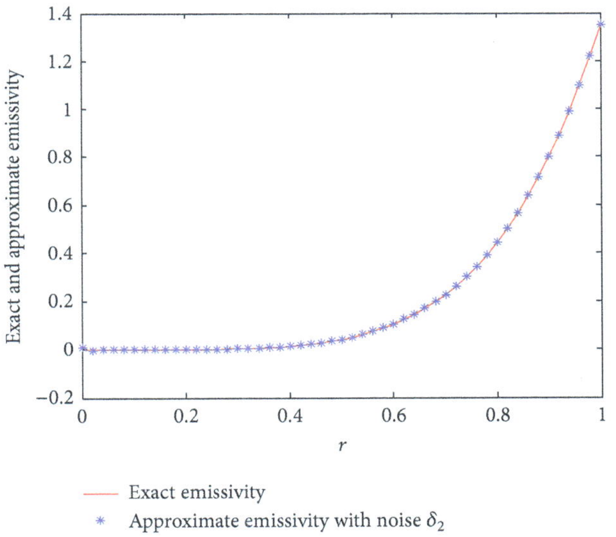

FIGURE 3: Exact and approximate emissivities with noise $\delta_2 = 1.6081 \times 10^{-4}$ for $n = 50$.

where

$$I_1(y) = \begin{cases} \frac{4}{3} y^{3/2}, & 0 \le y < \frac{1}{2}, \\ \frac{4}{3} y^{3/2} - \frac{8}{3}\left(y - \frac{1}{2}\right)^{3/2}, & \frac{1}{2} \le y \le 1. \end{cases} \tag{63}$$

The exact solution of the integral equation (62) is given by

$$\eta(r) = \begin{cases} r, & 0 \le r < \frac{1}{2}, \\ 1 - r, & \frac{1}{2} \le r \le 1. \end{cases} \tag{64}$$

In Table 4, the absolute errors for different noises have been shown. Various parameters used for Table 4 are

TABLE 4: The absolute errors $E_k(r)$, at different nodal points r, for $n = 1000$, in Example 8.

r	0.0	0.2	0.4	0.6	0.8	1.0
$E_0(r)$	0	4.4409×10^{-16}	5.5511×10^{-17}	2.7200×10^{-15}	2.1649×10^{-15}	3.3888×10^{-15}
$E_1(r)$	5.5359×10^{-14}	7.8271×10^{-15}	2.2093×10^{-14}	1.1768×10^{-14}	1.4433×10^{-14}	1.5344×10^{-14}
$E_2(r)$	0.0042764	0.0018814	3.4439×10^{-5}	0.0014343	0.0024313	0.00018694

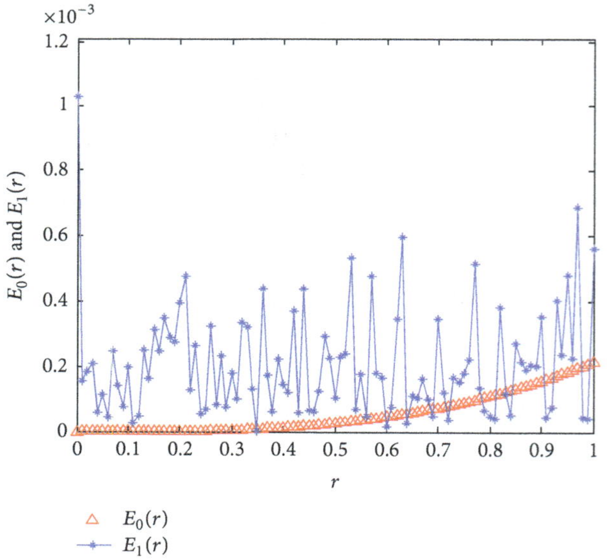

FIGURE 4: Absolute errors $E_0(r)$ and $E_1(r)$ for $\delta_1 = 8.2435 \times 10^{-5}$, $n = 100$.

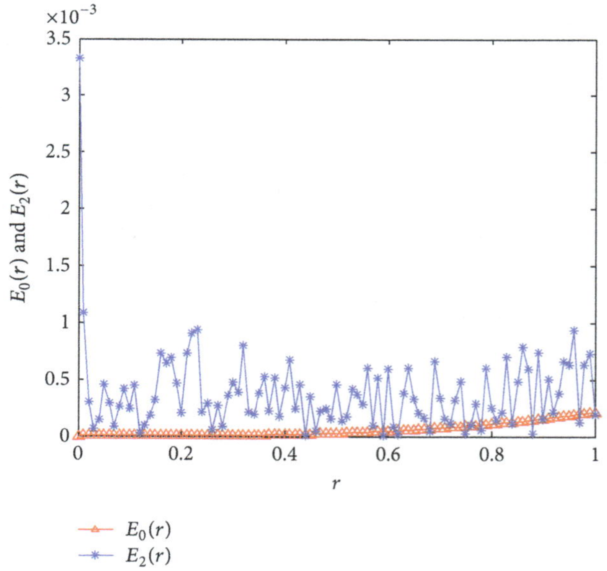

FIGURE 5: Absolute errors $E_0(r)$ and $E_2(r)$ for $\delta_2 = 1.5732 \times 10^{-4}$, $n = 100$.

$\sigma_{1001} = 1.8576\times10^{-15} = \delta_1, \mu_{1001} = 0.3446$, and $\delta_2 = 3.4462\times 10^{-4}$. Figure 6 shows the graph of exact and approximate emissivities $\eta(r)$ (without noise) for $n = 50$. Absolute errors $E_0(r)$ and $E_1(r)$, for $n = 50$ and $n = 100$, are shown in Figures 7 and 8, respectively.

Example 9. Consider the generalized Abel integral equation:

$$I_1(y) = e^y \sin(y) \int_0^y \frac{\eta(r)}{(y-r)^{1/3}} dr + e^{-y} \cos(y) \int_y^1 \frac{\eta(r)}{(r-y)^{1/3}} dr, \quad 0 < y < 1, \tag{65}$$

where $I_1(y) = e^y \sin(y)\ y^{5/3} + e^{-y}\ \cos(y)(1-y)^{2/3}(2+3y)$. The exact solution of (65) is $\eta(r) = 10r/9$. In Figure 9, the comparison between $E_0(r)$ and $E_1(r)$ is shown, for $n = 100$.

Example 10. For the pair [14, 15, 23]:

$$\eta(r) = (1-r)^2 (1+12r), \quad \text{for } 0 \le r \le 1,$$
$$I_1(y) = \left[\left(\frac{384}{35} y^{7/2} - \frac{368}{15} y^{5/2} + \frac{40}{3} y^{3/2} + 2\sqrt{y}\right) + \frac{16}{105}(1-y)^{5/2}(19+72y)\right], \tag{66}$$

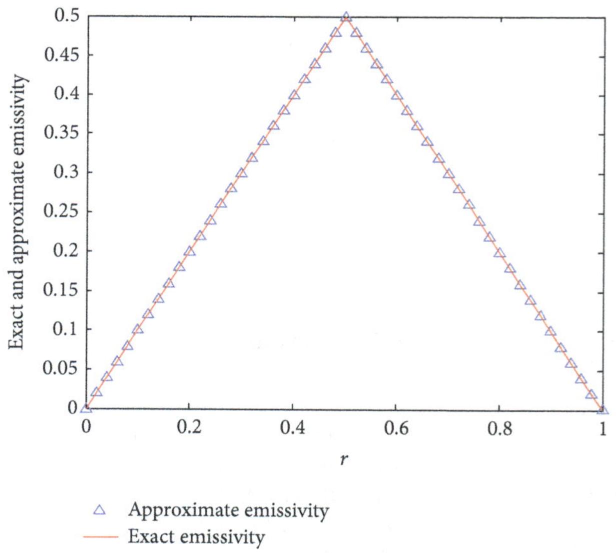

FIGURE 6: Exact and approximate solutions for $n = 50$.

with $a(y) = b(y) = 1$, and the various parameters are as follows:

$\delta_1 = 2.1111 \times 10^{-7}$ ($n = 3000$), $\delta_1 = 4.7384 \times 10^{-7}$ ($n = 2000$), $\delta_1 = 1.8852 \times 10^{-6}$ ($n = 1000$), and $\delta_2 = 0.0036$, for all the three chosen values of n.

TABLE 5: The absolute errors $E_k(r)$, at different nodal points r, for $n = 3000, 2000$, and 1000.

	n	$r = 0.0$	$r = 0.2$	$r = 0.4$	$r = 0.6$	$r = 0.8$	$r = 1.0$
$E_0(r)$	3000	5.4281×10^{-7}	2.8998×10^{-7}	1.5818×10^{-7}	2.6121×10^{-8}	1.0578×10^{-7}	2.9684×10^{-7}
	2000	1.2088×10^{-6}	6.5113×10^{-7}	3.5535×10^{-7}	5.887×10^{-8}	2.3717×10^{-7}	6.5839×10^{-7}
	1000	4.7364×10^{-6}	2.5925×10^{-6}	$1.4164e \times 10^{-6}$	2.3637×10^{-7}	9.4123×10^{-7}	2.559×10^{-6}
$E_1(r)$	3000	6.8261×10^{-6}	2.1337×10^{-6}	2.2043×10^{-6}	1.4765×10^{-6}	1.4642×10^{-6}	3.421×10^{-6}
	2000	1.589×10^{-6}	5.7191×10^{-6}	2.8577×10^{-6}	2.5629×10^{-6}	4.4183×10^{-6}	3.9074×10^{-8}
	1000	1.9156×10^{-5}	4.4716×10^{-6}	1.643×10^{-5}	1.712×10^{-5}	3.6022×10^{-6}	3.6477×10^{-6}
$E_2(r)$	3000	0.01495	0.020628	0.021094	0.065933	0.011844	0.095356
	2000	0.022427	0.010769	0.024946	0.014392	0.0068535	0.028831
	1000	0.056845	0.033342	0.013319	0.018905	0.028086	0.045314

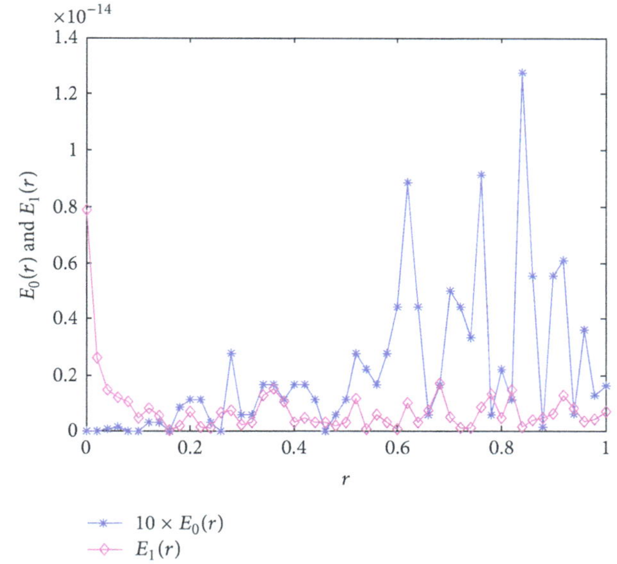

FIGURE 7: Absolute errors $E_0(r)$ and $E_1(r)$ for $\delta_1 = 3.4218 \times 10^{-16}$, $n = 50$.

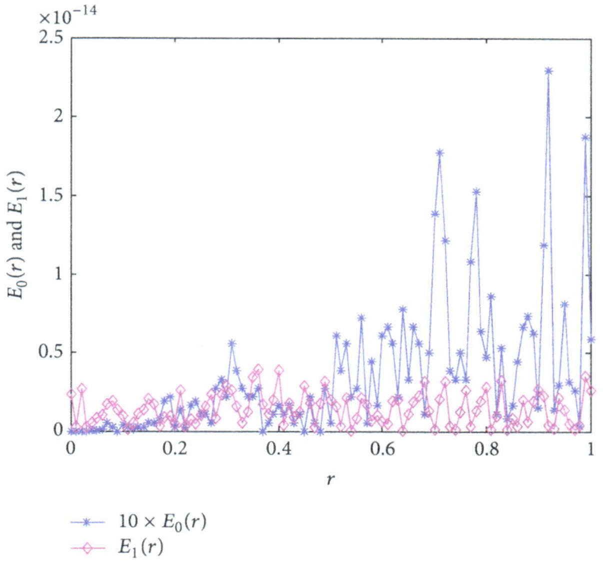

FIGURE 8: Absolute errors $E_0(r)$, $E_1(r)$ for $\delta_1 = 5.5794 \times 10^{-16}$, $n = 100$.

The absolute errors corresponding to different noises are given in Table 5. Figure 10 shows the exact and approximate emissivities (without noise and with noise $\delta_2 = 0.0036$) whereas, in Figure 11, a comparison between $E_0(r)$ and $E_1(r)$ is shown for $\delta_1 = 1.8232 \times 10^{-4}$, $n = 100$.

Example 11. Consider the generalized Abel integral equation (15) with $\gamma = 1/2$, $a(y) = (3/4)\exp(y)$, $b(y) = \exp(2y) + (1/\sqrt{2\pi})$, for the pair $\eta(r) = \sin r$ and

$$I_1(y) = \left[\begin{array}{c} \exp(y)\, y_1^{3/2} F_2\left(1, \frac{5}{4}, \frac{7}{4}, -\frac{y^2}{4}\right) + \sqrt{2\pi}\left(\exp(2y) + \frac{1}{\sqrt{2\pi}}\right) \\ \left(C\left(\frac{\sqrt{2-2y}}{\sqrt{\pi}}\right)\sin y + S\left(\frac{\sqrt{2-2y}}{\sqrt{\pi}}\right)\cos y\right) \end{array}\right], \quad (67)$$

where $C(z)$ and $S(z)$ in (67) are called Fresnel integrals. These are defined as

$$C(z) = \int_0^z \cos\left(\frac{\pi t^2}{2}\right) dt, \qquad S(z) = \int_0^z \sin\left(\frac{\pi t^2}{2}\right) dt. \quad (68)$$

For, $n = 3000$, different absolute errors are given in Table 6. The various parameters for $n = 3000$ are $\delta_1 = 0.0173$ and $\delta_2 = 7.2830 \times 10^{-4}$. In Figure 12, the exact and approximate emissivities (without noise) are shown for $n = 100$.

5. Conclusions

We have constructed operational matrices of integration based on extended hat functions and used them to propose a stable algorithm for the numerical inversion of the generalized Abel integral equation. The earlier numerical

TABLE 6: The absolute errors $E_k(r)$ for $n = 3000$, for Example 11.

r	0.0	0.2	0.4	0.6	0.8	0.9
$E_0(r)$	3.389×10^{-7}	4.7548×10^{-7}	7.3287×10^{-7}	1.3451×10^{-6}	3.7972×10^{-6}	1.0734×10^{-5}
$E_1(r)$	0.38495	0.079121	0.17527	0.096373	0.40061	0.050512
$E_2(r)$	0.0066376	0.0058214	0.0070449	0.018136	0.013987	0.0021626

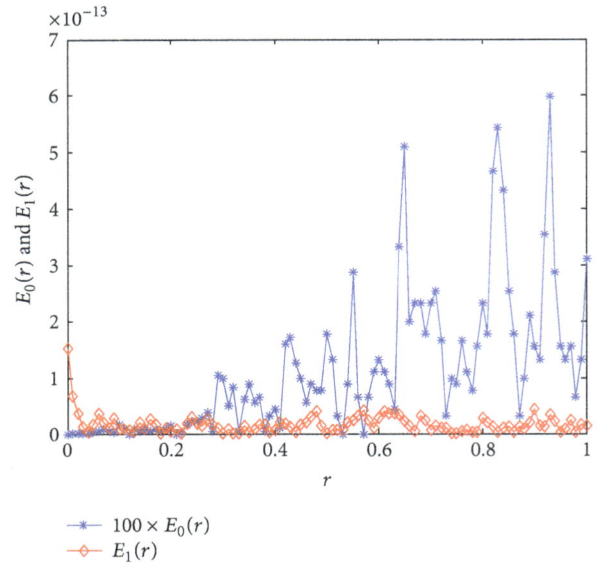

FIGURE 9: Comparison between $E_0(r)$ and $E_1(r)$ for $\delta_1 = 1.6911 \times 10^{-15}$, $n = 100$.

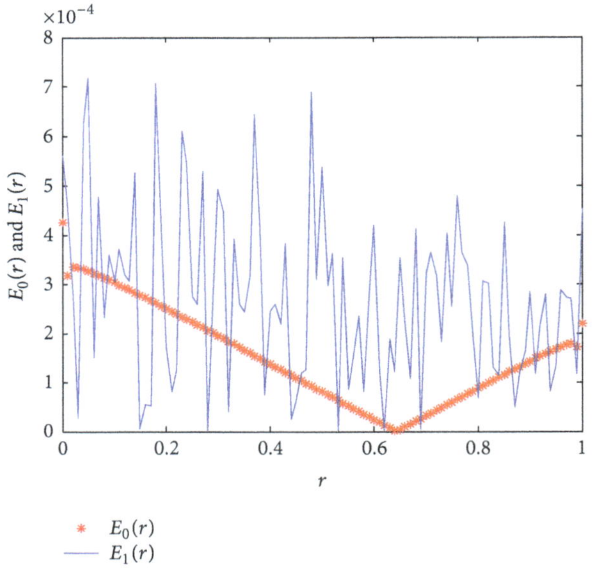

FIGURE 11: Comparison between $E_0(r)$ and $E_1(r)$ for $\delta_1 = 1.8232 \times 10^{-4}$, $n = 100$.

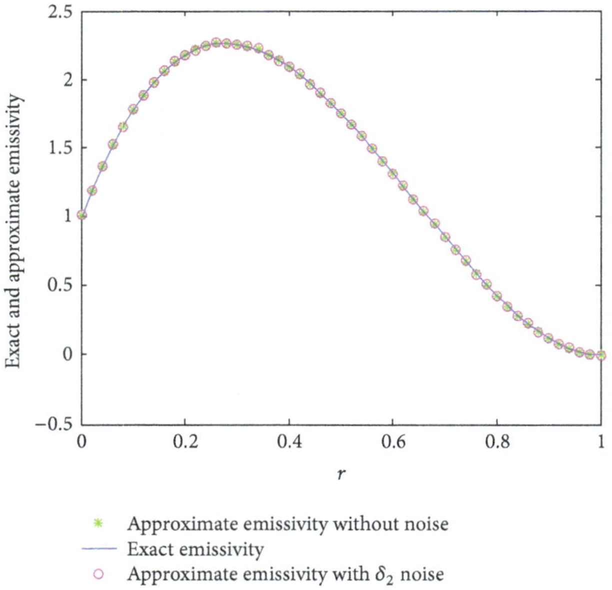

FIGURE 10: Exact and approximate emissivities (without noise and with noise $\delta_2 = 0.0036$) for $n = 50$.

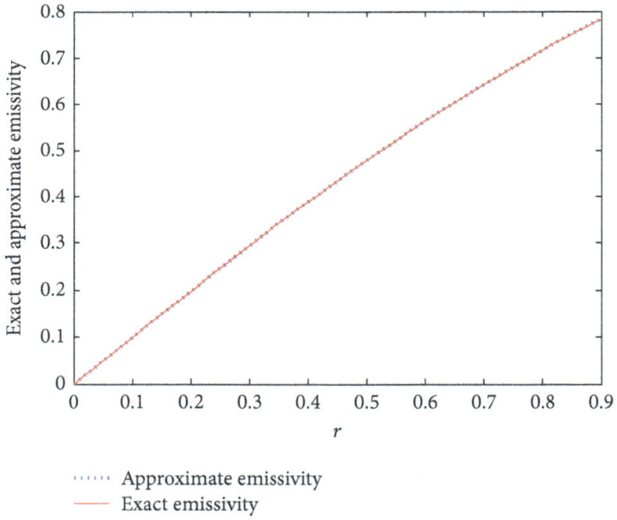

FIGURE 12: Exact and approximate emissivities (without noise) for $n = 100$.

inversions were restricted to a part of the general Abel's integral equation. We have extended the hat function beyond their domain $[0, 1]$ to avoid the singularity of the matrix at $t = 0, 1$. These operational matrices are given by the general formulae (39) and (44), thus making the evaluation of these matrices of any order extremely easy whereas in case of Bernstein operational matrices no such formula was available [19, 20]. The stability with respect to the data is restored and good accuracy is obtained even for high noise levels in the data. An error analysis and stability analysis are also given.

Acknowledgments

The authors are grateful to the learned reviewer for his valuable suggestions which have led to the improvement of the paper in the present form. Also, the first author acknowledges the financial support from UGC New-Delhi, India, under Faculty Improvement Program (FIP), whereas the second and the third authors acknowledge the financial support from UGC and CSIR New-Delhi, India, respectively, under JRF schemes.

References

[1] N. H. Abel, "Resolution d'un problem de mechanique," *Journal Für Die Reine Und Angewandte Mathematik*, vol. 1, pp. 153–157, 1826.

[2] L. Mach, *Wien. Akad. Ber. Math. Phys. Klasse*, vol. 105, p. 605, 1896.

[3] H. R. Griem, *Plasma Spectroscopy*, McGraw-Hill, New York, NY, USA, 1963.

[4] F. G. Tricomi, *Integral Equations*, Wiley-Interscience, New York, NY, USA, 1975.

[5] R. S. Anderssen, "Stable procedures for the inversion of Abel's equation," *Journal of the Institute of Mathematics and its Applications*, vol. 17, no. 3, pp. 329–342, 1976.

[6] I. Beniaminy and M. Deutsch, "ABEL: stable, high accuracy program for the inversion of Abel's integral equation," *Computer Physics Communications*, vol. 27, no. 4, pp. 415–422, 1982.

[7] S. Bhattacharya and B. N. Mandal, "Use of Bernstein polynomials in numerical solutions of Volterra integral equations," *Applied Mathematical Sciences*, vol. 2, no. 33-36, pp. 1773–1787, 2008.

[8] C. J. Cremers and R. C. Birkebak, "Application of the Abel integral equation to spectroscopic data," *Applied Optics*, vol. 5, pp. 1057–1064, 1966.

[9] M. Deutsch and I. Beniaminy, "Derivative-free inversion of Abel's integral equation," *Applied Physics Letters*, vol. 41, no. 1, pp. 27–28, 1982.

[10] M. Deutsch, "Abel inversion with a simple analytic representation for experimental data," *Applied Physics Letters*, vol. 42, no. 3, pp. 237–239, 1983.

[11] R. Grenflo, "Computation of rough solutions of Abel integral equation," in *Inverse ILL-Posed Problems*, H. W. Engel and C. W. Groetsch, Eds., pp. 195–210, Academic Press, NewYork, NY, USA, 1987.

[12] L. M. Ignjatović and A. A. Mihajlov, "The realization of Abel's inversion in the case of discharge with undetermined radius," *Journal of Quantitative Spectroscopy & Radiative Transfer*, vol. 72, no. 5, pp. 677–689, 2002.

[13] J.-P. Lanquart, "Error attenuation in Abel inversion," *Journal of Computational Physics*, vol. 47, no. 3, pp. 434–443, 1982.

[14] S. Ma, H. Gao, L. Wu, and G. Zhang, "Abel inversion using Legendre polynomials approximations," *Journal of Quantitative Spectroscopy & Radiative Transfer*, vol. 109, no. 10, pp. 1745–1757, 2008.

[15] S. Ma, H. Gao, G. Zhang, and L. Wu, "Abel inversion using Legendre wavelets expansion," *Journal of Quantitative Spectroscopy & Radiative Transfer*, vol. 107, no. 1, pp. 61–71, 2007.

[16] G. N. Minerbo and M. E. Levy, "Inversion of Abel's integral equation by means of orthogonal polynomials," *SIAM Journal on Numerical Analysis*, vol. 6, pp. 598–616, 1969.

[17] D. A. Murio, D. G. Hinestroza, and C. E. Mejía, "New stable numerical inversion of Abel's integral equation," *Computers and Mathematics with Applications*, vol. 23, no. 11, pp. 3–11, 1992.

[18] M. Sato, "Inversion of the Abel integral equation by use of simple interpolation formulas," *Contributions to Plasma Physics*, vol. 25, pp. 573–577, 1985.

[19] V. K. Singh, R. K. Pandey, and O. P. Singh, "New stable numerical solutions of singular integral equations of Abel type by using normalized Bernstein polynomials," *Applied Mathematical Sciences*, vol. 3, no. 5-8, pp. 241–255, 2009.

[20] O. P. Singh, V. K. Singh, and R. K. Pandey, "A stable numerical inversion of Abel's integral equation using almost Bernstein operational matrix," *Journal of Quantitative Spectroscopy & Radiative Transfer*, vol. 111, pp. 245–252, 2010.

[21] A. Chakrabarti, "Solution of the generalized Abel integral equation," *Journal of Integral Equations and Applications*, vol. 20, no. 1, pp. 1–11, 2008.

[22] L. Huang, Y. Huang, and X.-F. Li, "Approximate solution of Abel integral equation," *Computers and Mathematics with Applications*, vol. 56, no. 7, pp. 1748–1757, 2008.

[23] M. J. Buie, J. T. P. Pender, J. P. Holloway, T. Vincent, P. L. G. Ventzek, and M. L. Brake, "Abel's inversion applied to experimental spectroscopic data with off axis peaks," *Journal of Quantitative Spectroscopy & Radiative Transfer*, vol. 55, no. 2, pp. 231–243, 1996.

16

Some Existence and Convergence Theorems for Nonexpansive Type Mappings

S. N. Mishra,[1] **Rajendra Pant,**[2] **and R. Panicker**[1,3]

[1] *Department of Mathematics, Walter Sisulu University, Mthatha 5117, South Africa*
[2] *Department of Mathematics, Visvesvaraya National Institute of Technology, Nagpur 440010, India*
[3] *Department of Mathematics, Rhodes University, Grahamstown 6140, South Africa*

Correspondence should be addressed to Rajendra Pant; pant.rajendra@gmail.com

Academic Editor: Zhijun Qiao

Some existence and convergence theorems for a class of nonexpansive type mappings are obtained in a normed space. The results obtained herein generalize certain known results.

1. Introduction

Let C be a nonempty subset of a normed space X. A mapping $T : C \to C$ is said to be nonexpansive if

$$\|Tx - Ty\| \le \|x - y\|, \qquad (1)$$

for all $x, y \in C$. Suppose $S : C \to C$ is another mapping on C. Then the mapping T is said to be S-nonexpansive if

$$\|Tx - Ty\| \le \|Sx - Sy\|, \qquad (2)$$

for all $x, y \in C$. The class of S-nonexpansive mappings is more general than nonexpansive mappings [1–3].

We extend the above notion of S-nonexpansive mappings to a more general class of nonexpansive mappings.

Definition 1. Let X be a normed space, C a nonempty subset of X and $S, T : C \to C$. We say that T is a generalized S-nonexpansive type mapping if

$$\|Tx - Ty\| \le M(x, y), \qquad (3)$$

for all $x, y \in C$, where

$$M(x, y) = \max\left\{\|Sx - Sy\|, \frac{\|Sx - Tx\| + \|Sy - Ty\|}{2}, \frac{\|Sx - Ty\| + \|Sy - Tx\|}{2}\right\}. \qquad (4)$$

Further T will be called a generalized nonexpansive type mappings if

$$\|Tx - Ty\| \le m(x, y), \qquad (5)$$

for all $x, y \in C$, where

$$m(x, y) = \max\left\{\|x - y\|, \frac{\|x - Tx\| + \|y - Ty\|}{2}, \frac{\|x - Ty\| + \|y - Tx\|}{2}\right\}. \qquad (6)$$

In this paper we obtain some common fixed point theorems for generalized nonexpansive type mappings in normed spaces. Specifically, in Section 3, we obtain some coincidence and common fixed point theorems while in Section 4, the weak convergence of Mann iterations (cf. Mann [4]) to a common fixed point for the above class of mappings is discussed. The results obtained herein generalize certain results of Kim et al. [1], Rhoades and Temir [3], and Shazad [5] among others.

2. Preliminaries

Let C be a nonempty subset of a normed space X and $S, T : C \to X$. A point $z \in C$ is called a coincidence point of S and T if $Sz = Tz$ and a common fixed point if $Sz = Tz = z$.

Now onwards $\mathbb{N}$ will denote the set of naturals while $F(S)$ and $F(T)$ the set of fixed points of S and T, respectively.

Definition 2. Let X be a normed space, C a nonempty subset of X, and $S, T : C \to C$. The pair of mappings (S, T) is called

(i) commuting if $TSx = STx$ for all $x \in C$.

(ii) weakly commuting if for all $x \in C$, $\|STx - TSx\| \leq \|Sx - Tx\|$ (see [6]).

(iii) R-weakly commuting if for all $x \in C$, there exists $R > 0$ such that $\|STx - TSx\| \leq R\|Sx - Tx\|$ (see [7]).

The following example illustrates that weakly commuting mappings are R-weakly commuting but the converse is not true in general.

Example 3. Let $X = C = [1, \infty)$ be endowed with usual norm $\|x\| = |x|$. Let $S, T : C \to C$ be mappings defined by

$$Sx = x^2, \qquad Tx = 2x^2 - 1, \quad \forall x \in X. \tag{7}$$

Then

$$\begin{gathered} \|STx - TSx\| = \left\|2x^2 - 4x + 2\right\| = 2\left|x^2 - 2x + 1\right|, \\ \|Sx - Tx\| = \left|x^2 - 2x + 1\right|. \end{gathered} \tag{8}$$

Therefore $\|STx - TSx\| = 2|x^2 - 2x + 1| = 2\|Sx - Tx\|$ and the pair (S, T) is R-weakly commuting with $R = 2$ but not weakly commuting.

In general, commuting $\Rightarrow$ weakly commuting $\Rightarrow$ R-weakly commuting.

Definition 4 (cf. [1]). Let X be a normed space and C a nonempty subset of X. The set C is called q-starshaped with $q \in C$ if for all $x \in C$, the segment $[q, x] = \{(1 - t)q + tx\}$ joining q to x, is contained in C; where $0 \leq t \leq 1$.

Further if C is a nonempty q-starshaped subset of a normed space X, then the mapping $S : C \to C$ is said to be q-affine if

$$tSx + (1 - t)\, Sq \in C \tag{9}$$

for all $x \in C$ and $0 \leq t \leq 1$.

Definition 5 (cf. [1]). Let X be a normed space, C a nonempty subset of X, and $S, T : C \to C$ such that $F(S) \neq \varnothing$. Suppose $q \in F(S)$ and C is q-starshaped. Then the pair of mappings (S, T) is called R-subweakly commuting on C if for all $x \in C$, there exists a real number $R > 0$ such that

$$\|STx - TSx\| \leq R \operatorname{dist}(Sx, [q, Tx]), \tag{10}$$

where $\operatorname{dist}(Sx, [q, Tx]) = \inf\{\|Sx - y\| : y \in [q, Tx]\}$.

We note that R-subweakly commuting mappings are R-weakly commuting but the converse is not true in general.

Example 6. Let $X = \mathbb{R}$ (set of reals) with norm $\|x\| = |x|$ and $C = [0, 10]$. Define $S, T : C \to C$ by

$$Tx = \frac{x + 1}{2}, \qquad Sx = \frac{x}{2}. \tag{11}$$

Then

$$\begin{gathered} \|TSx - STx\| = \left|\frac{x + 2}{4} - \frac{x + 1}{4}\right| = \frac{1}{4}, \\ \|Tx - Sx\| = \left|\frac{x + 1}{2} - \frac{x}{2}\right| = \frac{1}{2}, \quad \forall x \in C. \end{gathered} \tag{12}$$

Therefore

$$\|TSx - STx\| = \frac{1}{4} = \frac{1}{2}\left(\frac{1}{2}\right) = R\,\|Tx - Sx\| \tag{13}$$

holds for $R = 1/2$ and T and S are R-weakly commuting on C.

On the other hand, $q = 0 \in F(S)$, and for all $x \in C$,

$$\|Sx - [Tx, q]\| = \left|\frac{x}{2} - \left[\frac{x + 1}{2}, 0\right]\right| = 0. \tag{14}$$

So, there does not exist any $R > 0$ such that for all $x \in C$,

$$\|TSx - STx\| = \frac{1}{4} \leq R\,\|Sx - [Tx, q]\| = 0 \tag{15}$$

holds. Thus T and S are not R-subweakly commuting on C.

Definition 7. Let C be a nonempty subset of a normed space X and $T : C \to C$. Let $\{x_n\}$ be a sequence in X. We denote the weak and strong convergence of $\{x_n\}$ to x by $x_n \rightharpoonup x$ and $x_n \to x$, respectively. The mapping T is said to be *demicontinuous* if $\{x_n\}$ is a sequence in X such that $x_n \to x$, then $Tx_n \rightharpoonup Tx$.

Definition 8. A Banach space X is said to satisfy the Opial's condition (see [8]), if whenever a sequence $\{x_n\}$ in X, converges weakly to x ($x_n \rightharpoonup x$), then

$$\lim_{n \to \infty} \inf \|x_n - x\| < \lim_{n \to \infty} \inf \|x_n - y\|, \tag{16}$$

for all $y \in X$, $y \neq x$.

We note that the L^p spaces, $p \neq 2$ do not satisfy Opial' condition while all l^p spaces $(1 < p < \infty)$ do (see for details Goebel and Kirk [9]).

3. Existence Results

The following common fixed point theorem is due to Shahzad [5, Theorem 2.1]. For related results we refer to [2, 10–12].

Theorem 9. *Let (X, d) be a metric space and C a nonempty subset of X. Let $S, T : C \to C$ be a pair of mappings such that*

(i) $T(C) \subseteq S(C)$;

(ii) $d(Tx, Ty) \leq k \max\{d(Sx, Sy), d(Sx, Tx), d(Sy, Ty), (d(Sx, Ty)+d(Sy, Tx))/2\}$, $k \in (0, 1)$;

(iii) *the pair (S, T) is R-weakly commuting on C.*

If $\mathrm{cl}(T(C))$ *(closure of $T(C)$), is complete and T is continuous, then $F(S) \cap F(T) \cap C$ is a singleton.*

Now we obtain a more general version of the above theorem, where the continuity condition on T has been dispensed with and the completeness of $\mathrm{cl}(T(C))$ has been replaced by completeness of $T(C)$.

Theorem 10. *Let (X, d) be a metric space and C a nonempty subset of X. Let $S, T : C \to C$ be a pair of mappings such that*

(i) $T(C) \subseteq S(C)$;

(ii) $d(Tx, Ty) \leq k \max\{d(Sx, Sy), d(Sx, Tx), d(Sy, Ty), (d(Sx, Ty)+d(Sy, Tx))/2\}$, $k \in (0, 1)$;

(iii) *the pair (S, T) is R-weakly commuting on C.*

Then we have the following:

(a) *$F(S) \cap F(T) \cap T(C)$ is a singleton if $T(C)$ is complete,*

(b) *$F(S) \cap F(T) \cap S(C)$ is a singleton if $S(C)$ is complete.*

Proof. Pick $x_0 \in C$. Since $T(C) \subseteq S(C)$, we can construct a sequence $\{x_n\}$ in C such that $Sx_n = Tx_{n-1}$ for all $n \in \mathbb{N}$ (see [12]). By (ii), we have

$$\begin{aligned}
&d(Sx_{n+1}, Sx_n)\\
&= d(Tx_n, Tx_{n-1})\\
&\leq k \max\Big\{d(Sx_n, Sx_{n-1}), d(Sx_n, Tx_n), d(Sx_{n-1}, Tx_{n-1}),\\
&\qquad \frac{d(Sx_n, Tx_{n-1})+d(Sx_{n-1}, Tx_n)}{2}\Big\}\\
&= k \max\Big\{d(Sx_n, Sx_{n-1}), d(Sx_n, Sx_{n+1}), d(Sx_{n-1}, Sx_n),\\
&\qquad \frac{d(Sx_n, Sx_n)+ d(Sx_{n-1}, Sx_{n+1})}{2}\Big\}\\
&= k \max\Big\{d(Sx_n, Sx_{n-1}), d(Sx_n, Sx_{n+1}), d(Sx_{n-1}, Sx_n),\\
&\qquad \frac{d(Sx_{n-1}, Sx_{n+1})}{2}\Big\}\\
&\leq k \max\Big\{d(Sx_n, Sx_{n-1}), d(Sx_n, Sx_{n+1}), d(Sx_{n-1}, Sx_n),\\
&\qquad \frac{d(Sx_{n-1}, Sx_n) + d(Sx_n, Sx_{n+1})}{2}\Big\},\\
&= k \max\Big\{d(Sx_n, Sx_{n-1}),\\
&\qquad \frac{d(Sx_n, Sx_{n+1})+d(Sx_{n-1}, Sx_n)}{2}\Big\}.
\end{aligned} \tag{17}$$

Now if

$$\max\left\{d(Sx_n, Sx_{n-1}), \frac{d(Sx_n, Sx_{n+1}) + d(Sx_{n-1}, Sx_n)}{2}\right\} = \frac{d(Sx_n, Sx_{n+1}) + d(Sx_{n-1}, Sx_n)}{2}, \tag{18}$$

then

$$d(Sx_{n+1}, Sx_n) \leq \frac{k}{2}\left[d(Sx_n, Sx_{n-1}) + d(Sx_{n+1}, Sx_n)\right]. \tag{19}$$

Therefore

$$d(Sx_{n+1}, Sx_n) \leq kd(Sx_n, Sx_{n-1}). \tag{20}$$

Since $k < 1$, $\{Sx_n\}$ is a Cauchy sequence in C (see [1, 5]).

(a) Suppose that $T(C)$ is complete. Then there exists a point $z \in T(C)$ such that $Tx_n \to z \in T(C)$. Thus, $Sx_n \to z$. Since $z \in T(C) \subseteq S(C)$, there exists $u \in C$ such that $z = Su$. Again by (ii), we have

$$\begin{aligned}
&d(Tu, Tx_n)\\
&\quad\leq k \max\Big\{d(Su, Sx_n), d(Su, Tu),\\
&\qquad d(Sx_n, Tx_n), \frac{d(Su, Tx_n) + d(Sx_n, Tu)}{2}\Big\}.
\end{aligned} \tag{21}$$

Making $n \to \infty$, yields

$$\begin{aligned}
d(Tu, Su) &\leq k \max\left\{0, d(Su, Tu), \frac{d(Su, Tu)}{2}\right\}\\
&= kd(Su, Tu) < d(Su, Tu),
\end{aligned} \tag{22}$$

a contradiction. Therefore $d(Tu, Su) = 0$ and $Su = Tu = z$. Since the pair (S, T) is R-weakly commuting on C, it follows that

$$d(STu, TSu) \leq Rd(Su, Tu) = Rd(z, z) = 0. \tag{23}$$

Therefore $d(STu, TSu) = 0$ and $Sz = Tz$. Again by (ii), we have

$$d(Tz, Tx_n) \le k \max\left\{d(Sz, Sx_n), d(Sz, Tz), d(Sx_n, Tx_n), \frac{d(Sz, Tx_n) + d(Sx_n, Tz)}{2}\right\}. \quad (24)$$

Making $n \to \infty$, yields

$$\begin{aligned} d(Tz, z) &\le k \max\left\{d(Sz, z), d(Sz, Tz), d(z, z), \frac{d(Sz, z) + d(z, Tz)}{2}\right\} \\ &= k \max\{d(Sz, z), 0, d(Sz, z)\} \\ &= kd(Sz, z) = kd(Tz, z), \end{aligned} \quad (25)$$

which implies $z = Tz = Sz$. Using (ii), uniqueness of z can be proved. Since $z \in T(C)$, we conclude that $F(S) \cap F(T) \cap T(C) = \{z\}$.

(b) Suppose $S(C)$ is complete. Then $Sx_n \to z$ for some $z \in S(C)$ and there exist $u \in C$ such that $z = Su$. As in part (a), we can show that $Sz = Tz = z$. Thus $F(S) \cap F(T) \cap S(C) = \{z\}$. □

Recently Kim et al. [1] obtained the following result for S-nonexpansive type mappings in a normed space.

Theorem 11. *Let C be a nonempty q-star shaped subset of a normed space X and $S, T : C \to C$ two mappings satisfying the following conditions:*

(i) *the mapping T is S-nonexpansive and S is q-affine with $q \in F(S)$;*
(ii) *$T(C) \subseteq S(C)$;*
(iii) *the pair (S, T) is R-subweakly commuting.*

Suppose $S(C)$ is compact. Then we have the following:

(a) S and T have a coincidence point $y \in S(C)$.
(b) If S or T is demicontinuous, then $y \in F(S) \cap F(T)$.

We extend the above theorem to a more general class of nonexpansive type mappings. In the sequel we will need the following Lemma 12 and Proposition 13.

Lemma 12. *Let (X, d) be a metric space and C a nonempty subset of X. Let $S, T : C \to C$ be a pair of mappings such that*

(i) *$T(C) \subseteq S(C)$;*
(ii) *$d(Tx, Ty) \le k \max\{d(Sx, Sy), (d(Sx, Tx) + d(Sy, Ty))/2, (d(Sx, Ty) + d(Sy, Tx))/2\}$, $k \in (0, 1)$;*
(iii) *the pair (S, T) is R-weakly commuting on C.*

Then we have the following:

(a) *$F(S) \cap F(T) \cap T(C)$ is a singleton if $T(C)$ is complete;*
(b) *$F(S) \cap F(T) \cap S(C)$ is a singleton if $S(C)$ is complete.*

Proof. The proof can be completed on the lines of the proof of Theorem 10, where

$$\max\left\{d(Sx, Sy), d(Sx, Tx), d(Sy, Ty), \frac{d(Sx, Ty) + d(Sy, Tx)}{2}\right\} \quad (26)$$

is replaced by

$$\max\left\{d(Sx, Sy), \frac{d(Sx, Tx) + d(Sy, Ty)}{2}, \frac{d(Sx, Ty) + d(Sy, Tx)}{2}\right\}. \quad (27)$$

□

Proposition 13. *Let C be a nonempty q-star shaped subset of a normed space X and $S, T : C \to C$ two mappings such that*

(i) *T is a generalized S-nonexpansive type mapping and S is q-affine with $q \in F(S)$;*
(ii) *$T(C) \subseteq S(C)$;*
(iii) *the pair (S, T) is R-subweakly commuting;*
(iv) *$S(C)$ is complete.*

Then there exist exactly one point x_λ such that

$$x_\lambda = Sx_\lambda = (1 - \lambda) q + \lambda Tx_\lambda, \quad (28)$$

for all $\lambda \in (0, 1)$.

Proof. Define $T_\lambda : C \to C$ by $T_\lambda x = (1 - \lambda)q + \lambda Tx$, for all $x \in C$ and for each $\lambda \in (0, 1)$.

Since (S, T) is R-subweakly commuting and S is q-affine, we have

$$\begin{aligned} \|ST_\lambda x - T_\lambda Sx\| &= \|[(1 - \lambda) q + \lambda STx] - [(1 - \lambda) q + \lambda TSx]\| \\ &= \lambda \|TSx - STx\| \\ &\le \lambda R \|Sx - T_\lambda x\|, \end{aligned} \quad (29)$$

for all $x \in C$. Thus, the pair (S, T_λ) is R-weakly commuting on C.

Also

$$\|T_\lambda x - T_\lambda y\| = \lambda \|Tx - Ty\| \le \lambda \max\left\{\|Sx - Sy\|, \frac{\|Sx - Tx\| + \|Sy - Ty\|}{2}, \frac{\|Sx - Ty\| + \|Sy - Tx\|}{2}\right\}, \quad (30)$$

for all $x, y \in C$. For $x \in C$, we have $Tx \in T(C) \subseteq S(C)$, that is, there exist a point $y \in C$ such that $Tx = Sy \in S(C)$.

Observe that

$$T_\lambda x = (1-\lambda) q + \lambda Tx = (1-\lambda) q + \lambda Sy \in S(C). \quad (31)$$

It follows that $T_\lambda(C) \subseteq S(C)$ for all $\lambda \in (0,1)$. Now for each $\lambda \in (0,1)$, we conclude that

(i)* $T_\lambda(C) \subseteq S(C)$,

(ii)* $\|T_\lambda x - T_\lambda y\| \le \lambda \max\{\|Sx - Sy\|, (\|Sx - Tx\| + \|Sy - Ty\|)/2, (\|Sx - Ty\| + \|Sy - Tx\|)/2\}$,

(iii)* $S(C)$ is complete,

(iv)* (S, T_λ) is R-weakly commuting on C.

Therefore by Lemma 12, there exist exactly one point $x_\lambda \in S(C)$ such that

$$x_\lambda = Sx_\lambda = T_\lambda x_\lambda, \quad (32)$$

which implies that $x_\lambda = Sx_\lambda = (1-\lambda)q + \lambda T x_\lambda$. □

Now we obtain a common fixed point theorem for generalized S-nonexpansive type mappings.

Theorem 14. *Let C be a nonempty subset of a normed space X. Let $S, T : C \to C$ be two mappings satisfying conditions (i)–(iii) of Proposition 13. Suppose $S(C)$ is compact. Then we have the following:*

(a) *S and T have a coincidence point $y \in S(C)$,*

(b) *If S or T is demicontinuous, then $y \in F(S) \cap F(T)$.*

Proof. Let $\{\lambda_n\}$ be a sequence in $(0,1)$ such that $\lambda_n \to 1$. By Proposition 13, there exists exactly one point $x_{\lambda_n} \in S(C)$ such that

$$x_{\lambda_n} = Sx_{\lambda_n} = (1-\lambda_n) q + \lambda_n T x_{\lambda_n}, \quad (33)$$

for all $n \in \mathbb{N}$.

Set $x_{\lambda_n} := y_n$. Since $S(C)$ is compact, there exist a subsequence $\{y_{n_j}\}$ of $\{y_n\}$ such that

$$\lim_{j\to\infty} Sy_{n_j} = y \in S(C). \quad (34)$$

Thus $y = Su$ for some $u \in C$.

The assumption (ii) implies that $\{Ty_{n_j}\}$ is bounded. It follows that

$$\begin{aligned}\left\|y_{n_j} - Ty_{n_j}\right\| &= \left\|(1-\lambda_m) q + \lambda_m T y_{n_j} - Ty_{n_j}\right\| \\ &= (1-\lambda_m)\left\|q - Ty_{n_j}\right\| \longrightarrow 0, \quad \text{as } j \longrightarrow \infty. \end{aligned} \quad (35)$$

Thus $\lim_{j\to\infty} Ty_{n_j} = \lim_{j\to\infty} y_{n_j} = y$. By the condition (3), we have

$$\begin{aligned}&\left\|Ty_{n_j} - Tu\right\| \\ &\le \max\left\{\left\|Sy_{n_j} - Su\right\|, \frac{\left\|Sy_{n_j} - Ty_{n_j}\right\| + \left\|Su - Tu\right\|}{2}, \frac{\left\|Sy_{n_j} - Tu\right\| + \left\|Su - Ty_{n_j}\right\|}{2}\right\} \\ &= \max\left\{\left\|Sy_{n_j} - y\right\|, \frac{\left\|Sy_{n_j} - Ty_{n_j}\right\| + \left\|y - Tu\right\|}{2}, \frac{\left\|Sy_{n_j} - Tu\right\| + \left\|y - Ty_{n_j}\right\|}{2}\right\}.\end{aligned} \quad (36)$$

Making $j \to \infty$, we get

$$\|y - Tu\| \le \max\left\{0, \frac{1}{2}\|y-Tu\|, \frac{1}{2}\|y-Tu\|\right\} = \frac{1}{2}\|y-Tu\|, \quad (37)$$

a contradiction. Therefore $\|y - Tu\| = 0$ and $Tu = y$.

(a) Since the pair (S, T) is R-subweakly commuting, we have

$$\begin{aligned}\|STu - TSu\| &\le R \operatorname{dist}(Su, [Tu, q]) \\ &= R\left\|Su - [(1-\lambda_n) q + \lambda_n Tu\right\| \\ &= R\|Su - Su\| = 0.\end{aligned} \quad (38)$$

Thus $Sy = Ty$.

(b) Suppose S is demicontinuous. Since $\lim_{m\to\infty} x_m = \lim_{m\to\infty} Sx_m = y$, it follows from the demicontinuity of S that $Sy = y$. But $Sy = Ty$. Thus we conclude that $y \in F(S) \cap F(T)$. Similarly we can prove that $y \in F(S) \cap F(T)$ when T is demicontinuous. □

The following example shows the generality of Theorem 14 over Theorem 11.

Example 15. Let $X = \mathbb{R}$ (set of reals) with norm $\|x\| = |x|$ and $C = [0,4]$. Define $S, T : C \to C$ by

$$Tx = \begin{cases} 1, & \text{if } x \in \{1,3,4\}, \\ 2, & \text{if } x = 2, \\ \frac{3}{2}, & \text{otherwise}, \end{cases} \qquad Sx = \begin{cases} 1, & \text{if } x = 1, \\ 4, & \text{if } x \in \{2,3\}, \\ 2, & \text{if } x = 4, \\ \frac{1}{2}, & \text{otherwise}. \end{cases} \quad (39)$$

For $x = 2$ and $y = 3$, we have

$$\|Tx - Ty\| = 1 > 0 = \|Sx - Sy\|, \quad (40)$$

and the condition (2) of Theorem 11 is not satisfied. Further, it can be easily verified that S and T satisfy all the hypotheses of Theorem 14 and $S1 = T1 = 1$, is a common fixed point of S and T.

4. Convergence Results

Recently, Rhoades and Temir [3] obtained the following theorem.

Theorem 16. *Let X be a Banach space and C a closed convex subset of X which satisfies Opial's condition. Let $S, T : C \to C$ be mappings such that*

(i) *T is S-nonexpansive;*

(ii) *S is nonexpansive.*

Suppose $\{k_n\}$ is a real sequence in $(0, 1)$. Then the sequence of Mann iterates defined for an arbitrary $x_0 \in C$ by

$$x_{n+1} = (1 - k_n) x_n + k_n T x_n, \quad n \in \{0\} \cup \mathbb{N} \tag{41}$$

converges weakly to a common fixed point of S and T.

The following theorem extends Theorem 16 to generalized S-nonexpansive type mappings.

Theorem 17. *Let X be a Banach space and C a closed convex subset of X which satisfies Opial's condition. Let $S, T : C \to C$ be such that*

(i) *T is generalized S-nonexpansive type;*

(ii) *S is nonexpansive.*

Suppose $\{k_n\}$ is a real sequence in $(0, 1)$. Then the sequence of Mann iterates defined for an arbitrary $x_0 \in C$ by

$$x_{n+1} = (1 - k_n) x_n + k_n T x_n, \quad n \in \{0\} \cup \mathbb{N} \tag{42}$$

converges weakly to a common fixed point of S and T.

Proof. If $F(T) \cap F(S)$ is nonempty and singleton, then the proof is complete. Assume that $F(T) \cap F(S)$ is nonempty and that $F(T) \cap F(S)$ is not a singleton. Let $z \in F(T) \cap F(S)$. Then

$$\begin{aligned} \|x_{n+1} - z\| &= \|(1 - k_n) x_n + k_n T x_n - z\| \\ &= \|(1 - k_n)(x_n - z) + k_n (T x_n - z)\|. \end{aligned} \tag{43}$$

Since T is generalized S-nonexpansive type, we have

$$\|x_{n+1} - z\| \le (1 - k_n)\|x_n - z\| + k_n M(x_n, z). \tag{44}$$

Now the following cases arise.

Case 1. ($M(x_n, z) = \|Sx_n - Sz\|$). Then

$$\|Tx_n - z\| \le \|Sx_n - Sz\|. \tag{45}$$

Since S is nonexpansive on C, the above inequality reduces to

$$\|Tx_n - z\| \le \|x_n - z\|. \tag{46}$$

Case 2. ($M(x_n, z) = (\|Sx_n - Tx_n\| + \|Sz - Tz\|)/2$). Then

$$\begin{aligned} \|Tx_n - z\| &\le \frac{\|Sx_n - Tx_n\| + \|Sz - Tz\|}{2} \\ &= \frac{\|Sx_n - Tx_n\|}{2} \\ &\le \frac{\|Sx_n - z\| + \|z - Tx_n\|}{2} \\ &= \frac{\|Sx_n - Sz\| + \|z - Tx_n\|}{2}, \\ \|Tx_n - z\| &\le \|Sx_n - Sz\|. \end{aligned} \tag{47}$$

Nonexpansiveness of S on C implies

$$\|Tx_n - z\| \le \|x_n - z\|. \tag{48}$$

Case 3. ($M(x_n, z) = (\|Sx_n - Tz\| + \|Sz - Tx_n\|)/2$). Then

$$\begin{aligned} \|Tx_n - z\| &\le \frac{\|Sx_n - Tz\| + \|Sz - Tx_n\|}{2} \\ &= \frac{\|Sx_n - Tz\| + \|Tx_n - z\|}{2} \\ &= \frac{\|Sx_n - Sz\| + \|Tx_n - z\|}{2}, \\ \|Tx_n - z\| &\le \|Sx_n - Sz\|. \end{aligned} \tag{49}$$

Again, since S is nonexpansive on C, it follows that

$$\|Tx_n - z\| \le \|x_n - z\|. \tag{50}$$

Therefore in all the cases, we get

$$\|Tx_n - z\| \le \|x_n - z\|. \tag{51}$$

By (44) and (51), we get

$$\begin{aligned} \|x_{n+1} - z\| &\le (1 - k_n)\|x_n - z\| + k_n \|x_n - z\| \\ &= \|x_n - z\|. \end{aligned} \tag{52}$$

Thus, for $k_n \neq 0$, $\{\|x_n - z\|\}$ is a nonincreasing sequence. Hence, $\lim_{n\to\infty}\|x_n - z\|$ exists. Now we show that $\{x_n\}$ converges weakly to a common fixed point of S and T. Let $\{x_{n_k}\}$ and $\{x_{m_k}\}$ be two subsequences of $\{x_n\}$ which converge weakly to z and $\tilde{z}$, respectively. We will show that $z = \tilde{z}$. Suppose the contrary. Since X satisfies Opial's condition and $\lim_{n\to\infty}\|x_n - z\|$ exists for any $z \in F(T) \cap F(S)$, we get

$$\begin{aligned} \lim_{n\to\infty}\|x_n - z\| &= \lim_{k\to\infty}\|x_{n_k} - z\| \\ &< \lim_{k\to\infty}\|x_{n_k} - \tilde{z}\| \\ &= \lim_{n\to\infty}\|x_n - \tilde{z}\| \\ &= \lim_{j\to\infty}\|x_{m_j} - \tilde{z}\| \\ &< \lim_{j\to\infty}\|x_{m_j} - z\| = \lim_{n\to\infty}\|x_n - z\|, \end{aligned} \tag{53}$$

a contradiction. Hence $z = \tilde{z}$. □

Corollary 18. *Theorem 16.*

Proof. It comes from Theorem 17, when $M(x, y) = \|Sx - Sy\|$. □

Corollary 19. *Let X be a Banach space and C a closed convex subset of X which satisfies Opial's condition. Let $T : C \to C$ be generalized nonexpansive type mapping.*

Suppose $\{k_n\}$ is a real sequence in $(0, 1)$. Then the sequence of Mann iterates defined for an arbitrary $x_0 \in C$ by

$$x_{n+1} = (1 - k_n)\, x_n + k_n T x_n, \quad n \in \{0\} \cup \mathbb{N} \tag{54}$$

converges weakly to a common fixed point of S and T.

Proof. It comes from Theorem 17, when S is an identity mapping on X. □

Corollary 20. *Let X be a Banach space and C a closed convex subset of X which satisfies Opial's condition. Let $T : C \to C$ be a nonexpansive mapping.*

Suppose $\{k_n\}$ is a real sequence in $(0, 1)$. Then the sequence of Mann iterates defined for arbitrary $x_0 \in C$ defined

$$x_{n+1} = (1 - k_n)\, x_n + k_n T x_n, \quad n \in \{0\} \cup \mathbb{N} \tag{55}$$

converges weakly to a common fixed point of S and T.

Proof. It comes from Corollary 19 when $m(x, y) = \|x - y\|$. □

Acknowledgments

The authors would like to thank Professor M. Abbas for his useful comments. The authors would like to thank the referees for their constructive comments and useful suggestions.

References

[1] J. K. Kim, D. R. Sahu, and S. Anwar, "Browder's type strong convergence theorem for S-nonexpansive mappings," *Bulletin of the Korean Mathematical Society*, vol. 47, no. 3, pp. 503–511, 2010.

[2] S. Park, "On f-nonexpansive maps," *Journal of the Korean Mathematical Society*, vol. 16, no. 1, pp. 29–38, 1979.

[3] B. E. Rhoades and S. Temir, "Convergence theorems for I-nonexpansive mapping," *International Journal of Mathematics and Mathematical Sciences*, vol. 2006, Article ID 63435, 4 pages, 2006.

[4] W. R. Mann, "Mean value methods in iteration," *Proceedings of the American Mathematical Society*, vol. 4, pp. 506–510, 1953.

[5] N. Shahzad, "Invariant approximations, generalized I-contractions, and R-subweakly commuting maps," *Fixed Point Theory and Applications*, vol. 2005, no. 1, pp. 79–86, 2005.

[6] S. Sessa, "On a weak commutativity condition of mappings in fixed point considerations," *Publications de l'Institut Mathématique*, vol. 32, no. 46, pp. 149–153, 1982.

[7] R. P. Pant, "R-weak commutativity and common fixed points," *Soochow Journal of Mathematics*, vol. 25, no. 1, pp. 37–42, 1999.

[8] Z. Opial, "Weak convergence of the sequence of successive approximations for nonexpansive mappings," *Bulletin of the American Mathematical Society*, vol. 73, pp. 591–597, 1967.

[9] K. Goebel and W. A. Kirk, *Topics in Metric Fixed Point Theory*, Cambridge University Press, Cambridge, UK, 1990.

[10] M. A. Al-Thagafi and N. Shahzad, "Generalized I-nonexpansive selfmaps and invariant approximations," *Acta Mathematica Sinica*, vol. 24, no. 5, pp. 867–876, 2008.

[11] S. Reich, "Weak convergence theorems for nonexpansive mappings in Banach spaces," *Journal of Mathematical Analysis and Applications*, vol. 67, no. 2, pp. 274–276, 1979.

[12] N. Shahzad, "Invariant approximations and R-subweakly commuting maps," *Journal of Mathematical Analysis and Applications*, vol. 257, no. 1, pp. 39–45, 2001.

17

Bounded Nonlinear Functional Derived by the Generalized Srivastava-Owa Fractional Differential Operator

Rabha W. Ibrahim

Institute of Mathematical Sciences, University of Malaya, 50603 Kuala Lumpur, Malaysia

Correspondence should be addressed to Rabha W. Ibrahim; rabhaibrahim@yahoo.com

Academic Editor: Alain Miranville

By making use of the generalized Srivastava-Owa fractional differential operator, a class of analytical functions is imposed. The sharp bound for the nonlinear functional associated with the Hankel determinant is computed. We consider a new technique to prove our results. Important properties such as inclusion, subordination, and Hadamard product are studied. Some recent results are included.

1. Introduction

Fractional calculus (real and complex) is a rapidly growing subject of interest for physicists and mathematicians. The reason for this is that problems may be discussed in a much more stringent and elegant way than using traditional methods. Fractional differential equations have emerged as a new branch of applied mathematics which has been used for many mathematical models in science and engineering. In fact, fractional differential equations are considered as an alternative model to nonlinear differential equations. Several different derivatives were introduced: Riemann-Liouville, Hadamard, Grunwald-Letnikov, Riesz, Erdelyi-Kober operators, and Caputo [1–7].

Recently, the theory of fractional calculus has found interesting applications in the theory of analytic functions. The classical definitions of fractional operators and their generalizations have fruitfully been employed for imposing, for example, the characterization properties, coefficient estimates [8], distortion inequalities [9], and convolution structures for various subclasses of analytic functions and the works in the research monographs. In [10], Srivastava and Owa defined the fractional operators (derivative and integral) in the complex z-plane $\mathbb{C}$ as follows.

Definition 1. The fractional derivative of order α is defined, for a function $f(z)$ by

$$D_z^{\alpha} f(z) := \frac{1}{\Gamma(1-\alpha)} \frac{d}{dz} \int_0^z \frac{f(\zeta)}{(z-\zeta)^{\alpha}} d\zeta; \quad 0 \le \alpha < 1, \quad (1)$$

where the function $f(z)$ is analytical in simply-connected region of the complex z-plane $\mathbb{C}$ containing the origin and the multiplicity of $(z-\zeta)^{-\alpha}$ is removed by requiring $\log(z-\zeta)$ to be real when $(z-\zeta) > 0$.

Definition 2. The fractional integral of order α is defined, for a function $f(z)$, by

$$I_z^{\alpha} f(z) := \frac{1}{\Gamma(\alpha)} \int_0^z f(\zeta)(z-\zeta)^{\alpha-1} d\zeta; \quad \alpha > 0, \quad (2)$$

where the function $f(z)$ is analytical in simply connected region of the complex z-plane $(\mathbb{C})$ containing the origin and the multiplicity of $(z-\zeta)^{\alpha-1}$ is removed by requiring $\log(z-\zeta)$ to be real when $(z-\zeta) > 0$.

In [11], the author generalized a formula for the fractional integral as follows: for natural $n \in \mathbb{N} = \{1, 2, \ldots\}$ and real μ, the n-fold integral of the form

$$I_z^{\alpha,\mu} f(z) = \int_0^z \zeta_1^\mu d\zeta_1 \int_0^{\zeta_1} \zeta_2^\mu d\zeta_2 \cdots \int_0^{\zeta_{n-1}} \zeta_n^\mu f(\zeta_n)\, d\zeta_n. \quad (3)$$

Employing the Dirichlet technique implies

$$\int_0^z \zeta_1^\mu d\zeta_1 \int_0^{\zeta_1} \zeta^\mu f(\zeta)\, d\zeta = \int_0^z \zeta^\mu f(\zeta)\, d\zeta \int_\zeta^z \zeta_1^\mu d\zeta_1 = \frac{1}{\mu+1}\int_0^z \left(z^{\mu+1} - \zeta^{\mu+1}\right)\zeta^\mu f(\zeta)\, d\zeta. \quad (4)$$

Repeating the above step $n-1$ times yields

$$\int_0^z \zeta_1^\mu d\zeta_1 \int_0^{\zeta_1} \zeta_2^\mu d\zeta_2 \cdots \int_0^{\zeta_{n-1}} \zeta_n^\mu f(\zeta_n)\, d\zeta_n = \frac{(\mu+1)^{1-n}}{(n-1)!}\int_0^z \left(z^{\mu+1} - \zeta^{\mu+1}\right)^{n-1}\zeta^\mu f(\zeta)\, d\zeta, \quad (5)$$

which imposes the fractional operator type

$$I_z^{\alpha,\mu} f(z) = \frac{(\mu+1)^{1-\alpha}}{\Gamma(\alpha)}\int_0^z \left(z^{\mu+1} - \zeta^{\mu+1}\right)^{\alpha-1}\zeta^\mu f(\zeta)\, d\zeta, \quad (6)$$

where α and $\mu \neq -1$ are real numbers and the function $f(z)$ is analytic in simply connected region of the complex z-plane $\mathbb{C}$ containing the origin and the multiplicity of $(z^{\mu+1} - \zeta^{\mu+1})^{-\alpha}$ is removed by requiring $\log(z^{\mu+1} - \zeta^{\mu+1})$ to be real when $(z^{\mu+1} - \zeta^{\mu+1}) > 0$. When $\mu = 0$, we arrive at the standard Srivastava-Owa fractional integral. Further information can be found in [11].

Corresponding to the fractional integral operator, the fractional differential operator is

$$D_z^{\alpha,\mu} f(z) := \frac{(\mu+1)^{\alpha}}{\Gamma(1-\alpha)}\frac{d}{dz}\int_0^z \frac{\zeta^\mu f(\zeta)}{\left(z^{\mu+1} - \zeta^{\mu+1}\right)^{\alpha}} d\zeta; \quad 0 \le \alpha < 1, \quad (7)$$

where the function $f(z)$ is analytical in simply connected region of the complex z-plane $\mathbb{C}$ containing the origin and the multiplicity of $(z^{\mu+1} - \zeta^{\mu+1})^{-\alpha}$ is removed by requiring $\log(z^{\mu+1} - \zeta^{\mu+1})$ to be real when $(z^{\mu+1} - \zeta^{\mu+1}) > 0$. We have

$$D_z^{\alpha,\mu} z^\nu = \frac{(\mu+1)^{\alpha-1}\Gamma((\nu/(\mu+1))+1)}{\Gamma((\nu/(\mu+1))+1-\alpha)} z^{(1-\alpha)(\mu+1)+\nu-1}. \quad (8)$$

Let $\mathcal{A}$ denote the class of functions $f(z)$ normalized by

$$f(z) = z + \sum_{n=2}^{\infty} a_n z^n, \quad z \in U. \quad (9)$$

Also, let $\mathcal{S}, \mathcal{S}^*$ and $\mathcal{C}$ denote the subclasses of $\mathcal{A}$ consisting of functions which are, respectively, univalent, starlike $\Re(zf'(z)/f(z)) > 0$, and convex $\Re(1 + (z^2 f''(z)/f'(z))) > 0$ in U. It is well known that, if the function $f(z)$ given by (9) is in the class $\mathcal{S}$, then $|a_n| \le n$, $n \in \mathbb{N} \setminus \{1\}$. Moreover, if the function $f(z)$ given by (9) is in the class $\mathcal{C}$, then $|a_n| \le 1, n \in \mathbb{N}$.

In our present investigation, we will also make use of the Fox-Wright generalization ${}_q\Psi_p[z]$ of the hypergeometric ${}_qF_p$ function defined by [12]

$$\begin{aligned} {}_q\Psi_p\left[\begin{matrix}(\alpha_1, A_1), \ldots, (\alpha_q, A_q); \\ (\beta_1, B_1), \ldots, (\beta_p, B_p);\end{matrix}\; z\right] &= {}_q\Psi_p\left[(\alpha_j, A_j)_{1,q}; (\beta_j, B_j)_{1,p}; z\right] \\ &:= \sum_{n=0}^{\infty} = \frac{\Gamma(\alpha_1 + nA_1)\cdots\Gamma(\alpha_q + nA_q)}{\Gamma(\beta_1 + nB_1)\cdots\Gamma(\beta_p + nB_p)}\frac{z^n}{n!} \\ &= \sum_{n=0}^{\infty}\frac{\prod_{j=1}^{q}\Gamma(\alpha_j + nA_j)}{\prod_{j=1}^{q}\Gamma(\beta_j + nB_j)}\frac{z^n}{n!}, \end{aligned} \quad (10)$$

where $A_j > 0$ for all $j = 1, \ldots, q$, $B_j > 0$ for all $j = 1, \ldots, p$, and $1 + \sum_{j=1}^{p} B_j - \sum_{j=1}^{q} A_j \ge 0$ for suitable values $|z| < 1$, and α_i, β_j are complex parameters.

It is well known that

$${}_q\Psi_p\left[\begin{matrix}(\alpha_1, 1), \ldots, (\alpha_q, 1); \\ (\beta_1, 1), \ldots, (\beta_p, 1);\end{matrix}\; z\right] = \Lambda^{-1}{}_qF_p(\alpha_1, \ldots, \alpha_q, \beta_1, \ldots, \beta_p; z), \quad (11)$$

where

$$\Lambda := \frac{\prod_{j=1}^{p}\Gamma(\beta_j)}{\prod_{i=1}^{q}\Gamma(\alpha_i)}, \quad (12)$$

and ${}_qF_p$ is the generalized hypergeometric function.

Now by making use of the operator (7), we introduce the following extension operator $\Phi^{\alpha,\mu} : \mathcal{A} \to \mathcal{A}$:

$$\begin{aligned} \Phi^{\alpha,\mu} f(z) &:= \frac{\Gamma((1/(\mu+1)) + 1 - \alpha)}{(\mu+1)^{\alpha-1}\Gamma((1/(\mu+1)) + 1)} \times z^{\alpha-\mu+\mu\alpha} D_z^{\alpha,\mu} f(z) \\ &= \frac{\Gamma((1/(\mu+1)) + 1 - \alpha)}{(\mu+1)^{\alpha-1}\Gamma((1/(\mu+1)) + 1)} z^{\alpha-\mu+\mu\alpha} D_z^{\alpha,\mu} \times \left(z + \sum_{n=2}^{\infty} a_n z^n\right) \\ &= \frac{\Gamma((1/(\mu+1)) + 1 - \alpha)}{(\mu+1)^{\alpha-1}\Gamma((1/(\mu+1)) + 1)} z^{\alpha-\mu+\mu\alpha} \end{aligned}$$

$$\times\left[\frac{(\mu+1)^{\alpha-1}\Gamma((1/(\mu+1))+1)}{\Gamma((1/(\mu+1))+1-\alpha)}z^{-\alpha+\mu-\mu\alpha+1}\right.$$
$$\left.+\sum_{n=2}^{\infty}\frac{\Gamma((n/(\mu+1))+1)}{\Gamma((n/(\mu+1))+1-\alpha)}a_n z^{n-\alpha+\mu-\mu\alpha}\right]$$
$$= z+\sum_{n=2}^{\infty}\frac{\Gamma((1/(\mu+1))+1-\alpha)}{\Gamma((1/(\mu+1))+1)}$$
$$\times\frac{\Gamma((n/(\mu+1))+1)}{\Gamma((n/(\mu+1))+1-\alpha)}a_n z^n$$
$$:= z+\sum_{n=2}^{\infty}\phi_n^{\alpha,\mu}a_n z^n. \tag{13}$$

Obviously, when $\mu = 0$, we have the extension fractional differential operator defined in [13] ([14] for recent work), which contains the Carlson-Shaffer operator. In term of the Fox-Wright generalized function,

$$\Phi^{\alpha,\mu}f(z) = z+\sum_{n=2}^{\infty}\frac{\Gamma(n+1)\Gamma((1/(\mu+1))+1-\alpha)}{\Gamma((1/(\mu+1))+1)}$$
$$\times\frac{\Gamma((n/(\mu+1))+1)}{\Gamma((n/(\mu+1))+1-\alpha)}\frac{a_n}{n!}z^n$$
$$=\sum_{n=0}^{\infty}\frac{\Gamma(n+1)\Gamma((1/(\mu+1))+1-\alpha)}{\Gamma((1/(\mu+1))+1)}$$
$$\times\frac{\Gamma((n/(\mu+1))+1)}{\Gamma((n/(\mu+1))+1-\alpha)}\frac{a_n}{n!}z^n$$
$$=\frac{\Gamma((1/(\mu+1))+1-\alpha)}{\Gamma((1/(\mu+1))+1)}$$
$$\times{}_2\Psi_1\left[\begin{matrix}(1,1),\left(1,\dfrac{1}{\mu+1}\right);\\ \left(1-\alpha,\dfrac{1}{\mu+1}\right);\end{matrix}\; z\right]*f(z)$$
$$=\frac{\Gamma((1/(\mu+1))+1-\alpha)}{\Gamma((1/(\mu+1))+1)}\,{}_2\Psi_1[z]*f(z)$$
$$:=\Psi(\alpha,\mu;z)*f(z), \tag{14}$$

where $a_0 = 0$, $a_1 = 1$, and $*$ is the Hadamard product. Note that

$$\Phi^{0,0}f(z)=f(z),\qquad \Phi^{1,0}f(z)=zf'(z),$$
$$\left(\Phi^{\alpha,0}f\right)(z)=(L(2,2-\alpha)f)(z), \tag{15}$$

where $L(a,c)f$ is the Carlson-Shaffer operator. Moreover, operator (14) can be viewed as a linear operator which is essentially analogous to the Dziok-Srivastava operator whenever used instead of the Fox-Wright generalization of the hypergeometric function.

Recently, various results, such as convolution and inclusion properties, distortion theorem, extreme points, and coefficient estimates, are proposed by many authors for the operators due to Srivastava involving the Wright function, generalized hypergeometric function, and Meijer's G-functions. These operators are Dziok-Srivastava, Srivastava-Wright, Cho-Kwon-Srivastava operator, Cho-Saigo-Srivastava operator, Jung-Kim-Srivastava, and Srivastava-Owa operators (see [15–24]). Going on in this generalization, we have finally the Erdelyi-Kober operator of fractional integration with three parameters used in [25].

Definition 3 (subordination principal). For two functions f and g analytical in U, we say that the function $f(z)$ is subordinated to $g(z)$ in U and write $f(z) \prec g(z) (z \in U)$, if there exists a Schwartz function $w(z)$ analytical in U with $w(0) = 0$, and $|w(z)| < 1$, such that $f(z) = g(w(z)), z \in U$. In particular, if the function $g(z)$ is univalent in U, the above subordination is equivalent to $f(0) = g(0)$ and $f(U) \subset g(U)$.

Definition 4. For the function f defined by (9), the Hankel determinant of f is defined by

$$\begin{vmatrix} a_n & a_{n+1} & \cdots & a_{n+q-1}\\ a_{n+1} & a_{n+2} & \cdots & a_{n+q}\\ \vdots & \vdots & \vdots & \vdots\\ a_{n+q-1} & a_{n+q} & \cdots & a_{n+2q-2}\end{vmatrix}. \tag{16}$$

Now we proceed to define a new class of analytic function involving the operator (13).

Definition 5. The function $f \in \mathscr{A}$ is said to be in the class $\mathscr{R}_{\alpha,\mu}(\theta,\rho)$, where $0 \le \alpha < 1$, $\mu \ge 0$, $|\theta| < \pi/2$, $0 \le \rho \le 1$, if it satisfies the inequality

$$\Re\left\{e^{i\theta}\frac{\Phi^{\alpha,\mu}f(z)}{z}\right\} > \rho\cos\theta,\quad (z\in U). \tag{17}$$

Consequently, from Definition 4, we have

$$f\in\mathscr{R}_{\alpha,\mu}(\theta,\rho)\Longleftrightarrow e^{i\theta}\frac{\Phi^{\alpha,\mu}f(z)}{z}$$
$$=[(1-\rho)p(z)+\rho]\cos\theta+i\sin\theta, \tag{18}$$

where $p(z) = 1+c_1z+c_2z^2+\cdots$, $z \in U$ satisfies the following properties [26]:

$$\text{(i)}\ |c_n|\le 2 \text{ and } \Re(p(z))>0,$$
$$\text{(ii)}\ 2c_2=c_1^2+w\left(4-c_1^2\right),$$
$$\text{(iii)}\ 4c_3=c_1^3+2\left(4-c_1^2\right)c_1w-c_1\left(4-c_1^2\right)w^2$$
$$+2\left(4-c_1^2\right)\left(1-|w|^2\right)z,\quad (|w|\le1,|z|\le1). \tag{19}$$

We denote this class by $\mathscr{P}$.

Note that

$$\mathscr{R}_{\alpha,0}(0,\rho)=\Re\left\{\frac{\Phi^{\alpha,0}f(z)}{z}\right\}>\rho \tag{20}$$

(see [27]),

$$\mathscr{R}_{\alpha,0}(\theta,\rho) = \mathbb{R}\left\{e^{i\theta}\frac{\Phi^{\alpha,0}f(z)}{z}\right\} > \rho\cos\theta \qquad (21)$$

(see [28]).

It is well known that, for the univalent function f of the form (9), the sharp inequality $|a_3 - a_2^2| \le 1$ holds. In the recent paper, we assume the Hankel determinant for $n = 2$, $q = 2$ and calculate the sharp bound for the functional $|a_2a_4 - a_3^2|$ for $f \in \mathscr{R}_{\alpha,\mu}(\theta,\rho)$. Properties of this class are illustrated, and some well-known results are generalized. For this purpose, we need the following preliminary in the sequel, which can be found in [29].

Lemma 6. *Let ϕ and ψ be univalent convex in U. Then, the Hadamard product $\phi * \psi$ is also univalent convex function in U.*

Lemma 7. *Let Φ and Ψ be univalent convex in U, and $\phi \prec \Phi$ and $\psi \prec \Psi$. Then, $\phi * \psi \prec \Phi * \Psi$.*

Lemma 8. *Let ϕ and ψ be starlike of order 1/2 then, for function Φ satisfying $\mathbb{R}(\Phi(z)) > \sigma(\sigma \in [0,1))$,*

$$\mathbb{R}\left(\frac{\phi(z) * \Phi(z)\psi(z)}{\phi(z) * \psi(z)}\right) > \sigma, \quad (z \in U). \qquad (22)$$

2. Main Results

We have the following result.

Theorem 9. *Let the function f be in the class $\mathscr{R}_{\alpha,\mu}(\theta,\rho)$. Then*

$$\left|a_2a_4 - a_3^2\right| \le \frac{(1-\rho)^2\cos^2\theta}{\mu_1^2\mu_2\mu_4\mu_3^2}\left(80\beta + 16\mu_3^2\right), \qquad (23)$$

where $\beta := |\mu_3^2 - \mu_2\mu_4|$ and

$$\mu_1 = \frac{\Gamma((1/(\mu+1)) + 1 - \alpha)}{\Gamma((1/(\mu+1)) + 1)},$$
$$\mu_2 = \frac{\Gamma((2/(\mu+1)) + 1)}{\Gamma((2/(\mu+1)) + 1 - \alpha)},$$
$$\mu_3 = \frac{\Gamma((3/(\mu+1)) + 1)}{\Gamma((3/(\mu+1)) + 1 - \alpha)},$$
$$\mu_4 = \frac{\Gamma((4/(\mu+1)) + 1)}{\Gamma((4/(\mu+1)) + 1 - \alpha)}. \qquad (24)$$

The estimate (23) is sharp.

Proof. Since $f \in \mathscr{R}_{\alpha,\mu}(\theta,\rho)$, then

$$e^{i\theta}\frac{\Phi^{\alpha,\mu}f(z)}{z} = \left[(1-\rho)\left(1 + c_1z + c_2z^2 + \cdots\right) + \rho\right]\cos\theta + i\sin\theta. \qquad (25)$$

Comparing the coefficients of (13) and (25), we receive

$$a_2 = \frac{(1-\rho)c_1\cos\theta}{e^{i\theta}\mu_1\mu_2}, \qquad a_3 = \frac{(1-\rho)c_2\cos\theta}{e^{i\theta}\mu_1\mu_3},$$
$$a_4 = \frac{(1-\rho)c_3\cos\theta}{e^{i\theta}\mu_1\mu_4}, \qquad (26)$$

where

$$\mu_1 = \frac{\Gamma((1/(\mu+1)) + 1 - \alpha)}{\Gamma((1/(\mu+1)) + 1)},$$
$$\mu_2 = \frac{\Gamma((2/(\mu+1)) + 1)}{\Gamma((2/(\mu+1)) + 1 - \alpha)},$$
$$\mu_3 = \frac{\Gamma((3/(\mu+1)) + 1)}{\Gamma((3/(\mu+1)) + 1 - \alpha)},$$
$$\mu_4 = \frac{\Gamma((4/(\mu+1)) + 1)}{\Gamma((4/(\mu+1)) + 1 - \alpha)}. \qquad (27)$$

Therefore, (26) implies

$$\left|a_2a_4 - a_3^2\right| = \frac{(1-\rho)^2\cos^2\theta}{\mu_1^2}\left|\frac{c_1c_3}{\mu_2\mu_4} - \frac{c_2^2}{\mu_3^2}\right|. \qquad (28)$$

By letting $c_1 := c$ and using (i)–(iii), we have

$$\begin{aligned}\left|a_2a_4 - a_3^2\right| = {} & \frac{(1-\rho)^2\cos^2\theta}{\mu_1^2\mu_2\mu_4\mu_3^2} \\ & \times\left|c^4\left(\mu_3^2 - \mu_2\mu_4\right) + 2wc^2\left(4-c^2\right)\right. \\ & \times\left(\mu_3^2 - \mu_2\mu_4\right) - w^2\left(4-c^2\right) \\ & \times\left(\mu_3^2c^2 + \mu_2\mu_4\left(4-c^2\right)\right) \\ & \left.+\left(1-|w|^2\right)z\left(2c\left(4-c^2\right)\mu_3^2\right)\right|.\end{aligned} \qquad (29)$$

By employing the triangle inequality and assuming $|w| := x$, $\beta := |\mu_3^2 - \mu_2\mu_4|$, $c > 0$, and $|z| \le 1$, we obtain

$$\begin{aligned}\left|a_2a_4 - a_3^2\right| \le {} & \frac{(1-\rho)^2\cos^2\theta}{\mu_1^2\mu_2\mu_4\mu_3^2} \\ & \times\left\{c^4\beta + 2xc^2\left(4-c^2\right)\beta\right. \\ & + x^2\left(4-c^2\right)\beta\left(c^2+2c\right) \\ & \left.+2c\left(4-c^2\right)\mu_3^2\right\} := F(x,c).\end{aligned} \qquad (30)$$

Our aim is to maximize F in the interior of the domain $\mathscr{D} = [0,1] \times [0,2]$. Since

$$\frac{\partial F(x,c)}{\partial x} = \frac{(1-\rho)^2\cos^2\theta}{\mu_1^2\mu_2\mu_4\mu_3^2}\left\{2c^2\left(4-c^2\right)\beta + 2x\left(4-c^2\right)\beta\left(c^2+2c\right)\right\} > 0, \qquad (31)$$

thus F cannot have a maximum in the interior of $\mathscr{D}$. Furthermore,

$$\max_{x\in[0,1]} F(x,c) = F(1,c) := H(c), \tag{32}$$

where

$$H(c) \le \frac{(1-\rho)^2\cos^2\theta}{\mu_1^2\mu_2\mu_4\mu_3^2}\left[16\beta + 16\beta\left(4-c^2\right) + 4\left(4-c^2\right)\mu_3^2\right] := G(c). \tag{33}$$

But

$$\max_{c\in[0,2]} G(c) = G(0); \tag{34}$$

hence the upper bound of (28) is

$$\left|a_2a_4 - a_3^2\right| \le \frac{(1-\rho)^2\cos^2\theta}{\mu_1^2\mu_2\mu_4\mu_3^2}\left(80\beta + 16\mu_3^2\right). \tag{35}$$

The equality holds for the functions

$$f(z) = \Psi(\alpha,\mu;z) * e^{i\theta}\left[z\left(\frac{1+(1-2\rho)z^2}{1-z^2}\cos\theta + i\sin\theta\right)\right]. \tag{36}$$

Remark 10. Letting $\mu = 0$, we receive a recent result due to Mishra and Gochhayat [28]; putting $\alpha \to 1$, $\mu = 0$, $\rho = 0$, we obtain a result given by Janteng et al. [30].

Theorem 11. *Assume that $\theta \in (-\pi/2,\pi/2)$, $\rho \in [0,1)$ and $\alpha_1,\alpha_2 \in [0,1)$, with $\alpha_1 < \alpha_2$. If the subordination*

$$z\mathscr{G}''\left(\frac{1}{\mu+1}+1-\alpha_2, \frac{1}{\mu+1}+1-\alpha_1; z\right) + \mathscr{G}'\left(\frac{1}{\mu+1}+1-\alpha_2, \frac{1}{\mu+1}+1-\alpha_1; z\right) \prec 1+z, \tag{37}$$

where

$$\mathscr{G}(a,c;z) := \sum_{n=0}^{\infty}\frac{(a)_n}{(c)_n}z^{n+1}, \quad (c \ne 0,-1,-2,\ldots), \tag{38}$$

holds, then

$$\mathscr{R}_{\alpha_2,\mu}(\theta,\rho) \subset \mathscr{R}_{\alpha_1,\mu}(\theta,\rho). \tag{39}$$

Proof. Let $f \in \mathscr{R}_{\alpha_2,\mu}(\theta,\rho)$. We rewrite

$$\begin{aligned}\Phi^{\alpha_1,\mu}f(z) &= \Psi(\alpha_1,\mu;z) * f(z) \\ &= \left(\Psi^{(-1)}(\alpha_2,\mu;z) * \Psi(\alpha_2,\mu;z)\right) * \Psi(\alpha_1,\mu;z) * f(z) \\ &= \left(\Psi^{(-1)}(\alpha_2,\mu;z) * \Psi(\alpha_1,\mu;z)\right) * \Psi(\alpha_2,\mu;z) * f(z) \\ &= \mathscr{G}\left(\frac{1}{\mu+1}+1-\alpha_2, \frac{1}{\mu+1}+1-\alpha_1; z\right) * \Phi^{\alpha_2,\mu}f(z),\end{aligned} \tag{40}$$

where

$$\begin{aligned}\Psi(\alpha_2,\mu;z) &= \sum_{n=0}^{\infty}\frac{\Gamma(n+1)\Gamma((1/(\mu+1))+1-\alpha)}{\Gamma((1/(\mu+1))+1)} \times \frac{\Gamma((n/(\mu+1))+1)}{\Gamma((n/(\mu+1))+1-\alpha)}\frac{1}{n!}z^n, \\ \Psi^{(-1)}(\alpha_2,\mu;z) &= \sum_{n=0}^{\infty}\frac{\Gamma((1/(\mu+1))+1)}{\Gamma(n+1)\Gamma((1/(\mu+1))+1-\alpha)} \times \frac{\Gamma((n/(\mu+1))+1-\alpha)}{\Gamma((n/(\mu+1))+1)}n!\,z^n, \\ \mathscr{G}(a,c;z) &= \sum_{n=0}^{\infty}\frac{(a)_n}{(c)_n}z^{n+1} \quad (c \ne 0,-1,-2,\ldots),\end{aligned} \tag{41}$$

and $(x)_k = \Gamma(x+k)/\Gamma(x)$ is a Pochhammer symbol. Therefore,

$$\frac{e^{i\theta}\Phi^{\alpha_1,\mu}f(z)}{z} = \frac{\mathscr{G}((1/(\mu+1))+1-\alpha_2,(1/(\mu+1))+1-\alpha_1;z) * \left(e^{i\theta}\Phi^{\alpha_2,\mu}f(z)/z\right)z}{\mathscr{G}((1/(\mu+1))+1-\alpha_2,(1/(\mu+1))+1-\alpha_1;z) * z}. \tag{42}$$

Assumption (37) implies that $\mathscr{G}((1/(\mu+1))+1-\alpha_2,(1/(\mu+1))+1-\alpha_1;z)$ is convex (see [31, Theorem 1.9]) and consequently $\mathscr{G}((1/(\mu+1))+1-\alpha_2,(1/(\mu+1))+1-\alpha_1;z) \in \mathscr{S}^*(1/2)$ (Marx-Strohhäcker Theorem [32]). Moreover, the function $\psi(z) = z$ is starlike of order $1/2$, then in view of Lemma 8, we obtain that

$$\Re\left(\frac{e^{i\theta}\Phi^{\alpha_1,\mu}f(z)}{z}\right) > \rho\cos\theta, \tag{43}$$

and consequently $\mathscr{R}_{\alpha_2,\mu}(\theta,\rho) \subset \mathscr{R}_{\alpha_1,\mu}(\theta,\rho)$.

Remark 12. Condition (37) can be replaced by another condition to obtain the convexity of the function $\mathscr{G}$, such that

$$\left|z\mathscr{G}''\left(\frac{1}{\mu+1}+1-\alpha_2, \frac{1}{\mu+1}+1-\alpha_1; z\right)\right| < \frac{1}{2} \tag{44}$$

yields that $\mathscr{G}((1/(\mu+1))+1-\alpha_2,(1/(\mu+1))+1-\alpha_1;z)$ is convex (see [31]).

Theorem 13. *Let $f \in \mathscr{S}^*$ and $h \in \mathscr{R}_{\alpha,\mu}(\theta,\rho)$ $(\rho \in [0,1], \theta \in (-\pi/2,\pi/2), \alpha \in [0,1), \mu \ge 0)$. Then $f * h \in \mathscr{R}_{\alpha,\mu}(\theta,\rho)$.*

Proof. By employing the properties of the Hadamard product, we receive

$$\Phi^{\alpha,\mu}(f * h)(z) = f(z) * \Phi^{\alpha,\mu}h(z). \quad (45)$$

Therefore,

$$\frac{e^{i\theta}\Phi^{\alpha,\mu}(f * h)(z)}{z} = \frac{f(z) * \left(\left(e^{i\theta}\Phi^{\alpha,\mu}h(z)\right)/z\right)z}{f(z) * z}. \quad (46)$$

In virtue of Lemma 8, we have

$$\Re\left(\frac{e^{i\theta}\Phi^{\alpha,\mu}(f * h)(z)}{z}\right) > \rho\cos\theta. \quad (47)$$

Hence $f * h \in \mathcal{R}_{\alpha,\mu}(\theta,\rho)$.

Theorem 14. *Let $f \in \mathcal{R}_{\alpha,\mu}(\theta,\rho)$ ($\rho \in [0,1]$, $\theta \in (-\pi/2,\pi/2)$, $\alpha \in [0,1)$, $\mu \geq 0$). Then the integral*

$$(\mathcal{J}f)(z) = \frac{\tau+1}{z^{\tau}}\int_0^z \xi^{\tau-1}f(\xi)\,d\xi \quad (z \in U, \tau > -1) \quad (48)$$

is also in $\mathcal{R}_{\alpha,\mu}(\theta,\rho)$.

Proof. It is easy to show that

$$\Phi^{\alpha,\mu}(\mathcal{J}f)(z) = \mathcal{G}(\tau+1,\tau+2) * \Phi^{\alpha,\mu}f(z). \quad (49)$$

Therefore,

$$\begin{aligned}&\frac{e^{i\theta}\Phi^{\alpha,\mu}(\mathcal{J}f)(z)}{z}\\ &= \frac{\mathcal{G}(\tau+1,\tau+2) * \left(\left(e^{i\theta}\Phi^{\alpha,\mu}f(z)\right)/z\right)z}{\mathcal{G}(\tau+1,\tau+2) * z}.\end{aligned} \quad (50)$$

But $\mathcal{G}(\tau+1,\tau+2) \in \mathcal{S}^*(1/2)$, thus in view of Lemma 8, the proof is complete.

Remark 15. When $\mu = 0$ in Theorems 13 and 14, we have the results given in [28].

Theorem 16. *Let $\rho \in [0,1]$, $\theta \in (-\pi/2,\pi/2)$, $\alpha \in [0,1)$, $\mu \geq 0$. If the subordination*

$$\begin{aligned}&\mathcal{G}''\left(\frac{1}{\mu+1}+1-\alpha, \frac{1}{\mu+1}+1; z\right)\\ &-\frac{2}{z}\mathcal{G}'\left(\frac{1}{\mu+1}+1-\alpha, \frac{1}{\mu+1}+1; z\right)\\ &+\frac{2}{z^2}\mathcal{G}\left(\frac{1}{\mu+1}+1-\alpha, \frac{1}{\mu+1}+1; z\right) \prec 1+z\end{aligned} \quad (51)$$

holds, then $g(z) := (e^{-i\theta}/z)(\mathcal{G}((1/(\mu+1))+1-\alpha, (1/(\mu+1))+1; z) * z(((1+(1-2\rho)z^2)/(1-z^2))\cos\theta + i\sin\theta))$ is univalent convex function.

Proof. Condition (51) yields that $\mathcal{G}(1/(\mu+1))+1-\alpha, (1/(\mu+1))+1; z/z$ is univalent convex (see [31, Theorem 1.9]) and consequently, in view of Lemma 6, $g(z)$ is univalent convex function.

Theorem 17. *Let $f \in \mathcal{R}_{\alpha_2,\mu}(\theta,\rho)$. Then*

$$\frac{f(z)}{z} \prec g(z), \quad (52)$$

where g is defined in Theorem 16.

Proof. Since $f \in \mathcal{R}_{\alpha_2,\mu}(\theta,\rho)$, then we have

$$\frac{\Phi^{\alpha,\mu}f(z)}{z} \prec e^{-i\theta}\left(\frac{1+(1-2\rho)z^2}{1-z^2}\cos\theta + i\sin\theta\right). \quad (53)$$

But g is univalent convex function (Theorem 16); thus by an application of Lemma 7, we obtain the desired assertion.

3. Conclusion

We defined a new fractional differential operator which generalized well-known linear and nonlinear operators such as Carlson-Shaffer operator and the Dziok-Srivastava (linear operators) and Srivastava-Owa fractional differential operators (nonlinear operator). By making uses this operator a generalized class of analytic functions is defined and studied. The sharp bound for nonlinear functional based on the second-order Hankel determinant $|a_2a_4 - a_3^2|$, involving the generalized fractional differential operator, is computed. Several properties, depending on the Hadamard product, are imposed. We have shown that some results are generalized by recent works due to Mishra-Gochhayat, Ling-Ding, and Janteng et al. Furthermore, a new approach is introduced in the proof of Theorems 11 and 16 based on the subordination concept and employing the result due to Ponnusamy and Singh.

References

[1] S. G. Samko, A. A. Kilbas, and O. I. Marichev, *Fractional Integrals and Derivatives: Theory and Applications*, Gordon and Breach Science, Yverdon, Switzerland, 1993.

[2] I. Podlubny, *Fractional Differential Equations*, vol. 198, Academic Press, San Diego, Calif, USA, 1999.

[3] R. Hilfer, *Applications of Fractional Calculus in Physics*, World Scientific, Singapore, 2000.

[4] K. Diethelm, *The Analysis of Fractional Differential Equations*, vol. 2004 of *Lecture Notes in Mathematics*, Springer, Berlin, Germany, 2010.

[5] A. A. Kilbas, H. M. Srivastava, and J. J. Trujillo, *Theory and Applications of Fractional Differential Equations*, vol. 204 of *North-Holland Mathematics Studies*, Elsevier Science, Amsterdam, The Netherland, 2006.

[6] J. Sabatier, O. P. Agrawal, and J. A. T. Machado, *Advance in Fractional Calculus: Theoretical Developments and Applications in Physics and Engineering*, Springer, New York, NY, USA, 2007.

[7] V. Lakshmikantham, S. Leela, and J. V. Devi, *Theory of Fractional Dynamic Systems*, Cambridge Scientific, Cambridge, UK, 2009.

[8] M. Darus and R. W. Ibrahim, "Radius estimates of a subclass of univalent functions," *Matematichki Vesnik*, vol. 63, no. 1, pp. 55–58, 2011.

[9] H. M. Srivastava, Y. Ling, and G. Bao, "Some distortion inequalities associated with the fractional derivatives of analytic and univalent functions," *Journal of Inequalities in Pure and Applied Mathematics*, vol. 2, no. 2, article 23, 6 pages, 2001.

[10] H. M. Srivastava and S. Owa, *Univalent Functions, Fractional Calculus, and Their Applications*, Halsted Press, John Wiley and Sons, New York, NY, USA, 1989.

[11] R. W. Ibrahim, "On generalized Srivastava-Owa fractional operators in the unit disk," *Advances in Difference Equations*, vol. 2011, article 55, 10 pages, 2011.

[12] H. M. Srivastava and P. W. Karlsson, *Multiple Gaussian Hypergeometric Series*, Halsted Press, John Wiley and Sons, New York, NY, USA, 1985.

[13] S. Owa and H. M. Srivastava, "Univalent and starlike generalized hypergeometric functions," *Canadian Journal of Mathematics*, vol. 39, no. 5, pp. 1057–1077, 1987.

[14] R. W. Ibrahim and M. Darus, "Differential operator generalized by fractional derivatives," *Miskolc Mathematical Notes*, vol. 12, no. 2, pp. 167–184, 2011.

[15] Y. Yang, Y.-Q. Tao, and J.-L. Liu, "Differential subordinations for certain meromorphically multivalent functions defined by Dziok-Srivastava operator," *Abstract and Applied Analysis*, vol. 2011, Article ID 726518, 9 pages, 2011.

[16] V. Kiryakova, "Criteria for univalence of the Dziok-Srivastava and the Srivastava-Wright operators in the class *A*," *Applied Mathematics and Computation*, vol. 218, no. 3, pp. 883–892, 2011.

[17] M. Darus and R. W. Ibrahim, "On the existence of univalent solutions for fractional integral equation of Volterra type in complex plane," *ROMAI Journal*, vol. 7, no. 1, pp. 77–86, 2011.

[18] H. M. Srivastava, M. Darus, and R. W. Ibrahim, "Classes of analytic functions with fractional powers defined by means of a certain linear operator," *Integral Transforms and Special Functions*, vol. 22, no. 1, pp. 17–28, 2011.

[19] K. Piejko and J. Sokól, "Subclasses of meromorphic functions associated with the Cho-Kwon-Srivastava operator," *Journal of Mathematical Analysis and Applications*, vol. 337, no. 2, pp. 1261–1266, 2008.

[20] J. Sokól, "On some applications of the Dziok-Srivastava operator," *Applied Mathematics and Computation*, vol. 201, no. 1-2, pp. 774–780, 2008.

[21] R. W. Ibrahim and M. Darus, "On analytic functions associated with the Dziok-Srivastava linear operator and Srivastava-Owa fractional integral operator," *Arabian Journal for Science and Engineering*, vol. 36, no. 3, pp. 441–450, 2011.

[22] B. A. Frasin, "New properties of the Jung-Kim-Srivastava integral operators," *Tamkang Journal of Mathematics*, vol. 42, no. 2, pp. 205–215, 2011.

[23] Z.-G. Wang, R. Aghalary, M. Darus, and R. W. Ibrahim, "Some properties of certain multivalent analytic functions involving the Cho-Kwon-Srivastava operator," *Mathematical and Computer Modelling*, vol. 49, no. 9-10, pp. 1969–1984, 2009.

[24] H. M. Srivastava, "Some Fox-Wright generalized hypergeometric functions and associated families of convolution operators," *Applicable Analysis and Discrete Mathematics*, vol. 1, no. 1, pp. 56–71, 2007, Proceedings of the International Conference on Topics in Mathematical Analysis and Graph Theory (2007).

[25] V. Kiryakova, "Convolutions of Erdélyi-Kober fractional integration operators," in *Complex Analysis and Applications '87*, pp. 273–283, The Bulgarian Academy of Sciences, Sofia, Bulgaria, 1989.

[26] R. J. Libera and E. J. Złotkiewicz, "Coefficient bounds for the inverse of a function with derivative in *P*," *Proceedings of the American Mathematical Society*, vol. 87, no. 2, pp. 251–257, 1983.

[27] L. Yi and D. Shusen, "A class of analytic functions defined by fractional derivation," *Journal of Mathematical Analysis and Applications*, vol. 186, no. 2, pp. 504–513, 1994.

[28] A. K. Mishra and P. Gochhayat, "Second Hankel determinant for a class of analytic functions defined by fractional derivative," *International Journal of Mathematics and Mathematical Sciences*, vol. 2008, Article ID 153280, 10 pages, 2008.

[29] St. Ruscheweyh and T. Sheil-Small, "Hadamard products of Schlicht functions and the Pólya-Schoenberg conjecture," *Commentarii Mathematici Helvetici*, vol. 48, pp. 119–135, 1973.

[30] A. Janteng, S. A. Halim, and M. Darus, "Coefficient inequality for a function whose derivative has a positive real part," *Journal of Inequalities in Pure and Applied Mathematics*, vol. 7, no. 2, article 50, 5 pages, 2006.

[31] S. Ponnusamy and V. Singh, "Criteria for univalent, starlike and convex functions," *Bulletin of the Belgian Mathematical Society*, vol. 9, no. 4, pp. 511–531, 2002.

[32] S. S. Miller and P. T. Mocanu, *Differential Subordinantions: Theory and Applications*, vol. 225 of *Pure and Applied Mathematics*, Dekker, New York, NY, USA, 2000.

18

On Polynomial Stability of Variational Nonautonomous Difference Equations in Banach Spaces

Mihail Megan,[1,2] Traian Ceauşu,[2] and Mihaela Aurelia Tomescu[3]

[1] *Academy of Romanian Scientists, Independenţei 54, 050094 Bucharest, Romania*
[2] *West University of Timişoara, Department of Mathematics, V. Pârvan Boulevard, No. 4, 300223 Timişoara, Romania*
[3] *University of Petroşani, Department of Mathematics, University Street 20, 332006 Petroşani, Romania*

Correspondence should be addressed to Mihaela Aurelia Tomescu; mtomescu@upet.ro

Academic Editor: Seenith Sivasundaram

Our goal in this paper is to give characterizations for some concepts of polynomial stability for variational nonautonomous difference equations. The obtained results can be considered generalizations for the case of variational nonautonomous difference equations of some theorems proved by Barbashin (1967), Datko (1973), and Lyapunov (1992), for evolution operators.

1. Introduction

In this paper we define and characterize two types of polynomial stability: (nonuniform) polynomial stability and strong polynomial stability for variational nonautonomous difference equations. These concepts are different from the concept of exponential stability studied for variational nonautonomous difference equations in [1], as shown in this paper.

In the case of evolution operators, the concept of nonuniform polynomial stability was studied by Barreira and Valls [2]. Moreover, characterizations for polynomial stability of evolution operators have been given in [3].

The variational nonautonomous difference equations considered in this paper generate discrete evolution cocycle over a discrete evolution semiflow. The concept of evolution cocycle was introduced by Megan and Stoica in [4].

We will consider the sets $\Delta = \{(m,n) \in \mathbb{N}^2$, with $m \geq n\}$ and $T = \{(m,n,p) \in \mathbb{N}^3$, with $m \geq n \geq p\}$, a metric space (X,d) and V a real or complex Banach space. The norm on V and on $\mathcal{B}(V)$ (the Banach algebra of all bounded linear operators on V) will be denoted by $\|\cdot\|$.

Definition 1. A mapping $\varphi : \Delta \times X \to X$ is called a discrete evolution semiflow on X if the following conditions hold:

(s_1) $\varphi(n,n,x) = x$, for all $(n,x) \in \mathbb{N} \times X$;

(s_2) $\varphi(m,n,\varphi(n,p,x)) = \varphi(m,p,x)$, for all $(m,n,p,x) \in T \times X$.

Given a sequence $(A_m)_{m\in\mathbb{N}}$ with $A_m : X \to \mathcal{B}(V)$ and a discrete evolution semiflow $\varphi : \Delta \times X \to X$, we consider the problem of existence of a sequence $(v_m)_{m\in\mathbb{N}}$ with $v_m : \mathbb{N} \times X \to X$ such that

$$v_{m+1}(n,x) = A_m(\varphi(m,n,x))\, v_m(n,x) \qquad (1)$$

for all $(m,n,x) \in \Delta \times X$. We will denote this problem by (A,φ) and we say that (A,φ) is a *variational (nonautonomous) discrete-time system*.

For $(m,n) \in \Delta$ we define the application $\Phi_m^n : X \to \mathcal{B}(V)$ by

$$\Phi_m^n(x)\,v = \begin{cases} A_{m-1}(\varphi(m-1,n,x)) \cdots A_{n+1}(\varphi(n+1,n,x))\, A_n(x)\, v, & \text{if } m > n \\ v, & \text{if } m = n. \end{cases} \qquad (2)$$

Remark 2. From the definitions of v_m and Φ_m^n it follows that

(c_1) $\Phi_m^m(x)v = v$, for all $(m, x, v) \in \mathbb{N} \times X \times V$;

(c_2) $\Phi_m^p(x) = \Phi_m^n(\varphi(n, p, x))\Phi_n^p(x)$, for all $(m, n, p, x) \in T \times X$;

(c_3) $v_m(n, x) = \Phi_m^n(x)v_n(n, x)$, for all $(m, n, x) \in \Delta \times X$.

Definition 3. A mapping $\Phi : \Delta \times X \to \mathcal{B}(V)$ is called a discrete evolution cocycle over the discrete evolution semiflow $\varphi : \Delta \times X \to X$ if the following properties hold:

(c_1) $\Phi(n, n, x) = I$ (the identity operator on V), for all $(n, x) \in \mathbb{N} \times X$,

(c_2) $\Phi(m, p, x) = \Phi(m, n, (\varphi(n, p, x))\Phi(n, p, x)$, for all $(m, n, p, x) \in T \times X$.

If Φ is a discrete evolution cocycle over the discrete evolution semiflow φ, then the pair $S = (\Phi, \varphi)$ is called a discrete skew-evolution semiflow on X.

Remark 4. From Remark 2 it results that the mapping

$$\Phi : \Delta \times X \to \mathcal{B}(V), \qquad \Phi(m, n, x)v = \Phi_m^n(x)v \tag{3}$$

is a discrete evolution cocycle over discrete evolution semiflow φ.

2. Polynomial Stability

Let (A, φ) be a discrete variational system associated with the discrete evolution semiflow $\varphi : \Delta \times X \to X$ and with the sequence of mappings $A = (A_m)$, where $A_m : X \to \mathcal{B}(V)$, for all $m \in \mathbb{N}$.

Definition 5. The system (A, φ) is said to be

(i) exponentially stable (and denoted as e.s.) if there exist the constants $N \geq 1$, $\alpha > 0$ and $\beta \geq 0$, such that

$$e^{\alpha(m-n)} \left\|\Phi_m^n(x)v\right\| \leq Ne^{\beta n}\|v\| \tag{4}$$

for all $(m, n, x, v) \in \Delta \times X \times V$;

(ii) polynomially stable (and denoted as p.s.) if there exist the constants $N \geq 1$, $\alpha > 0$ and $\beta \geq 0$ such that

$$(m+1)^{\alpha}\left\|\Phi_m^n(x)v\right\| \leq N(n+1)^{\alpha+\beta}\|v\| \tag{5}$$

for all $(m, n, x, v) \in \Delta \times X \times V$.

Remark 6. The system (A, φ) is

(i) exponentially stable if and only if there are $N \geq 1$, $\alpha > 0$, and $\beta \geq 0$ with

$$e^{\alpha(m-n)}\left\|\Phi_m^p(x)v\right\| \leq Ne^{\beta n}\left\|\Phi_n^p(x)v\right\| \tag{6}$$

for all $(m, n, p, x, v) \in T \times X \times V$;

(ii) polynomially stable if and only if there exist $N \geq 1$, $\alpha > 0$, and $\beta \geq 0$ with

$$(m+1)^{\alpha}\left\|\Phi_m^p(x)v\right\| \leq N(n+1)^{\alpha+\beta}\left\|\Phi_n^p(x)v\right\| \tag{7}$$

for all $(m, n, p, x, v) \in T \times X \times V$.

The connection between the two concepts of stability defined previously is established in the following.

Remark 7. It is obvious that

$$\text{e.s.} \Longrightarrow \text{p.s.} \tag{8}$$

The following example shows that the converse implication is not valid.

Example 8. Let $\mathcal{C} = \mathcal{C}(\mathbb{R}_+, \mathbb{R})$ be the metric space of all bounded continuous functions $x : \mathbb{R}_+ \to \mathbb{R}$, with the topology of uniform convergence. $\mathcal{C}$ is metrizable with respect to the metric $d(x_1, x_2) = \sup_{t\in\mathbb{R}_+}|x_1(t) - x_2(t)|$. Let $f : \mathbb{R}_+ \to (0, \infty)$ be a bounded decreasing function with the property that there exists $\lim_{t\to\infty} f(t) = l > 0$. We denote by X the closure in $\mathcal{C}$ of the set $\{f_t,\ t \in \mathbb{R}_+\}$, where $f_t(s) = f(t + s)$ for all $s \in \mathbb{R}_+$. The mapping $\varphi : \Delta \times X \to X$ defined by $\varphi(m, n, x) = x_{m-n}$ is a discrete evolution semiflow. Let us consider the Banach space V and let the sequence of mappings $A_m : X \to \mathcal{B}(V)$, defined by

$$A_m(x)v = \frac{u(m)x(\tau)}{u(m+1)x(\tau+1)}v \tag{9}$$

for all $(m, x, v) \in \mathbb{N} \times X \times V$, where the sequence $u : \mathbb{N} \to \mathbb{R}$ is given by

$$u(m) = (m+1)\left(m + 1 - m\cos\frac{m\pi}{2}\right)^2. \tag{10}$$

Then

$$\Phi_m^n(x)v = \frac{(n+1)(n+1-n\cos(n\pi/2))^2 x(\tau)}{(m+1)(m+1-m\cos(m\pi/2))^2 x(m-n+\tau)}v, \tag{11}$$

and it results that

$$\begin{aligned}\left\|\Phi_m^n(x)v\right\| &\leq (2n+1)^2\frac{(n+1)x(0)}{(m+1)l}\|v\| \\ &\leq N\frac{n+1}{m+1}(n+1)^2\|v\|\end{aligned} \tag{12}$$

for all $(m, n, x, v) \in \Delta \times X \times V$, where $N = 4x(0)/l$. Hence (A, φ) is p.s. Assume by a contradiction that (A, φ) is e.s. According to Definition 5, there are $N \geq 1$, $\alpha > 0$, and $\beta \geq 0$ such that

$$e^{\alpha(m-n)}\left\|\Phi_m^n(x)v\right\| \leq Ne^{\beta n}\|v\| \tag{13}$$

for all $(m, n, x, v) \in \Delta\times X\times V$. The previous inequality for the considered system becomes

$$e^{\alpha(m-n)} \times \frac{(n+1)(n+1-n\cos(n\pi/2))^2 x(\tau)}{(m+1)(m+1-m\cos(m\pi/2))^2 x(m-n+\tau)} \leq Ne^{\beta n} \tag{14}$$

for all $(m, n, x) \in \Delta \times X$. If we take $m = 4k^2 + 4$ and $n = 4k + 2$, $k \in \mathbb{N}$, then

$$e^{4\alpha k^2 + k(\alpha-\beta-2)+2} \frac{(4k+3)(8k+5)^2}{4k^2+5} \times \frac{x(\tau)}{x(4k^2 - 4k + 2 + \tau)} \le Ne^{2\beta}. \quad (15)$$

Passing to the limit for $k \to \infty$ we obtain a contradiction. We have shown that (A, φ) is not e.s.

Lemma 9. *The system (A, φ) is polynomially stable if and only if there are $N \ge 1$ and $0 < c \le d$ such that*

$$(m+1)^c \left\|\Phi_m^n(x)v\right\| \le N(n+1)^d \|v\| \quad (16)$$

for all $(m, n, x, v) \in \Delta \times X \times V$.

Proof. Necessity. If (A, φ) is p.s., then there are $N \ge 1$, $\alpha > 0$, and $\beta \ge 0$ such that

$$(m+1)^\alpha \left\|\Phi_m^n(x)v\right\| \le N(n+1)^{\alpha+\beta} \|v\| \quad (17)$$

for all $(m, n, x, v) \in \Delta \times X \times V$. Hence inequality (16) holds for $c = \alpha$ and $d = \alpha + \beta$.

Sufficiency. From the hypothesis it results that relation (5) of Definition 5 holds for $\alpha = c$ and $\beta = d - c$. □

A necessary condition for the polynomial stability property is presented by the following theorem.

Theorem 10. *If the system (A, φ) is polynomially stable, then there are $D \ge 1$, $d > 0$, $\omega > 0$, and $\gamma \ge 0$ such that*

$$\sum_{k=n}^{\infty} \frac{1}{k+1} \left(\frac{k+1}{n+1}\right)^d \left\|\Phi_k^n(x)v\right\| \le D(n+1)^\gamma \|v\| \quad (18)$$

for all $(n, x, v) \in \mathbb{N} \times X \times V$ and

$$\left\|\Phi_m^n(x)v\right\| \le D(m+1)^\omega (n+1)^{\gamma-\omega} \|v\| \quad (19)$$

for all $(m, n, x, v) \in \Delta \times X \times V$.

Proof. Let $N \ge 1$, $\alpha > 0$, and $\beta \ge 0$ as in Definition 5. Then, for every $d \in (0, \alpha)$ we have that

$$\begin{aligned}
&\sum_{k=n}^{\infty} \frac{1}{k+1} \left(\frac{k+1}{n+1}\right)^d \left\|\Phi_k^n(x)v\right\| \\
&\le N(n+1)^{\alpha+\beta-d} \|v\| \sum_{k=n}^{\infty} \frac{1}{(k+1)^{\alpha+1-d}} \\
&= N\|v\|(n+1)^{\alpha+\beta-d} \left(\frac{1}{(n+1)^{\alpha+1-d}} + \sum_{k=n+1}^{\infty} \frac{1}{(k+1)^{\alpha+1-d}}\right) \\
&\le N\|v\|(n+1)^{\alpha+\beta-d} \left(\frac{1}{(n+1)^{\alpha+1-d}} + \frac{1}{\alpha-d} \frac{1}{(n+1)^{\alpha-d}}\right) \\
&\le N\|v\|(n+1)^{\alpha+\beta-d} \left(1 + \frac{1}{\alpha-d}\right) \frac{1}{(n+1)^{\alpha-d}} \\
&= \frac{N(1+\alpha-d)}{\alpha-d} \|v\|(n+1)^\beta \le D\|v\|(n+1)^\gamma
\end{aligned} \quad (20)$$

for all $(n, x, v) \in \mathbb{N} \times X \times V$, where $D = N(1+\alpha-d)/(\alpha-d)$ and $\gamma = \beta$. In addition

$$\begin{aligned}
\left\|\Phi_m^n(x)v\right\| &\le N \frac{(n+1)^{\alpha+\beta}}{(m+1)^\alpha} \|v\| \\
&= N\left(\frac{m+1}{n+1}\right)^\omega \left(\frac{n+1}{m+1}\right)^{\alpha+\omega} (n+1)^\beta \|v\| \\
&\le D\left(\frac{m+1}{n+1}\right)^\omega (n+1)^\gamma \|v\|
\end{aligned} \quad (21)$$

for all $(m, n, x, v) \in \Delta \times X \times V$. □

Next, a sufficient condition for the polynomial stability property is presented by.

Theorem 11. *If there are $D \ge 1$ and $d > \omega > \gamma \ge 0$ such that*

$$\sum_{k=n}^{\infty} \frac{1}{k+1} \left(\frac{k+1}{n+1}\right)^d \left\|\Phi_k^n(x)v\right\| \le D(n+1)^\gamma \|v\| \quad (22)$$

for all $(n, x, v) \in \mathbb{N} \times X \times V$ and

$$\left\|\Phi_m^n(x)v\right\| \le D(m+1)^\omega (n+1)^{\gamma-\omega} \|v\| \quad (23)$$

for all $(m, n, x, v) \in \Delta \times X \times V$, then the system (A, φ) is polynomially stable.

Proof. From the hypothesis it results that

$$\left\|\Phi_m^n(x)v\right\| \le D(m+1)^\omega (k+1)^{\gamma-\omega} \left\|\Phi_k^n(x)v\right\| \quad (24)$$

for all $(m, k, n, x, v) \in T \times X \times V$. We suppose that $m \ge 2n$ and we denote by $j = [m/2]$. Then

$$\begin{aligned}
&\frac{m+2}{2} \frac{(m+1)^{d-\omega}}{(n+1)^{d-\omega+\gamma}} \left\|\Phi_m^n(x)v\right\| \\
&\le \sum_{k=j}^{m} \frac{(m+1)^{d-\omega}}{(n+1)^{d-\omega+\gamma}} \left\|\Phi_m^n(x)v\right\| \\
&\le D\sum_{k=j}^{m} \frac{(m+1)^{d-\omega}}{(n+1)^{d-\omega+\gamma}} \left(\frac{m+1}{k+1}\right)^\omega
\end{aligned}$$

$$\times (k+1)^{\gamma} \left\|\Phi_k^n(x)v\right\|$$
$$= D\sum_{k=j}^{m}\frac{1}{k+1}\left(\frac{k+1}{n+1}\right)^{d}\left\|\Phi_k^n(x)v\right\|$$
$$\times\left(\frac{m+1}{k+1}\right)^{d}\left(\frac{n+1}{k+1}\right)^{\omega-\gamma}(k+1)$$
$$\le D4^{d}(m+1)\sum_{k=j}^{m}\frac{1}{k+1}\left(\frac{k+1}{n+1}\right)^{d}\left\|\Phi_k^n(x)v\right\|$$
$$\le D4^{d}(m+1)\sum_{k=n}^{\infty}\frac{1}{k+1}\left(\frac{k+1}{n+1}\right)^{d}\left\|\Phi_k^n(x)v\right\|$$
$$\le D^{2}4^{d}(m+1)(n+1)^{\gamma}\left\|v\right\|. \tag{25}$$

Hence,

$$\left(\frac{m+1}{n+1}\right)^{d-\omega}\left\|\Phi_m^n(x)v\right\|$$
$$\le D^{2}4^{d}\frac{2(m+1)}{m+2}(n+1)^{2\gamma}\left\|v\right\| \tag{26}$$
$$\le D^{2}2^{2d+1}(n+1)^{2\gamma}\left\|v\right\|$$
$$= N(n+1)^{2\gamma}\left\|v\right\|$$

for all $(x,v)\in X\times V$, where $N = D^{2}2^{2d+1}$. If $n\le m<2n$, then

$$\left(\frac{m+1}{n+1}\right)^{d-\omega}\left\|\Phi_m^n(x)v\right\|$$
$$\le D\left(\frac{m+1}{n+1}\right)^{d-\omega}\left(\frac{m+1}{n+1}\right)^{\omega}(n+1)^{\gamma}\left\|v\right\| \tag{27}$$
$$\le D2^{d}(n+1)^{\gamma}\left\|v\right\|\le N(n+1)^{2\gamma}\left\|v\right\|$$

for all $(x,v)\in X\times V$ thus we have proved that (A,φ) is p.s. □

As a generalization of a theorem of Barbashin [5], we give the following characterization of the polynomial stability property.

Theorem 12. *The system (A,φ) is polynomially stable if and only if there are $B\ge 1$ and $a>b\ge 0$ such that*

$$\sum_{k=n}^{m}\frac{1}{k+1}\left(\frac{m+1}{k+1}\right)^{a}\left\|\Phi_m^k(x)v\right\|\le B(m+1)^{b}\left\|v\right\| \tag{28}$$

for all $(m,n,x,v)\in\Delta\times X\times V$.

Proof. Necessity. Let $N\ge 1$, $\alpha>0$, and $\beta\ge 0$ as in Definition 5. Then, for every $a>0$ with $0\le\beta<a<\alpha+\beta<a+1$ we have that

$$\sum_{k=n}^{m}\frac{1}{k+1}\left(\frac{m+1}{k+1}\right)^{a}\left\|\Phi_m^k(x)v\right\|$$
$$\le N\left\|v\right\|\sum_{k=n}^{m}\frac{1}{k+1}\left(\frac{m+1}{k+1}\right)^{a}\frac{(k+1)^{\alpha+\beta}}{(m+1)^{\alpha}}$$
$$= N\left\|v\right\|(m+1)^{a-\alpha}\sum_{k=n}^{m}\frac{1}{(k+1)^{a+1-\alpha-\beta}} \tag{29}$$
$$\le N\left\|v\right\|(m+1)^{a-\alpha}\frac{1}{\alpha+\beta-a}(m+1)^{\alpha+\beta-a}$$
$$\le D(m+1)^{b}\left\|v\right\|$$

for all $(m,n,x,v)\in\Delta\times X\times V$, where $D=N/(\alpha+\beta-a)$ and $b=\beta$.

Sufficiency. From the hypothesis we have

$$\frac{1}{n+1}\left(\frac{m+1}{n+1}\right)^{a}\left\|\Phi_m^n(x)v\right\|\le B(m+1)^{b}\left\|v\right\| \tag{30}$$

for all $(m,n,x,v)\in\Delta\times X\times V$. Hence

$$(m+1)^{a-b}\left\|\Phi_m^n(x)v\right\|\le B(n+1)^{a+1}\left\|v\right\|, \tag{31}$$

and relation (16) from Lemma 9 holds for $0<c=a-b\le a<a+1=d$. □

Definition 13. An application $L:\Delta\times X\times V\to\mathbb{R}_{+}$ is called a Lyapunov polynomial stability function for the system (A,φ) if there exists $l>0$ such that

$$L(m,p,x,v)+\sum_{k=n}^{m-1}\frac{1}{k+1}\left(\frac{k+1}{n+1}\right)^{l}\left\|\Phi_k^p(x)v\right\|\le L(n,p,x,v) \tag{32}$$

for all $(m,n,p,x,v)\in T\times X\times V$, with $m>n$.

The constant $l>0$ is called the order of the Lyapunov function L.

Theorem 14. *If the system (A,φ) is polynomially stable, then there are a Lyapunov polynomial stability function for the system (A,φ) and constants $K\ge 1$, $\nu>0$ and $\delta\ge 0$ such that*

$$L(m,n,x,v)\le K(n+1)^{\delta}\left\|v\right\|,$$
$$\left\|\Phi_m^n(x)v\right\|\le K(m+1)^{\nu}(n+1)^{\delta-\nu}\left\|v\right\| \tag{33}$$

for all $(m,n,x,v)\in\Delta\times X\times V$.

Proof. From Theorem 10 we have that there are $D\ge 1$, $d>0$, $\omega>0$, and $\gamma\ge 0$ such that

$$\sum_{k=n}^{\infty}\frac{1}{k+1}\left(\frac{k+1}{n+1}\right)^{d}\left\|\Phi_k^n(x)v\right\|\le D(n+1)^{\gamma}\left\|v\right\|,$$
$$\left\|\Phi_m^n(x)v\right\|\le D(m+1)^{\omega}(n+1)^{\gamma-\omega}\left\|v\right\| \tag{34}$$

for all $(m,n,x,v) \in \Delta \times X \times V$. We define the application $L : \Delta \times X \times V \to \mathbb{R}_+$ by

$$L(m,p,x,v) = \sum_{k=m}^{\infty} \frac{1}{k+1}\left(\frac{k+1}{m+1}\right)^d \left\|\Phi_k^p(x)v\right\| \quad (35)$$

for all $(m,p,x,v) \in \Delta \times X \times V$. Then, for all $(m,n,p,x,v) \in T \times X \times V$, with $m > n$ we have

$$\begin{aligned} L(m,p,x,v) &+ \sum_{k=n}^{m-1} \frac{1}{k+1}\left(\frac{k+1}{n+1}\right)^d \left\|\Phi_k^p(x)v\right\| \\ &= \sum_{k=m}^{\infty} \frac{1}{k+1}\left(\frac{k+1}{m+1}\right)^d \left\|\Phi_k^p(x)v\right\| \\ &\quad + \sum_{k=n}^{m-1} \frac{1}{k+1}\left(\frac{k+1}{n+1}\right)^d \left\|\Phi_k^p(x)v\right\| \\ &\le \sum_{k=n}^{\infty} \frac{1}{k+1}\left(\frac{k+1}{n+1}\right)^d \left\|\Phi_k^p(x)v\right\| \\ &= L(n,p,x,v). \end{aligned} \quad (36)$$

In addition, from Theorem 10 we have that

$$\begin{aligned} L(m,n,x,v) &= \sum_{k=m}^{\infty} \frac{1}{k+1}\left(\frac{k+1}{n+1}\right)^d \left\|\Phi_k^n(x)v\right\| \\ &\le \sum_{k=n}^{\infty} \frac{1}{k+1}\left(\frac{k+1}{n+1}\right)^d \left\|\Phi_k^n(x)v\right\| \\ &\le D(n+1)^\gamma \|v\|, \\ \left\|\Phi_m^n(x)v\right\| &\le D(m+1)^\omega (n+1)^{\gamma-\omega} \|v\|. \end{aligned} \quad (37)$$

Consequently, relations (33) are satisfied for $K = D$, $\delta = \gamma$ and $\nu = \omega$. □

Theorem 15. *If there exist a Lyapunov polynomial stability function with the order $l > 0$ for the system (A,φ) and the constants $K \ge 1$, $\nu > 0$, and $\delta \ge 0$ with $l > \nu > \delta \ge 0$ such that*

$$\begin{aligned} L(m,n,x,v) &\le K(n+1)^\delta \|v\|, \\ \left\|\Phi_m^n(x)v\right\| &\le K(m+1)^\nu (n+1)^{\delta-\nu} \|v\| \end{aligned} \quad (38)$$

for all $(m,n,x,v) \in \Delta \times X \times V$, then the system (A,φ) is polynomially stable.

Proof. From the hypothesis and Definition 13 we have that

$$\begin{aligned} \sum_{k=n}^{m-1} &\frac{1}{k+1}\left(\frac{k+1}{n+1}\right)^l \left\|\Phi_k^n(x)v\right\| \\ &\le L(n,n,x,v) - L(m,n,x,v) \\ &\le L(n,n,x,v) \le K(n+1)^\delta \|v\| \end{aligned} \quad (39)$$

for all $(m,n,x,v) \in \Delta \times X \times V$, with $m > n$. Passing to the limit for $m \to \infty$ we obtain that

$$\sum_{k=n}^{\infty} \frac{1}{k+1}\left(\frac{k+1}{n+1}\right)^l \left\|\Phi_k^n(x)v\right\| \le K(n+1)^\delta \|v\| \quad (40)$$

for all $(n,x,v) \in \mathbb{N} \times X \times V$. Now, from Theorem 11 the conclusion follows. □

3. Strong Polynomial Stability

Definition 16. The system (A,φ) is said to be strongly polynomially stable (and denoted as s.p.s.) if there are three constants $N \ge 1$ and $\alpha > \beta \ge 0$ such that

$$(m+1)^\alpha \left\|\Phi_m^n(x)v\right\| \le N(n+1)^{\alpha+\beta} \|v\| \quad (41)$$

for all $(m,n,x,v) \in \Delta \times X \times V$.

Remark 17. It is easy to see that (A,φ) is strongly polynomially stable if and only if there are $N \ge 1$ and $\alpha > \beta \ge 0$ with

$$(m+1)^\alpha \left\|\Phi_m^p(x)v\right\| \le N(n+1)^{\alpha+\beta} \left\|\Phi_n^p(x)v\right\| \quad (42)$$

for all $(m,n,p,x,v) \in T \times X \times V$.

Remark 18. It is obvious that

$$\text{s.p.s.} \Longrightarrow \text{p.s.} \quad (43)$$

The following example shows that the converse implication is not valid.

Example 19. Let (X,d) be the metric space, let V be a Banach space, and let φ be the evolution semiflow given as in Example 8. We define the sequence of mappings $A_m : X \to \mathscr{B}(\mathbb{R})$ by

$$A_m(x)v = \frac{u(m)x(\tau)}{u(m+1)x(\tau+1)}v \quad (44)$$

for all $(m,x,v) \in \mathbb{N} \times X \times V$, where the sequence $u : \mathbb{N} \to \mathbb{R}$ is given by

$$u(m) = (m+1)\left(m+1-m\cos\frac{m\pi}{2}\right). \quad (45)$$

Then

$$\Phi_m^n(x)v = \frac{(n+1)(n+1-n\cos(n\pi/2))x(\tau)}{(m+1)(m+1-m\cos(m\pi/2))x(m-n+\tau)}v, \quad (46)$$

and it follows that

$$\begin{aligned} \left\|\Phi_m^n(x)v\right\| &\le (2n+1)\frac{(n+1)x(0)}{(m+1)l}\|v\| \\ &\le N\frac{n+1}{m+1}(n+1)\|v\| \end{aligned} \quad (47)$$

for all $(m,n,x,v) \in \Delta \times X \times V$, where $N = 2x(0)/l$. Hence we have proved that (A,φ) is p.s. Let us suppose now that the system (A, φ) is s.p.s. According to Definition 16, there exist $N \geq 1$ and $\alpha > \beta \geq 0$ such that

$$\frac{(n+1)(n+1-n\cos(n\pi/2))\,x(\tau)}{(m+1)(m+1-m\cos(m\pi/2))\,x(m-n+\tau)} \leq N\left(\frac{n+1}{m+1}\right)^{\alpha}(n+1)^{\beta} \tag{48}$$

for all $(m, n, x) \in \Delta \times X$. If we take $n = 4k+2$ and $m = 4k^2+4$, $k \in \mathbb{N}$, we have that

$$\left(\frac{4k+3}{4k^2+5}\right)^{1-\alpha}\frac{8k+5}{4k+3}\frac{x(\tau)}{x(4k^2-4k+2+\tau)} \leq N(4k+3)^{\beta-1}. \tag{49}$$

It follows that $\alpha \leq 1$, $\beta \geq 1$, and $N \geq 2$ which implies that $0 < \alpha \leq 1 \leq \beta$, contradicting the fact that $\alpha > \beta \geq 0$. This proves that (A, φ) is not s.p.s.

Lemma 20. *The system (A, φ) is strongly polynomially stable if and only if there are $N \geq 1$ and $0 < c \leq d < 2c$ such that*

$$(m+1)^{c}\left\|\Phi_{m}^{n}(x)\,v\right\| \leq N(n+1)^{d}\|v\| \tag{50}$$

for all $(m, n, x, v) \in \Delta \times X \times V$.

Proof. It is similar to the proof of Lemma 9 with the condition $0 < \alpha = c \leq d = \alpha + \beta < 2\alpha = 2c$ in the case of the necessity and with $0 \leq \beta = d - c < c = \alpha$ for the sufficiency. □

Theorem 21. *If the system (A, φ) is strongly polynomially stable, then there are $D \geq 1$, $d > 0$, $\omega > 0$, and $\gamma \geq 0$ such that*

$$\sum_{k=n}^{\infty}\frac{1}{k+1}\left(\frac{k+1}{n+1}\right)^{d}\left\|\Phi_{k}^{n}(x)\,v\right\| \leq D(n+1)^{\gamma}\|v\| \tag{51}$$

for all $(n, x, v) \in \mathbb{N} \times X \times V$,

$$\left\|\Phi_{m}^{n}(x)\,v\right\| \leq D(m+1)^{\omega}(n+1)^{\gamma-\omega}\|v\| \tag{52}$$

for all $(m, n, x, v) \in \Delta \times X \times V$.

Proof. It results from the proof of Theorem 10 with $0 \leq \beta < d < \alpha$ for the first inequality and with $0 \leq \beta = \gamma < \omega < d < \alpha$ for the second inequality. □

Theorem 22. *If there are $D \geq 1$ and $d > \omega > \gamma \geq 0$ with $2\gamma + \omega < d$ such that*

$$\sum_{k=n}^{\infty}\frac{1}{k+1}\left(\frac{k+1}{n+1}\right)^{d}\left\|\Phi_{k}^{n}(x)\,v\right\| \leq D(n+1)^{\gamma}\|v\| \tag{53}$$

for all $(n, x, v) \in \mathbb{N} \times X \times V$ and

$$\left\|\Phi_{m}^{n}(x)\,v\right\| \leq D(m+1)^{\omega}(n+1)^{\gamma-\omega}\|v\| \tag{54}$$

for all $(m, n, x, v) \in \Delta \times X \times V$, then the system (A, φ) is strongly polynomially stable.

Proof. It is analogous to the proof of Theorem 11. □

Next, we present a generalization of a theorem due to Lyapunov for the case of strong polynomial stability of discrete variational systems.

Theorem 23. *If the system (A, φ) is strongly polynomially stable, then there exist a Lyapunov polynomial stability function for the system (A, φ) and constants $K \geq 1$ and $0 \leq \delta < \nu$ such that*

$$L(m, n, x, v) \leq K(n+1)^{\delta}\|v\|, \quad \left\|\Phi_{m}^{n}(x)\,v\right\| \leq K(m+1)^{\nu}(n+1)^{\delta-\nu}\|v\| \tag{55}$$

for all $(m, n, x, v) \in \Delta \times X \times V$.

Proof. Using the technique from the proof of Theorem 14 for $0 \leq \gamma < \omega < d$ we obtain the conclusion. □

Theorem 24. *If there are a Lyapunov polynomial stability function with the order $l > 0$ for the system (A, φ) and the constants $K \geq 1$, $\nu > 0$ and $\delta \geq 0$ with $l > \nu > \delta \geq 0$ and $2\delta + \nu < l$ such that:*

$$L(m, n, x, v) \leq K(n+1)^{\delta}\|v\|, \quad \left\|\Phi_{m}^{n}(x)\,v\right\| \leq K(m+1)^{\nu}(n+1)^{\delta-\nu}\|v\| \tag{56}$$

for all $(m, n, x, v) \in \Delta \times X \times V$, then the system (A, φ) is strongly polynomially stable.

Proof. It is the same as the proof of Theorem 15. □

Remark 25. If the system (A, φ) is strongly polynomially stable, then there are $B \geq 1$ and $a > b \geq 0$ such that

$$\sum_{k=n}^{m}\frac{1}{k+1}\left(\frac{m+1}{k+1}\right)^{a}\left\|\Phi_{m}^{k}(x)\,v\right\| \leq B(m+1)^{b}\|v\| \tag{57}$$

for all $(m, n, x, v) \in \Delta \times X \times V$.

Remark 26. If there are $B \geq 1$, $a > 0$, and $b \geq 0$, with $a > 2b + 1$ such that

$$\sum_{k=n}^{m}\frac{1}{k+1}\left(\frac{m+1}{k+1}\right)^{a}\left\|\Phi_{m}^{k}(x)\,v\right\| \leq B(m+1)^{b}\|v\| \tag{58}$$

for all $(m, n, x, v) \in \Delta \times X \times V$, then the system (A, φ) is strongly polynomially stable.

References

[1] M. Megan, T. Ceauşu, and M. A. Tomescu, "On exponential stability of variational nonautonomous difference equations in Banach spaces," *Annals of the Academy of Romanian Scientists*, vol. 4, no. 1, pp. 20–31, 2012.

[2] L. Barreira and C. Valls, "Polynomial growth rates," *Nonlinear Analysis: Theory, Methods & Applications*, vol. 71, no. 11, pp. 5208–5219, 2009.

[3] M. Megan, T. Ceaușu, and M. L. Ramneanțu, "Polynomial stability of evolution operators in Banach spaces," *Opuscula Mathematica*, vol. 31, no. 2, pp. 279–288, 2011.

[4] M. Megan and C. Stoica, "Discrete asymptotic behaviors for skew-evolution semi ows on Banach spaces," *Carpathian Journal of Mathematics*, vol. 24, pp. 348–355, 2008.

[5] E. A. Barbashin, *Introduction in Stability Theory*, Nauka, Moscow, Russia 1967.

Universality Properties of a Double Series by the Generalized Walsh System

Sergo A. Episkoposian

Faculty of Applied Mathematics, State Engineering University of Armenia, Teryan Street 105, 375049 Yerevan, Armenia

Correspondence should be addressed to Sergo A. Episkoposian; sergoep@ysu.am

Academic Editor: Frédéric Robert

We consider a question on existence of a double series by the generalized Walsh system, which is universal in weighted $L^1_\mu[0,1]^2$ spaces. In particular, we construct a weighted function $\mu(x,y)$ and a double series by generalized Walsh system of the form $\sum_{n,k=1}^{\infty} c_{n,k}\psi_n(x)\psi_k(y)$ with the $\sum_{n,k=1}^{\infty} |c_{n,k}|^q < \infty$ for all $q > 2$, which is universal in $L^1_\mu[0,1]^2$ concerning subseries with respect to convergence, in the sense of both spherical and rectangular partial sums.

1. Introduction

Let X be a Banach space.

Definition 1. A series

$$\sum_{k=1}^{\infty} f_k, \quad f_k \in X \tag{1}$$

is said to be universal in X with respect to rearrangements, if for any $f \in X$ the members of (1) can be rearranged so that the obtained series $\sum_{k=1}^{\infty} f_{\sigma(k)}$ converges to f by norm of X.

Definition 2. The series (1) is said to be universal (in X) concerning subseries, if for any $f \in X$, it is possible to choose a subseries $\sum_{k=1}^{\infty} f_{n_k}$ from (1), which converges to the f by norm of X.

Note that for one-dimensional case there are many papers that are devoted to the question on existence of various types of universal series in the sense of convergence almost everywhere and on a measure (see [1–10]).

Let $a \geq 2$ be a fixed integer and $\omega_a = e^{2\pi i/a}$. Recall the following definitions (see [11]).

The Rademacher system of order a is defined inductively as follows. For $n = 0$ let

$$\varphi_0(x) = \omega_a^k \quad \text{if } x \in \left[\frac{k}{a}, \frac{k+1}{a}\right), \quad k = 0, 1, \ldots, a-1, \tag{2}$$

and for $n \geq 1$ let

$$\varphi_n(x+1) = \varphi_n(x) = \varphi_0(a^n x). \tag{3}$$

The generalized Walsh system of order a is defined by

$$\psi_0(x) = 1, \tag{4}$$

and if $n = \alpha_1 a^{n_1} + \cdots + \alpha_s a^{n_s}$, where $n_1 > \cdots > n_s$, $0 \leq \alpha_j < a$, $j = 1, 2, \ldots, s$, then

$$\psi_n(x) = \varphi_{n_1}^{\alpha_1}(x) \cdot \ldots \cdot \varphi_{n_s}^{\alpha_s}(x). \tag{5}$$

We denote the generalized Walsh system of order a by Ψ_a. Not that Ψ_2 is the classical Walsh system. The basic properties of the generalized Walsh system of order a have been obtained by Chrestenson, Fine, Watari, Young, Vilenkin, and others (see [11–16]).

In [6–9], the existence of universal one-dimensional series by trigonometric and the classical Walsh system with respect to rearrangements and subseries in some weighted space $L^1_\mu[0,1]$. Some results for two-dimensional case for the classical Walsh system were obtained in [10] is proved. In this paper we consider the universality properties of a double series by the generalized Walsh system.

2. Preliminary Notes

Now we list some properties of Ψ_a, $a \geq 2$, which will be useful later.

(i) Each nth Rademacher function has period a^{-n} and

$$\varphi_n(x) = \text{const} \in \Omega_a = \left\{1, \omega_a, \omega_a^2, \ldots, \omega_a^{a-1}\right\}, \quad (6)$$

if $x \in \Delta_{n+1}^{(k)} = [k/a^{n+1}, (k+1)/a^{n+1})$, $k = 0, \ldots, a^{n+1} - 1$, $n = 1, 2, \ldots$.

(ii)

$$(\varphi_n(x))^k = (\varphi_n(x))^m, \quad \forall n,\ k \in \mathcal{N},\ m = k \pmod{a}. \quad (7)$$

(iii) $\psi_n(x)$ is a finite product of the Rademacher functions with values in Ω_a.

(iv)

$$\psi_{a^k+j}(x) = \varphi_k(x) \cdot \psi_j(x), \quad \text{if } 0 \leq j \leq a^k - 1. \quad (8)$$

(v) $\Psi_a, a \geq 2$ is a complete orthonormal system in $L^2[0,1)$ and it is a basis in $L^p[0,1]$ for $p > 1$.

The rectangular and spherical partial sums of the double series

$$\sum_{k,\nu=1}^{\infty} c_{k,\nu}\psi_k(x)\psi_\nu(y), \quad (x,y) \in T = [0,1)^2 \quad (9)$$

will be denoted by

$$S_{n,m}(x,y) = \sum_{k=1}^{n}\sum_{\nu=1}^{m} c_{k,\nu}\psi_k(x)\psi_\nu(y),$$
$$S_R(x,y) = \sum_{\nu^2+k^2 \leq R^2} c_{k,\nu}\psi_k(x)\psi_\nu(y). \quad (10)$$

If $g(x,y)$ is a continuous function on $T = [0,1]^2$, then we set

$$\|g(x,y)\|_C = \max_{(x,y)\in T} |g(x,y)|. \quad (11)$$

3. Main Results

Let us denote the generalized Walsh system of order a by Ψ_a, $a \geq 2$. These are the main results of the paper.

Theorem 3. *There exists a double series of the form*

$$\sum_{n,k=1}^{\infty} c_{n,k}\psi_n(x)\psi_k(y) \quad \text{with} \sum_{n,k=1}^{\infty} |c_{n,k}|^q < \infty \quad \forall q > 2 \quad (12)$$

with the following property: for any number $\varepsilon > 0$ a weighted function $\mu(x,y)$ satisfying

$$0 < \mu(x,y) \leq 1, \quad |\{(x,y) \in T : \mu(x,y) \neq 1\}| < \varepsilon \quad (13)$$

can be constructed so that the series (12) is universal in $L^1_\mu(T)$ concerning subseries with respect to convergence in the sense of both spherical and rectangular partial sums.

Theorem 4. *There exists a double series of the form (12) with the following property: for any number $\varepsilon > 0$ a weighted function $\mu(x,y)$ with (13) can be constructed, so that the series (12) is universal in $L^1_\mu(T)$ concerning rearrangements with respect to convergence in the sense of both spherical and rectangular partial sums.*

Repeating the reasoning of the proof of [17, Lemma 2] we will receive the following lemma.

Lemma 5. *For any given numbers $0 < \varepsilon < 1$, $N_0 > 2$ ($N_0 \in \mathcal{N}$) and a step function*

$$f(x) = \sum_{s=1}^{q} \gamma_s \cdot \chi_{\Delta_s}(x), \quad (14)$$

where Δ_s is an interval of the form $\Delta_m^{(i)} = [(i-1)/2^m, i/2^m]$, $1 \leq i \leq 2^m$, there exist a measurable set $E \subset [0,1]$ and a polynomial $P(x)$ of the form

$$P(x) = \sum_{k=N_0}^{N} c_k\psi_k(x) \quad (15)$$

which satisfy the following conditions:

(1)

$$P(x) = f(x) \text{ on } E, \quad (16)$$

(2)

$$|E| > (1-\varepsilon), \quad (17)$$

(3)

$$\sum_{k=N_0}^{N} |c_k|^{2+\varepsilon} < \varepsilon, \quad (18)$$

(4)

$$\max_{N_0 \leq m < N}\left[\int_e \left|\sum_{k=N_0}^{m} c_k\psi_k(x)\right| dx\right] < \varepsilon + \int_e |f(x)|\,dx, \quad (19)$$

for every measurable subset e of E.

Then applying this Lemma we get the next one.

Lemma 6. *For any numbers $\gamma \neq 0$, $0 < \delta < 1$, $N > 1$ and for any square $\Delta = \Delta_1 \times \Delta_2 \subset T$, there exists a measurable set $E \subset T$ and a polynomial $P(x,y)$ of the form*

$$P(x,y) = \sum_{k,s=N}^{M} c_{k,s}\psi_k(x)\cdot\psi_s(y), \tag{20}$$

with the following properties:

(1)

$$|E| > 1-\delta, \tag{21}$$

(2)

$$\sum_{k,s=N}^{M} \left|c_{k,s}\right|^{2+\delta} < \delta, \tag{22}$$

(3)

$$P(x,y) = \gamma\cdot\chi_\Delta(x,y) \quad \text{for } (x,y) \in E, \tag{23}$$

(4)

$$\begin{aligned} &\max_{N\le\overline{n},\overline{m}\le M}\left[\int\int_e \left|\sum_{k,s=N}^{\overline{n},\overline{m}} c_{k,s}\psi_k(x)\cdot\psi_s(y)\right| dx\,dy\right] \\ &+ \max_{\sqrt{2}N\le R\le\sqrt{2}M}\left[\int\int_e\left|\sum_{2N^2\le k^2+s^2\le R^2} c_{k,s}\psi_k(x)\cdot\psi_s(y)\right| dx\,dy\right] \\ &\le 16\cdot|\gamma|\cdot|\Delta|, \end{aligned} \tag{24}$$

for every measurable subset e of E.

Proof. We apply Lemma 5, setting

$$f(x) = \gamma\cdot\chi_{\Delta_1}(x), \qquad N_0 = N, \qquad \varepsilon = \frac{\delta}{2}. \tag{25}$$

Then we can define a measurable set $E_1 \subset [0,1]$ and a polynomial $P_1(x)$ of the form

$$P_1(x) = \sum_{k=N}^{N_1} a_k\psi_k(x) \tag{26}$$

which satisfy the following conditions:

(1^0)

$$P_1(x) = \gamma\cdot\chi_{\Delta_1}(x) \quad \text{for } x \in E_1, \tag{27}$$

(2^0)

$$\left|E_1\right| > 1 - \frac{\delta}{2}, \tag{28}$$

(3^0)

$$\sum_{k=N}^{N_1}\left|a_k\right|^{2+\delta} < \delta, \tag{29}$$

(4^0)

$$\max_{N\le\overline{n}\le N_1}\left[\int_{e_1}\left|\sum_{k=N}^{\overline{n}} a_k\psi_k(x)\right| dx\right] \le 2\cdot|\gamma|\cdot\left|\Delta_1\right|, \tag{30}$$

for every measurable subset e_1 of E_1.
Set

$$M_0 = 2\cdot\left(N_1^2+1\right) \tag{31}$$

and apply Lemma 5 again, setting

$$f(y) = \chi_{\Delta_2}(y), \qquad N_0 = M_0, \qquad \varepsilon = \frac{\delta}{2}. \tag{32}$$

Then we can define a measurable set $E_2 \subset [0,1]$ and a polynomial $P_2(y)$ of the form

$$P_2(y) = \sum_{s=M_0}^{M} b_s\psi_s(y), \tag{33}$$

which satisfy the following conditions:

(1^{00})

$$P_2(y) = \chi_{\Delta_2}(y) \quad \text{for } y \in E_2, \tag{34}$$

(2^{00})

$$\left|E_2\right| > 1-\frac{\delta}{2}, \tag{35}$$

(3^{00})

$$\sum_{s=M_0}^{M}\left|b_s\right|^{2+\delta} < \delta, \tag{36}$$

(4^{00})

$$\max_{M_0\le\overline{m}\le M}\left[\int_{e_2}\left|\sum_{s=M_0}^{\overline{m}} b_s\psi_s(y)\right| dy\right] \le 2\cdot\left|\Delta_2\right|, \tag{37}$$

for every measurable subset e_2 of E_2.
Set

$$\begin{gathered} E = E_1\times E_2, \\ P(x,y) = P_1(x)\cdot P_2(x) = \sum_{k,s=N}^{M} c_{k,s}\psi_k(x)\cdot\psi_s(y), \end{gathered} \tag{38}$$

where

$$\begin{gathered} c_{k,s} = a_k\cdot b_s, \quad \text{if } N\le k\le N_1,\ M_0\le s\le M, \\ c_{k,s} = 0, \quad \text{for other } k, s. \end{gathered} \tag{39}$$

By (1^0)–(3^0), (1^{00})–(3^{00}), and (38), (39), we obtain

$$|E| > 1 - \delta,$$

$$\sum_{k,s=N}^{M} |c_{k,s}|^{2+\delta} = \sum_{k=N}^{N_1} |a_k|^{2+\delta} \cdot \sum_{s=M_0}^{M} |b_s|^{2+\delta} < \delta, \quad (40)$$

$$P(x,y) = \gamma \cdot \chi_\Delta (x,y) \quad \text{for } (x,y) \in E.$$

Thus, the statements (1)–(3) of Lemma 6 are satisfied. Now we will check the fulfillment of statement (4).

Let $N^2 + M_0^2 < R^2 < N_1^2 + M^2$, then for some $m_0 > M_0$ we have $m_0 < R < m_0 + 1$ and from (31) it follows that $R^2 - N_1^2 > (m_0 - 1)^2$.

Consequently taking relations (4^0), (4^{00}), and (38), (39) for any measurable set $e \subset E$ ($e = e_1 \times e_2$, $e_1 \subset E_1$, $e_2 \subset E_2$) we obtain

$$\begin{aligned}
&\int\int_e \left| \sum_{N^2+M^2 \le k^2+s^2 \le R^2} c_{k,s} \psi_k(x) \cdot \psi_s(y) \right| dx\,dy \\
&\le \int\int_e \left| \sum_{k=N}^{N_1} \sum_{s=M_0}^{m_0-1} c_{k,s} \psi_k(x) \cdot \psi_s(y) \right| dx\,dy \\
&+ \max_{N<n\le N_1} \left[\int\int_e \left| \sum_{k=N}^{n} c_{k,m_0} \psi_k(x) \cdot \psi_{m_0}(y) \right| dx\,dy \right] \\
&\le \left[\int_{e_1} \left| \sum_{k=N}^{N_1} a_k \psi_k(x) \right| dx \right] \cdot \left[\int_{e_2} \left| \sum_{s=M_0}^{m_0-1} b_s \psi_s(y) \right| dy \right] \\
&+ |b_{m_0}| \cdot \left[\int_{e_2} |\psi_{m_0}(y)|\, dy \right] \\
&\cdot \max_{N<n\le N_1} \left[\int_{e_1} \left| \sum_{k=N}^{n} a_k \psi_k(x) \right| dx \right] \\
&\le 12 \cdot |\gamma| \cdot |\Delta|.
\end{aligned} \quad (41)$$

Similarly, for $N \le \overline{n} \le N_1$, $M_0 \le \overline{m} \le M$, we get

$$\int\int_e \left| \sum_{k,s=N}^{\overline{n},\overline{m}} c_{k,s} \psi_k(x) \cdot \psi_s(y) \right| dx\,dy \le 4 \cdot |\gamma| \cdot |\Delta|. \quad (42)$$

Lemma 6 is proved. □

Lemma 7. *For any numbers $\varepsilon > 0$, $N > 1$ and a step function*

$$f(x,y) = \sum_{\nu=1}^{\nu_0} \gamma_\nu \cdot \chi_{\Delta_\nu}(x,y), \quad (43)$$

there exists a measurable set $E \subset T$ and a polynomial $P(x,y)$ of the form

$$P(x,y) = \sum_{k,s=N}^{M} c_{k,s} \psi_k(x) \cdot \psi_s(y), \quad (44)$$

which satisfy the following conditions:

(1^0)

$$P(x,y) = f(x,y) \quad \textit{for } (x,y) \in E, \quad (45)$$

(2^0)

$$|E| > 1 - \varepsilon, \quad (46)$$

(3^0)

$$\sum_{k,s=N}^{M} |c_{k,s}|^{2+\varepsilon} < \varepsilon, \quad (47)$$

(4^0)

$$\begin{aligned}
&\max_{N \le \overline{n},\overline{m} < M} \left[\int\int_e \left| \sum_{k,s=N}^{\overline{n},\overline{m}} c_{k,s} \psi_k(x) \cdot \psi_s(y) \right| dx\,dy \right] \\
&+ \max_{\sqrt{2}N \le R \le \sqrt{2}M} \left[\int\int_e \left| \sum_{2N^2 \le k^2+s^2 \le R^2} c_{k,s} \psi_k(x) \cdot \psi_s(y) \right| dx\,dy \right] \\
&\le 2 \cdot \int\int_e |f(x,y)|\, dx\,dy + \varepsilon,
\end{aligned} \quad (48)$$

for every measurable subset e of E.

Proof. Without any loss of generality, we assume that

$$\max_{1 \le \nu \le \nu_0} (|\gamma_\nu| \cdot |\Delta_\nu|) < \frac{\varepsilon}{32}, \quad (49)$$

Δ_ν, $1 \le \nu \le \nu_0$ are the constancy rectangular domain of $f(x,y)$, that is, where the function $f(x,y)$ is constant.

Given an integer $1 \le \nu \le \nu_0$, by applying Lemma 6 with $\delta = \varepsilon/16\nu_0$, we find that there exists a measurable set $E_\nu \subset T$ and a polynomial $P_\nu(x,y)$ of the form

$$P_\nu(x,y) = \sum_{k,s=N_\nu}^{M_\nu} c_{k,s}^{(\nu)} \psi_k(x) \cdot \psi_s(y) \quad (50)$$

with the following properties:

$$|E_\nu| > 1 - \frac{\varepsilon}{2^\nu}, \quad (51)$$

$$\sum_{k,s=N_\nu}^{M_\nu} \left|c_{k,s}^{(\nu)}\right|^{2+\varepsilon} < \frac{\varepsilon}{\nu_0}, \quad (52)$$

$$P_\nu(x,y) = \gamma_\nu \cdot \chi_{\Delta_\nu}(x,y) \quad \text{for } (x,y) \in E_\nu, \quad (53)$$

$$\max_{N_\nu \le \overline{n},\overline{m} \le M_\nu} \left[\int\int_e \left| \sum_{k,s=N_\nu}^{\overline{n},\overline{m}} c_{k,s}^{(\nu)} \psi_k(x) \cdot \psi_s(y) \right| dx\,dy \right] + \max_{\sqrt{2}N_\nu \le R \le \sqrt{2}M_\nu} \left[\int\int_e \left| \sum_{2N_\nu^2 \le k^2+s^2 \le R^2} c_{k,s}^{(\nu)} \psi_k(x) \cdot \psi_s(y) \right| dx\,dy \right] \le 16 \cdot |\gamma_\nu| \cdot |\Delta_\nu| < \frac{\varepsilon}{2}, \quad (54)$$

for every measurable subset e of E_ν (see (49)).

Then we can take

$$N_1 = N, \qquad n_\nu = M_{\nu-1} + 1, \qquad 1 \le \nu \le \nu_0. \quad (55)$$

Set

$$E = \bigcap_{\nu=1}^{\nu_0} E_\nu, \quad (56)$$

$$P(x,y) = \sum_{\nu=1}^{\nu_0} P_\nu(x,y) = \sum_{k,s=N}^{M} c_{k,s} \psi_k(x) \cdot \psi_s(y), \qquad M = M_{\nu_0}, \quad (57)$$

where

$$c_{k,s} = c_{k,s}^{(\nu)}, \quad \text{for } N_\nu \le k, s \le M_\nu, \ 1 \le \nu \le \nu_0, \quad (58)$$

$$c_{k,s} = 0, \quad \text{for other } k, s. \quad (59)$$

From (51)–(53) and (56)–(58) we obtain

$$P(x,y) = f(x,y) \quad \text{for } (x,y) \in E, \qquad |E| > 1 - \varepsilon, \qquad \sum_{k,s=N}^{M} |c_{k,s}|^{2+\varepsilon} < \sum_{\nu=1}^{\nu_0} \left[\sum_{k,s=N_\nu}^{M_\nu} |c_{k,s}^{(\nu)}|^{2+\varepsilon} \right] < \varepsilon. \quad (60)$$

Then, let $R \in [\sqrt{2}N, \sqrt{2}M]$, then for some ν', $1 \le \nu' \le \nu_0$ we have $\sqrt{2}N_{\nu'} \le R \le \sqrt{2}N_{\nu'+1}$, consequently from (57) and (58) we have

$$\sum_{2N^2 \le k^2+s^2 \le R^2} c_{k,s} \psi_k(x) \cdot \psi_s(y) = \sum_{\nu=1}^{\nu'-1} P_\nu(x,y) + \sum_{2N_{\nu'}^2 \le k^2+s^2 \le R^2} c_{k,s}^{(\nu')} \psi_k(x) \cdot \psi_s(y). \quad (61)$$

In view of the conditions (51)–(54) and the equality $P(x,y) = f(x,y)$ on E, for any measurable set $e \subset E$ we obtain

$$\int\int_e \left| \sum_{2N^2 \le k^2+s^2 \le R^2} c_{k,s} \psi_k(x) \cdot \psi_s(y) \right| dx\,dy \le \int\int_e \left| \sum_{\nu=1}^{\nu'-1} P_\nu(x,y) \right| dx\,dy + \int\int_e \left| \sum_{2N_{\nu'}^2 \le k^2+s^2 \le R^2} c_{k,s}^{(\nu')} \psi_k(x) \cdot \psi_s(y) \right| dx\,dy \le \int\int_e |f(x,y)|\,dx\,dy + \frac{\varepsilon}{2}. \quad (62)$$

Similarly, for any $e \subset E$ we have

$$\max_{N \le \overline{n},\overline{m} \le M} \left[\int\int_e \left| \sum_{k,s=N}^{\overline{n},\overline{m}} c_{k,s} \psi_k(x) \cdot \psi_s(y) \right| dx\,dy \right] \le \int\int_e |f(x,y)|\,dx\,dy + \frac{\varepsilon}{2}. \quad (63)$$

Lemma 7 is proved. □

4. Proofs of the Theorems

Theorem 3 is proved similarly [10, Theorem 3], but for maintenance of integrity of this paper, here we will give the proof.

Proof of Theorem 3. Let

$$\{f_s(x,y)\}_{s=1}^{\infty}, \quad (x,y) \in T \quad (64)$$

be a sequence of all step functions, values, and constancy interval endpoints which are rational numbers. Applying Lemma 7 consecutively, we can find a sequence $\{E_s\}_{s=1}^{\infty}$ of sets and a sequence of polynomials

$$P_s(x,y) = \sum_{k,\nu=N_{s-1}}^{N_s-1} c_{k,\nu}^{(s)} \psi_k(x) \psi_\nu(y), \qquad 1 = N_0 < N_1 < \cdots < N_s < \cdots, \quad s = 1, 2, \ldots, \quad (65)$$

which satisfy the following conditions:

$$P_s(x,y) = f_s(x,y), \quad (x,y) \in E_s, \quad (66)$$

$$|E_s| > 1 - 2^{-2(s+1)}, \quad E_s \subset T, \quad (67)$$

$$\sum_{k,\nu=N_{s-1}}^{N_s-1} \left| c_{k,\nu}^{(s)} \right|^{2+2^{-2s}} < 2^{-2s}, \quad (68)$$

$$\max_{N_{s-1}\le \overline{n},\overline{m}<N_s}\left[\int\int_e\left|\sum_{k,\nu=N_{s-1}}^{\overline{n},\overline{m}} c_{k,\nu}^{(s)}\psi_k(x)\cdot\psi_\nu(y)\right|dx\,dy\right] + \max_{\sqrt{2}N_{s-1}\le R\le\sqrt{2}N_s}\left[\int\int_e\left|\sum_{2N_{s-1}^2\le k^2+\nu^2\le R^2} c_{k,\nu}^{(s)}\psi_k(x)\cdot\psi_\nu(y)\right|dx\,dy\right] \le 2\cdot\int\int_e|f_s(x,y)|\,dx\,dy + 2^{-2(s+1)}, \tag{69}$$

for every measurable subset e of E_s.

Denote

$$\sum_{k,\nu=1}^{\infty} c_{k,\nu}\psi_k(x)\psi_\nu(y) = \sum_{s=1}^{\infty}\left[\sum_{k,\nu=N_{s-1}}^{N_s-1} c_{k,\nu}^{(s)}\psi_k(x)\psi_\nu(y)\right], \tag{70}$$

where

$$c_{k,\nu} = c_{k,\nu}^{(s)}, \quad \text{for } N_{s-1}\le k,\ \nu<N_s,\ s=1,2,\ldots. \tag{71}$$

For an arbitrary number $\varepsilon>0$ we set

$$\Omega_n = \bigcap_{s=n}^{\infty}E_s, \quad n=1,2,\ldots,$$
$$E=\Omega_{n_0}=\bigcap_{s=n_0}^{\infty}E_s, \quad n_0=\left[\log_{1/1}\varepsilon\right]+1, \tag{72}$$
$$B=\bigcup_{n=n_0}^{\infty}\Omega_n=\Omega_{n_0}\bigcup\left(\bigcup_{n=n_0+1}^{\infty}\Omega_n\setminus\Omega_{n-1}\right).$$

It is obvious (see (67) and (72)) that $|B|=1$ and $|E|>1-\varepsilon$.

We define a function $\mu(x,y)$ in the following way:

$$\mu(x,y)=\begin{cases}1, & \text{for } (x,y)\in E\cup(T\setminus B);\\ \mu_n, & \text{for } (x,y)\in\Omega_n\setminus\Omega_{n-1},\ n\ge n_0+1,\end{cases} \tag{73}$$

where

$$\mu_n=\left[2^{2n}\cdot\prod_{s=1}^{n}h_s\right]^{-1},$$
$$h_s=\|f_s\|_C+\max_{N_{s-1}\le\overline{n},\overline{m}<N_s}\left\|\sum_{k,\nu=N_{s-1}}^{\overline{n},\overline{m}}c_{k,\nu}^{(s)}\psi_k(x)\cdot\psi_\nu(y)\right\|_C + \max_{\sqrt{2}N_{s-1}\le R\le\sqrt{2}N_s}\left\|\sum_{2N_{s-1}^2\le k^2+\nu^2\le R^2}c_{k,\nu}^{(s)}\psi_k(x)\cdot\psi_\nu(y)\right\|_C+1. \tag{74}$$

From (68) and (70)–(74) we obtain the following:

(A) $0<\mu(x,y)\le1$, $\mu(x,y)$ is a measurable function and

$$\left|\{(x,y)\in T:\mu(x,y)\ne1\}\right|<\varepsilon. \tag{75}$$

(B) Consider $\sum_{k,\nu=1}^{\infty}|c_{k,\nu}|^q<\infty$ for all $q>2$.

Hence, obviously we have (see (68) and (70))

$$\lim_{\min\{k,\nu\}\to\infty}c_{k,\nu}=0. \tag{76}$$

It follows from (72)–(74) that for all $s\ge n_0$ and $N_{s-1}\le\overline{n},\ \overline{m}<N_s$

$$\int\int_{T\setminus\Omega_s}\left|\sum_{k,\nu=N_{s-1}}^{\overline{n},\overline{m}}c_{k,\nu}^{(s)}\psi_k(x)\cdot\psi_\nu(y)\right|\mu(x,y)\,dx\,dy = \sum_{n=s+1}^{\infty}\left[\int\int_{\Omega_n\setminus\Omega_{n-1}}\left|\sum_{k,\nu=N_{s-1}}^{\overline{n},\overline{m}}c_{k,\nu}^{(s)}\psi_k(x)\cdot\psi_\nu(y)\right|\mu_n\,dx\,dy\right] \le \sum_{n=s+1}^{\infty}2^{-2n}\left[\int\int_T\left|\sum_{k,\nu=N_{s-1}}^{\overline{n},\overline{m}}c_{k,\nu}^{(s)}\psi_k(x)\cdot\psi_\nu(y)\right|h_s^{-1}dx\,dy\right] < \frac{1}{3}2^{-2s}. \tag{77}$$

Analogously for all $s\ge n_0$ and $\sqrt{2}N_{s-1}\le R\le\sqrt{2}N_s$ we have

$$\int\int_{T\setminus\Omega_s}\left|\sum_{2N_{s-1}^2\le k^2+\nu^2\le R^2}c_{k,\nu}^{(s)}\psi_k(x)\cdot\psi_\nu(y)\right| \times\mu(x,y)\,dx\,dy<\frac{1}{3}2^{-2s}. \tag{78}$$

By (65) and (72)–(74) for all $s\ge n_0$ we have

$$\int\int_T|P_s(x,y)-f_s(x,y)|\,\mu(x,y)\,dx\,dy = \int\int_{\Omega_s}|P_s(x,y)-f_s(x,y)|\,\mu(x,y)\,dx\,dy + \int\int_{T\setminus\Omega_s}|P_s(x,y)-f_s(x,y)|\,\mu(x,y)\,dx\,dy = \sum_{n=s+1}^{\infty}\left[\int\int_{\Omega_n\setminus\Omega_{n-1}}|P_s(x,y)-f_s(x,y)|\,\mu_n\,dx\,dy\right] \le \sum_{n=s+1}^{\infty}2^{-2n}\left[\int\int_T\left(|f_s(x,y)|+\sum_{k,\nu=N_{s-1}}^{N_s-1}c_{k,\nu}^{(s)}\psi_k(x)\cdot\psi_\nu(y)\right)h_s^{-1}dx\,dy\right] < \frac{1}{3}2^{-2s}<2^{-2s}. \tag{79}$$

By (69) and (72)–(77) for all $N_{s-1} \le \overline{n},\ \overline{m} < N_s$ and $s \ge n_0 + 1$ we obtain

$$\int\int_T \left| \sum_{k,\nu=N_{s-1}}^{\overline{n},\overline{m}} c_{k,\nu}^{(s)} \psi_k(x) \cdot \psi_\nu(y) \right| \mu(x,y)\, dx\, dy$$

$$\int\int_{\Omega_s} \left| \sum_{k,\nu=N_{s-1}}^{\overline{n},\overline{m}} c_{k,\nu}^{(s)} \psi_k(x) \cdot \psi_\nu(y) \right| \mu(x,y)\, dx\, dy$$

$$\int\int_{T\setminus\Omega_s} \left| \sum_{k,\nu=N_{s-1}}^{\overline{n},\overline{m}} c_{k,\nu}^{(s)} \psi_k(x) \cdot \psi_\nu(y) \right| \mu(x,y)\, dx\, dy$$

$$< \sum_{n=n_0+1}^{s} \left[\int\int_{\Omega_n\setminus\Omega_{n-1}} \left| \sum_{k,\nu=N_{s-1}}^{\overline{n},\overline{m}} c_{k,\nu}^{(s)} \psi_k(x) \cdot \psi_\nu(y) \right| \cdot \mu_n\, dx\, dy \right] + \frac{1}{3} 2^{-2s}$$

$$< \sum_{n=n_0+1}^{s} \left(2^{-2(s+1)} + 2 \cdot \int\int_{\Omega_n\setminus\Omega_{n-1}} |f_s(x,y)|\, dx\, dy \right) \cdot \mu_n + \frac{1}{3} 2^{-2s}$$

$$= 2^{-2(s+1)} \cdot \sum_{n=n_0+1}^{s} \mu_n + \int\int_{\Omega_s} |f_s(x,y)|\, \mu(x,y)\, dx\, dy + \frac{1}{3} 2^{-2s}$$

$$< 2 \cdot \int\int_T |f_s(x,y)|\, \mu(x,y)\, dx\, dy + 2^{-2s}. \tag{80}$$

Analogously for all $s \ge n_0$ and $\sqrt{2}N_{s-1} \le R \le \sqrt{2}N_s$ we have (see (78))

$$\int\int_T \left| \sum_{2N_{s-1}^2 \le k^2+\nu^2 \le R^2} c_{k,\nu}^{(s)} \psi_k(x) \cdot \psi_\nu(y) \right| \mu(x,y)\, dx\, dy < 2 \cdot \int\int_T |f_s(x,y)|\, \mu(x,y)\, dx\, dy + 2^{-2s}. \tag{81}$$

Now we will show that the series (70) is universal in $L_\mu^1(T)$ concerning subseries with respect to convergence by both spherical and rectangular partial sums.

Let $f(x,y) \in L_\mu^1(T)$, that is,

$$\int\int_T |f(x,y)|\, \mu(x,y)\, dx\, dy < \infty. \tag{82}$$

It is easy to see that we can choose a function $f_{n_1}(x,y)$ from the sequence (64) such that

$$\int\int_T |f(x,y) - f_{n_1}(x,y)|\, \mu(x,y)\, dx\, dy < 2^{-2}, \quad n_1 > n_0 + 1. \tag{83}$$

Hence, we have

$$\int\int_T |f_{n_1}(x,y)|\, \mu(x,y)\, dx\, dy < 2^{-2} + \int\int_T |f(x,y)|\, \mu(x,y)\, dx\, dy. \tag{84}$$

From (80) and (83) we get

$$\begin{aligned}\int\int_T |f(x,y) - P_{n_1}(x,y)|\, \mu(x,y)\, dx\, dy &\le \int\int_T |f(x,y) - f_{n_1}(x,y)|\, \mu(x,y)\, dx\, dy \\ &\quad + \int\int_T |f_{n_1}(x,y) - P_{n_1}(x,y)|\, \mu(x,y)\, dx\, dy \\ &< 2 \cdot 2^{-2}.\end{aligned} \tag{85}$$

Assume that numbers $n_1 < n_2 < \cdots < n_{q-1}$ are chosen in such a way that the following condition is satisfied:

$$\int\int_T \left| f(x,y) - \sum_{s=1}^{j} P_{n_s}(x,y) \right| \mu(x,y)\, dx\, dy < 2 \cdot 2^{-2j}, \quad 1 \le j \le q-1. \tag{86}$$

Now we choose a function $f_{n_q}(x,y)$ from the sequence (64) such that

$$\int\int_T \left| \left(f(x,y) - \sum_{s=1}^{q-1} P_{n_s}(x,y) \right) - f_{n_q}(x,y) \right| \mu(x,y)\, dx\, dy < 2 \cdot 2^{-2q}, \quad n_q > n_{q-1}. \tag{87}$$

This with (86) implies

$$\int\int_T |f_{n_q}(x,y)|\, \mu(x,y)\, dx\, dy < 2^{-2q} + 2 \cdot 2^{-2(q-1)} = 9 \cdot 2^{-2q}. \tag{88}$$

Hence and from (65) and (79)–(81) we obtain

$$\int\int_T |f_{n_q}(x,y) - P_{n_q}(x,y)|\, \mu(x,y)\, dx\, dy < 2^{-2n_q}, \tag{89}$$

where

$$P_{n_q}(x,y) = \sum_{k,\nu=N_{n_q-1}}^{N_{n_q}-1} c_{k,\nu}^{(n_q)} \psi_k(x) \psi_\nu(y), \tag{90}$$

$$\max_{N_{n_q-1} \le \overline{n},\overline{m} < N_{n_q}} \left[\int\int_T \left| \sum_{k,\nu=N_{n_q-1}}^{\overline{n},\overline{m}} c_{k,\nu}^{(n_q)} \psi_k(x) \cdot \psi_\nu(y) \right| \mu(x,y)\, dx\, dy \right] < 19 \cdot 2^{-2q}. \tag{91}$$

Analogously we have

$$\max_{\sqrt{2}N_{n_q-1}\le R\le\sqrt{2}N_{n_q}}\left[\int\int_T\left|\sum_{2N^2_{n_q-1}\le k^2+\nu^2\le R^2} c_{k,\nu}^{(n_q)}\psi_k(x)\cdot\psi_\nu(y)\right|\mu(x,y)\,dx\,dy\right] < 19\cdot 2^{-2q}. \quad (92)$$

In quality subseries of the theorem we will take

$$\sum_{q=1}^{\infty}P_{n_q}(x,y)=\sum_{q=1}^{\infty}\left[\sum_{k,\nu=N_{n_q-1}}^{N_{n_q}-1}c_{k,\nu}^{(n_q)}\psi_k(x)\psi_\nu(y)\right]. \quad (93)$$

From (87) and (88) we have

$$\begin{aligned}\int\int_T&\left|f(x,y)-\sum_{s=1}^{q}P_{n_s}(x,y)\right|\mu(x,y)\,dx\,dy\\ &\le\int\int_T\left|\left(f(x,y)-\sum_{s=1}^{q-1}P_{n_s}(x,y)\right)-f_{n_q}(x,y)\right|\\ &\times\mu(x,y)\,dx\,dy\\ &+\int\int_T\left|f_{n_q}(x,y)-P_{n_q}(x,y)\right|\mu(x,y)\,dx\,dy<2\cdot 2^{-2q}.\end{aligned} \quad (94)$$

Let $\overline{n}$ and $\overline{m}$ be arbitrary natural numbers. Then for some natural number q we have

$$N_{n_q-1}\le\min\{\overline{n},\overline{m}\}<N_{n_q}. \quad (95)$$

Taking into account (89) and (93) for rectangular partial sums $S_{\overline{n},\overline{m}}(x,y)$ of (91) we get

$$\begin{aligned}\int\int_T&\left|S_{\overline{n},\overline{m}}(x,y)-f(x,y)\right|\mu(x,y)\,dx\,dy\\ &\le\int\int_T\left|f(x,y)-\sum_{s=1}^{q}P_{n_s}(x,y)\right|\mu(x,y)\,dx\,dy\\ &+\max_{N_{n_q-1}\le\overline{n},\overline{m}<N_{n_q}}\left[\int\int_T\left|\sum_{k,\nu=N_{n_q-1}}^{\overline{n},\overline{m}}c_{k,\nu}^{(n_q)}\psi_k(x)\cdot\psi_\nu(y)\right|\mu(x,y)\,dx\,dy\right]\\ &<21\cdot 2^{-2q}.\end{aligned} \quad (96)$$

Analogously for $\sqrt{2}N_{n_q-1}\le R\le\sqrt{2}N_{n_q}$ we have

$$\int\int_T\left|S_R(x,y)-f(x,y)\right|\mu(x,y)\,dx\,dy<21\cdot 2^{-2q}, \quad (97)$$

where $S_R(x,y)$ is the spherical partial sums of (91).

From (96) and (97) we conclude that the series (70) is universal in $L^1_\mu(T)$ concerning subseries with respect to convergence by both spherical and rectangular partial sums (see Definition 2).

Theorem 3 is proved. □

Remark 8. We can prove Theorem 4 by the same method used in the proof of Theorem 3.

References

[1] D. E. Menshov, "On the partial summs of trigonometric series," *Studia Mathematica*, vol. 20, pp. 197–238, 1947 (Russian).

[2] V. Ya. Kozlov, "On the complete systems of orthogonal functions," *Matematicheskii Sbornik*, vol. 26, pp. 351-364, 1950 (Russian).

[3] A. A. Talalian, "On the universal series with respect to rearrangements," *Izvestiya Akademii Nauk SSSR*, vol. 24, pp. 567-604, 1960 (Russian).

[4] O. P. Dzangadze, "On the universal double series," *Bulletin of the Georgian Academy of Sciences*, vol. 34, pp. 225–228, 1964 (Russian).

[5] W. Orlicz, "Über die unabhangig von der Anordnung fast überallberall kniwergenten Reihen," *Bulletin de l'Academie Polonaise des Sciences*, vol. 81, pp. 117–125, 1927.

[6] M. G. Grigorian, "On the representation of functions by orthogonal series in weighted L^p spaces," *Studia Mathematica*, vol. 134, no. 3, pp. 211–237, 1999.

[7] S. A. Episkoposian, "On the series by Walsh system universal in weighted $L^1_\mu[0,1]$ spaces," *Izvestiya Natsional'noĭ Akademii Nauk Armenii*, 1999, English Translation in: *Journal of Contemporary Mathematical Analysis*, vol. 34, pp. 25–40, 1999.

[8] M.G. Grigorian and S.A. Episkoposian, "Representation of functions in weighted spaces $L^1_\mu[0,1]$ by trigonometric and Walsh series," *Analysis Mathematica*, vol. 27, pp. 267–277, 2001.

[9] S. A. Episkoposian, "On the existence of universal series by trigonometric system," *Journal of Functional Analysis*, vol. 230, no. 1, pp. 169–183, 2006.

[10] S. A. Episkoposian, "Existence of double Walsh series universal in weighted spaces," *International Journal of Modern Mathematics*, vol. 2, pp. 231–247, 2007.

[11] H. E. Chrestenson, "A class of generalized Walsh functions," *Pacific Journal of Mathematics*, vol. 5, pp. 17–31, 1955.

[12] N. Ja. Vilenkin, "On a class of complete orthonormal systems," *American Mathematical Society Translations*, vol. 28, pp. 1–35, 1963.

[13] R. E. A. C. Paley, "A remarkable set of orthogonal functions," *London Mathematical Society*, vol. 34, pp. 241–279, 1932.

[14] N. J. Fine, "The generalized Walsh functions," *Transactions of the American Mathematical Society*, vol. 69, pp. 66–77, 1950.

[15] W. S. Young, "Mean convergence of generalized Walsh-Fourier series," *Transactions of the American Mathematical Society*, vol. 218, pp. 311–320, 1976.

[16] C. Watari, "On generalized Walsh Fourier series," *The Tohoku Mathematical Journal*, vol. 10, pp. 211–241, 1958.

[17] S. A. Episkoposian, "L^1-convergence of greedy algorithm by generalized Walsh system," *Banach Journal of Mathematical Analysis*, vol. 6, no. 1, pp. 161–174, 2012.

A Simple Method for Obtaining Coupled Fixed Points of α-ψ-Contractive Type Mappings

Sh. Rezapour[1,2] **and J. Hasanzade Asl**[1]

[1] *Department of Mathematics, Science and Research Branch, Islamic Azad University, Tehran, Iran*
[2] *Department of Mathematics, Azarbaijan Shahid Madani University, Azarshahr, Tabriz, Iran*

Correspondence should be addressed to Sh. Rezapour; sh.rezapour@azaruniv.edu

Academic Editor: Ahmed Zayed

In 2012, the notion of α-ψ-contractive type mappings was introduced by Samet, C. Vetro, and P. Vetro. By using a simple method, we give some coupled fixed point results for α-ψ-contractive type mappings.

1. Introduction

In 1987, Guo and Lakshmikantham introduced the notion of coupled fixed points [1]. Then some authors proved some coupled fixed point results via some applications in the last decade of the previous century [2–6]. Later, this field was completed by some researchers by using different sights (see for example [7–18]). In 2012, Samet et al. introduced the notion of α-ψ-contractive type mappings [19]. Also, Amini-Harandi has provided a method for obtaining coupled fixed point results [20]. The aim of this paper is to provide a simple method for obtaining some coupled fixed point results for α-ψ-contractive type mappings.

Denote by Ψ the family of nondecreasing functions $\psi : [0,\infty) \to [0,\infty)$ such that $\sum_{n=1}^{\infty} \psi^n(t) < \infty$ for all $t > 0$, where ψ^n is the nth iterate of ψ. It is known that $\psi(t) < t$ for all $t > 0$ and $\psi \in \Psi$ [19]. Let (X, d) be a metric space and T a self-map on X. Then T is called a α-ψ-contraction mapping whenever there exist $\psi \in \Psi$ and $\alpha : X \times X \to [0,\infty)$ such that $\alpha(x, y)d(Tx, Ty) \le \psi(d(x, y))$ for all $x, y \in X$ [19]. Also, we say that T is α-admissible whenever $\alpha(x, y) \ge 1$ implies $\alpha(Tx, Ty) \ge 1$ [19]. Also, we say that X has the property (B) if $\{x_n\}$ is a sequence in X such that $\alpha(x_n, x_{n+1}) \ge 1$ for all $n \ge 1$ and $x_n \to x$, then $\alpha(x_n, x) \ge 1$ for all $n \ge 1$ [19]. Let (X, d) be a complete metric space and T a α-admissible α-ψ-contractive mapping on X. Suppose that there exists $x_0 \in X$ such that $\alpha(x_0, Tx_0) \ge 1$. If T is continuous or X has the property (B), then T has a fixed point (see [19]; Theorems 2.1 and 2.2). Also, we say that X has the property (H) whenever for each $x, y \in X$ there exists $z \in X$ such that $\alpha(x, z) \ge 1$ and $\alpha(y, z) \ge 1$. If X has the property (H) in the Theorems 2.1 and 2.2, then X has a unique fixed point ([19]; Theorem 2.3). It is considerable that the results of Samet et al. generalize similar ordered results in the literature (see the results of third section in [19]). Let $F : X \times X \to X$ be a mapping, where (X, d) is a metric space. We say that $(x_0, y_0) \in X \times X$ is a coupled fixed point of F whenever $F(x_0, y_0) = x_0$ and $F(y_0, x_0) = y_0$. Define $T_F : X\times X \to X\times X$ by $T_F(x, y) = (F(x, y), F(y, x))$ for all $(x, y) \in X \times X$. Then, it is easy to check that (x_0, y_0) is a coupled fixed point of F if and only if (x_0, y_0) is a fixed point of T_f.

2. Main Results

In this section, we define $\delta((x, y), (u, v)) = d(x, u) + d(y, v)$ and

$$
\begin{aligned}
m&((x, y), (u, v)) \\
&= \max\Big\{\delta((x, y), (u, v)), \delta((x, y), (F(x, y), F(y, x))), \\
&\qquad \delta((u, v), (F(u, v), F(v, u))),
\end{aligned}
$$

$$\frac{1}{2}\left[\delta\left((x,y),(F(u,v),F(v,u))\right)\right.$$
$$\left.\left.+\delta\left((u,v),(F(x,y),F(y,x))\right)\right]\right\} \quad (1)$$

for all $(x,y),(u,v) \in X \times X$, where (X,d) is a metric space and $F : X\times X \to X$ is a mapping. Now, we are ready to state and prove our main results.

Lemma 1. *Let (X,d) be a complete metric space, $\alpha : X\times X \to [0,\infty)$ a function, $\psi \in \Psi$ and T a self-map on X such that*

$$\begin{aligned}\alpha(x,y)\,d(Tx,Ty) &\le \psi\Big(\max\Big\{d(x,y),d(x,Tx),d(y,Ty),\\ &\qquad \frac{1}{2}\left[d(x,Ty)+d(y,Tx)\right]\Big\}\Big)\end{aligned} \quad (2)$$

for all $x, y \in X$. Suppose that T is α-admissible and there exists $x_0 \in X$ such that $\alpha(x_0,Tx_0) \ge 1$. If X has the property (B), then T has a fixed point.

Proof. Take $x_0 \in X$ such that $\alpha(x_0,Tx_0) \ge 1$ and define the sequence $\{x_n\}$ in X by $x_{n+1} = Tx_n$ for all $n \ge 0$. If $x_n = x_{n+1}$ for some n, then $x^* = x_n$ is a fixed point of T. Assume that $x_n \neq x_{n+1}$ for all n. Since T is α-admissible, it is easy to check that $\alpha(x_n,x_{n+1}) \ge 1$ for all $n \ge 1$. Thus, for each natural number n one has

$$\begin{aligned}d(x_n,x_{n+1}) &= d(Tx_{n-1},Tx_n) \le \alpha(x_{n-1},x_n)\,d(Tx_{n-1},Tx_n)\\ &\le \psi\Big(\max\Big\{d(x_n,x_{n-1}),d(x_n,x_{n+1}),d(x_{n-1},x_n),\\ &\qquad \frac{1}{2}\left[d(x_n,x_n)+d(x_{n-1},x_{n+1})\right]\Big\}\Big)\\ &\le \psi\Big(\max\Big\{d(x_n,x_{n-1}),d(x_n,x_{n+1}),\\ &\qquad \frac{1}{2}\left[d(x_n,x_{n-1})+d(x_n,x_{n+1})\right]\Big\}\Big)\\ &= \psi\left(\max\left\{d(x_n,x_{n-1}),d(x_n,x_{n+1})\right\}\right).\end{aligned} \quad (3)$$

If $\max\{d(x_n,x_{n-1}),d(x_n,x_{n+1})\} = d(x_n,x_{n+1})$, then

$$d(x_{n+1},x_n) \le \psi\left(d(x_n,x_{n+1})\right) < d(x_{n+1},x_n) \quad (4)$$

which is contradiction. Thus, $\max\{d(x_n,x_{n-1}), d(x_n,x_{n+1})\} = d(x_n,x_{n-1})$ for all n. Hence, $d(x_{n+1},x_n) \le \psi(d(x_n,x_{n-1}))$ and so $d(x_{n+1},x_n) \le \psi^n(d(x_1,x_0))$ for all n. It is easy to check that $\{x_n\}$ is a Cauchy sequence. Thus, there exists $x^* \in X$ such that $x_n \to x^*$. By using the assumption, we have $\alpha(x_n,x^*) \ge 1$ for all n. Thus,

$$\begin{aligned}d(Tx^*,x^*) &\le d(Tx^*,Tx_n)+d(x_{n+1},x^*)\\ &\le \alpha(x_n,x^*)\,d(Tx^*,Tx_n)+d(x_{n+1},x^*)\\ &\le \psi\Big(\max\Big\{d(x_n,x^*),d(x_n,x_{n+1}),d(x^*,Tx^*),\\ &\qquad \frac{1}{2}\left[d(x_n,Tx^*)+d(x^*,x_{n+1})\right]\Big\}\Big)\\ &\quad + d(x_{n+1},x^*)\\ &= \psi\left(\max\left\{d(x_n,x^*),d(x_n,x_{n+1})\right\}\right)+d(x_{n+1},x^*)\end{aligned} \quad (5)$$

for all n. Hence, $d(Tx^*,x^*) = 0$ and so $Tx^* = x^*$.

By using a similar argument we can prove the following results.

Lemma 2. *Let (X,d) be a complete metric space, $\alpha : X\times X \to [0,\infty)$ a function, $\psi \in \Psi$ and T a continuous self-map on X such that*

$$\begin{aligned}\alpha(x,y)\,d(Tx,Ty) &\le \psi\Big(\max\Big\{d(x,y),d(x,Tx),d(y,Ty),\\ &\qquad \frac{1}{2}\left[d(x,Ty)+d(y,Tx)\right]\Big\}\Big)\end{aligned} \quad (6)$$

for all $x, y \in X$. Suppose that T is α-admissible and there exists $x_0 \in X$ such that $\alpha(x_0,Tx_0) \ge 1$. Then T has a fixed point.

Lemma 3. *Let (X,d) be a complete metric space, $\alpha : X\times X \to [0,\infty)$ a function, $\psi \in \Psi$ and T a self-map on X such that*

$$\alpha(x,y)\,d(Tx,Ty) \le \psi\left(d(x,u)+d(y,v)\right) \quad (7)$$

for all $x, y \in X$. Suppose that T is α-admissible and there exists $x_0 \in X$ such that $\alpha(x_0,Tx_0) \ge 1$. If X has the property (B), then T has a fixed point.

Lemma 4. *Let (X,d) be a complete metric space, $\alpha : X\times X \to [0,\infty)$ a function, $\psi \in \Psi$ and T a continuous self-map on X such that*

$$\alpha(x,y)\,d(Tx,Ty) \le \psi\left(d(x,u)+d(y,v)\right) \quad (8)$$

for all $x, y \in X$. Suppose that T is α-admissible and there exists $x_0 \in X$ such that $\alpha(x_0,Tx_0) \ge 1$. Then T has a fixed point.

Theorem 5. *Let (X,d) be a complete metric space and $F : X \times X \to X$ a mapping. Suppose that there exist a mapping $\alpha : X^2 \times X^2 \to [0,+\infty)$ and $\psi \in \Psi$ such that $\alpha((x,y),(u,v))d(F(x,y),F(u,v)) \le (1/2)\psi(m((x,y),(u,v)))$ for all $(x,y),(u,v) \in X \times X$, $\alpha((x,y),(u,v)) \ge 1$ implies*

$$\alpha\left((F(x,y),F(y,x)),(F(u,v),F(v,u))\right) \ge 1 \quad (9)$$

for all $(x, y), (u, v) \in X \times X$ *and there exists* $(x_0, y_0) \in X \times X$ *such that*

$$\alpha((x_0, y_0), (F(x_0, y_0), F(y_0, x_0))) \geq 1 \tag{10}$$

and $\alpha((F(y_0, x_0), F(x_0, y_0)), (y_0, x_0)) \geq 1$. *Also, suppose that for each convergent sequences* $\{x_n\}$ *and* $\{y_n\}$ *with* $x_n \to x$, $y_n \to y$, $\alpha((x_n, y_n), (x_{n+1}, y_{n+1})) \geq 1$ *and* $\alpha((y_{n+1}, x_{n+1}), (y_n, x_n)) \geq 1$ *for all n, we have* $\alpha((x_n, y_n), (x, y)) \geq 1$ *and* $\alpha((y, x), (y_n, x_n)) \geq 1$ *for all n. Then F has a coupled fixed point.*

Proof. It is easy to check that $m((x, y), (u, v)) = m((v, u), (y, x))$ for all $(x, y), (u, v)$ in $X \times X$, the metric space $(X \times X, \delta)$ is complete and $\beta(\xi, \eta)\delta(T_F\xi, T_F\eta) \leq \psi(m(\xi, \eta))$ for all $\xi = (\xi_1, \xi_2), \eta = (\eta_1, \eta_2) \in X \times X$, where

$$\beta((\xi_1, \xi_2), (\eta_1, \eta_2)) = \min\{\alpha((\xi_1, \xi_2), (\eta_1, \eta_2)), \alpha((\eta_2, \eta_1), (\xi_2, \xi_1))\}. \tag{11}$$

Thus, T_F is a β-ψ-contractive mapping. It is easy to check that T_F is β-admissible and $\beta((x_0, y_0), T_F(x_0, y_0)) \geq 1$. Since $(X \times X, \delta)$ has the property (B), T_F has a fixed point by using Lemma 1 and so F has a coupled fixed point.

Theorem 6. *Let* (X, d) *be a complete metric space and* $F : X \times X \to X$ *a continuous mapping. Suppose that there exist a mapping* $\alpha : X^2 \times X^2 \to [0, +\infty)$ *and* $\psi \in \Psi$ *such that* $\alpha((x, y), (u, v))d(F(x, y), F(u, v)) \leq (1/2)\psi(m((x, y), (u, v)))$ *for all* $(x, y), (u, v) \in X \times X$, $\alpha((x, y), (u, v)) \geq 1$ *implies*

$$\alpha((F(x, y), F(y, x)), (F(u, v), F(v, u))) \geq 1 \tag{12}$$

for all $(x, y), (u, v) \in X \times X$ *and there exists* $(x_0, y_0) \in X \times X$ *such that*

$$\alpha((x_0, y_0), (F(x_0, y_0), F(y_0, x_0))) \geq 1 \tag{13}$$

and $\alpha((F(y_0, x_0), F(x_0, y_0)), (y_0, x_0)) \geq 1$. *Then F has a coupled fixed point.*

Proof. The metric space $(X \times X, \delta)$ is complete and $\beta(\xi, \eta)\delta(T_F\xi, T_F\eta) \leq \psi(m(\xi, \eta))$ for all $\xi = (\xi_1, \xi_2)$, $\eta = (\eta_1, \eta_2) \in X \times X$, where

$$\beta((\xi_1, \xi_2), (\eta_1, \eta_2)) = \min\{\alpha((\xi_1, \xi_2), (\eta_1, \eta_2)), \alpha((\eta_2, \eta_1), (\xi_2, \xi_1))\}. \tag{14}$$

Thus, T_F is a continuous β-ψ-contractive mapping. Also, it is easy to check that T_F is β-admissible and $\beta((x_0, y_0), T_F(x_0, y_0)) \geq 1$. Now by using Lemma 2, T_F has a fixed point and so F has a coupled fixed point.

Example 7. Let $X = [0, +\infty)$ and $d(x, y) = |x - y|$ for all $x, y \in X$. Define the mapping $F : X \times X \to X$ by $F(x, y) = (x - y)/4$ whenever $x \geq y$ and $F(x, y) = 0$ whenever $x < y$. Then F is continuous mapping. Also, define $\alpha : X^2 \times X^2 \to [0, +\infty)$ by

$$\alpha((x, y), (u, v)) = \begin{cases} 1, & x \geq y,\ u \geq v, \\ 0, & \text{otherwise}. \end{cases} \tag{15}$$

Thus, for each $(x, y), (u, v) \in X \times X$ we have

$$\begin{aligned} &\alpha((x, y), (u, v))\, d(F(x, y), F(u, v)) \\ &\quad \leq |F(x, y) - F(u, v)| = \left|\frac{x - y}{4} - \frac{u - v}{4}\right| \\ &\quad = \frac{1}{4}|(x - u) + (v - y)| \leq \frac{1}{4}|x - u| + |v - y| \\ &\quad = \frac{1}{4}(d(x, u) + d(y, v)) \\ &\quad = \frac{1}{4}\delta((x, y), (u, v)) \leq \frac{1}{2}\psi(m((x, y), (u, v))), \end{aligned} \tag{16}$$

where $\psi(t) = t/2$ for all $t \geq 0$. Also, it is easy to check that $\alpha((x, y), (u, v)) \geq 1$ implies $\alpha((F(x, y), F(y, x)), (F(u, v), F(v, u))) \geq 1$ for all $(x, y), (u, v) \in X \times X$. Also, $\alpha((x_0, y_0), (F(x_0, y_0), F(y_0, x_0))) \geq 1$ and $\alpha((F(y_0, x_0), F(x_0, y_0)), (y_0, x_0)) \geq 1$, where $(x_0, y_0) = (1, 1)$. Thus by using Theorem 6, F has a coupled fixed point.

Corollary 8. *Let* (X, d) *be a complete metric space,* $F : X \times X \to X$ *a mapping and* $\preceq$ *an order on* $X \times X$. *Suppose that there exist* $(x_0, y_0) \in X \times X$ *such that* (x_0, y_0) *and* $(F(x_0, y_0), F(y_0, x_0))$ *are comparable and also* (y_0, xy_0) *and* $(F(y_0, x_0), F(x_0, y_0))$ *are comparable, and a mapping* $\psi \in \Psi$ *such that*

$$d(F(x, y), F(u, v)) \leq \frac{1}{2}\psi(m((x, y), (u, v))) \tag{17}$$

for all comparable elements $(x, y), (u, v)$ *in* $X \times X$. *Suppose that* $(F(x, y), F(y, x))$ *is comparable with* $(F(u, v), F(v, u))$ *whenever* (x, y) *is comparable with* (u, v). *Also, suppose that for each convergent sequences* $\{x_n\}$ *and* $\{y_n\}$ *with* $x_n \to x$, $y_n \to y$, (x_n, y_n) *is comparable with* (x_{n+1}, y_{n+1}) *and* (y_{n+1}, x_{n+1}) *is comparable with* (y_n, x_n) *for all n, one gets that* (x_n, y_n) *is comparable with* (x, y) *and* (y, x) *is comparable with* (y_n, x_n) *for all n. Then F has a coupled fixed point.*

Proof. Define the mapping $\alpha : X^2 \times X^2 \to [0, +\infty)$ by $\alpha((x, y), (u, v)) = 1$ whenever (x, y) and (u, v) are comparable and $\alpha((x, y), (u, v)) = 0$ otherwise. Then by using Theorem 5, F has a coupled fixed point.

Corollary 9. *Let* (X, d) *be a complete metric space,* $F : X \times X \to X$ *a continuous mapping,* $(x^*, y^*) \in X \times X$ *a fixed element and* $\preceq$ *an order on* $X \times X$. *Suppose that there exist* $(x_0, y_0) \in X \times X$ *such that* (x_0, y_0), (y_0, x_0), $(F(x_0, y_0), F(y_0, x_0))$ *and* $(F(y_0, x_0), F(x_0, y_0))$ *are comparable with* (x^*, y^*), *and a mapping* $\psi \in \Psi$ *such that* $d(F(x, y), F(u, v)) \leq (1/2)\psi(m((x, y), (u, v)))$ *for all* (x, y) *and* (u, v) *in* $X \times X$ *which are comparable with* (x^*, y^*). *Assume that* $(F(x, y), F(y, x))$ *and* $(F(u, v), F(v, u))$ *are comparable with* (x^*, y^*) *whenever* (x, y) *and* (u, v) *so are. Then F has a coupled fixed point.*

Proof. Define the mapping $\alpha : X^2 \times X^2 \to [0,+\infty)$ by $\alpha((x,y),(u,v)) = 1$ whenever (x,y) and (u,v) are comparable with (x^*,y^*) and $\alpha((x,y),(u,v)) = 0$ otherwise. Then by using Theorem 6, F has a coupled fixed point.

Theorem 10. *Let (X,d) be a complete metric space and $F : X\times X \to X$ a mapping. Suppose that there exist a mapping $\alpha : X^2 \times X^2 \to [0,+\infty)$ and $\psi \in \Psi$ such that*

$$\alpha((x,y),(u,v))\, d(F(x,y),F(u,v)) \leq \frac{1}{2}\psi(\delta((x,y),(u,v))) \tag{18}$$

for all $(x,y),(u,v) \in X\times X$, $\alpha((x,y),(u,v)) \geq 1$ implies

$$\alpha((F(x,y),F(y,x)),(F(u,v),F(v,u))) \geq 1 \tag{19}$$

for all $(x,y),(u,v) \in X\times X$ and there exists $(x_0,y_0) \in X\times X$ such that

$$\alpha((x_0,y_0),(F(x_0,y_0),F(y_0,x_0))) \geq 1 \tag{20}$$

and $\alpha((F(y_0,x_0),F(x_0,y_0)),(y_0,x_0)) \geq 1$. Also, suppose that for each convergent sequences $\{x_n\}$ and $\{y_n\}$ with $x_n \to x$, $y_n \to y$, $\alpha((x_n,y_n),(x_{n+1},y_{n+1})) \geq 1$ and $\alpha((y_{n+1},x_{n+1}),(y_n,x_n)) \geq 1$ for all n, one has $\alpha((x_n,y_n),(x,y)) \geq 1$ and $\alpha((y,x),(y_n,x_n)) \geq 1$ for all n. Then F has a coupled fixed point.

Proof. The metric space $(X\times X,\delta)$ is complete and $\beta(\xi,\eta)\delta(T_F\xi,T_F\eta) \leq \psi(\delta(\xi,\eta))$ for all $\xi = (\xi_1,\xi_2)$, $\eta = (\eta_1,\eta_2) \in X\times X$, where

$$\beta((\xi_1,\xi_2),(\eta_1,\eta_2)) = \min\{\alpha((\xi_1,\xi_2),(\eta_1,\eta_2)),\alpha((\eta_2,\eta_1),(\xi_2,\xi_1))\}. \tag{21}$$

Thus, $\beta(\xi,\eta)d(T_F\xi,T_F\eta) \leq \psi(\delta(\xi,\eta))$ for all $\xi,\eta \in X\times X$. Also, it is easy to check that T_F is β-admissible and $\beta((x_0,y_0),T_F(x_0,y_0)) \geq 1$. Since $(X\times X,\delta)$ has the property (B), T_F has a fixed point by using Lemma 3, and so F has a coupled fixed point.

Theorem 11. *Let (X,d) be a complete metric space and $F : X\times X \to X$ a continuous mapping. Suppose that there exist a mapping $\alpha : X^2\times X^2 \to [0,+\infty)$ and $\psi \in \Psi$ such that $\alpha((x,y),(u,v))d(F(x,y),F(u,v)) \leq (1/2)\psi(\delta((x,y),(u,v)))$ for all $(x,y),(u,v) \in X\times X$, $\alpha((x,y),(u,v)) \geq 1$ implies*

$$\alpha((F(x,y),F(y,x)),(F(u,v),F(v,u))) \geq 1 \tag{22}$$

for all $(x,y),(u,v) \in X\times X$ and there exists $(x_0,y_0) \in X\times X$ such that

$$\alpha((x_0,y_0),(F(x_0,y_0),F(y_0,x_0))) \geq 1 \tag{23}$$

and $\alpha((F(y_0,x_0),F(x_0,y_0)),(y_0,x_0)) \geq 1$. Then F has a coupled fixed point.

Proof. The metric space $(X\times X,\delta)$ is complete and $\beta(\xi,\eta)\delta(T_F\xi,T_F\eta) \leq \psi(\delta(\xi,\eta))$ for all $\xi = (\xi_1,\xi_2)$, $\eta = (\eta_1,\eta_2) \in X\times X$, where

$$\beta((\xi_1,\xi_2),(\eta_1,\eta_2)) = \min\{\alpha((\xi_1,\xi_2),(\eta_1,\eta_2)),\alpha((\eta_2,\eta_1),(\xi_2,\xi_1))\}. \tag{24}$$

Thus, $\beta(\xi,\eta)d(T_F\xi,T_F\eta) \leq \psi(\delta(\xi,\eta))$ for all $\xi,\eta \in X\times X$. Also, it is easy to check that T_F is a continuous β-admissible mapping and $\beta((x_0,y_0),T_F(x_0,y_0)) \geq 1$. Now by using Lemma 4, T_F has a fixed point and so F has a coupled fixed point.

Example 12. Let $X = \mathbb{R}$ and $d(x,y) = |x-y|$ for all $x,y \in X$. Define the mapping $F : X\times X \to X$ by $F(x,y) = (x-y)/3+1$ whenever $x \geq y$ and $F(x,y) = 0$ whenever $x < y$. Then F is a discontinuous mapping. Define $\alpha : X^2\times X^2 \to [0,+\infty)$ by

$$\alpha((x,y),(u,v)) = \begin{cases} 1, & x \geq y,\ u \geq v, \\ 0, & \text{otherwise.} \end{cases} \tag{25}$$

Thus, for each $(x,y),(u,v) \in X\times X$ we have

$$\begin{aligned} &\alpha((x,y),(u,v))\,d(F(x,y),F(u,v)) \\ &\quad \leq |F(x,y)-F(u,v)| = \left|\frac{x-y}{3} - \frac{u-v}{3}\right| \\ &\quad = \frac{1}{3}|(x-u)+(v-y)| \leq \frac{1}{3}|(x-u)| + |(v-y)| \\ &\quad = \frac{1}{3}(d(x,u)+d(y,v)) \\ &\quad = \frac{1}{3}\delta((x,y),(u,v)) = \frac{1}{2}\psi(\delta((x,y),(u,v))), \end{aligned} \tag{26}$$

where $\psi(t) = 2t/3$ for all $t \geq 0$. Also, $\alpha((x,y),(u,v)) \geq 1$ implies

$$\alpha((F(x,y),F(y,x)),(F(u,v),F(v,u))) \geq 1 \tag{27}$$

for all $(x,y),(u,v) \in X\times X$. Finally, $\alpha((x_0,y_0),(F(x_0,y_0),F(y_0,x_0))) \geq 1$ and $\alpha((F(y_0,x_0),F(x_0,y_0)),(y_0,x_0)) \geq 1$, where $(x_0,y_0) = (2,2)$. Thus by using Theorem 10, F has a coupled fixed point.

Corollary 13. *Let (X,d) be a complete metric space, $F : X\times X \to X$ a mapping and $\preceq$ an order on $X\times X$. Suppose that there exists a mapping $\psi \in \Psi$ such that $d(F(x,y),F(u,v)) \leq (1/2)\psi(\delta((x,y),(u,v)))$ for all comparable elements (x,y) and (u,v) in $X\times X$. Suppose that there exists $(x_0,y_0) \in X\times X$ such that $(F(x_0,y_0), F(y_0,x_0))$ and (x_0,y_0) are comparable and also $(F(y_0,x_0), F(x_0,y_0))$ and (y_0,x_0) are comparable, $(F(x,y), F(y,x))$ is comparable with $(F(u,v), F(v,u))$ whenever (x,y) is comparable with (u,v) and for each sequences $\{x_n\}$ and $\{y_n\}$ in X such that $x_n \to x$, $y_n \to y$, (x_n,y_n) and (x_{n+1},y_{n+1}) are comparable elements of $X\times X$ for all n, (x_n,y_n) and (x,y) are comparable elements for all n. Then F has a coupled fixed point.*

Proof. Define the mapping $\alpha : X^2 \times X^2 \to [0, +\infty)$ by $\alpha((x, y), (u, v)) = 1$ whenever (x, y) and (u, v) are comparable and $\alpha((x, y), (u, v)) = 0$ otherwise. Then by using Theorem 10, F has a coupled fixed point.

Corollary 14. *Let (X, d) be a complete metric space, $F : X \times X \to X$ a continuous mapping, $(x^*, y^*) \in X \times X$ a fixed element and $\preceq$ an order on $X \times X$. Suppose that there exist $(x_0, y_0) \in X \times X$ such that (x_0, y_0), (y_0, x_0), $(F(x_0, y_0), F(y_0, x_0))$ and $(F(y_0, x_0), F(x_0, y_0))$ are comparable with (x^*, y^*), a mapping $\psi \in \Psi$ such that $d(F(x, y), F(u, v)) \le (1/2)\psi(\delta((x, y), (u, v)))$ for all (x, y) and (u, v) in $X \times X$ which are comparable with (x^*, y^*). Assume that $(F(x, y), F(y, x))$ and $(F(u, v), F(v, u))$ are comparable with (x^*, y^*) whenever (x, y) and (u, v) so are. Then F has a coupled fixed point.*

Proof. Define the mapping $\alpha : X^2 \times X^2 \to [0, +\infty)$ by $\alpha((x, y), (u, v)) = 1$ whenever (x, y) and (u, v) are comparable with (x^*, y^*) and $\alpha((x, y), (u, v)) = 0$ otherwise. Then by using Theorem 11, F has a coupled fixed point.

Now, we give an application of Lemma 1. In this way, we study the nonlinear fractional differential equation $D^{\alpha_0}x(t) + D^{\beta_0}x(t) = f_0(t, x(t))$ for $t \in [0, 1]$ via the two-point boundary value condition $x(0) = x(1) = 0$, where $f_0 : [0, 1] \times \mathbb{R} \to \mathbb{R}$ is a continuous function and $0 < \beta_0 < \alpha_0 < 1$. Recall that the Green function associated to the equation is given by $G(t) = t^{\alpha_0-1}E_{\alpha_0-\beta_0,\alpha_0}(-t^{\alpha_0-\beta_0})$, where E_{α_0,β_0} is the two-parametric Mittag-Leffler function defined by

$$E_{(\alpha_0,\beta_0)}(z) = \sum_{k=0}^{\infty} \frac{z^k}{\Gamma(k\alpha_0 + \beta_0)} \tag{28}$$

for $\alpha_0 > 0$ and $\beta_0 > 0$ (see [21, 22]). Let $X = C_{\mathbb{R}}([0, 1])$ and

$$d(f, g) = \sup_{x\in[0,1]} |f(x) - g(x)|. \tag{29}$$

Now by considering some conditions, we give the following result about existence of solution of the nonlinear fractional differential equation.

Theorem 15. *Suppose that*

(i) *there exist a function $\xi : \mathbb{R}^2 \to \mathbb{R}$ and $\psi \in \Psi$ such that*

$$|f_0(t, a) - f_0(t, b)| \le \alpha_0\psi(d(a, b)) \tag{30}$$

for all $t \in [0, 1]$ and $a, b \in \mathbb{R}$ with $\xi(a, b) \ge 0$,

(ii) *there exist $x_0 \in X$ such that $\xi(x_0(t), \int_0^t G(t - s)f_0(s, x_0(s))ds) \ge 0$ for all $t \in [0, 1]$, for each $t \in [0, 1]$ and $x, y \in X$, $\xi(x(t), y(t)) \ge 0$ implies*

$$\xi\left(\int_0^t G(t-s) f_0(s, x(s)), \int_0^t G(t-s) f_0(s, y(s))\, ds\right) \ge 0, \tag{31}$$

(iii) *if $\{x_n\}$ is a sequence in X with $x_n \to x$ and $\xi(x_n, x_{n+1}) \ge 0$ for all n, then $\xi(x_n, x) \ge 0$ for all n.*

Then the nonlinear fractional differential equation has at least one solution.

Proof. It is well known that $x \in X$ is a solution of the nonlinear fractional differential equation if and only if is a solution of the integral equation

$$x(t) = \int_0^t G(t-s) f_0(s, x(s))\, ds \tag{32}$$

for all $t \in [0, 1]$. Define the operator $F : X \to X$ by $Fx(t) = \int_0^t G(t - s)f_0(s, x(s))ds$ for all $t \in [0, 1]$. Thus, for finding a solution of the the nonlinear fractional differential equation it is sufficient we find a fixed point of the continuous operator F. Let $x, y \in X$ be such that $\xi(x(t), y(t)) \ge 0$ for all $t \in [0, 1]$. By using (i), we get

$$\begin{aligned} &|Fx(t) - Fy(t)| \\ &= \left|\int_0^t G(t-s)(f_0(s, x(s))) - f_0(s, y(s))\, ds\right| \\ &\le \int_0^t |G(t-s)|\,|f_0(s, x(s)) - f_0(s, y(s))|\, ds \\ &\le \int_0^t |G(t-s)|\,\alpha_0\psi(|x(s) - y(s)|)\, ds \\ &\le \alpha_0\psi(\|x - y\|_\infty) \sup_{t\in I} \int_0^t |G(t-s)|\, ds \\ &\le \psi(\|x - y\|_\infty). \end{aligned} \tag{33}$$

Note that, $G(t) = t^{\alpha_0-1}E_{\alpha_0-\beta_0,\alpha_0}(-t^{\alpha_0-\beta_0}) \le t^{\alpha_0-1}(1/(1+|-t^{\alpha_0-\beta_0}|)) \le t^{\alpha_0-1}$ for all $t \in [0, 1]$. Thus, $\sup_{t\in[0,1]} \int_0^t G(t - s)\, ds \le 1/\alpha_0$. Now, define $\alpha : X\times X \to [0, \infty)$ by $\alpha(x, y) = 1$ whenever $\xi(x(t), y(t)) \ge 0$ for all $t \in [0, 1]$ and $\alpha(x, y) = 0$ otherwise. Hence, $\alpha(x, y)\|Fx - Fy\|_\infty \le \psi(\|x - y\|_\infty)$ and so

$$\begin{aligned} &\alpha(x, y)\, d(Fx, Fy) \\ &\le \psi\Big(\max\Big\{d(x, y), d(x, Fx), d(y, Fy), \\ &\qquad \frac{1}{2}[d(x, Fy) + d(y, Fx)]\Big\}\Big) \end{aligned} \tag{34}$$

for all $x, y \in X$. Thus, F is an α-ψ-contractive mapping. By using (iii), $\alpha(x, y) \ge 1$ implies $\alpha(Fx, Fy) \ge 1$. Therefore, F is α-admissible. From (ii), there exists $x_0 \in C(I)$ such that $\alpha(x_0, Fx_0) \ge 1$. Now by using (iv) and Lemma 1, there exists $x^* \in X$ such that $Fx^* = x^*$.

Similar to the work of Samet et al. [19], we say that the space X has the property (H^*) whenever for each $(x, y), (u, v) \in X \times X$ there exists $(z_1, z_2) \in X \times X$ such that $\alpha((x, y), (z_1, z_2)) \ge 1$, $\alpha((z_2, z_1), (y, x)) \ge 1$, $\alpha((u, v), (z_1, z_2)) \ge 1$ and $\alpha((z_2, z_1), (v, u)) \ge 1$. It is easy to check that if X has the property (H^*) in before results, then the mapping has a unique coupled fixed point.

If there is an order $\leq$ on X, then one can construct the order $\preccurlyeq$ on $X \times X$ by $(x, y) \preccurlyeq (u, v)$ whenever $x \leq u$ and $v \leq y$. By using this idea, it has been provided some coupled fixed point results for mixed monotone mappings on ordered metric spaces [7]. Let $(X, \preccurlyeq)$ be a partially ordered set and $F : X \times X \to X$ a mapping. We say that F has the mixed monotone property whenever $x_1 \preccurlyeq x_2$ implies $F(x_1, y) \preccurlyeq F(x_2, y)$ and $y_1 \preccurlyeq y_2$ implies $F(x, y_2) \preccurlyeq F(x, y_1)$ for all $x, y \in X$ [7]. Thus by considering the provided corollaries and the explanations in the text-body, we can obtain some similar coupled fixed point results for mixed monotone mappings as special case of above results. Finally, it has been published interesting fixed point results on metric spaces with a graph (see e.g., [23–26]). There is a connection between fixed point results on ordered metric spaces and fixed point results on metric spaces with a graph [23, 27]. It is notable that one can get some similar coupled fixed point results on metric spaces with a graph.

Acknowledgments

The authors express their gratitude to the referees for their helpful suggestions which improved final version of this paper.

References

[1] D. Guo and V. Lakshmikantham, "Coupled fixed points of nonlinear operators with applications," *Nonlinear Analysis*, vol. 11, no. 5, pp. 623–632, 1987.

[2] S. S. Chang and Y. H. Ma, "Coupled fixed points for mixed monotone condensing operators and an existence theorem of the solutions for a class of functional equations arising in dynamic programming," *Journal of Mathematical Analysis and Applications*, vol. 160, no. 2, pp. 468–479, 1991.

[3] Y. Z. Chen, "Existence theorems of coupled fixed points," *Journal of Mathematical Analysis and Applications*, vol. 154, no. 1, pp. 142–150, 1991.

[4] S. Heikkilä and V. Lakshmikantham, "A unified theory for first-order discontinuous scalar differential equations," *Nonlinear Analysis*, vol. 26, no. 4, pp. 785–797, 1996.

[5] S. Jingxian and L. Lishan, "Iterative method for coupled quasi-solutions of mixed monotone operator equations," *Applied Mathematics and Computation*, vol. 52, no. 2-3, pp. 301–308, 1992.

[6] Y. Sun, "A fixed point theorem for mixed monotone operators with applications," *Journal of Mathematical Analysis and Applications*, vol. 156, no. 1, pp. 240–252, 1991.

[7] V. Berinde, "Generalized coupled fixed point theorems for mixed monotone mappings in partially ordered metric spaces," *Nonlinear Analysis*, vol. 74, no. 18, pp. 7347–7355, 2011.

[8] V. Berinde and M. Borcut, "Tripled fixed point theorems for contractive type mappings in partially ordered metric spaces," *Nonlinear Analysis*, vol. 74, no. 15, pp. 4889–4897, 2011.

[9] M. Borcut and V. Berinde, "Tripled coincidence theorems for contractive type mappings in partially ordered metric spaces," *Applied Mathematics and Computation*, vol. 218, no. 10, pp. 5929–5936, 2012.

[10] H. S. Ding, L. Li, and S. Radenovic, "Coupled coincidence point theorems for generalized nonlinear contraction in partially ordered metric spaces," *Fixed Point Theory and Applications*, vol. 2012, article 96, 10 pages, 2012.

[11] H. S. Ding and L. Li, "Coupled fixed point theorems in partially ordered cone metric spaces," *Filomat*, vol. 25, no. 2, pp. 137–149, 2011.

[12] W. S. Du, "Coupled fixed point theorems for nonlinear contractions satisfied Mizoguchi-Takahashi's condition in quasiordered metric spaces," *Fixed Point Theory and Applications*, vol. 2010, Article ID 876372, 9 pages, 2010.

[13] R. H. Haghi, S. Rezapour, and N. Shahzad, "Some fixed point generalizations are not real generalizations," *Nonlinear Analysis*, vol. 74, no. 5, pp. 1799–1803, 2011.

[14] X. Q. Hu and X. Y. Ma, "Coupled coincidence point theorems under contractive conditions in partially ordered probabilistic metric spaces," *Nonlinear Analysis*, vol. 74, no. 17, pp. 6451–6458, 2011.

[15] W. Long, B. E. Rhoades, and M. Rajovic, "Coupled coincidence points for two mappings in metric spaces and cone metric spaces," *Fixed Point Theory and Applications*, vol. 2012, article 66, 9 pages, 2012.

[16] N. V. Luong and N. X. Thuan, "Coupled fixed points in partially ordered metric spaces and application," *Nonlinear Analysis*, vol. 74, no. 3, pp. 983–992, 2011.

[17] H. K. Nashine and W. Shatanawi, "Coupled common fixed point theorems for a pair of commuting mappings in partially ordered complete metric spaces," *Computers and Mathematics with Applications*, vol. 62, no. 4, pp. 1984–1993, 2011.

[18] B. Samet and C. Vetro, "Coupled fixed point theorems for multi-valued nonlinear contraction mappings in partially ordered metric spaces," *Nonlinear Analysis*, vol. 74, no. 12, pp. 4260–4268, 2011.

[19] B. Samet, C. Vetro, and P. Vetro, "Fixed point theorems for α-ψ-contractive type mappings," *Nonlinear Analysis*, vol. 75, no. 4, pp. 2154–2165, 2012.

[20] A. Amini-Harandi, "Coupled and tripled fixed point theory in partially ordered metric spaces with application to initial value problem," *Mathematical and Computer Modelling*. In press.

[21] D. Baleanu, H. Mohammadi, and S. Rezapour, "Positive solutions of a boundary value problem for nonlinear fractional differential equations," *Journal of Mathematical Analysis and Applications*, vol. 311, no. 2, pp. 495–505, 2005.

[22] D. Baleanu, H. Mohammadi, and S. Rezapour, "Some existence results on nonlinear fractional differentialequations," *Philosophical Transactions of the Royal Society*. In press.

[23] S. M. A. Aleomraninejad, S. Rezapour, and N. Shahzad, "Some fixed point results on a metric space with a graph," *Topology and Its Applications*, vol. 159, no. 3, pp. 659–663, 2012.

[24] I. Beg, A. R. Butt, and S. Radojević, "The contraction principle for set valued mappings on a metric space with a graph," *Computers and Mathematics with Applications*, vol. 60, no. 5, pp. 1214–1219, 2010.

[25] J. Jachymski, "The contraction principle for mappings on a metric space with a graph," *Proceedings of the American Mathematical Society*, vol. 136, no. 4, pp. 1359–1373, 2008.

[26] A. Nicolae, D. O'Regan, and A. Petruşel, "Fixed point theorems for singlevalued and multivalued generalized contractions in metric spaces endowed with a graph," *Georgian Mathematical Journal*, vol. 18, no. 2, pp. 307–327, 2011.

[27] D. O'Regan and A. Petruşel, "Fixed point theorems for generalized contractions in ordered metric spaces," *Journal of Mathematical Analysis and Applications*, vol. 341, no. 2, pp. 1241–1252, 2008.

$\mathcal{I}^{\mathcal{K}}$-Convergence in the Topology Induced by Random 2-Normed Spaces

U. Yamancı and M. Gürdal

Department of Mathematics, Suleyman Demirel University, East Campus, 32260 Isparta, Turkey

Correspondence should be addressed to U. Yamancı; ulasyamanci@sdu.edu.tr

Academic Editor: Baruch Cahlon

We study $\mathcal{I}^{\mathcal{K}}$-convergence which is common generalization of the $\mathcal{I}^*$-convergence of sequences in the topology induced by random 2-normed spaces and prove some important results.

1. Introduction

Kostyrko et al. (cf. [1]; a similar concept was presented in [2]) introduced the concept of $\mathcal{I}$-convergence of sequences in a metric space and studied some properties of such convergence. Note that $\mathcal{I}$-convergence is an interesting generalization of statistical convergence. The notion of statistical convergence of sequences of real numbers was introduced by Fast in [3] and Steinhaus in [4].

Motivated by a result of Šalát [5] and Fridy [6] about statistically convergent sequences, in [7] Kostyrko et al. also defined so-called $\mathcal{I}^*$-convergence and asked for which ideals the notions of $\mathcal{I}$-convergence and $\mathcal{I}^*$-convergence coincide. This question was answered in [1] where the authors showed that these notions coincide if and only if the ideal $\mathcal{I}$ satisfies the property AP, which we call AP ($\mathcal{I}$, Fin) here (see also [8]).

Another important variant of ideal convergence is the notion of $\mathcal{I}^{\mathcal{K}}$-convergence introduced by Mačaj and Sleziak [9]. Recently, Eshaghi Gordji et al. [10] studied $\mathcal{I}^{\mathcal{K}}$-convergence in 2-normed spaces.

The concept of 2-normed spaces was initially introduced by Gähler [11] in the 1960s. Since then, many researchers have studied these subjects and obtained various results [12–17].

The theory of probabilistic normed (PN) spaces is an important area of research in functional analysis. Much work has been done in this theory and it has many important applications in real-world problems. PN spaces are the vector spaces in which the norm of each vector is an appropriate probability distribution function rather than a number. A PN space is a generalization of an ordinary normed linear space. In a PN space, the norms are represented by distance distribution functions. If x is an element of a PN space, then its norm is denoted by F_x, and the value $F_x(t)$ is interpreted as the probability that the norm of x is smaller than t. PN spaces were first introduced by Šerstnev in [18] by means of a definition that was closely modelled on the theory of normed spaces. In 1993, Alsina et al. [19] presented a new definition of a PN space which includes the definition of Šerstnev [20] as a special case. This new definition has naturally led to the definition of the principal class of PN spaces, the Menger spaces, and is compatible with various possible definitions of a probabilistic inner product space. It is based on the probabilistic generalization of a characterization of ordinary normed spaces by means of a betweenness relation and relies on the tools of the theory of probabilistic metric (PM) spaces (see [21]). This new definition quickly became the standard one and it has been adopted by many authors (e.g., [22–29]), who have investigated several properties of PN spaces. A detailed history and the development of the subject up to 2006 can be found in [30].

The notion of $\mathcal{I}^{\mathcal{K}}$-convergence of sequences has not been studied previously in the setting of random 2-normed spaces. Motivated by this fact, in this paper, as a variant of $\mathcal{I}$-convergence, the notion of $\mathcal{I}^{F_{\mathcal{K}}}$-convergence of sequences is introduced in a random 2-normed space as a common generalization of all these types of $\mathcal{I}^{F^*}$-convergence, and some important results are established.

2. Preliminaries

In this section we recall some of the basic concepts, which will be used in this paper.

The notion of a statistically convergent sequence can be defined using the asymptotic density of subsets of the set of positive integers $\mathbb{N} = \{1, 2, \ldots\}$. A subset K of $\mathbb{N}$ is said to have density $\delta(K)$ if $\delta(K) = \lim_{n\to\infty} n^{-1} \sum_{k=1}^{n} \chi_K(k)$ exists, where χ_K is the characteristic function of the set K [31]. The notion of statistical convergence was originally defined for sequences of numbers in the paper [3] and also in [32]. We say that a number sequence $x = (x_k)_{k\in\mathbb{N}}$ statistically converges to a point L if for each $\varepsilon > 0$ we have $\delta(K(\varepsilon)) = 0$, where $K(\varepsilon) = \{k \in \mathbb{N} : |x_k - L| \geq \varepsilon\}$ and in such situation we will write $L = \text{st-lim}\, x_k$.

The notion of statistical convergence was further generalized in the paper [1, 8] using the notion of an ideal of subsets of the set $\mathbb{N}$. We say that a nonempty family of sets $\mathcal{I} \subset \mathcal{P}(\mathbb{N})$ is an ideal on $\mathbb{N}$ if (a) $\emptyset \in \mathcal{I}$; (b) $A, B \in \mathcal{I}$ imply $A \cup B \in \mathcal{I}$; (c) $A \in \mathcal{I}, B \subset A$ imply $B \in \mathcal{I}$. An ideal $\mathcal{I}$ on $\mathbb{N}$ for which $\mathcal{I} \neq \mathcal{P}(\mathbb{N})$ is called a proper ideal. A proper ideal $\mathcal{I}$ is called admissible if $\mathcal{I}$ contains all finite subsets of $\mathbb{N}$. If not otherwise stated in the sequel $\mathcal{I}$ will denote an admissible ideal. Let $\mathcal{I} \subset \mathcal{P}(\mathbb{N})$ be a nontrivial ideal. A class $\mathcal{F}(\mathcal{I}) = \{M \subset \mathbb{N} : \exists A \in \mathcal{I} : M = \mathbb{N} \setminus A\}$, called the filter associated with the ideal $\mathcal{I}$, is a filter on $\mathbb{N}$.

Recall the generalization of statistical convergence from [1, 8].

Let $\mathcal{I}$ be an admissible ideal on $\mathbb{N}$ and $x = (x_k)_{k\in\mathbb{N}}$ a sequence of points in a metric space (X, ρ). We say that the sequence x is $\mathcal{I}$-convergent (or $\mathcal{I}$-converges) to a point $\xi \in X$, and we denote it by $\mathcal{I}\text{-lim}\, x = \xi$, if for each $\varepsilon > 0$ we have

$$A(\varepsilon) = \{k \in \mathbb{N} : \rho(x_k, \xi) \geq \varepsilon\} \in \mathcal{I}. \quad (1)$$

This generalizes the notion of usual convergence, which can be obtained when we take for $\mathcal{I}$ the ideal $\mathcal{I}_f$ (Fin the ideal) of all finite subsets of $\mathbb{N}$. A sequence is statistically convergent if and only if it is $\mathcal{I}_\delta$-convergent, where $\mathcal{I}_\delta := \{K \subset \mathbb{N} : \delta(K) = 0\}$ is the admissible ideal of the sets of zero asymptotic density.

Definition 1 (see [11]). Let X be a real vector space of dimension d, where $2 \leq d < \infty$. A 2-norm on X is a function $\|\cdot,\cdot\| : X \times X \to \mathbb{R}$ which satisfies (i) $\|x, y\| = 0$ if and only if x and y are linearly dependent; (ii) $\|x, y\| = \|y, x\|$; (iii) $\|\alpha x, y\| = |\alpha|\|x, y\|, \alpha \in \mathbb{R}$; (iv) $\|x, y + z\| \leq \|x, y\| + \|x, z\|$. The pair $(X, \|\cdot,\cdot\|)$ is then called a 2-normed space.

As an example of a 2-normed space, we may take $X = \mathbb{R}^2$ being equipped with the 2-norm $\|x, y\| :=$ the area of the parallelogram spanned by the vectors x and y, which may be given explicitly by the formula

$$\|x, y\| = |x_1 y_2 - x_2 y_1|, \quad x = (x_1, x_2),\ y = (y_1, y_2). \quad (2)$$

Observe that in any 2-normed space $(X, \|\cdot,\cdot\|)$ we have $\|x, y\| \geq 0$ and $\|x, y + \alpha x\| = \|x, y\|$ for all $x, y \in X$ and $\alpha \in \mathbb{R}$. Also, if x, y, and z are linearly dependent, then $\|x, y + z\| = \|x, y\| + \|x, z\|$ or $\|x, y - z\| = \|x, y\| + \|x, z\|$. Given a 2-normed space $(X, \|\cdot,\cdot\|)$, one can derive a topology for it via the following definition of the limit of a sequence: a sequence (x_n) in X is said to be convergent to x in X if $\lim_{n\to\infty} \|x_n - x, y\| = 0$ for every $y \in X$.

All the concepts listed below are studied in depth in the fundamental book by Schweizer and Sklar [21].

Definition 2. Let $\mathbb{R}$ denote the set of real numbers, $\mathbb{R}_+ = \{x \in \mathbb{R} : x \geq 0\}$, and $S = [0, 1]$ the closed unit interval. A mapping $f : \mathbb{R} \to S$ is called a distribution function if it is nondecreasing and left continuous with $\inf_{t\in\mathbb{R}} f(t) = 0$ and $\sup_{t\in\mathbb{R}} f(t) = 1$.

We denote the set of all distribution functions by D^+ such that $f(0) = 0$. If $a \in \mathbb{R}_+$, then $H_a \in D^+$, where

$$H_a(t) = \begin{cases} 1 & \text{if } t > a, \\ 0 & \text{if } t \leq a. \end{cases} \quad (3)$$

It is obvious that $H_0 \geq f$ for every $f \in D^+$.

Definition 3. A triangular norm (t-norm) is a continuous mapping $* : S \times S \to S$ such that $(S, *)$ is an abelian monoid with unit one and $c * d \leq a * b$ if $c \leq a$ and $d \leq b$ for all $a, b, c, d \in S$. A triangle function τ is a binary operation on D^+ which is commutative, associative, and $\tau(f, H_0) = f$ for every $f \in D^+$.

Definition 4. Let X be a linear space of dimension greater than one, τ a triangle function, and $F : X \times X \to D^+$. Then, F is called a probabilistic 2-norm and (X, F, τ) a probabilistic 2-normed space if the following conditions are satisfied.

(i) $F(x, y; t) = H_0(t)$ if x and y are linearly dependent, where $F(x, y; t)$ denotes the value of $F(x, y)$ at $t \in \mathbb{R}$.

(ii) $F(x, y; t) \neq H_0(t)$ if x and y are linearly independent.

(iii) $F(x, y; t) = F(y, x; t)$ for all $x, y \in X$.

(iv) $F(\alpha x, y; t) = F(x, y; t/|\alpha|)$ for every $t > 0, \alpha \neq 0$ and $x, y \in X$.

(v) $F(x + y, z; t) \geq \tau(F(x, z; t), F(y, z; t))$ whenever $x, y, z \in X$, and $t > 0$.

If (v) is replaced with

(vi) $F(x + y, z; t) \geq \sup\{F(x, z; t_1) * F(y, z; t_2) : t_1 + t_2 = t\}$ for all $x, y, z \in X$ and $t_1, t_2 \in \mathbb{R}_+$; then $(X, F, *)$ is called a random 2-normed space (for short, RTN space), where $*$ and τ are given by Definition 3.

As a standard example, we can give the following.

Example 5. Let $(X, \|\cdot,\cdot\|)$ be a 2-normed space, and let $a * b = \min\{a, b\}$ for all $a, b \in S$. For all $x, y \in X$ and every $t > 0$, consider

$$F(x, y; t) = H_0(t - \|x, y\|). \quad (4)$$

Then, observe that $(X, F, *)$ is a random 2-normed space.

Let $(X, F, *)$ be an RTN space. Since $*$ is a continuous t-norm, the system of (ε, λ)-neighborhoods of θ (the null vector in X)

$$\{\mathcal{N}_{\theta,z}(\varepsilon,\lambda) : \varepsilon > 0, \lambda \in (0,1), z \in X\}, \quad (5)$$

where

$$\mathcal{N}_{\theta,z}(\varepsilon,\lambda) = \{x, z \in X \times X : F_{x,z}(\varepsilon) > 1 - \lambda\} \quad (6)$$

determines a first countable Hausdorff topology on $X \times X$, called the F-topology. Thus, the F-topology can be completely specified by means of F-convergence of sequences. It is clear that $x - y \in \mathcal{N}_{\theta,z}$ means $y \in \mathcal{N}_{x,z}$ and vice versa.

A sequence $x = (x_k)$ in X is said to be F-convergence to $L \in X$ if for every $\varepsilon > 0$, $\lambda \in (0,1)$, and any nonzero $z \in X$ there exists a positive integer N such that

$$x_k, z - L \in \mathcal{N}_{\theta,z}(\varepsilon,\lambda) \quad \text{for each } k \geq N \quad (7)$$

or equivalently,

$$x_k, z \in \mathcal{N}_{L,z}(\varepsilon,\lambda) \quad \text{for each } k \geq N. \quad (8)$$

In this case, we write F-$\lim x_k, z = L$.

We also recall that the concept of $\mathcal{I}^F$-convergence and $\mathcal{I}^{F*}$-convergence of sequences in a random 2-normed space is studied in [24].

Definition 6. Let $(X, F, *)$ be an RTN space and $\mathcal{I}$ a proper ideal in $\mathbb{N}$. The sequence $x = (x_k)$ in X is said to be $\mathcal{I}^F$-convergent to $L \in X$ ($\mathcal{I}^F$-convergent to $L \in X$ with respect to F-topology) if for each $\varepsilon > 0$, $\lambda \in (0,1)$ and any nonzero $z \in X$,

$$\{k \in \mathbb{N} : x_k, z \notin \mathcal{N}_{L,z}(\varepsilon,\lambda)\} \in \mathcal{I}. \quad (9)$$

In this case the vector L is called the $\mathcal{I}^F$-limit of the sequence $x = (x_k)$ and we write $\mathcal{I}^F$-$\lim x, z = L$.

Definition 7. Let $(X, F, *)$ be an RTN space and $\mathcal{I}$ an admissible ideal in $\mathbb{N}$. We say that a sequence $x = (x_k)$ in X is said to be $\mathcal{I}^{F*}$-convergent to $L \in X$ with respect to the random 2-norm F if there exists a subset

$$M = \{m_k : m_1 < m_2 < \cdots\} \subset \mathbb{N} \quad (10)$$

such that $M \in \mathcal{F}(\mathcal{I})$ (i.e., $\mathbb{N} \setminus M \in \mathcal{I}$) and F-$\lim x_{m_k}, z = L$.

In this case we write $\mathcal{I}^{F*}$-$\lim x, z = L$ and L is called the $\mathcal{I}^{F*}$-limit of the sequence $x = (x_k)$.

3. $\mathcal{I}^{F_{\mathcal{K}}}$-Convergence in RTNS

In this section, we aim to generalize the notion of $\mathcal{I}^{F*}$-convergence of sequences in random 2-normed space.

We give a few basic facts concerning $\mathcal{I}^F$-convergence for future reference.

Lemma 8. *Let $(X, F, *)$, $(Y, F, *)$ be two RTN spaces and $\mathcal{I}$, $\mathcal{I}_1, \mathcal{I}_2$ be ideals on $\mathbb{N}$. Then,*

(i) *if $\mathcal{I}$ is not proper ideal, then every sequence $x = (x_k)$ in X is $\mathcal{I}^F$-convergent to each point of X with respect to the random 2-norm F;*

(ii) *If $\mathcal{I}_1 \subset \mathcal{I}_2$, then for every sequence $x = (x_k)$, one has*

$$\mathcal{I}_1^F\text{-}\lim x_k, z = L \Longrightarrow \mathcal{I}_2^F\text{-}\lim x_k, z = L. \quad (11)$$

Proof. (i) Let L be arbitrary element of X. Then for each $\varepsilon > 0$, $\lambda \in (0,1)$ and any nonzero $z \in X$

$$\{k \in \mathbb{N} : x_k, z \notin \mathcal{N}_{L,z}(\varepsilon,\lambda)\} \in \mathcal{I}. \quad (12)$$

(ii) Let $\mathcal{I}_1 \subset \mathcal{I}_2$, $\mathcal{I}_1^F$-$\lim x_k, z = L$. Then we have for each $\varepsilon > 0$, $\lambda \in (0,1)$ and any nonzero $z \in X$

$$\{k \in \mathbb{N} : x_k, z \notin \mathcal{N}_{L,z}(\varepsilon,\lambda)\} \in \mathcal{I}_1 \subset \mathcal{I}_2. \quad (13)$$

Hence, $\mathcal{I}_2^F$-$\lim x_k, z = L$. □

As we have previously mentioned, we aim to generalize the notion of $\mathcal{I}^{F*}$-convergence of sequences, introduced in [24]. Therefore, we modify this definition in the following way.

Definition 9. Let $(X, F, *)$ be an RTN space and $\mathcal{I}$ an ideal in $\mathbb{N}$. A sequence $(x_k)_{k\in\mathbb{N}}$ in X is called $\mathcal{I}^{F*}$-convergent to the point $L \in X$ with respect to the random 2-norm F if there exists a set $M \in \mathcal{F}(\mathcal{I})$ such that the sequence $(y_k)_{k\in\mathbb{N}}$ defined by

$$y_k = \begin{cases} x_k, & \text{if } k \in M \\ L, & \text{if } k \notin M \end{cases} \quad (14)$$

is Fin-convergent to L with respect to the random 2-norm F. If $(x_k)_{k\in\mathbb{N}}$ is $\mathcal{I}^{F*}$-convergent to L, we write $\mathcal{I}^{F*}$-$\lim x_k, z = L$.

We introduce the definition of $\mathcal{I}^{F_{\mathcal{K}}}$-convergence in $(X, F, *)$; we simply replace the ideal Fin with an arbitrary ideal on the set $\mathbb{N}$.

Definition 10. Let $(X, F, *)$ be an RTN space and let $\mathcal{K}$ and $\mathcal{I}$ be ideals on $\mathbb{N}$. The sequence $(x_k)_{k\in\mathbb{N}}$ in X is called $\mathcal{I}^{F_{\mathcal{K}}}$-convergent to the point $L \in X$ with respect to the random 2-norm F if there exists a set $M \in \mathcal{F}(\mathcal{I})$ such that the sequence $(y_k)_{k\in\mathbb{N}}$ given by

$$y_k = \begin{cases} x_k, & \text{if } k \in M \\ L, & \text{if } k \notin M \end{cases} \quad (15)$$

is $\mathcal{K}$-convergent to L with respect to the random 2-norm F. If $(x_k)_{k\in\mathbb{N}}$ is $\mathcal{I}^{F_{\mathcal{K}}}$-convergent to L, we write $\mathcal{I}^{F_{\mathcal{K}}}$-$\lim x_k, z = L$.

Remark 11. The definition of $\mathcal{I}^{F_{\mathcal{K}}}$-convergence can be reformulated in the form of decomposition theorem. A sequence $(x_k)_{k\in\mathbb{N}}$ is $\mathcal{I}^{F_{\mathcal{K}}}$-convergent if and only if it can be written as $(x_k)_{k\in\mathbb{N}} = (y_k)_{k\in\mathbb{N}} + (z_k)_{k\in\mathbb{N}}$, where $(y_k)_{k\in\mathbb{N}}$ is $\mathcal{K}$-convergent with respect to the random 2-norm F and $(z_k)_{k\in\mathbb{N}}$ is nonzero only on a set from $\mathcal{I}$.

Example 12. (i) Put $\mathcal{I}_0 = \mathcal{K}_0 = \{\emptyset\}$. $\mathcal{I}_0$ is the minimal ideal in $\mathbb{N}$. A sequence $(x_k)_{k\in\mathbb{N}}$ is $\mathcal{I}_0^{F_{\mathcal{K}_0}}$-convergent if and only if it is constant.

(ii) Let $\emptyset \neq M \subset \mathbb{N}$, $M \neq \mathbb{N}$. Let $\mathcal{K}$ be a proper ideal in $\mathbb{N}$. Let $\mathcal{I} = \{\emptyset\}$. A sequence $(x_k)_{k\in\mathbb{N}}$ is $\mathcal{I}^{F_{\mathcal{K}}}$-convergent if and only if it is constant on $\mathbb{N} \setminus M$.

(iii) Let $\mathcal{K}$ be an admissible ideal in $\mathbb{N}$ and $\mathcal{I}$ an arbitrary ideal. A sequence $(x_k)_{k\in\mathbb{N}}$ is $\mathcal{I}^{F_{\mathcal{K}}}$-convergent if there exists a set $M \in \mathcal{F}(\mathcal{I})$ and the sequence $(y_k)_{k\in\mathbb{N}}$ that is the usual F-convergences.

Theorem 13. *Let $(X, F, *)$ be an RTN space and let $\mathcal{K}$ and $\mathcal{I}$ be ideals on $\mathbb{N}$. $\mathcal{I}^{F_{\mathcal{K}}}$-limit of any sequence if exists is unique.*

Proof. Let $(x_k)_{k\in\mathbb{N}}$ be any sequence and suppose that $\mathcal{I}^{F_{\mathcal{K}}}$-$\lim x_k, z = \xi$, $\mathcal{I}^{F_{\mathcal{K}}}$-$\lim x_k, z = \eta$, where $\xi \neq \eta$. Since $\xi \neq \eta$, select $\varepsilon > 0$, $\lambda \in (0,1)$ and any nonzero $z \in X$ such that $\mathcal{N}_{\xi,z}(\varepsilon,\lambda)$ and $\mathcal{N}_{\eta,z}(\varepsilon,\lambda)$ are disjoint neighborhoods of ξ and η. Since $\mathcal{I}^{F_{\mathcal{K}}}$-$\lim x_k, z = \xi$, and $\mathcal{I}^{F_{\mathcal{K}}}$-$\lim x_k, z = \eta$, by the definition there exists $M_1, M_2 \in \mathcal{F}(\mathcal{I})$ such that the sequences $(y_k)_{k\in\mathbb{N}}, (z_k)_{k\in\mathbb{N}}$ given by

$$y_k = \begin{cases} x_k, & \text{if } k \in M_1 \\ \xi, & \text{if } k \notin M_1, \end{cases} \qquad z_k = \begin{cases} x_k, & \text{if } k \in M_2 \\ \eta, & \text{if } k \notin M_2 \end{cases} \tag{16}$$

having the following properties for each $\varepsilon > 0$, $\lambda \in (0,1)$ and any nonzero $z \in X$

$$A = \left\{k \in \mathbb{N} : y_k, z \notin \mathcal{N}_{\xi,z}(\varepsilon,\lambda)\right\}, \qquad B = \left\{k \in \mathbb{N} : z_k, z \notin \mathcal{N}_{\eta,z}(\varepsilon,\lambda)\right\} \tag{17}$$

belong to $\mathcal{K}$. This implies that sets $A^c = \{k \in \mathbb{N} : y_k, z \in \mathcal{N}_{\xi,z}(\varepsilon,\lambda)\}$ and $B^c = \{k \in \mathbb{N} : z_k, z \in \mathcal{N}_{\eta,z}(\varepsilon,\lambda)\}$ belong to $\mathcal{F}(\mathcal{I})$. Since $\mathcal{F}(\mathcal{I})$ is a filter in $\mathbb{N}$, we have that $A^c \cap B^c$ is a nonempty in $\mathcal{F}(\mathcal{I})$. In this way, we obtain a contradiction to the fact that the neighborhoods $\mathcal{N}_{\xi,z}(\varepsilon,\lambda)$ and $\mathcal{N}_{\eta,z}(\varepsilon,\lambda)$ of ξ and η are disjoints. Hence, we have $\xi = \eta$. This completes the proof. □

Lemma 14. *Let $(X, F, *)$ be a RTN space and let $\mathcal{K}$ and $\mathcal{I}$ be ideals on $\mathbb{N}$. If (x_k) and (y_k) are two sequences in $(X, F, *)$ with $a * a > a$ for every $a \in (0,1)$; then*

(i) *if $\mathcal{I}^{F_{\mathcal{K}}}$-$\lim x_k, z = \xi$ and $\mathcal{I}^{F_{\mathcal{K}}}$-$\lim y_k, z = \eta$, then $\mathcal{I}^{F_{\mathcal{K}}}$-$\lim(x_k + y_k), z = \xi + \eta$;*

(ii) *if $\mathcal{I}^{F_{\mathcal{K}}}$-$\lim x_k, z = \xi$ and $\alpha \in \mathbb{R}$, then $\mathcal{I}^{F_{\mathcal{K}}}$-$\lim \alpha x_k$, $z = \alpha\xi$;*

(iii) *if $\mathcal{I}^{F_{\mathcal{K}}}$-$\lim x_k, z = \xi$ and $\mathcal{I}^{F_{\mathcal{K}}}$-$\lim y_k, z = \eta$, then $\mathcal{I}^{F_{\mathcal{K}}}$-$\lim(x_k - y_k), z = \xi - \eta$.*

Proof. (i) Let $\varepsilon > 0$, $\lambda \in (0,1)$, and any nonzero $z \in X$. Since $\mathcal{I}^{F_{\mathcal{K}}}$-$\lim x_k, z = \xi$, and $\mathcal{I}^{F_{\mathcal{K}}}$-$\lim y_k, z = \eta$, there exists $M_1, M_2 \in \mathcal{F}(\mathcal{I})$ such that the sequences $(z_k)_{k\in\mathbb{N}}, (t_k)_{k\in\mathbb{N}}$ given by

$$z_k = \begin{cases} x_k, & \text{if } k \in M_1 \\ \xi, & \text{if } k \notin M_1, \end{cases} \qquad t_k = \begin{cases} y_k, & \text{if } k \in M_2 \\ \eta, & \text{if } k \notin M_2 \end{cases} \tag{18}$$

having the following conditions

$$A = \left\{k \in \mathbb{N} : z_k, z \notin \mathcal{N}_{\xi,z}\left(\frac{\varepsilon}{2},\lambda\right)\right\}, \qquad B = \left\{k \in \mathbb{N} : t_k, z \notin \mathcal{N}_{\eta,z}\left(\frac{\varepsilon}{2},\lambda\right)\right\} \tag{19}$$

belong to $\mathcal{K}$. Let $M = M_1 \cap M_2$ and

$$p_k = \begin{cases} x_k + y_k, & \text{if } k \in M \\ \xi + \eta, & \text{if } k \notin M \end{cases} \tag{20}$$

have the condition $C = \{k \in \mathbb{N} : p_k, z \notin \mathcal{N}_{\xi+\eta,z}(\varepsilon,\lambda)\}$. Since $\mathcal{K}$ is an ideal, it is sufficient to show that $C \subset A \cup B$. This is equivalent to showing that $C^c \supset A^c \cap B^c$ where A^c and B^c belong to $\mathcal{F}(\mathcal{I})$. Let $k \in A^c \cap B^c$, that is, $k \in A^c$ and $k \in B^c$, and we have

$$\begin{aligned} &F_{(p_k)-(\xi+\eta),z}(\varepsilon) \\ &= F_{(z_k+t_k)-(\xi+\eta),z}(\varepsilon) \\ &\geq \sup\left\{F_{z_k-\xi,z}\left(\frac{\varepsilon}{2}\right) * F_{t_k-\eta,z}\left(\frac{\varepsilon}{2}\right) : \frac{\varepsilon}{2} + \frac{\varepsilon}{2} = \varepsilon\right\} \\ &> \sup\{(1-\lambda) * (1-\lambda)\} \\ &> (1-\lambda). \end{aligned} \tag{21}$$

Since $k \in C^c \supset A^c \cap B^c \in \mathcal{F}(\mathcal{I})$, we have $C \subset A \cup B \in \mathcal{K}$.

(ii) Let $\varepsilon > 0, \lambda \in (0,1)$ and any nonzero $z \in X$. Since $\mathcal{I}^{F_{\mathcal{K}}}$-$\lim x_k, z = \xi$, there exists $M \in \mathcal{F}(\mathcal{I})$ such that the sequence $(y_k)_{k\in\mathbb{N}}$ given by

$$y_k = \begin{cases} x_k, & \text{if } k \in M \\ \xi, & \text{if } k \notin M \end{cases} \tag{22}$$

have the following condition

$$A = \left\{k \in \mathbb{N} : y_k, z \notin \mathcal{N}_{\xi,z}(\varepsilon,\lambda)\right\} \in \mathcal{K}. \tag{23}$$

This implies that $A^c = \{k \in \mathbb{N} : y_k, z \in \mathcal{N}_{\xi,z}(\varepsilon,\lambda)\} \in \mathcal{F}(\mathcal{I})$. Let $k \in A^c$. For the case $\alpha = 0$, we have

$$F_{0y_k - 0\xi,z}(\varepsilon) = F_{0,z}(\varepsilon) > (1-\lambda) \tag{24}$$

and for the case $\alpha \neq 0$

$$F_{\alpha y_k-\alpha\xi,z}(\varepsilon) = F_{y_k-\xi,z}\left(\frac{\varepsilon}{|\alpha|}\right) \geq \sup\left\{F_{y_k-\xi,z}(\varepsilon) * F_{0,z}\left(\frac{\varepsilon}{|\alpha|}-\varepsilon\right) : \varepsilon + \frac{\varepsilon}{|\alpha|} - \varepsilon = \varepsilon\right\} > \sup\{(1-\lambda) * 1\} = (1-\lambda). \tag{25}$$

So $\{k \in \mathbb{N} : \alpha y_k, z \notin \mathcal{N}_{\alpha\xi,z}(\varepsilon,\lambda)\} \in \mathcal{K}$. Hence $\mathcal{I}^{F_{\mathcal{K}}}$-$\lim \alpha x_k, z = \alpha\xi$.

(iii) The result follows from (i) and (ii). □

One can show easily directly the definitions that $\mathcal{K}$-convergence with respect to the random 2-norm F implies $\mathcal{I}^{F_{\mathcal{K}}}$-convergence.

Lemma 15. *Let $\mathcal{K}$ and $\mathcal{I}$ be ideals on $\mathbb{N}$. If $(X, F, *)$ is an RTN space and $(x_k)_{k\in\mathbb{N}}$ is a sequence in X such that $\mathcal{K}$-$\lim x_k, z = \xi$ with respect to the random 2-norm F, then $\mathcal{I}^{F_{\mathcal{K}}}$-$\lim x_k, z = \xi$.*

Lemma 16. *Let $(X, F, *)$ an RTN space and let $\mathcal{I}, \mathcal{I}_1, \mathcal{I}_2, \mathcal{K}, \mathcal{K}_1$, and $\mathcal{K}_2$ be ideals on $\mathbb{N}$ such that $\mathcal{I}_1 \subset \mathcal{I}_2$ and $\mathcal{K}_1 \subset \mathcal{K}_2$. Then, for every sequence $(x_k)_{k\in\mathbb{N}}$ in X, we have*

(i) $\mathcal{I}_1^{F_{\mathcal{K}}}$-$\lim x_k, z = \xi \Rightarrow \mathcal{I}_2^{F_{\mathcal{K}}}$-$\lim x_k, z = \xi$,

(ii) $\mathcal{I}^{F_{\mathcal{K}_1}}$-$\lim x_k, z = \xi \Rightarrow \mathcal{I}^{F_{\mathcal{K}_2}}$-$\lim x_k, z = \xi$.

Proof. (i) Let $\varepsilon > 0$, $\lambda \in (0, 1)$ and any nonzero $z \in X$ and suppose that $\mathcal{I}_1^{F_{\mathcal{K}}}$ -$\lim x_k, z = \xi$. Then there exists $M \in \mathcal{F}(\mathcal{I}_1)$ such that the sequence $(y_k)_{k\in\mathbb{N}}$ given by

$$y_k = \begin{cases} x_k, & \text{if } k \in M \\ \xi, & \text{if } k \notin M \end{cases} \tag{26}$$

has the following condition

$$A = \left\{k \in \mathbb{N} : y_k, z \notin \mathcal{N}_{\xi,z}(\varepsilon,\lambda)\right\} \in \mathcal{K}. \tag{27}$$

On the other hand, since $\mathcal{I}_1 \subset \mathcal{I}_2$, $M \in \mathcal{F}(\mathcal{I}_1) \subset \mathcal{F}(\mathcal{I}_2)$ and the by Definition 10, $\mathcal{I}_2^{F_{\mathcal{K}}}$-$\lim x_k, z = \xi$.

(ii) Let $\varepsilon > 0$, $\lambda \in (0, 1)$ and any nonzero $z \in X$. Since $\mathcal{I}^{F_{\mathcal{K}_1}}$-$\lim x_k, z = \xi$, there exists $M \in \mathcal{F}(\mathcal{I})$ such that the sequence $(y_k)_{k\in\mathbb{N}}$ given by

$$y_k = \begin{cases} x_k, & \text{if } k \in M \\ \xi, & \text{if } k \notin M \end{cases} \tag{28}$$

has the following condition

$$A = \left\{k \in \mathbb{N} : y_k, z \notin \mathcal{N}_{\xi,z}(\varepsilon,\lambda)\right\} \in \mathcal{K}_1 \subset \mathcal{K}_2. \tag{29}$$

Hence, $A \in \mathcal{K}_2$ and the proof is complete. □

In the next theorem, we show the relationship between the $\mathcal{I}^F$-convergence and $\mathcal{I}^{F_{\mathcal{K}}}$-convergence in random 2-normed spaces.

Theorem 17. *Let $(X, F, *)$ be an RTN space and let $\mathcal{K}$ and $\mathcal{I}$ be ideals on $\mathbb{N}$. Let $(x_k)_{k\in\mathbb{N}}$ be a sequence in X.*

(i) *If $\mathcal{I}^{F_{\mathcal{K}}}$-$\lim x_k, z = \xi$ implies $\mathcal{I}^F$-$\lim x_k, z = \xi$ holds for some $\xi \in X$, which has at least one neighborhood different from X, then $\mathcal{K} \subset \mathcal{I}$.*

(ii) *If $\mathcal{K} \subset \mathcal{I}$, then $\mathcal{I}^{F_{\mathcal{K}}}$-$\lim x_k, z = \xi$ implies $\mathcal{I}^F$-$\lim x_k, z = \xi$.*

Proof. (i) Let $\varepsilon > 0, \lambda \in (0, 1)$, and any nonzero $z \in X$. Suppose that $\mathcal{K} \not\subseteq \mathcal{I}$; that is, there exists a set $A \in \mathcal{K} \setminus \mathcal{I}$. Let $\xi \in X$ be a point with neighborhood $U \subset X$ such that $U \neq X$ and $y \in X \setminus U$. Let us define a sequence $\{x_k\}$ on X by

$$x_k = \begin{cases} \xi, & \text{if } k \notin A \\ y, & \text{otherwise.} \end{cases} \tag{30}$$

Clearly, $\mathcal{K}$-$\lim x_k, z = \xi$ and thus by Lemma 15 we get $\mathcal{I}^{F_{\mathcal{K}}}$-$\lim x_k, z = \xi$. As $\{k \in \mathbb{N} : x_k, z \notin \mathcal{N}_{\xi,z}(\varepsilon,\lambda)\} = A \notin \mathcal{I}$, the sequence $(x_k)_{k\in\mathbb{N}}$ is not $\mathcal{I}^{F_{\mathcal{K}}}$-convergent to ξ.

(ii) Let $(X, F, *)$ be an RTN space and $\xi \in X$ and let $(x_k)_{k\in\mathbb{N}}$ be a sequence on X. Assume that $\mathcal{K} \subset \mathcal{I}$ and $\mathcal{I}^{F_{\mathcal{K}}}$-$\lim x_k, z = \xi$. By the definition of $\mathcal{I}^{F_{\mathcal{K}}}$-convergence, there exists $M \in \mathcal{F}(\mathcal{I})$ such that the sequence $(y_k)_{k\in\mathbb{N}}$ given by

$$y_k = \begin{cases} x_k, & \text{if } k \in M \\ \xi, & \text{if } k \notin M \end{cases} \tag{31}$$

has the condition for every $\varepsilon > 0$, $\lambda \in (0, 1)$ and any nonzero $z \in X$

$$A = \left\{k \in \mathbb{N} : y_k, z \notin \mathcal{N}_{\xi,z}(\varepsilon,\lambda)\right\} = \left\{k \in M : x_k, z \notin \mathcal{N}_{\xi,z}(\varepsilon,\lambda)\right\} \in \mathcal{K}. \tag{32}$$

Hence, $A \cap M \in \mathcal{K} \subset \mathcal{I}$. Consequently,

$$\left\{k \in \mathbb{N} : x_k, z \notin \mathcal{N}_{\xi,z}(\varepsilon,\lambda)\right\} \subseteq (X \setminus M) \cup (A \cap M) \in \mathcal{I} \tag{33}$$

and thus $\mathcal{I}^F$-$\lim x_k, z = \xi$.

□

The following example shows that the converse of Theorem 17(ii) does need not to be true.

Example 18. Let $X = \mathbb{R}^2$, with the 2-norm $\|x, y\| := |x_1 y_2 - x_2 y_1|$ where $x = (x_1, x_2), y = (y_1, y_2) \in \mathbb{R}^2$, and let $a * b = ab$ for all $a, b \in S$. Now for all $x, y \in X$, and $t > 0$, let us define

$$F_{x,y}(t) = \frac{t}{t + \|x, y\|}. \quad (34)$$

Then, $(\mathbb{R}^2, F, *)$ is an RTN space. Let $\mathbb{N} = \cup_{i=1}^{\infty} D_i$ be a decomposition of $\mathbb{N}$ such that for any $n \in \mathbb{N}$ each D_i contains infinitely many i's where $i \geq j$ and $D_i \cap D_j = \emptyset$ for $i \neq j$. Denote by $\mathcal{I}$ the class of all $A \subseteq \mathbb{N}$ such that A intersects only a finite numbers of D_i. Let $\mathcal{K}$ be the family of all finite subsets of $\mathbb{N}$. We define a sequence $(x_k)_{k\in\mathbb{N}}$ as follows: $x_k = (1/i, 0) \in \mathbb{R}^2$ if $k \in D_i$. Then, for nonzero $z \in X$, we have

$$F_{x_k,z}(t) = \frac{t}{t + \|x_k, z\|} \longrightarrow 1 \quad (35)$$

as $k \to \infty$. Hence $\mathcal{I}^F\text{-}\lim x_k, z = \xi$.

Now assume that $\mathcal{I}^{F_{\mathcal{K}}}\text{-}\lim x_k, z = 0$. Then, there exists $M \in \mathcal{F}(\mathcal{I})$ such that the sequence $(y_k)_{k\in\mathbb{N}}$ given by

$$y_k = \begin{cases} x_k, & \text{if } k \in M \\ 0, & \text{if } k \notin M \end{cases} \quad (36)$$

has the condition for every $\varepsilon > 0$, $\lambda \in (0, 1)$ and any nonzero $z \in X$

$$\{k \in \mathbb{N} : y_k, z \notin \mathcal{N}_{\theta,z}(\varepsilon, \lambda)\} \in \mathcal{K}; \quad (37)$$

that is, $\mathcal{K}\text{-}\lim y_k, z = \xi$. Since $M \in \mathcal{F}(\mathcal{I})$, then there exists $A \in \mathcal{I}$ such that $M = \mathbb{N} \setminus A$. Now, by the definition of $\mathcal{I}$, there exists a $s \in \mathbb{N}$ such that $A \subset \cup_{i=1}^{s} D_i$. But $D_{s+1} \subset M$, and therefore $x_k = (1/(s+1), 0)$ for infinitely many k's from M which contradicts the assumption that $\mathcal{I}^{F_{\mathcal{K}}}\text{-}\lim x_k, z = 0$. Hence, the converse of the theorem does need not to be true.

Acknowledgment

The authors would like to express their appreciation to the referee for him/her valuable suggestions and corrections which help to provide better presentation of this paper.

References

[1] P. Kostyrko, T. Šalát, and W. Wilczyński, "$\mathcal{I}$-convergence," *Real Analysis Exchange*, vol. 26, no. 2, pp. 669–685, 2000.

[2] M. Katětov, "Products of filters," *Commentationes Mathematicae Universitatis Carolinae*, vol. 9, pp. 173–189, 1968.

[3] H. Fast, "Sur la convergence statistique," *Colloquium Mathematicum*, vol. 2, pp. 241–244, 1952.

[4] H. Steinhaus, "Sur la convergence ordinaire et la convergence asymptotique," *Colloquium Mathematicum*, vol. 2, pp. 73–74, 1951.

[5] T. Šalát, "On statistically convergent sequences of real numbers," *Mathematica Slovaca*, vol. 30, no. 2, pp. 139–150, 1980.

[6] J. A. Fridy, "On statistical convergence," *Analysis*, vol. 5, no. 4, pp. 301–313, 1985.

[7] P. Kostyrko, M. Macaj, and T. Salat, "Statistical convergence and $\mathcal{I}$-convergence," 1999, http://thales.doa.fmph.uniba.sk/macaj/ICON.pdf .

[8] P. Kostyrko, M. Mačaj, T. Šalát, and M. Sleziak, "$\mathcal{I}$-convergence and extremal $\mathcal{I}$-limit points," *Mathematica Slovaca*, vol. 55, no. 4, pp. 443–464, 2005.

[9] M. Mačaj and M. Sleziak, "$\mathcal{I}^{\mathcal{K}}$-convergence," *Real Analysis Exchange*, vol. 36, no. 1, pp. 177–193, 2010.

[10] M. Eshaghi Gordji, S. Sarabadan, and F. Amouei Arani, "$\mathcal{I}^{\mathcal{K}}$-convergence in 2-normed spaces," *Functional Analysis, Approximation and Computation*, vol. 4, no. 1, pp. 1–7, 2012.

[11] S. Gähler, "Lineare 2-normierte räume," *Mathematische Nachrichten*, vol. 28, no. 1-2, pp. 1–43, 1964.

[12] H. Gunawan and Mashadi, "On finite-dimensional 2-normed spaces," *Soochow Journal of Mathematics*, vol. 27, no. 3, pp. 321–329, 2001.

[13] M. Gürdal and S. Pehlivan, "The statistical convergence in 2-Banach spaces," *Thai Journal of Mathematics*, vol. 2, no. 1, pp. 107–113, 2004.

[14] M. Gürdal and I. Açık, "On $\mathcal{I}$-Cauchy sequences in 2-normed spaces," *Mathematical Inequalities & Applications*, vol. 11, no. 2, pp. 349–354, 2008.

[15] M. Gürdal and M. B. Huban, "$\mathcal{I}$-limit Points in Random 2-normed Spaces," *Theory and Applications of Mathematics and Computer Science*, vol. 2, no. 1, pp. 15–22, 2012.

[16] A. H. Siddiqi, "I^K-normed spaces," *The Aligarh Bulletin of Mathematics*, vol. 7, pp. 53–70, 1980.

[17] A. Şahiner, M. Gürdal, S. Saltan, and H. Gunawan, "Ideal convergence in 2-normed spaces," *Taiwanese Journal of Mathematics*, vol. 11, no. 5, pp. 1477–1484, 2007.

[18] A. N. Šerstnev, "Random normed spaces. Questions of completeness," *Kazanskii Gosudarstvennyi Universitet. Uchenye Zapiski*, vol. 122, pp. 3–20, 1962.

[19] C. Alsina, B. Schweizer, and A. Sklar, "On the definition of a probabilistic normed space," *Aequationes Mathematicae*, vol. 46, no. 1-2, pp. 91–98, 1993.

[20] A. N. Šerstnev, "On the concept of a stochastic normalized space," *Doklady Akademii Nauk SSSR*, vol. 149, pp. 280–283, 1963.

[21] B. Schweizer and A. Sklar, *Probabilistic Metric Spaces*, Elsevier Science Publishing, New York, NY, USA, 1983.

[22] C. Alsina, B. Schweizer, and A. Sklar, "Continuity properties of probabilistic norms," *Journal of Mathematical Analysis and Applications*, vol. 208, no. 2, pp. 446–452, 1997.

[23] I. Goleţ, "On probabilistic 2-normed spaces," *Novi Sad Journal of Mathematics*, vol. 35, no. 1, pp. 95–102, 2005.

[24] M. Gürdal and M. B. Huban, "On $\mathcal{I}$-convergence of double sequences in the topology induced by random 2-norms," *Matematicki Vesnik*. In press.

[25] M. Mursaleen and S. A. Mohiuddine, "On ideal convergence of double sequences in probabilistic normed spaces," *Mathematical Reports*, vol. 12, no. 4, pp. 359–371, 2010.

[26] M. Mursaleen and A. Alotaibi, "On $\mathcal{I}$-convergence in random 2-normed spaces," *Mathematica Slovaca*, vol. 61, no. 6, pp. 933–940, 2011.

[27] M. Mursaleen and S. A. Mohiuddine, "On ideal convergence in probabilistic normed spaces," *Mathematica Slovaca*, vol. 62, no. 1, pp. 49–62, 2012.

[28] M. R. S. Rahmat and K. K. Harikrishnan, "On I-convergence in the topology induced by probabilistic norms," *European Journal of Pure and Applied Mathematics*, vol. 2, no. 2, pp. 195–212, 2009.

[29] B. C. Tripathy, M. Sen, and S. Nath, "$\mathcal{I}$-convergence in probabilistic n-normed space," *Soft Computing*, vol. 16, pp. 1021–1027, 2012.

[30] C. Sempi, "A short and partial history of probabilistic normed spaces," *Mediterranean Journal of Mathematics*, vol. 3, no. 2, pp. 283–300, 2006.

[31] A. R. Freedman and J. J. Sember, "Densities and summability," *Pacific Journal of Mathematics*, vol. 95, no. 2, pp. 293–305, 1981.

[32] I. J. Schoenberg, "The integrability of certain functions and related summability methods," *The American Mathematical Monthly*, vol. 66, pp. 361–375, 1959.

Global Attractivity Results on Complete Ordered Metric Spaces for Third-Order Difference Equations

Mujahid Abbas[1] **and Maher Berzig**[2]

[1] *Department of Mathematics and Applied Mathematics, University of Pretoria, Lynnwood road, Pretoria 0002, South Africa*
[2] *University of Tunisia, Tunis College of Sciences and Techniques, 5 Avenue Taha Hussein, BP 56, Bab Manara, Tunis, Tunisia*

Correspondence should be addressed to Maher Berzig; maher.berzig@gmail.com

Academic Editor: Jacques Liandrat

We establish fixed-point theorems for mixed monotone mappings in the setting of ordered metric spaces which satisfy a contractive condition for all points that are related by a given ordering. We also give a global attractivity result for all solutions of the difference equation $u_{n+1} = F(u_n, u_{n-1}, u_{n-2}), n = 2, 3, \dots$, where F satisfies certain monotonicity conditions with respect to the given ordering. As an application of our obtained results, we present some iterative algorithms to solve a class of matrix equations. A numerical example is also presented to test the validity of the algorithms.

1. Introduction

The following global attractivity result from [1] (see also [2]) is very useful in establishing convergence results in many situations.

Theorem 1 (see [1]). *Let $[a, b]$ be a closed and bounded interval of real numbers and let $F \in C([a, b]^3, [a, b])$ satisfy the following conditions.*

(i) *The function $F(z_1, z_2, z_3)$ is monotonic in each of its arguments.*

(ii) *For each $m, M \in [a, b]$ and for each $i \in \{1, 2, 3\}$, one defines*

$$M_i(m, M) = \begin{cases} M, & \text{if } F \text{ is increasing in } z_i, \\ m, & \text{if } F \text{ is decreasing in } z_i, \end{cases} \quad (1)$$

$$m_i(m, M) = M_i(M, m)$$

and assume that if (m, M) is a solution of the system

$$\begin{aligned} M &= F(M_1(m, M), M_2(m, M), M_3(m, M)), \\ m &= F(m_1(m, M), m_2(m, M), m_3(m, M)), \end{aligned} \quad (2)$$

then $M = m$.
Then there exists exactly one equilibrium $\overline{x}$ of the equation

$$x_{n+1} = F(x_n, x_{n-1}, x_{n-2}), \quad n = 0, 1, \dots \quad (3)$$

and every solution of the above equation converges to $\overline{x}$.

The above result in Theorem 1 attracted considerable attention of the leading specialists in difference equations and discrete dynamic systems and has been generalized and extended to the case of maps in $\mathbb{R}^n$, see [3], and maps in Banach space with cone, see [4–6].

Moreover, there has been recent interest in establishing fixed-point theorems in partially ordered complete metric spaces with a contractivity condition which holds for all points that are related by partial ordering, (see, e.g., [7–19]). These fixed-point results have been applied mainly to the existence of solutions of boundary value problems for differential equations, namely [18], has been applied for solving a class of matrix equations.

In [20], Burgić et al. obtained the following global attractivity result for mixed monotone mappings in partially ordered complete metric spaces (see also Gnana Bhaskar and Lakshmikantham [10]).

Theorem 2 (see Burgić et al. [20]). *Let $(X, \preceq)$ be a partially ordered set and suppose there is a metric d on X such that (X, d) is a complete metric space. Let $F : X \times X \to X$ be a map such that $F(x, y)$ is nonincreasing in x for all $y \in X$ and nondecreasing in y for all $x \in X$. Suppose that the following conditions hold.*

(i) *There exists $k \in [0, 1)$ with*

$$d\left(F\left(x,y\right),F\left(u,v\right)\right) \leq \frac{k}{2}\left[d\left(x,u\right)+d\left(y,v\right)\right] \quad \forall x \preceq u, y \succeq v. \tag{4}$$

(ii) *There exists $x_0, y_0 \in X$ such that the following condition holds:*

$$x_0 \preceq F\left(x_0, y_0\right), \qquad y_0 \succeq F\left(y_0, x_0\right). \tag{5}$$

(iii) *If $\{x_n\} \subset X$ is a nondecreasing convergent sequence such that $\lim_{n\to\infty} x_n = x$, then $x_n \preceq x$, for all $n \in \mathbb{N}$ and if $\{y_n\} \subset X$ is a nonincreasing convergent sequence such that $\lim_{n\to\infty} y_n = y$, then $y_n \succeq y$, for all $n \in \mathbb{N}$; if $x_n \preceq y_n$ for every n, then $\lim_{n\to\infty} x_n \preceq \lim_{n\to\infty} y_n$.*

Then one has the following.

(a) *For every initial point $x_0, y_0 \in X$ such that condition (ii) holds, $F^n(x_0, y_0) \to x$, $F^n(y_0, x_0) \to y$, $n \to \infty$, where $x, y \in X$ satisfy*

$$x = F\left(x,y\right), \qquad y = F\left(y,x\right). \tag{6}$$

(b) *If $x_0 \preceq y_0$ in condition (ii), then $x = y$. If in addition $x = y$, then $\{x_n\}$, $\{y_n\}$ converge to the equilibrium of the equation*

$$x_{n+1} = F\left(x_n, y_n\right), \quad y_{n+1} = F\left(y_n, x_n\right), \quad n = 1, 2, \ldots. \tag{7}$$

(c) *In particular, every solution $\{z_n\}$ of*

$$z_{n+1} = F\left(z_n, z_{n-1}\right), \quad n = 2, 3, \ldots \tag{8}$$

such that $x_0 \preceq z_0$, $z_1 \preceq y_0$ converges to the equilibrium of (7).

In this paper, motivated by the results and ideas in a recent work of Berinde and Borcut [9], we extend Theorem 2 to mappings $F : X \times X \times X \to X$. Such extension allows us to study the third-order difference equation

$$u_{n+1} = F\left(u_n, u_{n-1}, u_{n-2}\right), \quad n = 2, 3, \ldots. \tag{9}$$

The presented theorems also extend and generalize the work in [9]. We use our obtained results to build some iterative algorithms to solve a class of matrix equations. A numerical example is also presented to test the validity of the algorithms.

Now we introduce the following concepts.

Definition 3. Let X be a nonempty set and $F : X \times X \times X \to X$ a given mapping. One says that $(x, y, z) \in X \times X \times X$ is a fixed-point of the third order if

$$F\left(x,y,x\right) = x, \qquad F\left(y,x,y\right) = y, \qquad F\left(z,y,z\right) = z. \tag{10}$$

Definition 4. Let$(X, \preceq)$ be a partially ordered set and $F : X \times X \times X \to X$ a given mapping. one says that F has the mixed monotone property if $F(x, y, z)$ is monotonously increasing in x and z and is monotonously decreasing in y; that is, for any $x, y, z \in X$,

(i) $x_1, x_2 \in X, x_1 \preceq x_2 \Rightarrow F(x_1, y, z) \preceq F(x_2, y, z)$,
(ii) $y_1, y_2 \in X, y_1 \preceq y_2 \Rightarrow F(x, y_1, z) \succeq F(x, y_2, z)$, and
(iii) $z_1, z_2 \in X, z_1 \preceq z_2 \Rightarrow F(x, y, z_1) \preceq F(x, y, z_2)$.

Through this paper we will use the following notations. Let $(X, \preceq)$ be a partially ordered set.

(i) For $x, y, z \in X \times X \times X$ the notation $x \in [y, z]$ means that $x \succeq y$ and $x \preceq z$.

(ii) We endow $X \times X \times X$ with the partial order that we denote also by $\preceq$, defined by

$$\begin{gathered}(x,y,z),(u,v,w) \in X \times X \times X,\\ (x,y,z) \preceq (u,v,w) \Longleftrightarrow x \preceq u, y \succeq v, z \preceq w.\end{gathered} \tag{11}$$

(iii) Let $F : X \times X \times X \to X$ be a given mapping. For all $x, y, z \in X$, we denote

$$\begin{gathered}F^2\left(x,y,z\right) = F\left(F\left(x,y,x\right),F\left(y,x,y\right),F\left(z,y,z\right)\right),\\ F^{p+1}\left(x,y,z\right) = F\left(F^p\left(x,y,x\right),F^p\left(y,x,y\right),F^p\left(z,y,z\right)\right),\\ \forall p \geq 2.\end{gathered} \tag{12}$$

2. Main Result

Our first result is the following.

Theorem 5. *Let $(X, \preceq)$ be a partially ordered set and suppose there is a metric d on X such that (X, d) is a complete metric space. Let $F : X \times X \times X \to X$ be a mapping having the mixed monotone property on X. Suppose that the following conditions hold.*

(i) *There exists $\lambda \in [0, 1)$ with*

$$\begin{aligned}&d\left(F\left(x,y,z\right),F\left(u,v,w\right)\right)\\ &\quad\leq \lambda \max\{d\left(x,u\right),d\left(y,v\right),d\left(z,w\right),d\left(F\left(x,y,x\right),x\right),\\ &\qquad d\left(F\left(x,y,x\right),u\right),d\left(F\left(y,x,y\right),y\right),\\ &\qquad d\left(F\left(y,x,y\right),v\right),d\left(F\left(z,y,z\right),z\right),\\ &\qquad d\left(F\left(z,y,z\right),w\right)\}\end{aligned} \tag{13}$$

for all $(x, y, z) \preceq (u, v, w)$.

(ii) *There exist* $x_0, y_0, z_0 \in X$ *such that*

$$x_0 \preceq F(x_0, y_0, x_0), \qquad y_0 \succeq F(y_0, x_0, y_0), \qquad z_0 \preceq F(z_0, y_0, z_0). \tag{14}$$

(iii) *If* $\{x_n\} \subset X$ *is a nondecreasing convergent sequence such that* $\lim_{n\to\infty} x_n = x$, *then* $x_n \preceq x$ *for all* $n \in \mathbb{N}$, *if* $\{y_n\} \subset X$ *is a nonincreasing convergent sequence such that* $\lim_{n\to\infty} y_n = y$, *then* $y_n \succeq y$ *for every* $n \in \mathbb{N}$, *and if* $x_n \preceq y_n$ *for every* n, *then* $\lim_{n\to\infty} x_n \preceq \lim_{n\to\infty} y_n$.

Then one has the following.

(a) *For every initial point* $(x_0, y_0, z_0) \in X \times X \times X$ *such that condition* (14) *holds:*

$$x_n = F^n(x_0, y_0, x_0) \longrightarrow x, \qquad y_n = F^n(y_0, x_0, y_0) \longrightarrow y, \qquad z_n = F^n(z_0, y_0, z_0) \longrightarrow z \quad \text{as } n \to \infty, \tag{15}$$

where x, y, *and* z *satisfy*

$$x = F(x, y, x), \qquad y = F(y, x, y), \qquad z = F(z, y, z). \tag{16}$$

If $x_0 \preceq y_0$ *and* $z_0 \preceq y_0$ *in condition* (14), *then* $x = y = z$ *and* $\{x_n\}$, $\{y_n\}$, *and* $\{z_n\}$ *converge to the equilibrium of the equation*

$$x_{n+1} = F(x_n, y_n, x_n), \qquad y_{n+1} = F(y_n, x_n, y_n), \qquad z_{n+1} = F(z_n, y_n, z_n), \quad n = 0, 1, 2 \ldots. \tag{17}$$

(b) *In particular, every solution* $\{u_n\}$ *of*

$$u_{n+1} = F(u_n, u_{n-1}, u_{n-2}), \quad n = 2, 3, \ldots, \tag{18}$$

such that $u_0, u_1, u_2 \in [x_0, y_0]$ *(or* $[z_0, y_0]$*) converges to the equilibrium of* (18).

(c) *The following estimates hold:*

$$\begin{aligned}
&d(x, F^n(x_0, y_0, x_0)) \\
&\quad \le \frac{\lambda^n}{1-\lambda} \max\{d(F(x_0, y_0, x_0), x_0), \\
&\qquad d(F(y_0, x_0, y_0), y_0), \\
&\qquad d(z_0, F(z_0, y_0, z_0))\}, \\
&d(y, F^n(y_0, x_0, y_0)) \\
&\quad \le \frac{\lambda^n}{1-\lambda} \max\{d(F(x_0, y_0, x_0), x_0), \\
&\qquad d(F(y_0, x_0, y_0), y_0), \\
&\qquad d(z_0, F(z_0, y_0, z_0))\}, \\
&d(z, F^n(z_0, y_0, z_0)) \\
&\quad \le \frac{\lambda^n}{1-\lambda} \max\{d(F(x_0, y_0, x_0), x_0), \\
&\qquad d(F(y_0, x_0, y_0), y_0), \\
&\qquad d(F(z_0, y_0, z_0), z_0)\}.
\end{aligned} \tag{19}$$

Proof. Let $x_0, y_0, z_0 \in X$ such that condition (14) is satisfied. Denote $x_1 = F(x_0, y_0, x_0)$, $y_1 = F(y_0, x_0, y_0)$, and $z_1 = F(z_0, y_0, z_0)$. Since $x_0 \preceq F(x_0, y_0, x_0) = x_1$, $y_0 \succeq F(y_0, x_0, y_0) = y_1$, and $z_0 \preceq F(z_0, y_0, z_0) = y_1$, from the mixed monotone property of F, we get that

$$\begin{aligned}
x_0 &\preceq x_1 = F(x_0, y_0, x_0) \preceq x_2 = F(x_1, y_1, x_1), \\
y_0 &\succeq y_1 = F(y_0, x_0, y_0) \succeq y_2 = F(y_1, x_1, y_1), \\
z_0 &\preceq z_1 = F(z_0, y_0, z_0) \preceq z_2 = F(z_1, y_1, z_1).
\end{aligned} \tag{20}$$

Consider the sequences $\{x_n\}$, $\{y_n\}$, and $\{z_n\}$ defined by (17). By induction and using the mixed monotone property of F, we obtain easily that

$$x_0 \preceq x_1 \preceq \cdots \preceq x_n \preceq \cdots; \qquad y_0 \succeq y_1 \succeq \cdots \succeq y_n \succeq \cdots; \qquad z_0 \preceq z_1 \preceq \cdots \preceq z_n \preceq \cdots. \tag{21}$$

For the sake of clarity, for all $n \ge 1$, denote

$$D_n^x = d(x_{n-1}, x_n), \qquad D_n^y = d(y_{n-1}, y_n), \qquad D_n^z = d(z_{n-1}, z_n). \tag{22}$$

We claim that, for all $n \ge 1$, we have

$$\begin{aligned}
D_{n+1}^x &\le \lambda^n \max\{D_1^x, D_1^y, D_1^z\}, \\
D_{n+1}^y &\le \lambda^n \max\{D_1^x, D_1^y, D_1^z\}, \\
D_{n+1}^z &\le \lambda^n \max\{D_1^x, D_1^y, D_1^z\}.
\end{aligned} \tag{23}$$

By (13) and (21), we obtain

$$\begin{aligned}
D_2^x &= d(x_1, x_2) = d(F(x_0, y_0, x_0), F(x_1, y_1, x_1)) \\
&\le \lambda \max\{d(x_0, x_1), d(y_0, y_1), d(x_0, x_1), \\
&\quad d(F(x_0, y_0, x_0), x_0), d(F(x_0, y_0, x_0), x_1), \\
&\quad d(F(y_0, x_0, y_0), y_0), d(F(y_0, x_0, y_0), y_1), \\
&\quad d(F(x_0, y_0, x_0), x_0), d(F(x_0, y_0, x_0), x_1)\} \\
&= \lambda \max\{d(x_0, x_1), d(y_0, y_1)\}, \\
D_2^y &= d(y_1, y_2) = d(F(y_0, x_0, y_0), F(y_1, x_1, y_1)) \\
&\le \lambda \max\{d(y_0, y_1), d(x_0, x_1), d(y_0, y_1), \\
&\quad d(F(y_0, x_0, y_0), y_0), d(F(y_0, x_0, y_0), y_1), \\
&\quad d(F(x_0, y_0, x_0), x_0), d(F(x_0, y_0, x_0), x_1), \\
&\quad d(F(y_0, x_0, y_0), y_0), d(F(y_0, x_0, y_0), y_1)\} \\
&= \lambda \max\{d(y_0, y_1), d(x_0, x_1)\}, \\
D_2^z &= d(z_1, z_2) = d(F(z_0, y_0, z_0), F(z_1, y_1, z_1)) \\
&\le \lambda \max\{d(z_0, z_1), d(y_0, y_1), d(z_0, z_1), \\
&\quad d(F(z_0, y_0, z_0), z_0), d(F(z_0, y_0, z_0), z_1), \\
&\quad d(F(y_0, z_0, y_0), y_0), d(F(y_0, z_0, y_0), y_1), \\
&\quad d(F(z_0, y_0, z_0), z_0), d(F(z_0, y_0, z_0), z_1)\} \\
&= \lambda \max\{d(z_0, z_1), d(y_0, y_1)\}.
\end{aligned} \tag{24}$$

Thus we get that

$$D_2^x \le \lambda \max\{D_1^x, D_1^y, D_1^z\}, \qquad D_2^y \le \lambda \max\{D_1^x, D_1^y, D_1^z\}, \qquad D_2^z \le \lambda \max\{D_1^x, D_1^y, D_1^z\}. \tag{25}$$

Then our claim holds for $n = 1$. Suppose now that (23) holds for some fixed $n \ge 1$. Similarly, by (13) and (21), we obtain

$$\begin{aligned}
D_{n+2}^x &= d(x_{n+1}, x_{n+2}) \\
&= d(F(x_n, y_n, x_n), F(x_{n+1}, y_{n+1}, x_{n+1})) \\
&\le \lambda \max\{D_{n+1}^x, D_{n+1}^y\} \\
&\le \lambda^{n+1} \max\{D_1^x, D_1^y, D_1^y\}.
\end{aligned} \tag{26}$$

Similarly, one can show that

$$\begin{aligned}
D_{n+2}^y &\le \lambda^{n+1} \max\{D_1^x, D_1^y, D_1^z\}, \\
D_{n+2}^z &\le \lambda^{n+1} \max\{D_1^x, D_1^y, D_1^z\}.
\end{aligned} \tag{27}$$

Then by the induction principle, (23) holds for all $n \ge 1$.

Now we will prove that $\{x_n\}$, $\{y_n\}$, and $\{z_n\}$ are Cauchy sequences in the metric space (X, d). Using (23) and the triangular inequality, for $m > n$, we have

$$\begin{aligned}
d(x_m, x_n) &\le \sum_{i=n}^{m-1} d(x_i, x_{i+1}) = \sum_{i=n}^{m-1} D_{i+1}^x \\
&\le \sum_{i=n}^{m-1} \lambda^i \max\{D_1^x, D_1^y, D_1^z\} \\
&\le \left(\sum_{i=n}^{m-1} \lambda^i\right) \max\{D_1^x, D_1^y, D_1^z\} \\
&\le \frac{\lambda^n}{1-\lambda} \max\{D_1^x, D_1^y, D_1^z\} \longrightarrow 0 \quad \text{as } n \longrightarrow \infty.
\end{aligned} \tag{28}$$

This implies that $\{x_n\}$ is a Cauchy sequence. Similarly, one can prove that $\{y_n\}$ and $\{z_n\}$ are Cauchy sequences.

Since (X, d) is complete, there exist $x, y, z \in X$ such that

$$x_n \longrightarrow x, \qquad y_n \longrightarrow y, \qquad z_n \longrightarrow z \quad \text{as } n \longrightarrow \infty. \tag{29}$$

From condition (iii) and (21), we get that

$$x_n \preceq x, \qquad y_n \succeq y, \qquad z_n \preceq z, \qquad \forall n. \tag{30}$$

We claim that $x = F(x, y, x)$. Indeed, we have

$$\begin{aligned}
&d(x, F(x, y, x)) \\
&\le d(x, x_{n+1}) + d(x_{n+1}, F(x, y, x)) \\
&= d(x, x_{n+1}) + d(F(x_n, y_n, x_n), F(x, y, x)) \\
&\le d(x, x_{n+1}) \\
&\quad + \lambda \max\{d(x_n, x), d(y_n, y), \\
&\qquad d(x_n, x), d(F(x_n, y_n, x_n), x_n), \\
&\qquad d(F(x_n, y_n, x_n), x), d(F(y_n, x_n, y_n), y_n), \\
&\qquad d(F(y_n, x_n, y_n), y), d(F(x_n, y_n, x_n), x_n), \\
&\qquad d(F(x_n, y_n, x_n), x)\} \\
&= d(x, x_{n+1}) + \lambda \max\{d(x_n, x), d(y_n, y)\} \\
&\longrightarrow 0 \quad \text{as } n \longrightarrow \infty \quad \text{by (29)}.
\end{aligned} \tag{31}$$

Then, our claim holds; that is, $x = F(x, y, x)$. Similarly, one can show that $y = F(y, x, y)$ and $z = F(z, y, z)$. Thus we proved (16).

On the other hand, from (28), for $m > n$ with n being fixed, we have

$$\begin{aligned}
d(x_m, x_n) &= d(x_m, F^n(x_0, y_0, x_0)) \\
&\le \frac{\lambda^n}{1-\lambda} \max\{D_1^x, D_1^y, D_1^z\}.
\end{aligned} \tag{32}$$

Letting $m \to \infty$ in the above inequality, we obtain

$$d(x, F^n(x_0, y_0, x_0)) \leq \frac{\lambda^n}{1-\lambda} \max\{d(x_0, x_1), d(y_0, y_1), d(z_0, z_1)\}; \tag{33}$$

that is,

$$\begin{aligned} d(x, F^n(x_0, y_0, x_0)) \leq \frac{\lambda^n}{1-\lambda} \max\{&d(x_0, F(x_0, y_0, x_0)), \\ &d(y_0, F(y_0, x_0, y_0)), \\ &d(z_0, F(z_0, y_0, z_0))\}. \end{aligned} \tag{34}$$

Similarly, one can show that

$$\begin{aligned} d(y, F^n(y_0, x_0, y_0)) \leq \frac{\lambda^n}{1-\lambda} \max\{&d(x_0, F(x_0, y_0, x_0)), \\ &d(y_0, F(y_0, x_0, y_0)), \\ &d(z_0, F(z_0, y_0, z_0))\}, \\ d(z, F^n(z_0, y_0, z_0)) \leq \frac{\lambda^n}{1-\lambda} \max\{&d(x_0, F(x_0, y_0, x_0)), \\ &d(y_0, F(y_0, x_0, y_0)), \\ &d(z_0, F(z_0, y_0, z_0))\}. \end{aligned} \tag{35}$$

Thus we proved (19).

Now, if $x_0 \preceq y_0$ and $z_0 \preceq y_0$, we claim that, for all $n \in \mathbb{N}$, $x_n \preceq y_n$ and $z_n \preceq y_n$. Indeed, by the mixed monotone property of F,

$$\begin{aligned} x_1 &= F(x_0, y_0, x_0) \preceq F(y_0, x_0, y_0) = y_1, \\ z_1 &= F(z_0, y_0, z_0) \preceq F(y_0, x_0, y_0) = y_1. \end{aligned} \tag{36}$$

Assume that $x_n \preceq y_n$ and $z_n \preceq y_n$ for some n. Then,

$$x_{n+1} = F(x_n, y_n, x_n) \preceq F(y_n, x_n, y_n) = y_{n+1}, \tag{37}$$

and similarly for y_n and z_n. Thus we proved that

$$x_n \preceq y_n, \quad z_n \preceq y_n, \quad \forall n \in \mathbb{N}. \tag{38}$$

Next, from (38), we have

$$\begin{aligned} d(x, y) &\leq d(x, x_{n+1}) + d(x_{n+1}, y_{n+1}) + d(y_{n+1}, y) \\ &\leq d(x, x_{n+1}) + d(F(x_n, y_n, x_n), F(y_n, x_n, y_n)) \\ &\quad + d(y_{n+1}, y) \\ &\leq d(x, x_{n+1}) + \lambda d(x_n, z_n) + d(y_{n+1}, y) \\ &\vdots \\ &\leq d(x, x_{n+1}) + \lambda^n d(x_0, y_0) + d(y_{n+1}, y) \\ &\longrightarrow 0 \quad \text{as } n \longrightarrow \infty, \end{aligned} \tag{39}$$

which implies that $x = y$. Similarly, we obtain that $d(y, z) = 0$, that is, $y = z$. Then $x = y = z$.

Now, assume that $x_0 \preceq y_0$. Then, in view of the monotonicity of F, we have

$$\begin{aligned} x_1 = F(x_0, y_0, x_0) \preceq x_2 = F(x_1, y_1, x_1) \\ \preceq F(y_1, x_1, y_1) = y_2 \preceq F(y_0, x_0, y_0) = y_1. \end{aligned} \tag{40}$$

Continuing this process, one can show that

$$x_n \preceq x_{n+1} \preceq y_{n+1} \preceq y_n, \quad \forall n. \tag{41}$$

If we assume that $u_0, u_1, u_2 \in [x_0, y_0]$, in view of the monotonicity of F, we have

$$\begin{aligned} x_1 &= F(x_0, y_0, x_0) \preceq F(u_2, u_1, u_0) \\ &= u_3 \preceq F(y_0, x_0, y_0) = y_1, \\ x_1 &= F(x_0, y_0, x_0) \preceq F(u_3, u_2, u_1) \\ &= u_4 \preceq F(y_0, x_0, y_0) = y_1, \\ x_1 &= F(x_0, y_0, x_0) \preceq F(u_4, u_3, u_2) \\ &= u_5 \preceq F(y_0, x_0, y_0) = y_1. \end{aligned} \tag{42}$$

Continuing in a similar way, we can prove that

$$x_k \preceq u_{3k}, u_{3k+1}, u_{3k+2} \preceq y_k, \quad \forall k \geq 1. \tag{43}$$

Letting $k \to \infty$ and using (iii) and (29), we get that $u_n \to x$ as $n \to \infty$, where x is the equilibrium of (17). Similarly, if $z_0 \preceq y_0$ and $u_0, u_1, u_2 \in [z_0, y_0]$, we obtain also the same result. □

Remark 6. If we replace the condition, if $\{x_n\} \subset X$ is a nondecreasing convergent sequence such that $\lim_{n\to\infty} x_n = x$, then $x_n \preceq x$ for all $n \in \mathbb{N}$ and if $\{y_n\} \subset X$ is a nonincreasing convergent sequence such that $\lim_{n\to\infty} y_n = y$, then $y_n \succeq y$ for every $n \in \mathbb{N}$, by the continuity of F, we can check easily that the result of Theorem 5 holds also in this case.

Theorem 7. *In addition to the hypotheses of Theorem 5, suppose that for every (x, y, z), $(x', y', z') \in X \times X \times X$, there* *exists (u, v, w) such that $(x, y, z) \preceq (u, v, w)$ and $(x', y', z') \preceq (u, v, w)$. Then one obtains the uniqueness of the fixed point of the third order.*

Proof. From (a) of Theorem 5, we know that F admits a fixed point of the third order $(x, y, z) \in X \times X \times X$; that is,

$$x = F(x, y, x), \qquad y = F(y, x, y), \qquad z = F(z, y, z), \tag{44}$$

where

$$x_n = F^n(x_0, y_0, x_0) \longrightarrow x, \qquad y_n = F^n(y_0, x_0, y_0) \longrightarrow y, \qquad z_n = F^n(z_0, y_0, z_0) \longrightarrow z \quad \text{as } n \longrightarrow \infty. \tag{45}$$

Suppose that $(x^*, y^*, z^*) \in X \times X \times X$ is another fixed point of the third order of F; that is,

$$x^* = F(x^*, y^*, x^*), \qquad y^* = F(y^*, x^*, y^*), \qquad z^* = F(z^*, y^*, z^*). \tag{46}$$

We will prove that

$$\eta((x, y, z), (x^*, y^*, z^*)) = 0, \tag{47}$$

where

$$\eta((x, y, z), (x^*, y^*, z^*)) = \max\{d(x, x^*), d(y, y^*), d(z, z^*)\}. \tag{48}$$

From the hypothesis of Theorem 7, there exists (u, v, w) such that

$$(x, y, z) \preceq (u, v, w), \qquad (x^*, y^*, z^*) \preceq (u, v, w). \tag{49}$$

Since F is a mixed monotone operator, we have

$$\begin{aligned} &(F^n(x, y, x), F^n(y, x, y), F^n(z, y, z)) \\ &\quad = (x, y, z) \preceq (F^n(u, v, w), F^n(v, u, v), F^n(w, v, w)), \\ &(F^n(x^*, y^*, x^*), F^n(y^*, x^*, y^*), F^n(z^*, y^*, z^*)) \\ &\quad = (x^*, y^*, z^*) \preceq (F^n(u, v, w), F^n(v, u, v), \\ &\qquad F^n(w, v, w)). \end{aligned} \tag{50}$$

We have

$$\begin{aligned} &\eta((x, y, z), (x^*, y^*, z^*)) \\ &\quad = \eta((F^n(x, y, x), F^n(y, x, y), F^n(z, y, z)), \\ &\qquad (F^n(x^*, y^*, x^*), F^n(y^*, x^*, y^*), \\ &\qquad F^n(z^*, y^*, z^*))) \\ &\quad \le \eta((F^n(x, y, x), F^n(y, x, y), F^n(z, y, z)), \\ &\qquad (F^n(u, v, u), F^n(v, u, v), F^n(w, v, w))) \\ &\qquad + \eta((F^n(u, v, u), F^n(v, u, v), F^n(w, v, w)), \\ &\qquad (F^n(x^*, y^*, x^*), F^n(y^*, x^*, y^*), \\ &\qquad F^n(z^*, y^*, z^*))) \\ &\quad \le \lambda^n \max\{d(x, u), d(y, v), d(z, w)\} \\ &\qquad + \lambda^n \max\{d(u, x^*), d(v, y^*), d(w, z^*)\} \\ &\quad \longrightarrow 0 \quad \text{as } n \to \infty, \end{aligned} \tag{51}$$

which implies that

$$\eta((x, y, z), (x^*, y^*, z^*)) = 0; \tag{52}$$

that is,

$$x = x^*, \qquad y = y^*, \qquad z = z^*. \tag{53}$$

Example 8. Let $X = [0, 1]$ be an ordered set with the natural ordering of real numbers and d a usual metric on X. Let $F : X \times X \times X \to X$ be defined by

$$F(x, y, z) = \frac{1 + x - y + 2z}{8} \quad \forall x, y, z \in X. \tag{54}$$

It is easy to check that F has the mixed monotone property on X. For $x, y, z, u, v, w \in X$ with $(x, y, z) \preceq (u, v, w)$, we have

$$\begin{aligned} &d(F(x, y, z), F(u, v, w)) \\ &\quad = \frac{1}{8}|x - u - (y - v) + 2(z - w)| \\ &\quad \le \frac{1}{8}[|x - u| + |y - v| + 2|z - w|] \\ &\quad \le \frac{3}{4}\max\{|x - u|, |y - v|, |z - w|\} \\ &\quad = \frac{3}{4}\max\{d(x, u), d(y, v), d(z, w)\} \\ &\quad \le \frac{3}{4}\max\{d(x, u), d(y, v), d(z, w), \\ &\qquad d(F(x, y, x), x), d(F(x, y, x), u), \\ &\qquad d(F(y, x, y), y), d(F(y, x, y), v), \\ &\qquad d(F(z, y, z), z), d(F(z, y, z), w)\}. \end{aligned} \tag{55}$$

Thus (13) is satisfied for $\lambda = 3/4 \in [0,1)$. Thus all the conditions of Theorems 5 and 7 are satisfied. Moreover, there exists a unique $(x, y, z) = (1/6, 1/6, 1/6)$ in $X \times X \times X$ such that

$$F(x, y, x) = x, \qquad F(y, x, y) = y, \qquad F(z, y, z) = z. \tag{56}$$

□

Corollary 9. *Let $(X, \preceq)$ be a partially ordered set and suppose there is a metric d on X such that (X, d) is a complete metric space. Let $F : X \times X \times X \to X$ be a mapping having the mixed monotone property on X. Suppose that the following conditions hold.*

(1) *There exists $a_{i's} \geq 0$ for $i = 1, 2, \ldots, 9$ and $\sum_{i=1}^{9} a_i < 1$ with*

$$\begin{aligned} d(F(x, y, z), F(u, v, w)) &\leq a_1 d(x, u) + a_2 d(y, v) + a_3 d(z, w) \\ &\quad + a_4 d(F(x, y, x), x) + a_5 d(F(x, y, x), u) \\ &\quad + a_6 d(F(y, x, y), y) + a_7 d(F(y, x, y), v) \\ &\quad + a_8 d(F(z, y, z), z) + a_9 d(F(z, y, z), w) \end{aligned} \tag{57}$$

for all $(x, y, z) \preceq (u, v, w)$.

(2) *There exist $x_0, y_0, z_0 \in X$ such that*

$$x_0 \preceq F(x_0, y_0, x_0), \qquad y_0 \succeq F(y_0, x_0, y_0), \qquad z_0 \preceq F(z_0, y_0, z_0). \tag{58}$$

(3) *If $\{x_n\} \subset X$ is a nondecreasing convergent sequence such that $\lim_{n\to\infty} x_n = x$, then $x_n \preceq x$ for all $n \in \mathbb{N}$, if $\{y_n\} \subset X$ is a nonincreasing convergent sequence such that $\lim_{n\to\infty} y_n = y$, then $y_n \succeq y$ for every $n \in \mathbb{N}$, and if $x_n \preceq y_n$ for every $n \in \mathbb{N}$, then $\lim_{n\to\infty} x_n \preceq \lim_{n\to\infty} y_n$.*

Then one has the following.

(a) *For every initial point $(x_0, y_0, z_0) \in X \times X \times X$ such that condition (14) holds,*

$$x_n = F^n(x_0, y_0, x_0) \longrightarrow x, \qquad y_n = F^n(y_0, x_0, y_0) \longrightarrow y, \qquad z_n = F^n(z_0, y_0, z_0) \longrightarrow z \quad \text{as } n \longrightarrow \infty, \tag{59}$$

where x, y, and z satisfy

$$x = F(x, y, x), \qquad y = F(y, x, y), \qquad z = F(z, y, z). \tag{60}$$

If $x_0 \preceq y_0$ and $z_0 \preceq y_0$ in condition (14), then $x = y = z$ and $\{x_n\}$, $\{y_n\}$, and $\{z_n\}$ converge to the equilibrium of the equation

$$x_{n+1} = F(x_n, y_n, x_n), \qquad y_{n+1} = F(y_n, x_n, y_n), \qquad z_{n+1} = F(z_n, y_n, z_n), \quad n = 0, 1, 2 \ldots. \tag{61}$$

(b) *In particular, every solution $\{u_n\}$ of*

$$u_{n+1} = F(u_n, u_{n-1}, u_{n-2}), \quad n = 2, 3, \ldots \tag{62}$$

such that $u_0, u_1, u_2 \in [x_0, y_0]$ (or $[z_0, y_0]$) converges to the equilibrium of (18).

Proof. Note that (14) implies that

$$\begin{aligned} d(F(x, y, z), F(u, v, w)) &\leq h \max\{d(x, u), d(y, v), d(z, w), \\ &\quad d(F(x, y, x), x), d(F(x, y, x), u), \\ &\quad d(F(y, x, y), y), d(F(y, x, y), v), \\ &\quad d(F(z, y, z), z), d(F(z, y, z), w)\} \end{aligned} \tag{63}$$

for all $(x, y, z) \preceq (u, v, w)$, where $h = \sum_{i=1}^{9} a_i < 1$ and the result follows from Theorem 5. □

Corollary 10. *In addition to the hypotheses of Corollary 9, suppose that for every (x, y, z), $(x', y', z') \in X \times X \times X$, there exists (u, v, w) such that $(x, y, z) \preceq (u, v, w)$ and $(x', y', z') \preceq (u, v, w)$. Then one obtains the uniqueness of the fixed point of the third order.*

Corollary 11. *Let $(X, \preceq)$ be a partially ordered set and suppose there is a metric d on X such that (X, d) is a complete metric space. Let $F : X \times X \times X \to X$ be a mapping having the mixed monotone property on X. Suppose that the following conditions hold.*

(1) *There exists $\lambda \in [0, 1)$ with*

$$d(F(x, y, z), F(u, v, w)) \leq \lambda \max\{d(x, u), d(y, v), d(z, w)\} \quad \forall (x, y, z) \preceq (u, v, w). \tag{64}$$

(2) *There exist $x_0, y_0, z_0 \in X$ such that*

$$x_0 \preceq F(x_0, y_0, x_0), \qquad y_0 \succeq F(y_0, x_0, y_0), \qquad z_0 \preceq F(z_0, y_0, z_0). \tag{65}$$

(3) *If $\{x_n\} \subset X$ is a nondecreasing convergent sequence such that $\lim_{n\to\infty} x_n = x$, then $x_n \preceq x$ for all $n \in \mathbb{N}$; if $\{y_n\} \subset X$ is a nonincreasing convergent sequence such that $\lim_{n\to\infty} y_n = y$, then $y_n \succeq y$ for every $n \in \mathbb{N}$, and if $x_n \preceq y_n$ for every n, then $\lim_{n\to\infty} x_n \preceq \lim_{n\to\infty} y_n$.*

Then one has the following.

(a) *For every initial point $(x_0, y_0, z_0) \in X \times X \times X$ such that condition (14) holds,*

$$x_n = F^n(x_0, y_0, x_0) \longrightarrow x, \qquad y_n = F^n(y_0, x_0, y_0) \longrightarrow y, \qquad z_n = F^n(z_0, y_0, z_0) \longrightarrow z \quad \text{as } n \longrightarrow \infty, \tag{66}$$

where x, y, and z satisfy

$$x = F(x, y, x), \qquad y = F(y, x, y), \qquad z = F(z, y, z). \tag{67}$$

If $x_0 \preceq y_0$ and $z_0 \preceq y_0$ in condition (14), then $x = y = z$ and $\{x_n\}$, $\{y_n\}$, and $\{z_n\}$ converge to the equilibrium of the equation

$$\begin{gathered} x_{n+1} = F(x_n, y_n, x_n), \qquad y_{n+1} = F(y_n, x_n, y_n), \\ z_{n+1} = F(z_n, y_n, z_n), \quad n = 0, 1, 2 \ldots. \end{gathered} \tag{68}$$

(b) *In particular, every solution $\{u_n\}$ of*

$$u_{n+1} = F(u_n, u_{n-1}, u_{n-2}), \quad n = 2, 3, \ldots \tag{69}$$

such that $u_0, u_1, u_2 \in [x_0, y_0]$ (or $[z_0, y_0]$) converges to the equilibrium of (18).

(c) *The following estimates hold:*

$$\begin{aligned} &d(x, F^n(x_0, y_0, x_0)) \\ &\quad \le \frac{\lambda^n}{1-\lambda} \max\{d(F(x_0, y_0, x_0), x_0), \\ &\qquad d(F(y_0, x_0, y_0), y_0), \\ &\qquad d(z_0, F(z_0, y_0, z_0))\}, \\ &d(y, F^n(y_0, x_0, y_0)) \\ &\quad \le \frac{\lambda^n}{1-\lambda} \max\{d(F(x_0, y_0, x_0), x_0), \\ &\qquad d(F(y_0, x_0, y_0), y_0), \\ &\qquad d(z_0, F(z_0, y_0, z_0))\}, \\ &d(z, F^n(z_0, y_0, z_0)) \\ &\quad \le \frac{\lambda^n}{1-\lambda} \max\{d(F(x_0, y_0, x_0), x_0), \\ &\qquad d(F(y_0, x_0, y_0), y_0), \\ &\qquad d(F(z_0, y_0, z_0), z_0)\}. \end{aligned} \tag{70}$$

Corollary 12. *In addition to the hypotheses of Corollary 11, suppose that for every (x, y, z), $(x', y', z') \in X \times X \times X$, there exists (u, v, w) such that $(x, y, z) \preceq (u, v, w)$ and $(x', y', z') \preceq (u, v, w)$. Then one obtains the uniqueness of the fixed point of the third order.*

3. Application: Solving a Class of Third-Order Difference Matrix Equations

In this section, we apply our main results to the study of a class of third-order difference matrix equations. At first, we start by fixing some notations and recalling some preliminaries.

We will use the symbol $H(N)$ for the set of all $N \times N$ Hermitian matrices. We denote by $P(N)$ the set of all $N \times N$ Hermitian positive definite matrices. Instead of $X \in P(N)$ we will also write $X \succ 0$. Furthermore, $X \succeq 0$ means that X is positive semidefinite. As a different notation for $X - Y \succeq 0$, $X - Z \succeq 0$ and $X \preceq Z \preceq Y$ we will use, respectively, $X \succeq Y$, $X \succeq Z$, and $Z \in [X, Y]$. The symbol $\|\cdot\|$ denotes the spectral norm, that is, $\|A\| = \lambda_{\max}^{1/2}(A^*A)$, the largest eigenvalue of A^*A. We denote by $\|\cdot\|_1$ the Ky Fan norm defined by $\|A\|_1 = \sum_{j=1}^{N} s_j(A)$, where $s_j(A)$, $j = 1, \ldots, N$ are the singular values of A. For a given $Q \in P(N)$, we define the modified norm $\|\cdot\|_{1,Q}$ given by $\|A\|_{1,Q} = \|Q^{1/2}AQ^{1/2}\|_1$. The set $H(N)$ endowed with this norm is a complete metric space for any positive definite matrix Q. For any $N \times N$ matrix A, we denote by $\operatorname{tr}(A)$ the trace of the matrix A.

The following lemmas will be useful later.

Lemma 13 (see [18]). *Let $A \succeq 0$ and $B \succeq 0$ be $N \times N$ matrices, then*

$$0 \le \operatorname{tr}(AB) \le \|A\| \operatorname{tr}(B). \tag{71}$$

Lemma 14 (See [21]). *Let $A \in H(N)$ satisfy $-I \prec A \prec I$, then* $\|A\| < 1$.

Finally, we recall the well-known Schauder fixed-point theorem.

Theorem 15 (The Schauder fixed-point theorem). *Let S be a nonempty, compact, and convex subset of a normed vector space. Every continuous function $f : S \to S$ mapping S into itself has a fixed point.*

Now, we consider the class of third-order difference matrix equations:

$$\begin{aligned} U_{n+1} = Q &+ \sum_{i=1}^{m} A_i U_n A_i - \sum_{i=1}^{m} B_i U_{n-1} B_i \\ &+ \sum_{i=1}^{m} C_i U_{n-2} C_i, \quad n = 2, 3, \ldots \end{aligned} \tag{72}$$

for given $U_0, U_1, U_2 \succ 0$, where $Q \succ 0$ and A_i, B_i, and C_i are $N \times N$ Hermitian matrices. This type of difference equations often arises from many areas such as ladder networks [22, 23], dynamic programming [24, 25], and control theory [26, 27].

3.1. A Convergence Result

Theorem 16. *Suppose that*

$$\begin{gathered} \sum_{i=1}^{m} A_i Q A_i \prec \frac{Q}{4}, \qquad \sum_{i=1}^{m} B_i Q B_i \prec \frac{Q}{2}, \\ \sum_{i=1}^{m} C_i Q C_i \prec \frac{Q}{4}. \end{gathered} \tag{73}$$

Then, one has the following.

(i) *Equation* (72) *has one and only one equilibrium point* $\widehat{X} > 0$.

(ii) $\widehat{X} \in [F(0, Q, 0), F(2Q, 0, 2Q)]$, *where*

$$F(0, 2Q, 0) = Q - 2\sum_{i=1}^{m} B_i Q B_i,$$
$$F(2Q, 0, 2Q) = Q + 2\sum_{i=1}^{m} A_i Q A_i + 2\sum_{i=1}^{m} C_i Q C_i. \quad (74)$$

(iii) *The sequences* (X_k) *and* (Y_k) *defined by* $X_0 = 0 = Z_0$, $Y_0 = 2Q$, *and*

$$X_{k+1} = Q + \sum_{i=1}^{m} A_i X_k A_i - \sum_{i=1}^{m} B_i Y_k B_i + \sum_{i=1}^{m} A_i X_k A_i,$$
$$Y_{k+1} = Q + \sum_{i=1}^{m} A_i Y_k A_i - \sum_{i=1}^{m} B_i X_k B_i + \sum_{i=1}^{m} A_i Y_k A_i, \quad (75)$$
$$Z_{k+1} = Q + \sum_{i=1}^{m} A_i Z_k A_i - \sum_{i=1}^{m} B_i Y_k B_i + \sum_{i=1}^{m} A_i Z_k A_i,$$
$$k = 0, 1, 2, \ldots$$

converge to $\widehat{X}$, *and the error estimation is given by*

$$\max\left\{\left\|X_{k+1} - \widehat{X}\right\|_{1,Q}, \left\|Y_{k+1} - \widehat{X}\right\|_{1,Q}, \left\|Z_{k+1} - \widehat{Z}\right\|_{1,Q}\right\}$$
$$\leq \frac{\lambda^k}{(1-k)} \max\left\{\left\|X_1 - X_0\right\|_{1,Q}, \left\|Y_1 - Y_0\right\|_{1,Q}, \left\|Z_1 - Z_0\right\|_{1,Q}\right\}, \quad (76)$$

for all $k = 0, 1, \ldots$, *where* λ *is a certain constant in* $[0, 1)$.

(iv) *For every* $U_0, U_1, U_2 \in [0, 2Q]$, *every solution* $\{U_n\}$ *of* (72) *converges to* $\widehat{X}$.

Proof. In order to make the proof easy, we divide it into several steps.

Step 1. We claim that there exists a unique $(X, Y, Z) \in H(N) \times H(N) \times H(N)$ solution to the system

$$X = Q + \sum_{i=1}^{m} A_i X A_i - \sum_{i=1}^{m} B_i Y B_i + \sum_{i=1}^{m} C_i X C_i,$$
$$Y = Q + \sum_{i=1}^{m} A_i Y A_i - \sum_{i=1}^{m} B_i X B_i + \sum_{i=1}^{m} C_i Y C_i, \quad (77)$$
$$Z = Q + \sum_{i=1}^{m} A_i Z A_i - \sum_{i=1}^{m} B_i Y B_i + \sum_{i=1}^{m} C_i Z C_i.$$

Consider the mapping $F : H(N) \times H(N) \times H(N) \to H(N)$ defined by

$$F(X, Y, Z) = Q + \sum_{i=1}^{m} A_i X A_i - \sum_{i=1}^{m} B_i Y B_i + \sum_{i=1}^{m} C_i Z C_i, \quad (78)$$

for all $X, Y, Z \in H(N)$. It is clear that F is a mapping having the mixed monotone property with respect to the partial order $\preceq$. Let $X, Y, Z, U, V, W \in H(N)$ such that $X \succeq U$, $Y \preceq V$, and $Z \succeq W$. Using Lemma 13, we have

$$\|F(X, Y, Z) - F(U, V, W)\|_{1,Q}$$
$$= \operatorname{tr}\left(Q^{1/2}(F(X, Y, Z) - F(U, V, W))Q^{1/2}\right)$$
$$= \operatorname{tr}\left(Q^{1/2}\left(\sum_{i=1}^{m} A_i X A_i - \sum_{i=1}^{m} B_i Y B_i + \sum_{i=1}^{m} C_i Z C_i - \sum_{i=1}^{m} A_i U A_i + \sum_{i=1}^{m} B_i V B_i + \sum_{i=1}^{m} C_i W C_i\right) Q^{1/2}\right)$$
$$= \sum_{i=1}^{m} \operatorname{tr}\left(Q^{1/2}(A_i(X - U)A_i + B_i(V - Y)B_i + C_i(Z - W)C_i)Q^{1/2}\right)$$
$$= \sum_{i=1}^{m} \operatorname{tr}\left(Q^{1/2}(A_i(X - U)A_i)Q^{1/2}\right) + \operatorname{tr}\left(Q^{1/2}(B_i(V - Y)B_i)Q^{1/2}\right) + \operatorname{tr}\left(Q^{1/2}(C_i(Z - W)C_i)Q^{1/2}\right)$$
$$= \sum_{i=1}^{m} \operatorname{tr}(A_i Q A_i(X - U)) + \operatorname{tr}(B_i Q B_i(V - Y)) + \operatorname{tr}(C_i Q C_i(Z - W))$$
$$= \sum_{i=1}^{m} \operatorname{tr}\left(A_i Q A_i Q^{-1/2} Q^{1/2}(X - U)Q^{1/2} Q^{-1/2}\right) + \operatorname{tr}\left(B_i Q B_i Q^{-1/2} Q^{1/2}(V - Y)Q^{1/2} Q^{-1/2}\right) + \operatorname{tr}\left(C_i Q C_i Q^{-1/2} Q^{1/2}(Z - W)Q^{1/2} Q^{-1/2}\right)$$
$$= \sum_{i=1}^{m} \operatorname{tr}\left(Q^{-1/2} A_i Q A_i Q^{-1/2} Q^{1/2}(X - U)Q^{1/2}\right) + \operatorname{tr}\left(Q^{-1/2} B_i Q B_i Q^{-1/2} Q^{1/2}(V - Y)Q^{1/2}\right) + \operatorname{tr}\left(Q^{-1/2} C_i Q C_i Q^{-1/2} Q^{1/2}(Z - W)Q^{1/2}\right)$$

$$
\begin{aligned}
&= \operatorname{tr}\left(\sum_{i=1}^{m} Q^{-1/2}A_iQA_iQ^{-1/2}Q^{1/2}(X-U)Q^{1/2}\right) \\
&\quad + \operatorname{tr}\left(\sum_{i=1}^{m} Q^{-1/2}B_iQB_iQ^{-1/2}Q^{1/2}(V-Y)Q^{1/2}\right) \\
&\quad + \operatorname{tr}\left(\sum_{i=1}^{m} Q^{-1/2}C_iQC_iQ^{-1/2}Q^{1/2}(Z-W)Q^{1/2}\right) \\
&\le \left\|\sum_{i=1}^{m} Q^{-1/2}A_iQA_iQ^{-1/2}\right\| \|X-U\|_{1,Q} \\
&\quad + \left\|\sum_{i=1}^{m} Q^{-1/2}B_iQB_iQ^{-1/2}\right\| \|V-Y\|_{1,Q} \\
&\quad + \left\|\sum_{i=1}^{m} Q^{-1/2}C_iQC_iQ^{-1/2}\right\| \|Z-W\|_{1,Q} \\
&\le \delta \max\left\{\|X-U\|_{1,Q}, \|V-Y\|_{1,Q}, \|X-U\|_{1,Q}\right\},
\end{aligned} \tag{79}
$$

where

$$
\delta = \max\left\{\left\|\sum_{i=1}^{m} Q^{-1/2}A_iQA_iQ^{-1/2}\right\|, \left\|\sum_{i=1}^{m} Q^{-1/2}B_iQB_iQ^{-1/2}\right\|, \left\|\sum_{i=1}^{m} Q^{-1/2}C_iQC_iQ^{-1/2}\right\|\right\}. \tag{80}
$$

From (73) and Lemma 14, we have $\delta < 1$. Thus, the contractive condition of Theorem 5 is satisfied for all $X, Y, Z, U, V, W \in H(N)$ with $X \succeq U$, $Y \preceq V$, and $Z \succeq W$. Moreover, from (73), we have $X_0 = Z_0 = 0 \prec F(0, 2Q, 0)$ and $F(2Q, 0, 2Q) \prec 2Q = Y_0$.

Now, all the hypotheses of Theorem 5 are satisfied. Consequently, there exists $(X, Y, Z) \in H(N)\times H(N)\times H(N)$ solution to (77). Since for every $X, Y, Z \in H(N)$ there is a greatest lower bound and a least upper bound with respect to the partial order $\preceq$, we deduce from Theorem 7 the uniqueness of the solution to (77). Then, our claim holds.

Step 2. We claim that $X = Y = Z = \widehat{X}$.

Since $X_0 = Z_0 = 0 \prec Y_0 = 2Q$ (note that Q is a positive definite matrix), applying Theorem 5, we obtain the equality $X = Y = Z$. This proves our claim.

From Steps 1 and 2, we know that (72) has a unique equilibrium point $\widehat{X} \in H(N)$. Now, we need to prove that $\widehat{X} \in P(N)$. This is the goal of the next step.

Step 3 (proof of (i)–(iii)). Define the mapping $G : [F(0, 2Q, 0), F(2Q, 0, 2Q)] \to H(N)$ by

$$
GX = Q + \sum_{i=1}^{m} A_iXA_i - \sum_{i=1}^{m} B_iXB_i + \sum_{i=1}^{m} C_iXC_i, \tag{81}
$$

for all $X \in [F(0, 2Q, 0), F(2Q, 0, 2Q)]$.

From (73), one can show that $G([F(0, 2Q, 0), F(2Q, 0, 2Q)]) \subseteq [F(0, 2Q, 0), F(2Q, 0, 2Q)]$. Now, applying the Schauder fixed-point theorem (see Theorem 15), we deduce the existence of a fixed point of G in $[F(0, 2Q, 0), F(2Q, 0, 2Q)]$. But a fixed point of G is an equilibrium point of (72), and from Step 2, we know that (72) has a unique equilibrium point in $H(N)$. Consequently, $0 \prec F(0, 2Q, 0) \preceq \widehat{X} \preceq F(2Q, 0, 2Q)$. This proves (i) and (ii). For the proof of (iii), we have only to apply the inequalities (19).

Step 4. Let $U_0 = 0$, $U_1 = Q$, and $U_2 = 2Q$. Then, we have $U_0, U_1, U_2 \in [0, 2Q]$, thus by (b) Theorem 5, every solution $\{U_n\}$ to (72) converges to $\widehat{X}$. □

3.2. Numerical Example. Consider a third-order difference matrix equation

$$
U_{n+1} = Q + AU_nA - BU_{n-1}B + CU_{n-2}C, \tag{82}
$$

where Q, A, B, C, U_0, U_1, and U_2 are given by

$$
\begin{gathered}
Q = \begin{pmatrix} 1.5 & 0.3 & 0.3 \\ 0.3 & 1.5 & 0.3 \\ 0.3 & 0.3 & 1.5 \end{pmatrix}, \qquad A = \begin{pmatrix} 0.3 & 0.05 & 0.05 \\ 0.05 & 0.3 & 0.05 \\ 0.05 & 0.05 & 0.3 \end{pmatrix}, \\
B = \begin{pmatrix} 0.1 & 0.2 & 0.1 \\ 0.2 & 0.1 & 0.2 \\ 0.1 & 0.2 & 0.1 \end{pmatrix}, \qquad C = \begin{pmatrix} 0.1 & 0.1 & 0.1 \\ 0.1 & 0.1 & 0.1 \\ 0.1 & 0.1 & 0.1 \end{pmatrix}, \\
U_0 = 0, \qquad U_1 = Q, \qquad U_2 = 2Q,
\end{gathered} \tag{83}
$$

We are interested to approximate the unique positive equilibrium to (82).

We use Algorithm (iii) of Theorem 16 with $X_0 = 0 = Z_0$ and $Y_0 = 2Q$.

After 35 iterations, we get the unique solution $\widehat{X}$ given by

$$
\begin{gathered}
\widehat{X} = X_{35} = Y_{35} = Z_{35} = \begin{pmatrix} 1.615643487571 & 0.31874620339355 & 0.33564348757095 \\ 0.31874620339355 & 1.5374221323462 & 0.31874620339355 \\ 0.33564348757095 & 0.31874620339355 & 1.615643487571 \end{pmatrix}, \\
U_{35} = \begin{pmatrix} 1.6156434875817 & 0.31874620340583 & 0.33564348758171 \\ 0.31874620340583 & 1.5374221323604 & 0.31874620340583 \\ 0.33564348758171 & 0.31874620340583 & 1.6156434875817 \end{pmatrix}.
\end{gathered} \tag{84}
$$

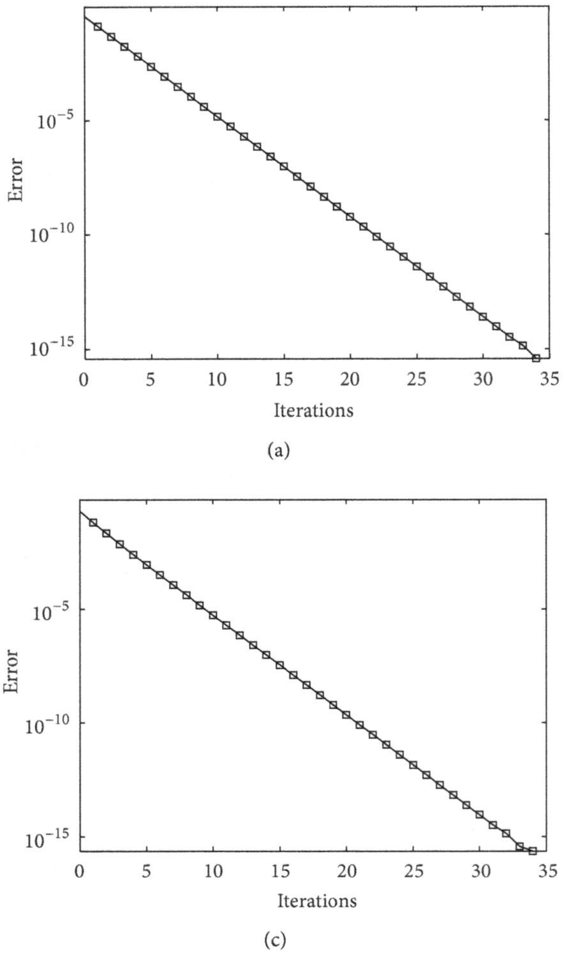

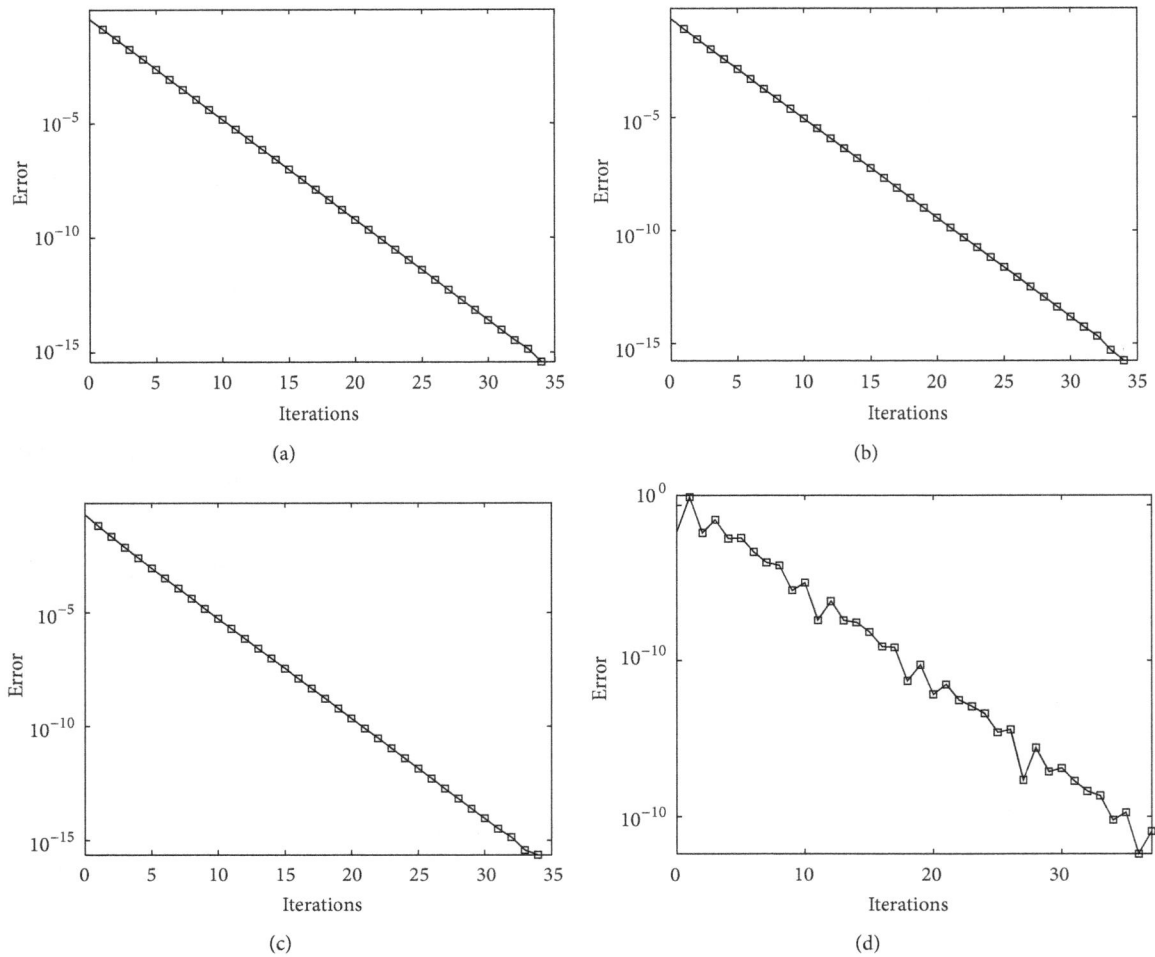

FIGURE 1: Convergence history for (82).

The residual errors are

$$
\begin{aligned}
R(X)_{35} &= \|X_{35} - (Q + A_1X_{35}A_1 + A_2X_{35}A_2 \\
&\quad -B_1X_{35}B_1 - B_2X_{35}B_2)\| \\
&= 3.792 \times 10^{-16}, \\
R(Y)_{35} &= \|Y_{35} - (Q + A_1Y_{35}A_1 + A_2Y_{35}A_2 \\
&\quad -B_1Y_{35}B_1 - B_2Y_{35}B_2)\| \\
&= 1.796 \times 10^{-16}, \\
R(Z)_{35} &= \|Z_{35} - (Q + A_1Z_{35}A_1 + A_2Z_{35}A_2 \\
&\quad -B_1Z_{35}B_1 - B_2Z_{35}B_2)\| \\
&= 2.351 \times 10^{-16}, \\
R(U)_{35} &= \|U_{35} - (Q + A_1U_{35}A_1 + A_2U_{35}A_2 \\
&\quad - B_1U_{35}B_1 - B_2U_{35}B_2)\| \\
&= 3.352 \times 10^{-11}.
\end{aligned}
\tag{85}
$$

The convergence history is given by Figure 1, where (a) corresponds to $\|X_k - F(X_k, X_k, X_k)\|$, (b) corresponds to $\|Y_k - F(Y_k, Y_k, Y_k)\|$, (c) corresponds to $\|Z_k - F(Z_k, Z_k, Z_k)\|$, and (d) corresponds to $\|U_k - F(U_k, U_k, U_k)\|$.

References

[1] M. R. S. Kulenović and G. Ladas, *Dynamics of Second Order Rational Difference Equations*, Chapman & Hall/CRC, Boca Raton, Fla, USA, 2002, With open problems and conjectures.

[2] E. Camouzis and G. Ladas, *Dynamics of third-order rational difference equations with open problems and conjectures*, vol. 5 of

Advances in Discrete Mathematics and Applications, Chapman Hall/CRC, Boca Raton, Fla, USA, 2008.

[3] M. R. S. Kulenović and O. Merino, "A global attractivity result for maps with invariant boxes," *Discrete and Continuous Dynamical Systems B*, vol. 6, no. 1, pp. 97–110, 2006.

[4] R. D. Nussbaum, "Global stability, two conjectures and Maple," *Nonlinear Analysis: Theory, Methods & Applications*, vol. 66, no. 5, pp. 1064–1090, 2007.

[5] H. L. Smith, "The discrete dynamics of monotonically decomposable maps," *Journal of Mathematical Biology*, vol. 53, no. 4, pp. 747–758, 2006.

[6] H. L. Smith, "Global stability for mixed monotone systems," *Journal of Difference Equations and Applications*, vol. 14, no. 10-11, pp. 1159–1164, 2008.

[7] R. P. Agarwal, M. A. El-Gebeily, and D. O'Regan, "Generalized contractions in partially ordered metric spaces," *Applicable Analysis*, vol. 87, no. 1, pp. 109–116, 2008.

[8] I. Altun and H. Simsek, "Some fixed point theorems on ordered metric spaces and application," *Fixed Point Theory and Applications*, vol. 2010, Article ID 621469, 17 pages, 2010.

[9] V. Berinde and M. Borcut, "Tripled fixed point theorems for contractive type mappings in partially ordered metric spaces," *Nonlinear Analysis: Theory, Methods & Applications*, vol. 74, no. 15, pp. 4889–4897, 2011.

[10] T. Gnana Bhaskar and V. Lakshmikantham, "Fixed point theorems in partially ordered metric spaces and applications," *Nonlinear Analysis: Theory, Methods & Applications*, vol. 65, no. 7, pp. 1379–1393, 2006.

[11] A. Brett and M. R. S. Kulenović, "Basins of attraction of equilibrium points of monotone difference equations," *Sarajevo Journal of Mathematics*, vol. 5, no. 2, pp. 211–233, 2009.

[12] D. Burgić, S. Kalabušić, and M. R. S. Kulenović, "Period-two trichotomies of a difference equation of order higher than two," *Sarajevo Journal of Mathematics*, vol. 4, no. 1, pp. 73–90, 2008.

[13] L. Cirić, N. Cakić, M. Rajović, and J. S. Ume, "Monotone generalized nonlinear contractions in partially ordered metric spaces," *Fixed Point Theory and Applications*, vol. 2008, Article ID 131294, 11 pages, 2008.

[14] E. Karapınar, "Couple fixed point theorems for nonlinear contractions in cone metric spaces," *Computers & Mathematics with Applications*, vol. 59, no. 12, pp. 3656–3668, 2010.

[15] V. Lakshmikantham and L. Cirić, "Coupled fixed point theorems for nonlinear contractions in partially ordered metric spaces," *Nonlinear Analysis: Theory, Methods & Applications*, vol. 70, no. 12, pp. 4341–4349, 2009.

[16] J. J. Nieto and R. Rodríguez-López, "Contractive mapping theorems in partially ordered sets and applications to ordinary differential equations," *Order*, vol. 22, no. 3, pp. 223–239, 2005.

[17] J. J. Nieto and R. Rodríguez-López, "Existence and uniqueness of fixed point in partially ordered sets and applications to ordinary differential equations," *Acta Mathematica Sinica*, vol. 23, no. 12, pp. 2205–2212, 2007.

[18] A. C. M. Ran and M. C. B. Reurings, "A fixed point theorem in partially ordered sets and some applications to matrix equations," *Proceedings of the American Mathematical Society*, vol. 132, no. 5, pp. 1435–1443, 2004.

[19] B. Samet, "Coupled fixed point theorems for a generalized Meir-Keeler contraction in partially ordered metric spaces," *Nonlinear Analysis: Theory, Methods & Applications*, vol. 72, no. 12, pp. 4508–4517, 2010.

[20] D. Burgić, S. Kalabušić, and M. R. S. Kulenović, "Global attractivity results for mixed-monotone mappings in partially ordered complete metric spaces," *Fixed Point Theory and Applications*, vol. 2009, Article ID 762478, 17 pages, 2009.

[21] R. Bhatia, *Matrix Analysis*, vol. 169 of *Graduate Texts in Mathematics*, Springer, New York, NY, USA, 1997.

[22] T. Ando, "Limit of iterates of cascade addition of matrices," *Numerical Functional Analysis and Optimization*, vol. 2, no. 7-8, pp. 579–289, 1980.

[23] W. N. Anderson, T. D. Morley, and G. E. Trapp, "Ladder networks, fixed points and the geometric mean," *Circuits, Systems and Signal Processing*, vol. 2, no. 3, pp. 259–268, 1983.

[24] J. C. Engwerda, "On the existence of a positive definite solution of the matrix equation $X + A^T X^{-1} A = I$," *Linear Algebra and its Applications*, vol. 194, pp. 91–108, 1993.

[25] W. Pusz and S. L. Woronowicz, "Functional calculus for sesquilinear forms and the purification map," *Reports on Mathematical Physics*, vol. 8, no. 2, pp. 159–170, 1975.

[26] B. L. Buzbee, G. H. Golub, and C. W. Nielson, "On direct methods for solving Poisson's equations," *SIAM Journal on Numerical Analysis*, vol. 7, pp. 627–656, 1970.

[27] W. L. Green and E. W. Kamen, "Stabilizability of linear systems over a commutative normed algebra with applications to spatially-distributed and parameter-dependent systems," *SIAM Journal on Control and Optimization*, vol. 23, no. 1, pp. 1–18, 1985.

23

Uniqueness of Meromorphic Functions Sharing Fixed Point

Subhas S. Bhoosnurmath and Veena L. Pujari

Department of Mathematics, Karnatak University, Dharwad 580003, India

Correspondence should be addressed to Subhas S. Bhoosnurmath; ssbmath@gmail.com

Academic Editor: Alexandre Timonov

We study the uniqueness of meromorphic functions concerning differential polynomials sharing fixed point and obtain some significant results , which improve the results due to Lin and Yi (2004).

1. Introduction and Main Results

Let $f(z)$ be a nonconstant meromorphic function in the whole complex plane $\mathbb{C}$. We will use the following standard notations of value distribution theory: $T(r,f), m(r,f)$, $N(r,f), \overline{N}(r,f), \ldots$ (see [1, 2]). We denote by $S(r,f)$ any function satisfying

$$S(r,f) = o\{T(r,f)\}, \quad \text{as } r \longrightarrow +\infty, \tag{1}$$

possibly outside of a set with finite measure.

Let a be a finite complex number and k a positive integer. We denote by $N_{k)}(r, 1/(f-a))$ the counting function for the zeros of $f(z) - a$ in $|z| \leq r$ with multiplicity $\leq k$ and by $\overline{N}_{k)}(r, 1/(f-a))$ the corresponding one for which multiplicity is not counted. Let $N_{(k}(r, 1/(f-a))$ be the counting function for the zeros of $f(z) - a$ in $|z| \leq r$ with multiplicity $\geq k$ and $\overline{N}_{(k}(r, 1/(f-a))$ the corresponding one for which multiplicity is not counted. Set

$$N_k\left(r, \frac{1}{f-a}\right) = \overline{N}\left(r, \frac{1}{f-a}\right) + \overline{N}_{(2}\left(r, \frac{1}{f-a}\right) + \cdots + \overline{N}_{(k}\left(r, \frac{1}{f-a}\right). \tag{2}$$

Let $g(z)$ be a nonconstant meromorphic function. We denote by $\overline{N}_L(r, 1/(f-a))$ the counting function for a-points of both $f(z)$ and $g(z)$ about which $f(z)$ has larger multiplicity than $g(z)$, where multiplicity is not counted. Similarly, we have notation $\overline{N}_L(r, 1/(g-a))$.

We say that f and g share a CM (counting multiplicity) if $f - a$ and $g - a$ have same zeros with the same multiplicities. Similarly, we say that f and g share a IM (ignoring multiplicity) if $f - a$ and $g - a$ have same zeros with ignoring multiplicities.

In 2004, Lin and Yi [3] obtained the following results.

Theorem A. *Let f and g be two transcendental meromorphic functions, $n \geq 12$ an integer. If $f^n(f-1)f'$ and $g^n(g-1)g'$ share z CM, then either $f(z) \equiv g(z)$ or*

$$g = \frac{(n+2)\left(1-h^{n+1}\right)}{(n+1)\left(1-h^{n+2}\right)}, \qquad f = \frac{(n+2)\,h\left(1-h^{n+1}\right)}{(n+1)\left(1-h^{n+2}\right)}, \tag{3}$$

where h is a nonconstant meromorphic function.

Theorem B. *Let f and g be two transcendental meromorphic functions, $n \geq 13$ an integer. If $f^n(f-1)^2 f'$ and $g^n(g-1)^2 g'$ share z CM, then $f(z) \equiv g(z)$.*

In this paper, we study the uniqueness problems of entire or meromorphic functions concerning differential polynomials sharing fixed point, which improves Theorems A and B.

1.1. Main Results

Theorem 1. *Let f and g be two nonconstant meromorphic functions, $n \geq 11$ a positive integer. If $f^n(f-1)f'$ and*

$g^n(g-1)g'$ share z CM, f and g share ∞ IM, then either $f(z) \equiv g(z)$ or

$$g = \frac{(n+2)\left(1-h^{n+1}\right)}{(n+1)\left(1-h^{n+2}\right)}, \qquad f = \frac{(n+2)h\left(1-h^{n+1}\right)}{(n+1)\left(1-h^{n+2}\right)}, \quad (4)$$

where h is a nonconstant meromorphic function.

Theorem 2. *Let f and g be two nonconstant meromorphic functions, $n \geq 12$ a positive integer. If $f^n(f-1)^2 f'$ and $g^n(g-1)^2 g'$ share z CM, f and g share ∞ IM, then $f(z) \equiv g(z)$.*

Theorem 3. *Let f and g be two nonconstant entire functions, $n \geq 7$ an integer. If $f^n(f-1)f'$ and $g^n(g-1)g'$ share z CM, then $f(z) \equiv g(z)$.*

2. Some Lemmas

Lemma 4 (see [4]). *Let f_1, f_2, and f_3 be nonconstant meromorphic functions such that $f_1 + f_2 + f_3 = 1$. If f_1, f_2, and f_3 are linearly independent, then*

$$T(r, f_1) < \sum_{i=1}^{3} N_2\left(r, \frac{1}{f_i}\right) + \sum_{i=1}^{3} \overline{N}(r, f_i) + o(T(r)), \quad (5)$$

where $T(r) = \max_{1\leq i\leq 3}\{T(r, f_i)\}$ and $r \notin E$.

Lemma 5 (see [1]). *Let f_1 and f_2 be two nonconstant meromorphic functions. If $c_1 f_1 + c_2 f_2 = c_3$, where c_1, c_2, and c_3 are non-zero constants, then*

$$T(r, f_1) \leq \overline{N}(r, f_1) + \overline{N}\left(r, \frac{1}{f_1}\right) + \overline{N}\left(r, \frac{1}{f_2}\right) + S(r, f_1). \quad (6)$$

Lemmas 4 and 5 play a very important role in proving our theorems.

Lemma 6 (see [1]). *Let f be a nonconstant meromorphic function and let k be a nonnegative integer, then*

$$N\left(r, \frac{1}{f^{(k)}}\right) \leq N\left(r, \frac{1}{f}\right) + k\overline{N}(r, f) + S(r, f). \quad (7)$$

The following lemmas play a cardinal role in proving our results.

Lemma 7. *Let f and g be nonconstant meromorphic functions. If $f^n(f-1)f'$ and $g^n(g-1)g'$ share z CM and $n > 6$, then*

$$T(r, g) \leq \left(\frac{n+3}{n-6}\right) T(r, f) + \log r + S(r, g). \quad (8)$$

Proof. Applying Nevanlinna's second fundamental theorem to $g^n(g-1)g'$, we have

$$\begin{aligned} T\left(r, g^n(g-1)g'\right) &\leq \overline{N}\left(r, g^n(g-1)g'\right) \\ &\quad + \overline{N}\left(r, \frac{1}{g^n(g-1)g'}\right) \\ &\quad + \overline{N}\left(r, \frac{1}{g^n(g-1)g' - z}\right) + S(r, g) \\ &\leq \overline{N}(r, g) + \overline{N}\left(r, \frac{1}{g}\right) \\ &\quad + \overline{N}\left(r, \frac{1}{g-1}\right) + \overline{N}\left(r, \frac{1}{g'}\right) \\ &\quad + \overline{N}\left(r, \frac{1}{f^n(f-1)f' - z}\right) + S(r, g). \end{aligned} \quad (9)$$

By the first fundamental theorem and (9), we have

$$\begin{aligned} (n+1)T(r, g) &\leq T\left(r, g^n(g-1)\right) + S(r, g) \\ &\leq T\left(r, g^n(g-1)g'\right) + T\left(r, \frac{1}{g'}\right) + S(r, g) \\ &\leq \overline{N}(r, g) + \overline{N}\left(r, \frac{1}{g}\right) \\ &\quad + \overline{N}\left(r, \frac{1}{g-1}\right) + \overline{N}\left(r, \frac{1}{g'}\right) \\ &\quad + \overline{N}\left(r, \frac{1}{f^n(f-1)f' - z}\right) \\ &\quad + T\left(r, g'\right) + S(r, g). \end{aligned} \quad (10)$$

We know that

$$\begin{aligned} \overline{N}\left(r, \frac{1}{f^n(f-1)f' - z}\right) &\leq T\left(r, \frac{1}{f^n(f-1)f' - z}\right) \\ &= T\left(r, f^n(f-1)f' - z\right) + O(1) \\ &\leq T\left(r, f^n(f-1)f'\right) \\ &\quad + T(r, z) + O(1) \\ &\leq nT(r, f) + T(r, f-1) \\ &\quad + T\left(r, f'\right) + \log r + O(1) \\ &\leq (n+3)T(r, f) + \log r + O(1). \end{aligned} \quad (11)$$

Therefore using Lemma 6, (10) becomes

$$(n+1)T(r,g) \le \overline{N}(r,g) + \overline{N}\left(r,\frac{1}{g}\right) + \overline{N}\left(r,\frac{1}{g-1}\right) + \overline{N}\left(r,\frac{1}{g'}\right) + (n+3)T(r,f) + T(r,g') + \log r + S(r,g)$$
$$\le 7T(r,g) + (n+3)T(r,f) + \log r + S(r,g) \tag{12}$$

$$\Longrightarrow (n-6)T(r,g) \le (n+3)T(r,f) + \log r + S(r,g); \tag{13}$$

since $n > 6$, we have

$$T(r,g) \le \left(\frac{n+3}{n-6}\right)T(r,f) + \log r + S(r,g) \tag{14}$$

This completes the proof of Lemma 7. □

Lemma 8. *Let f and g be nonconstant entire functions. If $f^n(f-1)f'$ and $g^n(g-1)g'$ share z CM and $n > 3$, then*

$$T(r,g) \le \left(\frac{n+2}{n-3}\right)T(r,f) + \log r + S(r,g). \tag{15}$$

Proof. Applying Nevanlinna's second fundamental theorem to $g^n(g-1)g'$, we have

$$T(r,g^n(g-1)g') \le \overline{N}(r,g^n(g-1)g') + \overline{N}\left(r,\frac{1}{g^n(g-1)g'}\right) + \overline{N}\left(r,\frac{1}{g^n(g-1)g'-z}\right) + S(r,g)$$
$$\le \overline{N}(r,g) + \overline{N}\left(r,\frac{1}{g}\right) + \overline{N}\left(r,\frac{1}{g-1}\right) + \overline{N}\left(r,\frac{1}{g'}\right) + \overline{N}\left(r,\frac{1}{f^n(f-1)f'-z}\right) + S(r,g). \tag{16}$$

Since g is an entire function, we have $\overline{N}(r,g) = 0$ and the above equation becomes

$$T(r,g^n(g-1)g') \le \overline{N}\left(r,\frac{1}{g}\right) + \overline{N}\left(r,\frac{1}{g-1}\right) + \overline{N}\left(r,\frac{1}{g'}\right) + \overline{N}\left(r,\frac{1}{f^n(f-1)f'-z}\right) + S(r,g). \tag{17}$$

By the first fundamental theorem and (17), we have

$$(n+1)T(r,g) \le T(r,g^n(g-1)) + S(r,g)$$
$$\le T(r,g^n(g-1)g') + T\left(r,\frac{1}{g'}\right) + S(r,g)$$
$$\le \overline{N}\left(r,\frac{1}{g}\right) + \overline{N}\left(r,\frac{1}{g-1}\right) + \overline{N}\left(r,\frac{1}{g'}\right) + \overline{N}\left(r,\frac{1}{f^n(f-1)f'-z}\right) + T(r,g') + S(r,g). \tag{18}$$

We know that

$$\overline{N}\left(r,\frac{1}{f^n(f-1)f'-z}\right) \le T\left(r,\frac{1}{f^n(f-1)f'-z}\right)$$
$$= T(r,f^n(f-1)f'-z) + O(1)$$
$$\le T(r,f^n(f-1)f') + T(r,z) + O(1)$$
$$\le nT(r,f) + T(r,f-1) + T(r,f') + \log r + O(1)$$
$$\le (n+2)T(r,f) + \log r + O(1). \tag{19}$$

Therefore using Lemma 6, (18) becomes

$$(n+1)T(r,g) \le \overline{N}\left(r,\frac{1}{g}\right) + \overline{N}\left(r,\frac{1}{g-1}\right) + \overline{N}\left(r,\frac{1}{g'}\right) + (n+2)T(r,f) + T(r,g') + \log r + S(r,g)$$
$$\le 4T(r,g) + (n+2)T(r,f) + \log r + S(r,g) \tag{20}$$

or

$$(n-3)T(r,g) \le (n+2)T(r,f) + \log r + S(r,g); \tag{21}$$

since $n > 3$, we have

$$T(r,g) \le \left(\frac{n+2}{n-3}\right)T(r,f) + \log r + S(r,g). \tag{22}$$

This completes the proof of Lemma 8. □

Lemma 9 (see [5]). *Suppose that $f(z)$ is a meromorphic function in the complex plane and $P(f) = a_0f^n + a_1f^{n-1} + \cdots + a_n$, where $a_0(\not\equiv 0), a_1, \ldots, a_n$ are small meromorphic functions of $f(z)$. Then*

$$T(r,P(f)) = nT(r,f) + S(r,f). \tag{23}$$

Lemma 10 (see [6]). *Let f_1, f_2, and f_3 be three meromorphic functions satisfying $\sum_{j=1}^{3} f_j = 1$, let $g_1 = -f_3/f_2$, $g_2 = 1/f_2$, and $g_3 = -f_1/f_2$. If f_1, f_2, and f_3 are linearly independent, then g_1, g_2, and g_3 are linearly independent.*

3. Proof of Theorems

Proof of Theorem 1. By assumption, $f^n(f-1)f'$ and $g^n(g-1)g'$ share z CM, and f and g share ∞ IM. Let

$$H = \frac{f^n(f-1)f' - z}{g^n(g-1)g' - z}. \tag{24}$$

Then, H is a meromorphic function satisfying

$$\begin{aligned} T(r,H) &= T\left(r, \frac{f^n(f-1)f' - z}{g^n(g-1)g' - z}\right) \\ &\leq T\left(r, f^n(f-1)f' - z\right) \\ &\quad + T\left(r, g^n(g-1)g' - z\right) + O(1) \\ &\leq (n+3)(T(r,f) + T(r,g)) + O(\log r). \end{aligned} \tag{25}$$

Therefore,

$$T(r,H) = O(T(r,f) + T(r,g)). \tag{26}$$

From (24), we easily see that the zeros and poles of H are multiple and satisfy

$$\overline{N}(r,H) \leq \overline{N}_L(r,f), \qquad \overline{N}\left(r, \frac{1}{H}\right) \leq \overline{N}_L(r,g). \tag{27}$$

Let

$$f_1 = \frac{f^n(f-1)f'}{z}, \qquad f_2 = H, \qquad f_3 = \frac{-Hg^n(g-1)g'}{z}. \tag{28}$$

Then, $f_1 + f_2 + f_3 = 1$ and $T(r)$ denote the maximum of $T(r, f_j)$, $j = 1, 2, 3$.

We have

$$\begin{aligned} T(r,f_1) &= O(T(r,f)), \\ T(r,f_2) &= O(T(r,f) + T(r,g)), \\ T(r,f_3) &= O(T(r,f) + T(r,g)). \end{aligned} \tag{29}$$

Therefore, $T(r) = O(T(r,f) + T(r,g))$, and thus

$$S(r,f) + S(r,g) = o(T(r)). \tag{30}$$

Now, we discuss the following three cases.

Case 1. Suppose that neither f_2 nor f_3 is a constant.

If f_1, f_2, and f_3 are linearly independent, then by Lemma 4 and (28), we have

$$\begin{aligned} T(r,f_1) &< \sum_{i=1}^{3} N_2\left(r, \frac{1}{f_i}\right) + \sum_{i=1}^{3} \overline{N}(r,f_i) + o(T(r)) \\ &\leq N_2\left(r, \frac{1}{f_1}\right) + N_2\left(r, \frac{1}{f_2}\right) + N_2\left(r, \frac{1}{f_3}\right) \\ &\quad + \overline{N}(r,f_1) + \overline{N}(r,f_2) + \overline{N}(r,f_3) + o(T(r)) \\ &\leq N_2\left(r, \frac{z}{f^n(f-1)f'}\right) + N_2\left(r, \frac{1}{H}\right) \\ &\quad + N_2\left(r, \frac{z}{Hg^n(g-1)g'}\right) \\ &\quad + \overline{N}\left(r, \frac{f^n(f-1)f'}{z}\right) \\ &\quad + \overline{N}(r,H) + \overline{N}\left(r, \frac{Hg^n(g-1)g'}{z}\right) + o(T(r)) \\ &= N_2\left(r, \frac{1}{f^n(f-1)f'}\right) + N_2\left(r, \frac{1}{H}\right) \\ &\quad + N_2\left(r, \frac{1}{Hg^n(g-1)g'}\right) \\ &\quad + \overline{N}\left(r, f^n(f-1)f'\right) \\ &\quad + \overline{N}(r,H) + \overline{N}\left(r, Hg^n(g-1)g'\right) \\ &\quad + 2\log r + o(T(r)). \end{aligned} \tag{31}$$

Using (27), we note that

$$\begin{aligned} N_2\left(r, \frac{1}{Hg^n(g-1)g'}\right) &\leq N_2\left(r, \frac{1}{H}\right) + N_2\left(r, \frac{1}{g^n(g-1)g'}\right) \\ &\leq 2\overline{N}\left(r, \frac{1}{H}\right) + N_2\left(r, \frac{1}{g^n(g-1)g'}\right) \\ &\leq 2\overline{N}_L(r,g) + N_2\left(r, \frac{1}{g^n(g-1)g'}\right). \end{aligned} \tag{32}$$

Since $\overline{N}_L(r,g) = 0$, we obtain that

$$N_2\left(r, \frac{1}{Hg^n(g-1)g'}\right) \leq N_2\left(r, \frac{1}{g^n(g-1)g'}\right), \tag{33}$$

$$\begin{aligned} \overline{N}\left(r, Hg^n(g-1)g'\right) &\leq \overline{N}(r,H) + \overline{N}\left(r, g^n(g-1)g'\right) \\ &\leq \overline{N}_L(r,f) + \overline{N}(r,g). \end{aligned} \tag{34}$$

But $\overline{N}_L(r, f) = 0$, so we get

$$\overline{N}\left(r, Hg^n(g-1)g'\right) \le \overline{N}(r, g). \quad (35)$$

Using (33) and (35) in (31), we get

$$\begin{aligned} T(r, f_1) \le\ & N_2\left(r, \frac{1}{f^n(f-1)f'}\right) + N_2\left(r, \frac{1}{H}\right) \\ & + N_2\left(r, \frac{1}{g^n(g-1)g'}\right) + \overline{N}(r, f) \\ & + \overline{N}(r, H) + \overline{N}(r, g) + 2\log r + o(T(r)). \end{aligned} \quad (36)$$

Since f and g share ∞ IM, we have $\overline{N}(r, f) = \overline{N}(r, g)$.
Using this with (27), we get

$$\begin{aligned} T(r, f_1) \le\ & N_2\left(r, \frac{1}{f^n(f-1)f'}\right) + 2\overline{N}_L(r, g) \\ & + N_2\left(r, \frac{1}{g^n(g-1)g'}\right) + \overline{N}_L(r, f) \\ & + 2\overline{N}(r, f) + 2\log r + o(T(r)) \\ \le\ & N\left(r, \frac{1}{f^n(f-1)f'}\right) \\ & - \left[N_{(3}\left(r, \frac{1}{f^n(f-1)f'}\right) - 2\overline{N}_{(3}\left(r, \frac{1}{f^n(f-1)f'}\right)\right] \\ & + N\left(r, \frac{1}{g^n(g-1)g'}\right) \\ & - \left[N_{(3}\left(r, \frac{1}{g^n(g-1)g'}\right) - 2\overline{N}_{(3}\left(r, \frac{1}{g^n(g-1)g'}\right)\right] \\ & + 2\overline{N}_L(r, g) + \overline{N}_L(r, f) + 2\overline{N}(r, f) \\ & + 2\log r + o(T(r)). \end{aligned} \quad (37)$$

If z_0 is a zero of f with multiplicity p, then z_0 is a zero of $f^n(f-1)f'$ with multiplicity $np + p - 1 \ge 3$; we have

$$\left[N_{(3}\left(r, \frac{1}{f^n(f-1)f'}\right) - 2\overline{N}_{(3}\left(r, \frac{1}{f^n(f-1)f'}\right)\right] \ge (n-2)N\left(r, \frac{1}{f}\right). \quad (38)$$

Similarly,

$$\left[N_{(3}\left(r, \frac{1}{g^n(g-1)g'}\right) - 2\overline{N}_{(3}\left(r, \frac{1}{g^n(g-1)g'}\right)\right] \ge (n-2)N\left(r, \frac{1}{g}\right). \quad (39)$$

Let

$$f_1^* = \frac{f^{n+2}}{n+2} - \frac{f^{n+1}}{n+1}. \quad (40)$$

By Lemma 9, we have $T(r, f_1^*) = (n+2)T(r, f) + S(r, f)$.
Since $(f_1^*)' = zf_1$, we have

$$\begin{aligned} m\left(r, \frac{1}{f_1^*}\right) & \le m\left(r, \frac{1}{zf_1}\right) + m\left(r, \frac{(f_1^*)'}{f_1^*}\right) \\ & \le m\left(r, \frac{1}{f_1}\right) + \log r + S(r, f). \end{aligned} \quad (41)$$

By the first fundamental theorem, we have

$$\begin{aligned} T(r, f_1^*) \le\ & T(r, f_1) + N\left(r, \frac{1}{f_1^*}\right) - N\left(r, \frac{1}{f_1}\right) \\ & + \log r + S(r, f); \end{aligned} \quad (42)$$

we have

$$N\left(r, \frac{1}{f_1^*}\right) = (n+1)N\left(r, \frac{1}{f}\right) + N\left(r, \frac{1}{f - (n+2)/(n+1)}\right). \quad (43)$$

From (37)–(43), we get

$$\begin{aligned} T(r, f_1^*) \le\ & N\left(r, \frac{1}{f^n(f-1)f'}\right) - (n-2)N\left(r, \frac{1}{f}\right) \\ & + N\left(r, \frac{1}{g^n(g-1)g'}\right) - (n-2)N\left(r, \frac{1}{g}\right) \\ & + 2\overline{N}_L(r, g) \\ & + \overline{N}_L(r, f) + 2\overline{N}(r, f) + (n+1)N\left(r, \frac{1}{f}\right) \\ & + N\left(r, \frac{1}{f - (n+2)/(n+1)}\right) \\ & - N\left(r, \frac{1}{f^n(f-1)f'}\right) + 3\log r + o(T(r)). \end{aligned} \quad (44)$$

Using Lemma 6, we get

$$
\begin{aligned}
T\left(r, f_1^*\right) \le\ & 3N\left(r, \frac{1}{f}\right) + 3N\left(r, \frac{1}{g}\right) \\
& + 2\overline{N}(r, f) + \overline{N}(r, g) + 2\overline{N}_L(r, g) \\
& + \overline{N}_L(r, f) + N\left(r, \frac{1}{g-1}\right) \\
& + N\left(r, \frac{1}{f-(n+2)/(n+1)}\right) \\
& + 3\log r + o(T(r))
\end{aligned}
$$

$$
\begin{aligned}
(n+2)\, T(r, f) \le\ & 3N\left(r, \frac{1}{f}\right) + 3N\left(r, \frac{1}{g}\right) \\
& + 2\overline{N}(r, f) + \overline{N}(r, g) \\
& + 2\overline{N}_L(r, g) + \overline{N}_L(r, f) + N\left(r, \frac{1}{g-1}\right) \\
& + N\left(r, \frac{1}{f-(n+2)/(n+1)}\right) \\
& + 3\log r + o(T(r)) .
\end{aligned} \tag{45}
$$

Let

$$
g_1 = -\frac{f_3}{f_2} = \frac{g^n(g-1)g'}{z}, \qquad g_2 = \frac{1}{f_2} = \frac{1}{H}, \qquad g_3 = -\frac{f_1}{f_2} = \frac{f^n(f-1)f'}{zH}. \tag{46}
$$

Then $g_1+g_2+g_3 = 1$. By Lemma 10, g_1, g_2, and g_3 are linearly independent. In the same manner as above, we get

$$
\begin{aligned}
(n+2)\, T(r, g) \le\ & 3N\left(r, \frac{1}{g}\right) + 3N\left(r, \frac{1}{f}\right) \\
& + 2\overline{N}(r, g) + \overline{N}(r, f) + 2\overline{N}_L(r, f) \\
& + \overline{N}_L(r, g) + N\left(r, \frac{1}{f-1}\right) \\
& + N\left(r, \frac{1}{g-(n+2)/(n+1)}\right) \\
& + 3\log r + o(T(r)) .
\end{aligned} \tag{47}
$$

Note that

$$
\overline{N}_L(r, f) + \overline{N}_L(r, g) \le \overline{N}(r, f) = \overline{N}(r, g) . \tag{48}
$$

Adding (45) and (47) gives

$$
\begin{aligned}
& (n+2)\left(T(r, f) + T(r, g)\right) \\
& \quad \le 6\left(N\left(r, \frac{1}{f}\right) + N\left(r, \frac{1}{g}\right)\right) + 3\left(\overline{N}(r, f) + \overline{N}(r, g)\right) \\
& \qquad + \left(\overline{N}_L(r, f) + \overline{N}_L(r, g)\right) \\
& \qquad + \left(N\left(r, \frac{1}{f-1}\right) + N\left(r, \frac{1}{g-1}\right)\right) \\
& \qquad + N\left(r, \frac{1}{f-(n+2)/(n+1)}\right) \\
& \qquad + N\left(r, \frac{1}{g-(n+2)/(n+1)}\right) + 6\log r + o(T(r)) \\
& (n-6)\left(T(r, f) + T(r, g)\right) \\
& \quad \le 3\left(\overline{N}(r, f) + \overline{N}(r, g)\right) \\
& \qquad + 3\left(\overline{N}_L(r, f) + \overline{N}_L(r, g)\right) + 6\log r + o(T(r)) .
\end{aligned} \tag{49}
$$

Using (48), we get

$$
\begin{aligned}
& (n-6)\left(T(r, f) + T(r, g)\right) \\
& \quad \le 3\overline{N}(r, f) + 6\overline{N}(r, g) + 6\log r + o(T(r))
\end{aligned} \tag{50}
$$

or

$$
\begin{aligned}
& (n-6)\left(T(r, f) + T(r, g)\right) \\
& \quad \le 6\overline{N}(r, f) + 3\overline{N}(r, g) + 6\log r + o(T(r)) .
\end{aligned} \tag{51}
$$

Combining (50) and (51), we get

$$
\begin{aligned}
(n-6)\left(T(r, f) + T(r, g)\right) \le\ & \frac{9}{2}\left(\overline{N}(r, f) + \overline{N}(r, g)\right) \\
& + 6\log r + o(T(r))
\end{aligned} \tag{52}
$$

$$
\left(n - \frac{21}{2}\right)\left(T(r, f) + T(r, g)\right) \le 6\log r + o(T(r)) .
$$

By $n \ge 11$ and (30), we get a contradiction. Thus f_1, f_2, and f_3 are linearly dependent. Then, there exists three constants $(c_1, c_2, c_3) \ne (0,0,0)$ such that

$$
c_1 f_1 + c_2 f_2 + c_3 f_3 = 0. \tag{53}
$$

If $c_1 = 0$, from (53) $c_2 \ne 0$, $c_3 \ne 0$, and

$$
f_3 = -\frac{c_2}{c_3} f_2; \tag{54}
$$

$$
\Longrightarrow g^n(g-1)g' = \frac{c_2}{c_3} z. \tag{55}
$$

On integrating, we get

$$\frac{g^{n+2}}{n+2} - \frac{g^{n+1}}{n+1} = \frac{c_2}{c_3}\frac{z^2}{2} + k, \quad k \text{ is a constant},$$

$$T\left(r, \frac{g^{n+2}}{n+2} - \frac{g^{n+1}}{n+1}\right) \le T\left(r, z^2\right) + O(1), \tag{56}$$

$$(n+2)\,T(r,g) \le 2\log r + O(1);$$

since $n \ge 11$, we get a contradiction.

Thus $c_1 \ne 0$, and by (53) we have

$$c_1 f_1 = -c_2 f_2 - c_3 f_3$$
$$\Longrightarrow f_1 = -\frac{c_2}{c_1} f_2 - \frac{c_3}{c_1} f_3. \tag{57}$$

Substituting this in $f_1 + f_2 + f_3 = 1$, we get

$$-\frac{c_2}{c_1} f_2 - \frac{c_3}{c_1} f_3 + f_2 + f_3 = 1; \tag{58}$$

that is,

$$\left(1 - \frac{c_2}{c_1}\right) f_2 + \left(1 - \frac{c_3}{c_1}\right) f_3 = 1, \quad \text{where } c_1 \ne c_3,\ c_2 \ne c_3. \tag{59}$$

From (28), we obtain

$$\left(1 - \frac{c_2}{c_1}\right) H + \left(1 - \frac{c_3}{c_1}\right)\left(\frac{-Hg^n(g-1)g'}{z}\right) = 1$$
$$\Longrightarrow \left(1 - \frac{c_2}{c_1}\right) - \left(1 - \frac{c_3}{c_1}\right)\left(\frac{g^n(g-1)g'}{z}\right) = 1 \tag{60}$$
$$\Longrightarrow \left(1 - \frac{c_3}{c_1}\right)\left(\frac{g^n(g-1)g'}{z}\right) + \frac{1}{H} = \left(1 - \frac{c_2}{c_1}\right).$$

Applying Lemma 5 to the above equation, we get

$$T\left(r, \frac{g^n(g-1)g'}{z}\right) \le \overline{N}\left(r, \frac{g^n(g-1)g'}{z}\right) + \overline{N}\left(r, \frac{z}{g^n(g-1)g'}\right) + \overline{N}(r,H) + S(r,g). \tag{61}$$

Note that

$$T\left(r, g^n(g-1)g'\right) \le T\left(r, \frac{g^n(g-1)g'}{z}\right) + T(r,z) \le T\left(r, \frac{g^n(g-1)g'}{z}\right) + \log r. \tag{62}$$

Using (61), we get

$$T\left(r, g^n(g-1)g'\right) \le \overline{N}(r,g) + \overline{N}\left(r, \frac{1}{g^n(g-1)g'}\right) + \overline{N}(r,g) + 2\log r + S(r,g)$$
$$= \overline{N}\left(r, \frac{1}{g^n(g-1)g'}\right) + 2\overline{N}(r,g) + 2\log r + S(r,g). \tag{63}$$

By Lemmas 9 and 6 and (63), we have

$$(n+1)\,T(r,g) = T\left(r, g^n(g-1)\right) + S(r,g)$$
$$\le T\left(r, g^n(g-1)g'\right) + T\left(r, \frac{1}{g'}\right) + S(r,g)$$
$$\le \overline{N}\left(r, \frac{1}{g^n(g-1)g'}\right) + 2\overline{N}(r,g) + 2\log r + 2T(r,g) + S(r,g)$$
$$\le \overline{N}\left(r, \frac{1}{g}\right) + \overline{N}\left(r, \frac{1}{g-1}\right) + N\left(r, \frac{1}{g'}\right) + 2\overline{N}(r,g) + 2T(r,g) + 2\log r + S(r,g)$$
$$\le 8T(r,g) + 2\log r + S(r,g)$$
$$\Longrightarrow (n-7)\,T(r,g) \le 2\log r + S(r,g); \tag{64}$$

we obtain $n \le 7$, which contradicts $n \ge 11$.

Case 2. Suppose that $f_2 = c\ (\ne 0)$, where c is a constant.

If $c \ne 1$, then we have

$$f_1 + f_2 + f_3 = 1$$
$$\Longrightarrow \frac{f^n(f-1)f'}{z} + c - \frac{cg^n(g-1)g'}{z} = 1; \tag{65}$$
$$\Longrightarrow \frac{f^n(f-1)f'}{z} - \frac{cg^n(g-1)g'}{z} = 1 - c. \tag{66}$$

Applying Lemma 5 to the above equation, we have

$$T\left(r, \frac{f^n(f-1)f'}{z}\right) \le \overline{N}\left(r, \frac{f^n(f-1)f'}{z}\right) + \overline{N}\left(r, \frac{z}{g^n(g-1)g'}\right) + \overline{N}\left(r, \frac{z}{f^n(f-1)f'}\right) + S(r,f)$$

$$\leq \overline{N}(r,f) + \overline{N}\left(r, \frac{1}{g^n(g-1)g'}\right) + \overline{N}\left(r, \frac{1}{f^n(f-1)f'}\right) + \log r + S(r,f). \tag{67}$$

Note that

$$T\left(r, f^n(f-1)f'\right) \leq T\left(r, \frac{f^n(f-1)f'}{z}\right) + \log r. \tag{68}$$

Therefore,

$$T\left(r, f^n(f-1)f'\right) \leq \overline{N}(r,f) + \overline{N}\left(r, \frac{1}{g^n(g-1)g'}\right) + \overline{N}\left(r, \frac{1}{f^n(f-1)f'}\right) + 2\log r + S(r,f). \tag{69}$$

Using Lemmas 9 and 6 and (69), we have

$$\begin{aligned}(n+1)T(r,f) &= T(r, f^n(f-1)) + S(r,f) \\ &\leq T\left(r, f^n(f-1)f'\right) + T\left(r, \frac{1}{f'}\right) + S(r,f) \\ &\leq \overline{N}(r,f) + \overline{N}\left(r, \frac{1}{f^n(f-1)f'}\right) + \overline{N}\left(r, \frac{1}{g^n(g-1)g'}\right) + T\left(r, f'\right) + 2\log r + S(r,f) \\ &\leq \overline{N}(r,f) + \overline{N}\left(r, \frac{1}{f}\right) + \overline{N}\left(r, \frac{1}{f-1}\right) + N\left(r, \frac{1}{f'}\right) + \overline{N}\left(r, \frac{1}{g}\right) + \overline{N}\left(r, \frac{1}{g-1}\right) + N\left(r, \frac{1}{g'}\right) + 2T(r,f) + 2\log r + S(r,f) \\ &\leq 7T(r,f) + 4T(r,g) + 2\log r + S(r,f) \\ \Longrightarrow (n-6)T(r,f) &\leq 4T(r,g) + 2\log r + S(r,f).\end{aligned} \tag{70}$$

Using Lemma 7, we get

$$(n-6)T(r,f) \leq 4\left(\frac{n+3}{n-6}\right)T(r,f) + 2\log r + S(r,f); \tag{71}$$

since $n \geq 11$, we get a contradiction.

Therefore $c = 1$, and by (27) and (24) we have

$$f^n(f-1)f' = g^n(g-1)g'. \tag{72}$$

On integrating, we get

$$\begin{gathered}\frac{f^{n+2}}{n+2} - \frac{f^{n+1}}{n+1} = \frac{g^{n+2}}{n+2} - \frac{g^{n+1}}{n+1} + k \\ F^* = G^* + k, \quad \text{where } k \text{ is a constant.}\end{gathered} \tag{73}$$

We claim that $k = 0$. Suppose that $k \neq 0$, then

$$\begin{aligned}&\Theta(0,F^*) + \Theta(k,F^*) + \Theta(\infty,F^*) \\ &\quad = \Theta(0,F^*) + \Theta(0,G^*) + \Theta(\infty,F^*).\end{aligned} \tag{74}$$

We have

$$\overline{N}\left(r, \frac{1}{F^*}\right) = \overline{N}\left(r, \frac{1}{f}\right) + \overline{N}\left(r, \frac{1}{f-(n+2)/(n+1)}\right) \leq 2T(r,f). \tag{75}$$

Similarly,

$$\begin{gathered}\overline{N}\left(r, \frac{1}{G^*}\right) \leq 2T(r,g), \\ \overline{N}(r,F^*) = \overline{N}(r,f) \leq T(r,f).\end{gathered} \tag{76}$$

Using Lemma 9, we have

$$\begin{gathered}T(r,F^*) = (n+2)T(r,f) + S(r,f), \\ T(r,G^*) = (n+2)T(r,g) + S(r,g).\end{gathered} \tag{77}$$

Thus,

$$\Theta(0,F^*) = 1 - \overline{\lim_{r\to\infty}} \frac{\overline{N}(r,1/F^*)}{T(r,F^*)} \geq 1 - \frac{2}{n+2}. \tag{78}$$

Similarly,

$$\Theta(0, G^*) \geq 1 - \frac{2}{n+2},$$
$$\Theta(\infty, F^*) = 1 - \varlimsup_{r\to\infty} \frac{\overline{N}(r, F^*)}{T(r, F^*)} \geq 1 - \frac{1}{n+2}. \quad (79)$$

Therefore, (74) becomes

$$\begin{aligned}\Theta(0, F^*) &+ \Theta(k, F^*) + \Theta(\infty, F^*) \\ &\geq 2\left(1 - \frac{2}{n+2}\right) + 1 - \frac{1}{n+2} \\ &= \frac{3n+1}{n+2} > 2 \quad \text{for } n \geq 11,\end{aligned} \quad (80)$$

which contradicts $\sum_{a\in\overline{\mathscr{C}}} \Theta(a, f) \leq 2$. Thus, we have

$$\frac{f^{n+2}}{n+2} - \frac{f^{n+1}}{n+1} = \frac{g^{n+2}}{n+2} - \frac{g^{n+1}}{n+1}. \quad (81)$$

Let $h = f/g$. If $h \not\equiv 1$, then we easily obtain that

$$g = \frac{(n+2)(h^{n+1}-1)}{(n+1)(h^{n+2}-1)}, \qquad f = \frac{(n+2)h(h^{n+1}-1)}{(n+1)(h^{n+2}-1)}. \quad (82)$$

If $h \equiv 1$, that is, $f \equiv g$.

Case 3. Suppose that $f_3 = c\,(\neq 0)$, where c is a constant.
If $c \neq 1$, then we have

$$\begin{aligned} f_1 + f_2 + f_3 &= 1 \\ f_1 + f_2 &= 1 - c \\ \frac{f^n(f-1)f'}{z} - \frac{cz}{g^n(g-1)g'} &= 1 - c. \end{aligned} \quad (83)$$

Applying Lemma 5 to the above equation, we have

$$\begin{aligned} T\left(r, \frac{f^n(f-1)f'}{z}\right) &\leq \overline{N}\left(r, \frac{f^n(f-1)f'}{z}\right) \\ &\quad + \overline{N}\left(r, \frac{z}{f^n(f-1)f'}\right) \\ &\quad + \overline{N}\left(r, \frac{g^n(g-1)g'}{z}\right) + S(r,f) \\ &\leq \overline{N}(r,f) + \overline{N}\left(r, \frac{1}{f^n(f-1)f'}\right) \\ &\quad + \overline{N}(r,g) + 2\log r + S(r,f). \end{aligned} \quad (84)$$

Note that

$$\begin{aligned} T(r, f^n(f-1)f') &\leq T\left(r, \frac{f^n(f-1)f'}{z}\right) + T(r,z) \\ &\leq T\left(r, \frac{f^n(f-1)f'}{z}\right) + \log r. \end{aligned} \quad (85)$$

Therefore using (84), we have

$$\begin{aligned} T(r, f^n(f-1)f') &\leq \overline{N}(r,f) + \overline{N}\left(r, \frac{1}{f^n(f-1)f'}\right) \\ &\quad + \overline{N}(r,g) + 3\log r + S(r,f). \end{aligned} \quad (86)$$

Using Lemmas 9 and 6 and (86), we have

$$\begin{aligned} (n+1)T(r,f) &= T(r, f^n(f-1)) + S(r,f) \\ &\leq T(r, f^n(f-1)f') \\ &\quad + T\left(r, \frac{1}{f'}\right) + S(r,f) \\ &\leq \overline{N}(r,f) + \overline{N}\left(r, \frac{1}{f^n(f-1)f'}\right) \\ &\quad + \overline{N}(r,g) + T(r,f') + 3\log r + S(r,f) \\ &\leq \overline{N}(r,f) + \overline{N}\left(r, \frac{1}{f}\right) + \overline{N}\left(r, \frac{1}{f-1}\right) \\ &\quad + N\left(r, \frac{1}{f'}\right) + \overline{N}(r,g) + 2T(r,f) \\ &\quad + 3\log r + S(r,f) \\ &\leq 7T(r,f) + T(r,g) + 3\log r + S(r,f) \\ \Longrightarrow (n-6)T(r,f) &\leq T(r,g) + 3\log r + S(r,f). \end{aligned} \quad (87)$$

Using Lemma 7, we get

$$(n-6)T(r,f) \leq \frac{(n+3)}{(n-6)}T(r,f) + 3\log r + S(r,f); \quad (88)$$

since $n \geq 11$, we get a contradiction. Thus, $c = 1$. Hence,

$$\begin{aligned} \frac{f^n(f-1)f'}{z} - \frac{z}{g^n(g-1)g'} &= 0 \\ \Longrightarrow f^n(f-1)f'g^n(g-1)g' &= z^2. \end{aligned} \quad (89)$$

Let z_0 be a zero of f of order p. From (89), we know that z_0 is a pole of g. Suppose that z_0 is a pole of g of order q. From (89), we obtain

$$\begin{aligned} np + p - 1 &= nq + 2q + 1 \\ \Longrightarrow (n+1)(p-q) &= q + 2, \end{aligned} \quad (90)$$

which implies that $p \geq q + 1$ and $q + 2 \geq n + 1$. Hence,

$$p \geq n. \quad (91)$$

Let z_1 be a zero of $(f-1)$ of order p_1, then from (89) z_1 is a pole of g (say order q_1). By (89), we get

$$\begin{aligned} p_1+p_1-1&=nq_1+2q_1+1\\ 2p_1-1&\geq n+3\\ 2p_1\geq n+4&\Longrightarrow p_1\geq\frac{n+4}{2}. \end{aligned}\tag{92}$$

Let z_2 be a zero of f' of order p_2 that is not zero of $f(f-1)$, then from (89), z_2 is a pole of g of order q_2. Again by (89), we get

$$\begin{aligned} p_2&=nq_2+2q_2+1\\ &\Longrightarrow p_2\geq n+3. \end{aligned}\tag{93}$$

In the same manner as above, we have similar results for the zeros of $g^n(g-1)g'$. From (89)–(93), we have

$$\overline{N}\left(r,f^n(f-1)f'\right)=\overline{N}\left(r,\frac{z^2}{g^n(g-1)g'}\right);\tag{94}$$

that is,

$$\begin{aligned} \overline{N}(r,f)&\leq\overline{N}\left(r,\frac{1}{g}\right)+\overline{N}\left(r,\frac{1}{g-1}\right)+N\left(r,\frac{1}{g'}\right)\\ &\leq\frac{1}{11}N\left(r,\frac{1}{g}\right)+\frac{2}{15}N\left(r,\frac{1}{g-1}\right)+\frac{1}{14}N\left(r,\frac{1}{g'}\right)\\ &\leq\frac{1}{11}T(r,g)+\frac{2}{15}T(r,g)+\frac{2}{14}T(r,g)+s(r,g)\\ &=\left(\frac{1}{11}+\frac{2}{15}+\frac{2}{7}\right)T(r,g)+S(r,g)\\ &\Longrightarrow\quad\overline{N}(r,f)<\frac{2}{3}T(r,g)+S(r,g). \end{aligned}\tag{95}$$

By Nevanlinna's second fundamental theorem, we have from (91), (92), and (95) that

$$\begin{aligned} T(r,f)&\leq\overline{N}\left(r,\frac{1}{f}\right)+\overline{N}\left(r,\frac{1}{f-1}\right)+\overline{N}(r,f)+S(r,f)\\ &<\frac{1}{11}T(r,f)+\frac{2}{15}T(r,f)+\frac{2}{3}T(r,g)\\ &\quad+S(r,f)+S(r,g)\\ &\Longrightarrow T(r,f)\\ &\leq\frac{37}{165}T(r,f)+\frac{2}{3}T(r,g)+S(r,f)+S(r,g). \end{aligned}\tag{96}$$

Similarly,

$$T(r,g)\leq\frac{37}{165}T(r,g)+\frac{2}{3}T(r,f)+S(r,f)+S(r,g).\tag{97}$$

From (96) and (97), we get

$$\begin{aligned} &T(r,f)+T(r,g)\\ &\quad\leq\frac{37}{165}\left(T(r,f)+T(r,g)\right)\\ &\qquad+\frac{2}{3}\left(T(r,f)+T(r,g)\right)+S(r,f)+S(r,g) \end{aligned}\tag{98}$$

$$\frac{18}{165}\left(T(r,f)+T(r,g)\right)\leq S(r,f)+S(r,g);$$

since $n\geq 11$, we get a contradiction.

This completes the proof of Theorem 1. □

Using the same argument as in the proof of Theorem 1, we can prove Theorem 2.

Proof of Theorem 3. By the assumption of the theorem, we know that either both f and g are two transcendental entire functions or both f and g are polynomials.

If f and g are transcendental entire functions, putting $\overline{N}(r,f)=0$, $\overline{N}(r,g)=0$ and using similar arguments as in the proof of Theorem 1, we easily obtain Theorem 3.

If f and g are polynomials, $f^n(f-1)f'$ and $g^n(g-1)g'$ share z CM, we get

$$f^n(f-1)f'-z=k\left(g^n(g-1)g'-z\right),\tag{99}$$

where k is a nonzero constant. Suppose that $k\neq 1$, (99) can be written as

$$\frac{f^n(f-1)f'}{z}-k\frac{g^n(g-1)g'}{z}=1-k.\tag{100}$$

Applying Lemma 5 to the above equation, we have

$$\begin{aligned} T\left(r,\frac{f^n(f-1)f'}{z}\right)&\leq\overline{N}\left(r,\frac{f^n(f-1)f'}{z}\right)\\ &\quad+\overline{N}\left(r,\frac{z}{g^n(g-1)g'}\right)\\ &\quad+\overline{N}\left(r,\frac{z}{f^n(f-1)f'}\right)+S(r,f). \end{aligned}\tag{101}$$

Since f is a polynomial, so it does not have any poles. Thus, we have,

$$\begin{aligned} &T\left(r,\frac{f^n(f-1)f'}{z}\right)\\ &\quad\leq\overline{N}\left(r,\frac{1}{g^n(g-1)g'}\right)\\ &\qquad+\overline{N}\left(r,\frac{1}{f^n(f-1)f'}\right)+2\log r+S(r,f). \end{aligned}\tag{102}$$

Note that

$$T\left(r,f^n(f-1)f'\right)\leq T\left(r,\frac{f^n(f-1)f'}{z}\right)+\log r.\tag{103}$$

Therefore,

$$\begin{aligned} &T\left(r, f^n(f-1)f'\right) \\ &\quad \le \overline{N}\left(r, \frac{1}{g^n(g-1)g'}\right) \\ &\quad + \overline{N}\left(r, \frac{1}{f^n(f-1)f'}\right) + 3\log r + S(r,f). \end{aligned} \tag{104}$$

Using Lemmas 9 and 6 and (104), we have

$$\begin{aligned} (n+1)T(r,f) &= T\left(r, f^n(f-1)\right) + S(r,f) \\ &\le T\left(r, f^n(f-1)f'\right) \\ &\quad + T\left(r, \frac{1}{f'}\right) + S(r,f) \\ &\le \overline{N}\left(r, \frac{1}{f^n(f-1)f'}\right) \\ &\quad + \overline{N}\left(r, \frac{1}{g^n(g-1)g'}\right) \\ &\quad + T\left(r, f'\right) + 3\log r + S(r,f) \\ &\le \overline{N}\left(r, \frac{1}{f}\right) + \overline{N}\left(r, \frac{1}{f-1}\right) + N\left(r, \frac{1}{f'}\right) \\ &\quad + \overline{N}\left(r, \frac{1}{g}\right) + \overline{N}\left(r, \frac{1}{g-1}\right) \\ &\quad + N\left(r, \frac{1}{g'}\right) + T(r,f) \\ &\quad + 3\log r + S(r,f) \\ &\le 4T(r,f) + 3T(r,g) + 3\log r + S(r,f) \\ \Longrightarrow (n-3)T(r,f) &\le 3T(r,g) + 3\log r + S(r,f). \end{aligned} \tag{105}$$

Using Lemma 8, we get

$$(n-3)T(r,f) \le 3\left(\frac{n+2}{n-3}\right)T(r,f) + 3\log r + S(r,f); \tag{106}$$

since $n \ge 7$, we get a contradiction.

Therefore, $k = 1$; so (99) becomes

$$f^n(f-1)f' = g^n(g-1)g'. \tag{107}$$

On integrating, we get

$$\begin{gathered} \frac{f^{n+2}}{n+2} - \frac{f^{n+1}}{n+1} = \frac{g^{n+2}}{n+2} - \frac{g^{n+1}}{n+1} + c \\ F^* = G^* + c, \quad \text{where } c \text{ is a constant.} \end{gathered} \tag{108}$$

We claim that $c = 0$. Suppose that $c \ne 0$, then

$$\Theta(0, F^*) + \Theta(c, F^*) = \Theta(0, F^*) + \Theta(0, G^*). \tag{109}$$

We have

$$\begin{aligned} &\overline{N}\left(r, \frac{1}{F^*}\right) \\ &\quad = \overline{N}\left(r, \frac{1}{f}\right) + \overline{N}\left(r, \frac{1}{f-(n+2)/(n+1)}\right) \le 2T(r,f). \end{aligned} \tag{110}$$

Similarly,

$$\overline{N}\left(r, \frac{1}{G^*}\right) \le 2T(r,g). \tag{111}$$

Using Lemma 9, we have

$$\begin{aligned} T(r,F^*) &= (n+2)T(r,f) + S(r,f), \\ T(r,G^*) &= (n+2)T(r,g) + S(r,g). \end{aligned} \tag{112}$$

Thus,

$$\Theta(0,F^*) = 1 - \overline{\lim_{r\to\infty}} \frac{\overline{N}(r,1/F^*)}{T(r,F^*)} \ge 1 - \frac{2}{n+2}. \tag{113}$$

Similarly,

$$\Theta(0,G^*) \ge 1 - \frac{2}{n+2}. \tag{114}$$

Therefore, (109) becomes

$$\begin{aligned} \Theta(0,F^*) + \Theta(c,F^*) &\ge 2\left(1 - \frac{2}{n+2}\right) \\ &= \frac{2n}{n+2} \ge \frac{14}{9} > 1 \quad \text{for } n \ge 7, \end{aligned} \tag{115}$$

which contradicts $\sum_{a\in\mathscr{C}} \Theta(a,f) \le 1$. Thus, we have

$$\frac{f^{n+2}}{n+2} - \frac{f^{n+1}}{n+1} = \frac{g^{n+2}}{n+2} - \frac{g^{n+1}}{n+1}. \tag{116}$$

Let $h = f/g$. If $h \not\equiv 1$, then by (116) we have

$$g = \frac{(n+2)\left(1+h+h^2+\cdots+h^n\right)}{(n+1)\left(1+h+h^2+\cdots+h^{n+1}\right)}. \tag{117}$$

By Picard's theorem, $h(z)$ is a constant. Hence, g is a constant, which is a contradiction. Therefore, $h(z) \equiv 1$, that is, $f(z) \equiv g(z)$. □

4. Remarks

If the condition "$f^n(f-1)f'$ and $g^n(g-1)g'$ share z CM" is replaced by the condition "$f^n(f-1)f'$ and $g^n(g-1)g'$ share $\alpha(z)$ CM," where α is a meromorphic function such that $\alpha \not\equiv 0, \infty$ and $T(r,\alpha) = o\{T(r,f), T(r,g)\}$; the conclusion of Theorems 1, 2, and 3 still holds. We, thus, obtain the following results.

Theorem 11. *Let f and g be two nonconstant meromorphic functions, $n \geq 11$ a positive integer. If $f^n(f-1)f'$ and $g^n(g-1)g'$ share $\alpha(z)$ CM, f and g share ∞ IM, then either $f(z) \equiv g(z)$ or*

$$g = \frac{(n+2)\left(1-h^{n+1}\right)}{(n+1)\left(1-h^{n+2}\right)}, \qquad f = \frac{(n+2)\,h\left(1-h^{n+1}\right)}{(n+1)\left(1-h^{n+2}\right)}, \tag{118}$$

where h is a nonconstant meromorphic function.

Theorem 12. *Let f and g be two nonconstant meromorphic functions, $n \geq 12$ a positive integer. If $f^n(f-1)^2 f'$ and $g^n(g-1)^2 g'$ share $\alpha(z)$ CM, f and g share ∞ IM, then $f(z) \equiv g(z)$.*

Theorem 13. *Let f and g be two nonconstant entire functions, $n \geq 7$ an integer. If $f^n(f-1)f'$ and $g^n(g-1)g'$ share $\alpha(z)$ CM, then $f(z) \equiv g(z)$.*

Acknowledgments

The authors thank the referees for their valuable suggestions. This research work is supported by the Department of Science and Technology, Government of India, Ministry of Science and Technology, Technology Bhavan, New Delhi, India (no. SR/S4/MS: 520/08).

References

[1] C.-C. Yang and H.-X. Yi, *Uniqueness Theory of Meromorphic Functions*, vol. 557 of *Mathematics and Its Applications*, Kluwer Academic, Dordrecht, The Netherlands, 2003.

[2] L. Yang, *Value Distribution Theory*, Springer, Berlin, Germany, 1993.

[3] W. Lin and H. Yi, "Uniqueness theorems for meromorphic functions concerning fixed-points," *Complex Variables*, vol. 49, no. 11, pp. 793–806, 2004.

[4] J.-F. Xu, F. Lü, and H.-X. Yi, "Fixed-points and uniqueness of meromorphic functions," *Computers and Mathematics with Applications*, vol. 59, no. 1, pp. 9–17, 2010.

[5] C. C. Yang, "On deficiencies of differential polynomials. II," *Mathematische Zeitschrift*, vol. 125, pp. 107–112, 1972.

[6] H. X. Yi, "Meromorphic functions that share two or three values," *Kodai Mathematical Journal*, vol. 13, no. 3, pp. 363–372, 1990.

Moment Problems on Bounded and Unbounded Domains

Octav Olteanu

Department of Mathematics-Informatics, Faculty of Applied Sciences, University Politehnica of Bucharest, 060042 Bucharest, Romania

Correspondence should be addressed to Octav Olteanu; olteanuoctav@yahoo.ie

Academic Editor: Zhijun Qiao

Using approximation results, we characterize the existence of the solution for a two-dimensional moment problem in the first quadrant, in terms of quadratic forms, similar to the one-dimensional case. For the bounded domain case, one considers a space of complex analytic functions in a disk and a space of continuous functions on a compact interval. The latter result seems to give sufficient (and necessary) conditions for the existence of a multiplicative solution.

1. Introduction

Applying the extension Hahn-Banach type results in existence questions concerning the moment problem is a well-known technique [1–11]. One of the most useful results is lemma of the majorizing subspace (see [12, Section 5.1.2] for the proof of the lattice-version of this lemma; see also [13]). It says that *if f is a linear positive operator on a subspace S of the ordered vector space X, the target space being an order complete vector lattice Y, and for each $x \in X$ there is $s \in S, x \leq s$, then f has a linear positive extension $F : X \to Y$*. Another geometric remark is that in the real case, the sublinear functional from the Hahn-Banach theorem can be replaced by a convex one. The theorem remains valid when the convex dominating functional is defined on a convex subset A with some qualities with respect to the subspace S (for instance, $S \cap \mathrm{ri}(A) \neq \Phi$), where $\mathrm{ri}(A)$ is the relative interior of A). Here we recall an answer published without proof in 1978 [14], without losing convexity, but strongly generalizing the classical result. The first detailed proof was published in 1983 [15]. The proof of a similar result, in terms of the moment problem, was published in [10]. Here we recall the general statement from [14]. One of the reasons is that many other results are consequences of this theorem, including Bauer's theorem [13], Namiokas's theorem, and abstract moment problem-results published firstly in [9]. Part of these generalizations of the Hahn-Banach principle are applied in the present work too. Throughout this first part, X will be a real vector space, Y an order-complete vector lattice, $A, B \subset X$ convex subsets, $q : A \to Y$ a concave operator, $p : B \to Y$ a convex operator, $S \subset X$ a vector subspace, and $f : S \to Y$ a linear operator.

Theorem 1. *Assume that*

$$f|_{S\cap A} \geq q|_{S\cap A}, \qquad f|_{S\cap B} \leq p|_{S\cap B}. \tag{1}$$

The following assertions are equivalent:

(a) *there is a linear extension $F : X \to Y$ of the operator f such that*

$$F|_A \geq q, \qquad F|_B \leq p; \tag{2}$$

(b) *there are $p_1 : A \to Y$ convex and $q_1 : B \to Y$ concave operators such that for all*

$$(\rho, t, \lambda', a_1, a', b_1, b', v) \in [0,1]^2 \times (0,\infty) \times A^2 \times B^2 \times S, \tag{3}$$

one has

$$\begin{aligned}(1-t)\,a_1 - tb_1 &= v + \lambda'\left[(1-\rho)\,a' - \rho b'\right] \\ &\Longrightarrow (1-t)\,p_1(a_1) - tq_1(b_1) \\ &\geq f(v) + \lambda'\left[(1-\rho)\,q(a') - \rho p(b')\right].\end{aligned} \tag{4}$$

The minus-sign appears to construct a convex operator in the left-hand side member and a concave operator in the right side. The idea of sandwich theorem on arbitrary convex subsets A, B is clear. Most of the applications hold for linear positive operators on linear ordered spaces (X, X_+), when we take $A = X_+$, $q \equiv 0$, $B = X$, p a suitable convex operator (a vector-valued norm, a sublinear operator), which "measures the continuity" of the extension F. One obtains the following result related to the theorem of H. Bauer ([13, Section 5.4]).

Theorem 2. *Let X be a preordered vector space with its positive cone X_+, $p : X \to Y$ a convex operator, $S \subset X$ a vector subspace, $f : S \to Y$ a linear positive operator. The following assertions are equivalent:*

(a) *there is a linear positive extension $F : X \to Y$ of f such that $F(x) \le p(x)$ for all $x \in X$;*

(b) *$f(x') \le p(x)$ for all $(x', x) \in S \times X$ such that $x' \le x$.*

Now we can deduce the main results on the abstract moment problem [9].

Theorem 3. *Let X, Y, $p : X \to Y$ be as in Theorem 2, $\{x_j\}_{j\in J} \subset X$, $\{y_j\}_{j\in J} \subset Y$ given families. The following assertions are equivalent:*

(a) *there is a linear positive operator $F : X \to Y$ such that*

$$F(x_j) = y_j \quad \forall j \in J, \qquad F(x) \le p(x) \quad \forall x \in X; \tag{5}$$

(b) *for any finite subset $J_0 \subset J$ and any $\{\lambda_j\}_{j\in J_0} \subset R$, one has:*

$$\sum_{j\in J_0} \lambda_j x_j \le x \Longrightarrow \sum_{j\in J_0} \lambda_j y_j \le p(x). \tag{6}$$

A clearer sandwich-moment problem variant is the following one.

Theorem 4. *Let $X, Y, \{x_j\}_{j\in J}, \{y_j\}_{j\in J}$ be as in Theorem 3 and $F_1, F_2 \in L(X, Y)$ two linear operators. The following statements are equivalent:*

(a) *there is a linear operator $F \in L(X, Y)$ such that*

$$\begin{gathered} F_1(x) \le F(x) \le F_2(x) \quad \forall x \in X_+, \\ F(x_j) = y_j, \quad \forall j \in J; \end{gathered} \tag{7}$$

(b) *for any finite subset $J_0 \subset J$ and any $\{\lambda_j\}_{j\in J_0} \subset R$, one has:*

$$\begin{gathered} \left(\sum_{j\in J_0} \lambda_j x_j = \varphi_2 - \varphi_1, \varphi_1, \varphi_2 \in X_+\right) \\ \Longrightarrow \sum_{j\in J_0} \lambda_j y_j \le F_2(\varphi_2) - F_1(\varphi_1). \end{gathered} \tag{8}$$

We recall the following approximation lemma on an unbounded bi-dimensional subset.

Lemma 5 (see [7] and [8, Lemma 1.3(d)]). *If $x \in C_0([0,\infty) \times [0,\infty))$ is a nonnegative continuous function with compact support, then there exists a sequence $(p_m)_m$ of positive polynomials on $[0,\infty) \times [0,\infty)$, such that*

$$p_m(t) > x(t) \quad \forall t \ge 0, \; \forall m \in Z_+, \; p_m \longrightarrow x, \tag{9}$$

uniformly on compact subsets of $[0,\infty) \times [0,\infty)$.

The idea of the proof is to add the ∞ point and to apply the Stone-Weierstrass theorem to the subalgebra generated by the functions $\exp(-mt_1 - nt_2)$, $m, n \in Z_+$. Then one uses for each such exp-function suitable majorizing or minorizing partial sums polynomials.

Note that Lemma 1.4 [8] asserts the density of positive polynomials in $(L^1_\nu(A))_+$, for any closed subset A of a finite dimensional space, ν being a positive regular Borel M-determinate measure. The results of this work are generally applications of the theorems stated above.

For uniqueness of the solution see [1, 3, 4, 16–19]. Several other interesting related results are contained in [20] (application of fixed-point principle, iterative methods) [6], (construction of a solution), [11, 21].

2. Moment Problems on Unbounded Subsets

The first result of this section concerns the multidimensional moment problem, being a consequence of the Stone Weierstrass theorem and of other usual results on positive polynomials on $[0,\infty)$. It is known that in several dimensions there are positive polynomials on R^n_+ which cannot be written involving sums of squares, as in the one-dimensional case. However, the approximation results from below seem to show that we can work with limits of tensor products of polynomials of one variable, for which such representation hold in each variable. This leads to characterizations of the existence of the solutions for multidimensional moment problems in terms of quadratic forms (Theorem 6(b)). Let H be a Hilbert space, A_1, A_2 two positive commuting self-adjoint operators acting on H, with spectrums $\sigma(A_j)$, $j = 1, 2$. We introduce the commutative algebra $Y = Y(A_1, A_2)$ of self-adjoint operators [5, 12], which is also an order-complete vector lattice:

$$\begin{gathered} Y_1 = \left\{T \in A(H); TA_j = A_j T, j = 1, 2\right\}, \\ Y = \left\{U \in Y_1; UT = TU \; \forall T \in Y_1\right\}. \end{gathered} \tag{10}$$

We denote by X the space of all continuous functions $x : [0,\infty)^2 \to R$, with the modulus dominated by a polynomial at each point of $[0,\infty)^2$, and by $x_{(j,k)}$ the elements of the base of polynomials, namely, $x_{(j,k)}(t_1, t_2) = t_1^j t_2^k$, $t_j \ge 0$.

Theorem 6 (see also [8, Theorem 3.2]). *Let* $(B_{(j,k)})_{(j,k)\in Z_+^2} \subset Y$. *The following assertions are equivalent:*

(a) *there is a linear operator* $F \in L(X,Y)$ *such that*

$$F\left(x_{(j,k)}\right) = B_{(j,k)} \quad \forall (j,k) \in Z_+^2,$$

$$0 \le F(x) \le \int_{\sigma(A_1)\times\sigma(A_2)} x(t_1,t_2)\, dE_{A_1} dE_{A_2} \quad \forall x \in X_+,$$

$$|F(\varphi)| \le \int_{\sigma(A_1)\times\sigma(A_2)} |\varphi(t_1,t_2)|\, dE_{A_1} dE_{A_2} \quad \forall \varphi \in X; \tag{11}$$

(b) *for any finite subsets* $J_1, J_2 \subset Z_+$ *and any* $\{\alpha_j\}_{j\in J_1}$, $\{\beta_k\}_{k\in J_2}$, *one has:*

$$\begin{aligned}
0 &\le \sum_{i,j\in J_1,k,l\in J_2} \alpha_i\alpha_j\beta_k\beta_l B_{(i+j,k+l)} \\
&\le \sum_{i,j\in J_1,k,l\in J_2} \alpha_i\alpha_j\beta_k\beta_l A_1^{i+j} A_2^{k+l}, \\
0 &\le \sum_{i,j\in J_1,k,l\in J_2} \alpha_i\alpha_j\beta_k\beta_l B_{(i+j+1,k+l)} \\
&\le \sum_{i,j\in J_1,k,l\in J_2} \alpha_i\alpha_j\beta_k\beta_l A_1^{i+j+1} A_2^{k+l}, \\
0 &\le \sum_{i,j\in J_1,k,l\in J_2} \alpha_i\alpha_j\beta_k\beta_l B_{(i+j,k+l+1)} \\
&\le \sum_{i,j\in J_1,k,l\in J_2} \alpha_i\alpha_j\beta_k\beta_l A_1^{i+j} A_2^{k+l+1}, \\
0 &\le \sum_{i,j\in J_1,k,l\in J_2} \alpha_i\alpha_j\beta_k\beta_l B_{(i+j+1,k+l+1)} \\
&\le \sum_{i,j\in J_1,k,l\in J_2} \alpha_i\alpha_j\beta_k\beta_l A_1^{i+j+1} A_2^{k+l+1}.
\end{aligned} \tag{12}$$

Proof. From the Stone-Weierstrass theorem, we can infer that the space $S = C_0(\sigma(A_1)) \otimes C_0(\sigma(A_2))$ is dense in the space $C_0(\sigma(A_1) \times \sigma(A_2))$. It follows that any continuous nonnegative function x on the product of spectrums can be uniformly approximated from above with elements of S. We extend all the functions involved with zero outside the compacts where they were defined, using then Luzin's theorem. Thus we obtain positive continuous approximations with compact support of x, defined on $[0,\infty)^2$. Each such approximation is the limit of a sequence of positive functions from $C_0([0,\infty)) \otimes C_0([0,\infty))$. Let $s = x_1 \otimes x_2$ be such a function. For each of the separate variables t_1, t_2, there is a sequence of positive polynomials on the whole nonnegative semi-axes such that

$$p_{m,j} > x_j \ge 0, \quad \forall m \in Z_+,\ p_{m,j} \longrightarrow x_j,\ m \longrightarrow \infty,\ j = 1,2, \tag{13}$$

the convergence being uniform on compact subsets. Because the polynomials involved are positive on the whole interval $[0,\infty)$, from [1] we know their form:

$$p_{m,j}\left(t_j\right) = q_{m,j}^2\left(t_j\right) + t_j r_{m,j}^2\left(t_j\right), \quad j = 1,2,\ m \in Z_+. \tag{14}$$

We define a linear operator on the subspace $P_1 \otimes P_2$ of products of polynomials in separate variables, such that the moment conditions from the statement holds:

$$\begin{aligned}
f\left(\sum_{j,k} \alpha_j\beta_k t_1^j t_2^k\right) &= \sum_{j,k} \alpha_j\beta_k B_{(j,k)} \Longrightarrow f\left(x_{(j,k)}\right) \\
&= B_{(j,k)} \quad \forall (j,k) \in Z_+^2.
\end{aligned} \tag{15}$$

We have already seen that these polynomials are dense in $C_0([0,\infty)^2)$. Using the above arguments, the assertion (b) says that we have

$$\begin{aligned}
0 &\le f(p_1 \otimes p_2) \\
&\le \iint_{\sigma(A_1)\times\sigma(A_2)} p_1(t_1)\, p_2(t_2)\, dE_{A_1} dE_{A_2}, \\
&\qquad \forall p_j \in \left(R\left[t_j\right]\right)_+, \quad j = 1,2.
\end{aligned} \tag{16}$$

An application of the majorizing subspace lemma [12] leads to the existence of a positive linear extension $F : X \to Y$ of f. Using the uniform convergence of the special polynomials on the product of spectrums, we obtain

$$\begin{aligned}
F(x) &= \lim F\left(p_{m,1} \otimes p_{m,2}\right) \\
&= \lim f\left(p_{m,1} \otimes p_{m,2}\right) \\
&\le \lim \iint_{\sigma(A_1)\times\sigma(A_2)} \left(p_{m,1} \otimes p_{m,2}\right) dE_{A_1} dE_{A_2} \\
&= \iint_{\sigma(A_1)\times\sigma(A_2)} x\,(t_1,t_2)\, dE_{A_1} dE_{A_2}, \quad \forall x \in X_+.
\end{aligned} \tag{17}$$

This relation leads to the last conclusion (a). Thus the proof of (b) $\Rightarrow$ (a) is finished. Since the converse is obvious, the theorem is proved.

Sometimes it is useful to find sufficient or necessary conditions for the existence of the solution in terms of some positive related sequences (assertion (b) of the next theorem).

Theorem 7. *Let us denote* $d\nu = \exp(-t)dt$ *on* $[0,\infty)$, $X = L_\nu^1 = L_\nu^1([0,\infty))$, *and let* $(y_j)_{j=0}^{\infty}$ *be a sequence of real numbers. The following assertions are equivalent:*

(a) *there exists a unique* $g \in L_\nu^\infty([0,\infty))$ *such that*

$$\begin{aligned}
&\int_o^\infty t^j g(t) \exp(-t)\, dt = y_j \\
&\forall j \in Z_+,\ 0 \le g(t) \le 1 \text{ a.e.};
\end{aligned} \tag{18}$$

(b) $\sum_{j=0}^{n} \alpha_j t^j \geq 0$ *for all* $t \geq 0 \Rightarrow 0 \leq \sum_{j=0}^{n} \alpha_j y_j \leq \sum_{j=0}^{n} \alpha_j \cdot \Gamma(j+1)$, *for all* $n \in Z_+$.

(c) *for any finite family* $(\lambda_j)_{j=0}^{n}$ *of real numbers, one has*

$$0 \leq \sum_{i,j=0}^{n} \lambda_i \lambda_j y_{i+j} \leq \sum_{i,j=0}^{n} \lambda_i \lambda_j \Gamma(i+j+1),$$
$$0 \leq \sum_{i,j=0}^{n} \lambda_i \lambda_j y_{i+j+1} \leq \sum_{i,j=0}^{n} \lambda_i \lambda_j \Gamma(i+j+2). \tag{19}$$

Proof ((b) $\Rightarrow$ *(a)).* Denote $x_j(t) = t^j$, $j \in Z_+$. One applies Theorem 4, (b) $\Rightarrow$ (a). Let $n \in Z_+$and $(\lambda_j)_{j=0}^{n}$ in R such that

$$\sum_{j=0}^{n} \lambda_j x_j = \varphi_2 - \varphi_1, \quad \varphi_l \in X_+ \bigcap C_0([0,\infty)) \; l = 1,2,$$
$$X = \left\{x \in L_\nu^1;\ \exists p \in P, |x| \leq p\right\}, P \tag{20}$$

being the subspace of polynomials. From the approximation lemma mentioned in the end of the introduction (see [7, Lemma 1.4]), and using (b), we infer that:

$$\sum_{j=0}^{n} \lambda_j x_j \leq p_m = \sum_{j=0}^{m} \beta_j x_j \longrightarrow \varphi_2 \Longrightarrow f\left(\sum_{j=0}^{n} \lambda_j x_j\right) := \sum_{j=0}^{n} \lambda_j y_j$$
$$\leq \sum_{j=0}^{m} \beta_j y_j = f(p_m) \leq \sum_{j=0}^{m} \beta_j \Gamma(j+1)$$
$$\longrightarrow \int_0^\infty \varphi_2(t) \exp(-t)\, dt, \quad m \longrightarrow \infty. \tag{21}$$

It follows that the linear form defined on polynomials is positive and dominated on $C_0([0,\infty))$ by the linear form F_2 defined below (note that p_m are positive polynomials on $[0,\infty)$):

$$F_2(\varphi) := \int_0^\infty \varphi(u) \exp(-u)\, du,$$
$$F_1 \equiv 0 \Longrightarrow \sum_{j=0}^{n} \lambda_j y_j \leq \sum_{j=0}^{m} \beta_j y_j$$
$$\longrightarrow \int_0^\infty \varphi_2(t) \exp(-t)\, dt$$
$$= F_2(\varphi_2) - F_1(\varphi_1), \quad m \longrightarrow \infty. \tag{22}$$

Application of Theorem 4 shows that there exists a linear positive form $F \in C_0^*$ satisfying the moment conditions and being majorized by F_2 on the positive cone X_+. Because of the density of $C_c([0,\infty)) \subset X$ in L_ν^1, and of the continuity of F_2 with respect to L_ν^1-norm, there is an extension $\widetilde{F} \in (L_\nu^1)^* \cong L^\infty$ of F. It follows that there exists a function in $g \in L_\nu^\infty([0,\infty))$ representing $\widetilde{F}$:

$$y_j = f(x_j) = \widetilde{F}(x_j) = \int_0^\infty t^j g(t)\, d\nu$$
$$= \int_0^\infty t^j g(t) \exp(-t)\, dt, \quad j \in Z_+,$$
$$0 \leq \int_0^\infty (\varphi \cdot g)(t) \cdot \exp(-t)\, dt$$
$$\leq \int_0^\infty \varphi(t) \cdot \exp(-t)\, dt, \quad \varphi \in X_+. \tag{23}$$

For $\varphi = \chi_E$, E Borel subset, one obtains

$$0 \leq \int_E g(t)\, d\nu \leq \int_E d\nu = \nu(E)$$
$$\forall E \in \Sigma \Longrightarrow 0 \leq g \leq 1 \text{ a.e.} \tag{24}$$

Since a direct computation and application of the Carleman criterion [18] show that $d\nu = \exp(-t)dt$ is an M-determinate measure, it follows from the same criterion that $g \cdot d\nu$ is M-determinate too. Thus the proof of (a) $\Leftrightarrow$ (b) is finished. The equivalence (b) $\Leftrightarrow$ (c) follows easily by the aid of the form of positive polynomials on the nonnegative semi-axes.

Corollary 8. *The set of all sequences* $(y_j)_j$ *verifying one of the assertions of Theorem 7 is a convex subset of* R^{Z_+}, *with a natural topology which makes it compact, and its extreme points are the sequences associated to those* $g \in U_1(L_\nu^\infty([0,\infty))) \cap (L_\nu^\infty([0,\infty)))_+$ *which take essentially only the values* 0 *and (or)* 1. *For each extreme point* g *with bounded support, one has*

$$\sum_{j=0}^{\infty} \frac{y_j}{j!} = m(\operatorname{Supp}(g)), \tag{25}$$

where m *denotes the Lebesgue measure.*

3. Moment Problems on Bounded Subsets

The next statement concerns a space of complex analytic functions with real coefficients, considered as a real ordered vector space. We denote by X_+ the convex cone of all absolutely convergent power series in the open disk $D = \{|z| < b\}$, with continuous sum

$$\sum_{j \in Z_+} \alpha_j z^j = \sum_{j \in Z_+} \alpha_j \varphi_j(z), \tag{26}$$

in the closed disk $\overline{D}$, and with nonnegative coefficients. Let $X = X_+ - X_+$. Let $(y_j)_{j \in Z_+}$ be a sequence of real numbers.

Theorem 9. *If $a \in (0, b)$, consider the following assertions:*

(a) *there is a real linear functional F on X such that*

$$F(\varphi_j) = y_j, \quad j \in Z_+,$$
$$|F(\varphi)| \leq \frac{b}{b-a}\varphi(b), \quad \forall \varphi \in X_+; \tag{27}$$

(b) *one has $|y_j| \leq a^j$, $j \in Z_+$.*

Then (b) implies (a) holds.

Proof. One observes that for all $\varphi \in X_+$, we have $\|\varphi\|_\infty = \varphi(b) = \varepsilon_b(\varphi)$. To apply Theorem 4, let us consider the following relation and its derivates:

$$\sum_{j=0}^{n} \lambda_j \varphi_j = \sum_{j=0}^{\infty} \alpha_j \varphi_j - \sum_{j=0}^{\infty} \beta_j \varphi_j$$
$$= \sum_{j=0}^{\infty} (\alpha_j - \beta_j)\varphi_j, \quad \alpha_j, \beta_j \geq 0. \tag{28}$$

Identification of the corresponding coefficients, Cauchy's inequalities and the hypothesis (b) lead to

$$-\frac{\varphi_1(b)}{b^j} = -\frac{\|\varphi_1\|_\infty}{b^j}$$
$$\leq -\beta_j \leq \lambda_j = \alpha_j - \beta_j \leq \alpha_j$$
$$\leq \frac{\|\varphi_2\|_\infty}{b^j} = \frac{\varphi_2(b)}{b^j}, \Longrightarrow$$
$$|\lambda_j| \leq \frac{\varphi_1(b) + \varphi_2(b)}{b^j}, \quad j = 0, \ldots, n \Longrightarrow \sum_{j=0}^{n} \lambda_j y_j$$
$$\leq \sum_{j=0}^{n} |\lambda_j| \cdot |y_j| \leq (\varphi_1(b) + \varphi_2(b)) \left(\sum_{j=0}^{\infty} \frac{a^j}{b^j}\right)$$
$$= \frac{b}{b-a}\varphi_2(b) - \left(-\frac{b}{b-a}\varphi_1(b)\right)$$
$$= F_2(\varphi_2) - F_1(\varphi_1),$$
$$F_2(\varphi) := \frac{b}{b-a}\varphi(b), \quad F_1 = -F_2. \tag{29}$$

Now from Theorem 4 we infer that there is a linear functional F satisfying the moment conditions, such that

$$-F_2(\varphi) \leq F(\varphi) \leq F_2(\varphi) \quad \forall \varphi \in X_+. \tag{30}$$

The conclusion follows.

Remark 10. For fixed $0 < a < b = 1$, Theorem 9 gives an upper bound and a lower bound for the convex set of solutions, both of them being Dirac measures multiplied with a constant. These bounds are realized if and only if we have $y_j = \pm 1/(1-a)$ for some $j \in Z_+$. The last results are applications of Theorem 3. Let T be a nonempty set, X an algebra of functions on T and X_+ the cone of pointwise nonnegative functions on T. Let Y be an order complete lattice, which is also a commutative algebra. Let

$$\{p_\alpha\}_{\alpha \in A} \subset X, \quad \{\varphi_\beta\}_{\beta \in B} \subset X,$$
$$\{u_\alpha\}_{\alpha \in A} \subset Y, \quad \{v_\beta\}_{\beta \in B} \subset Y \tag{31}$$

be which $S = Sp\{p_\alpha \varphi_\beta\}_{(\alpha,\beta) \in A \times B}$ is a majorizing subspace of X.

Theorem 11. *The following assertions are equivalent:*

(a) *there exists a linear positive operator $F \in L_+(X, Y)$ such that*

$$F(p_\alpha \varphi_\beta) = u_\alpha v_\beta \quad \forall (\alpha, \beta) \in A \times B,$$
$$x \leq \sum_{(\alpha,\beta)} \gamma_{(\alpha,\beta)} p_\alpha \varphi_\beta \Longrightarrow F(x) \leq \sum_{(\alpha,\beta)} \gamma_{(\alpha,\beta)} u_\alpha v_\beta, \tag{32}$$

where all the sums are finite;

(b) *the following implication holds:*

$$\sum_{(\alpha,\beta)} \lambda_{(\alpha,\beta)} p_\alpha \varphi_\beta \leq \sum_{(\alpha',\beta')} \gamma_{(\alpha',\beta')} p_{\alpha'} \varphi_{\beta'}$$
$$\Longrightarrow \sum_{(\alpha,\beta)} \lambda_{(\alpha,\beta)} u_\alpha v_\beta \tag{33}$$
$$\leq \sum_{(\alpha',\beta')} \gamma_{(\alpha',\beta')} u_{\alpha'} v_{\beta'}.$$

Proof. The implication (a) $\Rightarrow$ (b) is obvious. For the converse observe that any element of $x \in X$ is majorized by an element from S. Writing this for $-x$ too, we get that any $x \in X$ is between two elements $s_1(x) \leq x \leq s_2(x)$, $s_j(x) \in S$, $j = 1, 2$. We denote

$$x_{(\alpha,\beta)} = p_\alpha \varphi_\beta, \quad y_{(\alpha,\beta)} = u_\alpha v_\beta, \quad (\alpha, \beta) \in A \times B. \tag{34}$$

Define $p : X \to Y$,

$$p(x) = \inf \left\{ \sum_{(\alpha,\beta')} \gamma_{(\alpha' \cdot \beta')} u_{\alpha'} v_{\beta'}; \right.$$
$$\left. x \leq \sum_{(\alpha,\beta')} \gamma_{(\alpha' \cdot \beta')} p_{\alpha'} \varphi_{\beta'} \right\}, \quad x \in X. \tag{35}$$

If $f : S \to Y$ is the linear operator verifying the moment conditions, then (b) says that f is linear and positive on S, so that for any $s_1 \leq x$, $s_1 \in S$ we have

$$f(s_1) \leq p(x). \tag{36}$$

It follows that p is well defined and a standard computation shows that it is a convex operator. Moreover, these remarks and the definition of p lead to the following implication:

$$\begin{aligned}\sum_{(\alpha,\beta)} \lambda_{(\alpha,\beta)} p_\alpha \varphi_\beta = \sum_{(\alpha,\beta)} \lambda_{(\alpha,\beta)} x_{(\alpha,\beta)} \le x \\ \Longrightarrow \sum_{(\alpha,\beta)} \lambda_{(\alpha,\beta)} u_\alpha v_\beta \\ = \sum_{(\alpha,\beta)} \lambda_{(\alpha,\beta)} y_{(\alpha,\beta)} \le p(x).\end{aligned} \tag{37}$$

Hence the condition of Theorem 3, is accomplished, and the conclusion follows.

In the following result related to the preceding one, Y will be as above; additionally assume that it is a Banach algebra with multiplication unit u_0, which is a strong order unit too. An example is the algebra of self-adjoint operators preceding Theorem 6, when $u_0 = I$. In the sequel, this will be the target algebra of our operators. Let

$$\begin{aligned}X = C([a,b]), \quad p_\alpha(t) = \varphi_\alpha(t) = t^\alpha, \\ \forall \alpha \in A = B = Z_+, \quad \{u_\alpha\}_{\alpha\in Z_+} \subset Y, \\ \{v_\alpha\}_{\alpha\in Z_+} \subset Y, \quad u_0 = v_0 = I.\end{aligned} \tag{38}$$

Theorem 12. *Under the above assumptions, the following assertions are equivalent:*

(a) *there is a nondecreasing mapping* $\sigma : [a,b] \to Y$ *such that:*

$$\begin{aligned}\int_a^b t^{\alpha+\beta} d\sigma = u_\alpha v_\beta \quad \forall (\alpha,\beta) \in Z_+^2, \\ x \le \sum_{(\alpha,\beta)} y_{(\alpha,\beta)} p_{\alpha+\beta} \\ \Longrightarrow \int_a^b x(t)\, d\sigma \le \sum_{(\alpha,\beta)} y_{(\alpha,\beta)} u_\alpha v_\beta, \\ \int_a^b x(t) \cdot y(t)\, d\sigma = \int_a^b x(t)\, d\sigma \\ \cdot \int_a^b y(t)\, d\sigma \forall x, \quad y \in C([a,b]),\end{aligned} \tag{39}$$

(b) *the following implication holds:*

$$\begin{aligned}\sum_{(\alpha,\beta)} \lambda_{(\alpha,\beta)} p_{\alpha+\beta} \le \sum_{(\alpha',\beta'')} y_{(\alpha',\beta')} p_{\alpha'+\beta'} \\ \Longrightarrow \sum_{(\alpha,\beta)} \lambda_{(\alpha,\beta)} u_\alpha v_\beta \le \sum_{(\alpha',\beta'')} y_{(\alpha',\beta')} u_{\alpha'} v_{\beta'} \\ u_0 = v_0 = I.\end{aligned} \tag{40}$$

Proof. The implication (a) ⇒ (b) is almost obvious: the linear operator defined by the vector measure $d\sigma$ is multiplicative and positive on X. To prove the converse, we use the preceding result of Theorem 11. There is a linear positive operator $F \in L_+(X,Y)$ such that

$$\begin{aligned}F(p_\alpha \varphi_\beta) = F(p_{\alpha+\beta}) = u_\alpha v_\beta, \quad \forall (\alpha,\beta) \in Z_+^2, \\ F(x) \le p(x) \quad \forall x \in X.\end{aligned} \tag{41}$$

Since F applies topological bounded subsets of $C([a,b])$ into order-bounded subsets of Y, the theorem in [12, page 272] asserts that there exists σ representing mapping. Since F is positive, σ is nondecreasing. We have

$$\begin{aligned}F(p_{\alpha+\beta}) = \int_a^b t^{\alpha+\beta} d\sigma = u_\alpha v_\beta = (u_\alpha v_0) \cdot (v_\beta v_0) \\ = \int_a^b t^\alpha d\sigma \cdot \int_a^b t^\beta d\sigma \\ = F(p_\alpha) \cdot F(p_\beta), \quad (\alpha,\beta) \in Z_+^2.\end{aligned} \tag{42}$$

The conclusion is that on basic monomials, F is multiplicative. Now, a straightforward computation shows that F is multiplicative on the subspace of polynomials. From the density of the latter in $C([a,b])$ and using also the continuity of the product operation on $Y \times Y$, we deduce that F is multiplicative on the space of continuous functions. The proof is finished.

Corollary 13. *If F is the solution from Theorem 12, then denoting $B_1 = F(u_1) = \int_a^b t\, d\sigma$, one must has*

$$F(p_m) = F(\varphi_m) = u_m = v_m = B_1^m \quad \forall m \in Z_+ \tag{43}$$

so that the measure $d\sigma$ leads to the spectral measure of the operator B_1.

References

[1] N. I. Akhiezer, *The Classical Moment Problem and Some Related Questions in Analysis*, Oliver and Boyd, Edinburgh, UK, 1965.

[2] C. Ambrozie and O. Olteanu, "A sandwich theorem, the moment problem, finite-simplicial sets and some inequalities," *Revue Roumaine de Mathématique Pures et Appliquées*, vol. 49, pp. 189–210, 2004.

[3] G. Choquet, "Le problème des moments," in *Séminaire d'Initiation à l'Analise*, Institut H. Poicaré, Paris, France, 1962.

[4] M. G. Krein and A. A. Nudelman, *Markov Moment Problem and Extremal Problems*, American Mathematical Society, Providence, RI, USA, 1977.

[5] L. Lemnete-Ninulescu, "Using the solution of an abstract moment problem to solve some classical complex moment problems," *Romanian Journal of Pure and Applied Mathematics*, vol. 51, pp. 703–711, 2006.

[6] J. M. Mihăilă, O. Olteanu, and C. Udrişte, "Markov-type and operator-valued multidimensional moment problems, with some applications," *Romanian Journal of Pure and Applied Mathematics*, vol. 52, pp. 405–428, 2007.

[7] J. M. Mihăilă, O. Olteanu, and C. Udrişte, "Markov-type moment problems for arbitrary compact and for some non-compact Borel subsets of R^n," *Romanian Journal of Pure and Applied Mathematics*, vol. 52, pp. 655–664, 2007.

[8] A. Olteanu and O. Olteanu, "Some unexpected problems of the moment problem," in *Proceedings of the 6th Congress of Romanian Mathematicians*, vol. 1, pp. 347–355, Academiei, 2010.

[9] O. Olteanu, "Application de théorèmes de prolongement d'opérateurs linéaires au problème des moments et à une generalisation d'un théorème de Mazur-Orlicz," *Comptes Rendus de l'Académie des Sciences Série I*, vol. 313, pp. 739–742, 1991.

[10] O. Olteanu, "Applications of a general sandwich theorem for operators to the moment problem," *Revue Roumaine de Mathématique Pures et Appliquées*, no. 2, pp. 513–521, 1996.

[11] O. Olteanu, "New aspects of the classical moment problem," *Romanian Journal of Pure and Applied Mathematics*, vol. 49, pp. 63–77, 2004.

[12] R. Cristescu, *Ordered Vector Spaces and Linear Operators*, Academiei, Bucharest and Abacus Press, Kent, UK, 1976.

[13] H. H. Schaefer, *Topological Vector Spaces*, Springer, New York, NY, USA, 1971.

[14] O. Olteanu, "Convexité et prolongement d'opérateurs linéaires," *Comptes Rendus de l'Académie des Sciences, Série A*, vol. 286, pp. 511–514, 1978.

[15] O. Olteanu, "Théorèmes de prolongement d'opérateurs linéaires," *Revue Roumaine de Mathématique Pures et Appliquées*, vol. 28, no. 10, pp. 953–983, 1983.

[16] C. Berg, J. P. R. Christensen, and C. U. Jensen, *Harmonic Analysis on Semigroups: Theory of Positive Definite and Related Functions*, Springer, New York, NY, USA, 1984.

[17] B. Fuglede, "The multidimensional moment problem," *Expositiones Mathematicae*, vol. 1, pp. 47–65, 1983.

[18] C. Kleiber and J. Stoyanov, "Multivariate distributions and the moment problem," *Journal of Multivariate Analysis*, vol. 113, pp. 7–18, 2013.

[19] J. Stoyanov, "Stieltjes classes for moment-indeterminate probability distributions," *Journal of Applied Probability*, vol. 41, no. 1, pp. 281–294, 2004.

[20] C. Berg and A. J. Durán, "The fixed point for a transformation of Hausdorff moment sequences and iteration of a rational function," *Mathematica Scandinavica*, vol. 103, no. 1, pp. 11–39, 2008.

[21] M. Putinar and F. H. Vasilescu, "Problème des moments sur les compacts semi-algébriques," *Comptes Rendus de l'Académie des Sciences Série I*, vol. 323, pp. 787–791, 1996.

Permissions

The contributors of this book come from diverse backgrounds, making this book a truly international effort. This book will bring forth new frontiers with its revolutionizing research information and detailed analysis of the nascent developments around the world.

We would like to thank all the contributing authors for lending their expertise to make the book truly unique. They have played a crucial role in the development of this book. Without their invaluable contributions this book wouldn't have been possible. They have made vital efforts to compile up to date information on the varied aspects of this subject to make this book a valuable addition to the collection of many professionals and students.

This book was conceptualized with the vision of imparting up-to-date information and advanced data in this field. To ensure the same, a matchless editorial board was set up. Every individual on the board went through rigorous rounds of assessment to prove their worth. After which they invested a large part of their time researching and compiling the most relevant data for our readers. Conferences and sessions were held from time to time between the editorial board and the contributing authors to present the data in the most comprehensible form. The editorial team has worked tirelessly to provide valuable and valid information to help people across the globe.

Every chapter published in this book has been scrutinized by our experts. Their significance has been extensively debated. The topics covered herein carry significant findings which will fuel the growth of the discipline. They may even be implemented as practical applications or may be referred to as a beginning point for another development. Chapters in this book were first published by Hindawi Publishing Corporation; hereby published with permission under the Creative Commons Attribution License or equivalent.

The editorial board has been involved in producing this book since its inception. They have spent rigorous hours researching and exploring the diverse topics which have resulted in the successful publishing of this book. They have passed on their knowledge of decades through this book. To expedite this challenging task, the publisher supported the team at every step. A small team of assistant editors was also appointed to further simplify the editing procedure and attain best results for the readers.

Our editorial team has been hand-picked from every corner of the world. Their multi-ethnicity adds dynamic inputs to the discussions which result in innovative outcomes. These outcomes are then further discussed with the researchers and contributors who give their valuable feedback and opinion regarding the same. The feedback is then collaborated with the researches and they are edited in a comprehensive manner to aid the understanding of the subject.

Apart from the editorial board, the designing team has also invested a significant amount of their time in understanding the subject and creating the most relevant covers. They scrutinized every image to scout for the most suitable representation of the subject and create an appropriate cover for the book.

The publishing team has been involved in this book since its early stages. They were actively engaged in every process, be it collecting the data, connecting with the contributors or procuring relevant information. The team has been an ardent support to the editorial, designing and production team. Their endless efforts to recruit the best for this project, has resulted in the accomplishment of this book. They are a veteran in the field of academics and their pool of knowledge is as vast as their experience in printing. Their expertise and guidance has proved useful at every step. Their uncompromising quality standards have made this book an exceptional effort. Their encouragement from time to time has been an inspiration for everyone.

The publisher and the editorial board hope that this book will prove to be a valuable piece of knowledge for researchers, students, practitioners and scholars across the globe.

List of Contributors

Casey Johnson and Shusen Ding
Department of Mathematics, Seattle University, Seattle, WA 98122, USA

Vakeel A. Khan and Nazneen Khan
Department of Mathematics, A.M.U., Aligarh 202002, India

Deepak B. Pachpatte
Department of Mathematics, Dr. B.A.M. University, Aurangabad, Maharashtra 431004, India

R. M. El-Ashwah
Department of Mathematics, Faculty of Science, Damietta University, New Damietta 34517, Egypt

M. K. Aouf
Department of Mathematics, Faculty of Science, Mansoura University, Mansoura 33516, Egypt

A. A. M. Hassan and A. H. Hassan
Department of Mathematics, Faculty of Science, Zagazig University, Zagazig 44519, Egypt

R. A. Rashwan and Dhekra Mohammed Al-Baqeri
Department of Mathematics, University of Assiut, Assiut 71516, Egypt

P. K. Jhade
Department of Mathematics, NRI Institute of Information Science & Technology, Bhopal, Madhya Pradesh 462021, India

Pin-Lin Liu
Department of Automation, Engineering Institute of Mechatronoptic System, Chienkuo Technology University, Changhua 500, Taiwan

Chirasak Mongkolkeha and Poom Kumam
Department of Mathematics, Faculty of Science, King Mongkut's University of Technology Thonburi, Bang Mod, Thun Khru, Bangkok 10140,Thailand

H. H. G. Hashem
Faculty of Science, Alexandria University, Alexandria, Egypt
College of Science & Arts, Qassim University, P.O. Box 6644 Buriadah 81999, Saudi Arabia

A. R. Al-Rwaily
College of Science & Arts, Qassim University, P.O. Box 6644 Buriadah 81999, Saudi Arabia

Muhammad Arshad
Department of Mathematics, International Islamic University, H-10,Islamabad 44000, Pakistan

Jamshaid Ahmad
Department of Mathematics, COMSATS Institute of Information Technology, Chak Shahzad, Islamabad 44000, Pakistan

Erdal Karapinar
Department of Mathematics, Atilim University, ˙Incek, 06836 Ankara, Turkey

Daniela Paesano and Pasquale Vetro
Dipartimento di Matematica e Informatica, Universit`a Degli Studi di Palermo, Via Archirafi, 34-90123, Palermo, Italy

Subuhi Khan and Nusrat Raza
Department of Mathematics, Aligarh Muslim University, Aligarh 202002, India

T. M. Seoudy
Department of Mathematics, Faculty of Science, Fayoum University, Fayoum 63514, Egypt

Sébastien Gaboury and Richard Tremblay
Department of Mathematics and Computer Science, University of Quebec at Chicoutimi, Chicoutimi, QC, Canada G7H 2B1

Christian Märkl
Institut f¨ur Numerische und Angewandte Mathematik, Universit¨at Stuttgart, Pfaffenwaldring 57, 70569 Stuttgart, Germany

Gülnihal Meral
Department of Mathematics, Faculty of Arts and Sciences, B¨ulent Ecevit University, 67100 Zonguldak, Turkey

Christina Surulescu
Technische Universit¨at Kaiserslautern, Felix Klein Zentrum f¨ur Mathematik, Paul Ehrlich Strasse, 67663 Kaiserslautern, Germany

Manoj P. Tripathi
Department of Applied Mathematics, Indian Institute of Technology, Banaras Hindu University, Varanasi 221005, India
Department of Mathematics, Udai Pratap Autonomous College, Varanasi 221002, India

Ram K. Pandey, Vipul K. Baranwal and Om P. Singh
Department of Mathematics, Udai Pratap Autonomous College, Varanasi 221002, India

S. N. Mishra
Department of Mathematics, Walter Sisulu University, Mthatha 5117, South Africa

Rajendra Pant
Department of Mathematics, Visvesvaraya National Institute of Technology, Nagpur 440010, India

R. Panicker
Department of Mathematics, Walter Sisulu University, Mthatha 5117, South Africa
Department of Mathematics, Rhodes University, Grahamstown 6140, South Africa

Rabha W. Ibrahim
Institute of Mathematical Sciences, University of Malaya, 50603 Kuala Lumpur, Malaysia

Mihail Megan
Academy of Romanian Scientists, Independent¸ei 54, 050094 Bucharest, Romania
West University of Timis¸oara, Department of Mathematics, V. P^arvan Boulevard, No. 4, 300223 Timis¸oara, Romania

Traian Ceausu
West University of Timis¸oara, Department of Mathematics, V. P^arvan Boulevard, No. 4, 300223 Timis¸oara, Romania

Mihaela Aurelia Tomescu
University of Petros¸ani, Department of Mathematics, University Street 20, 332006 Petros¸ani, Romania

Sergo A. Episkoposian
Faculty of Applied Mathematics, State Engineering University of Armenia, Teryan Street 105, 375049 Yerevan, Armenia

Sh. Rezapour
Department of Mathematics, Science and Research Branch, Islamic Azad University, Tehran, Iran
Department of Mathematics, Azarbaijan Shahid Madani University, Azarshahr, Tabriz, Iran

J. Hasanzade Asl
Department of Mathematics, Science and Research Branch, Islamic Azad University, Tehran, Iran

U. Yamanci and M. Gürdal
Department of Mathematics, Suleyman Demirel University, East Campus, 32260 Isparta, Turkey

Mujahid Abbas
Department of Mathematics and Applied Mathematics, University of Pretoria, Lynnwood road, Pretoria 0002, South Africa

Maher Berzig
University of Tunisia, Tunis College of Sciences and Techniques, 5 Avenue Taha Hussein, BP 56, Bab Manara, Tunis, Tunisia

Subhas S. Bhoosnurmath and Veena L. Pujari
Department of Mathematics, Karnatak University, Dharwad 580003, India

Octav Olteanu
Department of Mathematics-Informatics, Faculty of Applied Sciences, University Politehnica of Bucharest, 060042 Bucharest, Romania

www.ingramcontent.com/pod-product-compliance
Lightning Source LLC
Chambersburg PA
CBHW080648200326
41458CB00013B/4778
9781632403360